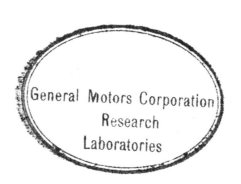

PURIFICATION
OF
LABORATORY CHEMICALS

Second Edition

Some Other Pergamon Titles of Interest

AMERICAN CHEMICAL SOCIETY AUDIO COURSES

Chemical Abstracts
Chemical Reaction Mechanisms
Electroorganic Synthesis
Laboratory Techniques in Organic Chemistry
Industrial Organic Chemistry
Modern Organic Synthesis, Parts I and II
Polymer Synthesis
Principles of Heterocyclic Chemistry
Use of the Chemical Literature

BARTON & OLLIS: Comprehensive Organic Chemistry, 6-Volume Set

PERRIN: Stability Constants of Metal-Ion Complexes, Part B, Organic
       Ligands

REUTOV et al: CH Acids

RIGAUDY: Nomenclature of Organic Chemistry

SERJEANT AND DEMPSEY: Ionisation Constants of Organic Acids in
       Aqueous Solution

STEPHEN et al: Solubilities of Inorganic and Organic Compounds,
       3-Volume Set

*Write to your nearest Pergamon office for further details about any of the above
publications.*

PURIFICATION
OF
LABORATORY CHEMICALS

Second Edition

D. D. PERRIN and W. L. F. ARMAREGO

Medical Chemistry Group
Australian National University, Canberra

and

D. R. PERRIN

Division of Plant Industry
Commonwealth Scientific and Industrial Research Organization
Canberra

PERGAMON PRESS

OXFORD · NEW YORK · TORONTO · SYDNEY · PARIS · FRANKFURT

U.K.	Pergamon Press Ltd., Headington Hill Hall, Oxford OX3 0BW, England
U.S.A.	Pergamon Press Inc., Maxwell House, Fairview Park, Elmsford, New York 10523, U.S.A.
CANADA	Pergamon of Canada, Suite 104, 150 Consumers Road, Willowdale, Ontario M2J 1P9, Canada
AUSTRALIA	Pergamon Press (Aust.) Pty. Ltd., P.O. Box 544, Potts Point, N.S.W. 2011, Australia
FRANCE	Pergamon Press SARL, 24 rue des Ecoles. 75240 Paris, Cedex 05, France
FEDERAL REPUBLIC OF GERMANY	Pergamon Press GmbH, 6242 Kronberg-Taunus, Pferdstrasse 1, Federal Republic of Germany

First edition 1966
Second edition 1980

British Library Cataloguing in Publication Data

Perrin, Douglas Dalzell
Purification of laboratory chemicals. - 2nd ed.
1. Chemicals - Purification
I. Title II. Armarego, W L F
III. Perrin, Dawn R
542   TP156.P83   79-41708
ISBN 0-08-022961-1

In order to make this volume available as economically and as rapidly as possible the authors' typescripts have been reproduced in their original forms. This method has its typographical limitations but it is hoped that they in no way distract the reader.

Printed in Great Britain by A. Wheaton & Co., Ltd., Exeter

Contents

CLASSES OF COMPOUNDS (continued)

Preface to First Edition

WE BELIEVE that a need exists for a book to help the chemist or bio-chemist who wishes to purify the reagents he uses. This need is emphasized by the previous lack of any satisfactory central source of references dealing with individual substances. Such a lack must undoubtedly have been a great deterrent to many busy research workers who have been left to decide whether to purify at all, to improvise possible methods, or to take a chance on finding, somewhere in the chemical literature, methods used by some previous investigators.

Although commercially available laboratory chemicals are usually satisfactory, as supplied, for most purposes in scientific and tech-nological work, it is also true that for many applications further purification is essential.

With this thought in mind, the present volume sets out, firstly, to tabulate methods, taken from the literature, for purifying some thousands of individual commercially available chemicals. To help in applying this information, two chapters describe the more common processes currently used for purification in chemical laboratories and give fuller details of new methods which appear likely to find increasing application for the same purpose. Finally, for dealing with substances not separately listed, a chapter is included setting out the usual methods for purifying specific classes of compounds.

To keep this book to a convenient size, and bearing in mind that its most likely users will be laboratory-trained, we have omitted manip-ulative details with which they can be assumed to be familiar, and also detailed theoretical discussion. Both are readily available elsewhere, for example in Vogel's very useful book Practical Organic Chemistry (Longmans, London, 3rd ed., 1956), or Fieser's Experiments in Organic Chemistry (Heath, Boston, 3rd ed., 1957).

For the same reason, only limited mention is made of the kinds of impurities likely to be present, and of tests for detecting them. In many cases, this information can be obtained readily from existing monographs.

By its nature, the present treatment is not exhaustive, nor do we claim that any of the methods taken from the literature are the best possible. Nevertheless, we feel that the information contained in this book is likely to be helpful to a wide range of laboratory workers, including physical and inorganic chemists, research students, bio-chemists, and biologists. We hope that it will also be of use, although perhaps to only a limited extent, to experienced organic chemists.

We are grateful to Professor A. Albert and Dr D.J. Brown for helpful comment on the manuscript.

Preface to Second Edition

SINCE the publication of the first edition of this book there have
been major advances in purification procedures. Sensitive methods
have been developed for the detection and elimination of progressively lower levels of impurities. Increasingly stringent requirements
for reagent purity have gone hand-in-hand with developments in
semiconductor technology, in the preparation of special alloys and
in the isolation of highly biologically active substances. The need
to eliminate trace impurities at the micro- and nanogram levels has
placed greater emphasis on ultrapurification technique. To meet
these demands the range of purities of laboratory chemicals has
become correspondingly extended. Purification of individual chemicals
thus depends more and more critically on the answers to two questions
-- Purification from what, and to what permissible level of contamination. Where these questions can be specifically answered, suitable
methods of purification can usually be devised.

Several periodicals devoted to ultrapurification and separations
have been started. These include "Progress in Separation and
Purification" Ed. (vol. 1) E. S. Perry, Wiley-Interscience, New York,
vols. 1-4, 1968-1971, and "Separation and Purification Methods",
Ed. E. S. Perry and C. J. van Oss, Marcel Dekker, New York, vol. 1 -,
1973 -. Nevertheless, there still remains a broad area in which a
general improvement in the level of purity of many compounds can be
achieved by applying more or less conventional procedures. The need
for a convenient source of information on methods of purifying
available laboratory chemicals was indicated by the continuing
demand for copies of this book even though it had been out of print
for several years.

We have sought to revise and update this volume, deleting sections
that have become more familiar or less important, and incorporating
more topical material. The number of compounds in Chapters 3 and 4
have been increased appreciably. Also, further details on purification and physical constants are given for many compounds that were
listed in the first edition.

We take this opportunity to thank users of the first edition who
pointed out errors and omissions, or otherwise suggested improve-
ments or additional material that should be included. We are
indebted to Mrs S. M. Schenk who emerged from retirement to type
this manuscript.

CHAPTER 1

COMMON PHYSICAL TECHNIQUES
USED IN PURIFICATION

<u>GENERAL REMARKS</u>

Purity is a matter of degree. Quite apart from any adventitious contaminants such as dust, scraps of paper, wax, cork, etc., that may
have been incorporated into a sample during manufacture, all commercially available chemical substances are in some measure impure. The
important question, then, is not whether a substance is pure but
whether a particular sample is sufficiently pure for some intended
purpose. That is, are its contaminants likely to interfere in the
process or measurement that is to be studied. By suitable manipulations it is often possible to reduce levels of impurities to acceptable
limits, but absolute purity is an ideal which, no matter how closely
approached, can never be shown to be attained. A "negative" physical
or chemical test indicates only that the amount of an impurity in a
substance lies below a certain level; no test can demonstrate that a
specified impurity is entirely absent.

When setting out to purify a laboratory chemical, it is desirable that
the starting material should be of the best grade commercially available. Particularly among organic solvents there is a range of qualit-
ies varying from "laboratory chemical" to "spectrophotometric",
"chromatographic" and "electronic" grades. Many of these are suitable
for use as received. With many of the commoner reagents it is possible
to obtain from current literature some indications of likely impurities, their probable concentrations and methods for detecting them.
See, for example, <u>Reagent Chemicals</u> (American Chemical Society Specifications, 5th edn., 1974), the American Society for Testing Materials D56-36, D92-46, and national pharmacopoeias. Other useful sources
include <u>Analar Standards for Laboratory Chemicals</u>, 7th edn., 1977
(The British Drug Houses Ltd., and Hopkin and Williams Ltd.), and
<u>Reagent Chemicals and Standards</u>, J. Rosin, 5th edn., 1967 (Van
Nostrand, New York). However, in many cases complete analyses are not
given so that significant concentrations of unspecified impurities may
be present.

Solvents and substances that are specified as "pure" for a particular
purpose may, in fact, be quite impure for other uses. Absolute ethanol
may contain traces of benzene, which makes it unsuitable for ultraviolet spectroscopy, or plasticizers which make it unsuitable for use
in solvent extraction.

1

Irrespective of the grade of material to be purified, it is essential that some criteria exist for assessing the degree of purity of the final product. The more common of these include:

1. Examination of physical properties such as

   (a) Melting point, freezing point, boiling point, and the freezing curve (i.e. the variation, with time, in the freezing point of a substance that is being slowly and continuously frozen).

   (b) Density.

   (c) Refractive index at a specified temperature and wavelength. The sodium D line, at 5892.6 $\overset{o}{A}$ (weighted mean of D_1 and D_2 lines) is the usual standard of wavelength but results from other wavelengths can often be interpolated from a plot of refractive index versus $1/(\text{wavelength})^2$.

   (d) Absorption spectra (ultraviolet, visible, infrared, and nuclear magnetic resonance).

   (e) Specific conductivity. (This can be used to detect, for example, water, salts, inorganic and organic acids and bases, in non-electrolytes.)

   (f) Optical rotation and circular dichroism.

   (g) Mass spectra.

2. Empirical analysis, for C, H, N, ash, etc.

3. Chemical tests for particular types of impurities, e.g. for peroxides in aliphatic ethers (with acidified KI), or for water in solvents (quantitatively by the Karl Fischer method).

4. Physical tests for particular types of impurities.

   (a) Emission and atomic absorption spectroscopy for detecting and determining metal ions.

   (b) Chromatography, including paper, thin layer, liquid and vapour phase.

   (c) Electron spin resonance for detecting free radicals.

   (d) X-ray spectroscopy.

   (e) Mass spectroscopy.

   (f) Fluorimetry.

5. Electrochemical methods.

6. Nuclear methods.

A substance is usually taken to be of an acceptable purity when the measured property is unchanged by further treatment (especially if it agrees with a recorded value). In general, at least two different methods, such as recrystallization and distillation, should be used in order to ensure maximum purification. Crystallization may be repeated until the substance has a constant melting point or absorption spectrum, and a substance may be redistilled in a fractionating column until it distils repeatedly within a narrow, specified, temperature range.

With liquids, the refractive index at a specified temperature and wavelength is a sensitive test of purity. Under favourable conditions, freezing curve studies are sensitive to impurity levels of as little as 0.001 moles per cent. (See, for example, Mair, Glasgow and Rossini, J.Res.Nat.Bur.Stand. 26, 591 (1941).) Analogous fusion curve or heat capacity measurements can be up to ten times as sensitive as this. (See, for example, Aston and Fink, Anal.Chem. 19, 218 (1947).) However, with these exceptions, most of the above methods are rather insensitive, especially if the impurities and the substances in which they occur are chemically similar. In some cases, even an impurity comprising many parts per million of a sample may escape detection.

The common methods of purification, discussed below, comprise distillation (including fractional distillation, distillation under reduced pressure, sublimation and steam distillation), crystallization, extraction and chromatographic adsorption. In some cases, volatile impurities (including water) can be removed by heating. Impurities can also sometimes be eliminated by the formation of derivatives from which the purified material is recovered.

Safety in the Chemical Laboratory

Although most of the manipulations involved in purifying laboratory chemicals are inherently safe, it remains true that care is necessary if hazards are to be avoided in the chemical laboratory. In particular there are dangers inherent in the inhalation of vapours and absorption of liquids through the skin. To the toxicity of solvents must be added the risk of their flammability and the possibility of eye damage. Chemicals, particularly in admixture, may be explosive. Compounds may be carcinogenic or otherwise deleterious to health. The use of radio-isotopic labelling poses problems of exposure and of disposal of laboratory waste.

At the least the laboratory should be well ventilated and safety glasses should be worn, particularly during distillation and manipulations carried out under reduced pressure or elevated temperatures. With this in mind we have endeavoured to warn users of this book whenever greater than usual care is needed in handling chemicals. As a general rule, however, all chemicals which are unfamiliar to the user should be treated with great care and assumed to be highly flammable and toxic.

Trace Impurities in Solvents

Some of the more obvious sources of contamination of solvents arise from storage in metal drums and plastic containers and from contact with grease. Many solvents contain water. Others have traces of acidic materials such as hydrochloric acid in chloroform. In both cases this leads to corrosion of the drum and contamination of the solvent by traces of metal ions, especially Fe^{3+}. Grease, for example on the stopcocks of separating funnels and other apparatus is also likely to contaminate solvents during extractions and other chemical manipulation.

A much more general source of contamination that has not received the consideration it merits comes from the use of plastics for tubing and containers. Plasticizers can readily be extracted by organic solvents from PVC and other plastics, so that most solvents, irrespective of their grade (including spectrograde and ultrapure) have been reported to contain 0.1 to 5 p.p.m. of plasticizer (de Zeeuw, Jonkman and van Mansvelt, Anal.Biochem., 67, 339 (1975)). Where large quantities of solvent are used for extraction (particularly of small amounts of compounds), followed by evaporation, this can introduce significant amounts of impurity, even exceeding the weight of the genuine extract and giving rise to spurious peaks in gas chromatography (for example, of fatty acid methyl esters (Pascaud, Anal.Biochem., 18, 570 (1967)). Likely contaminants are di(2-ethylhexyl)phthalate and dibutyl phthalate, but upwards of 20 different phthalic acid esters are listed as plasticizers as well as adipates, azelates, phosphates, epoxides, polyesters, trimellitates, and various heterocyclic compounds. These plasticizers would enter the solvent during passage through plastic tubing or containers or from plastic coatings used on cap liners for bottles. Such contamination could arise at any point in the manufacture or distribution of a solvent. The trouble with cap liners is avoidable by using corks wrapped in aluminium foil.

Solutions in contact with polyvinyl chloride may become contaminated with trace amounts of lead, titanium, tin, zinc, iron, magnesium or cadmium from additives used in the manufacture and moulding of PVC.

N-Phenyl-2-naphthylamine is a contaminant of solvents and biological materials that have been in contact with black rubber or neoprene (in which it is used as an antioxidant). Although it was only an artefact of the separation procedure it has been isolated as an apparent component of vitamin K preparations, extracts of plant lipids, algae, livers, butter, eye tissue and kidney tissue (Brown, Chem.Brit. 3, 524 (1967)).

Most of the above impurities can be removed by prior distillation of the solvent, but care should be taken to avoid plastic or black rubber as much as possible.

On Cleaning Apparatus

Laboratory glassware and Teflon equipment can be cleaned satisfactorily for most purposes by treating initially with a solution of sodium dichromate in concentrated sulphuric acid, draining, and rinsing copiously with distilled water. Where traces of chromium (adsorbed on the glass) must be avoided, a 1:1 mixture of concentrated sulphuric and nitric acid is a useful alternative. (Used in a fumehood to remove vapour and with adequate face protection. Acid washing is also suitable for polyethylene ware but prolonged contact (some weeks) leads to severe deterioration of the plastic.) For much glassware, washing with hot detergent solution, using tap water, followed by rinsing with distilled water and acetone, and heating to 200-300°

overnight, is adequate. (Volumetric apparatus should not be heated: after washing it is rinsed with acetone, then hexane, and air-dried. Prior to use, equipment can be rinsed with acetone, then with petroleum ether or hexane, to remove the last traces of contaminants.) Teflon equipment should be soaked, first in acetone, then in petroleum ether or hexane for 10 minutes prior to use.

For trace metal analyses, prolonged soaking of equipment in 1M nitric acid may be needed to remove adsorbed metal ions.

Soxhlet thimbles and filter papers contain traces of lipid-like material; for manipulations with highly pure materials, as in trace-pesticide analysis, they should be extracted before use using hexane.

Trace impurities in silica gel for TLC can be removed by heating at 300° for 16 hours or by Soxhlet extraction for 3 hours with redistilled chloroform, followed by 4 hours extraction with redistilled hexane.

DISTILLATION

One of the most widely applicable and most commonly used methods of purification (especially of organic chemicals) is fractional distillation at atmospheric, or some lower, pressure. Almost without exception, this method can be assumed to be suitable for all organic liquids and most of the low-melting organic solids. For this reason it has been possible in Chapter 3 to omit procedures for purification of organic chemicals when only a simple fractional distillation is involved—the suitability of such a procedure is implied from the boiling point.

The boiling point of a liquid varies with the atmospheric pressure to which it is exposed. A liquid boils when its vapour pressure is the same as the external pressure on its surface, its normal boiling point being the temperature at which its vapour pressure is equal to that of a standard atmosphere (760 mm Hg). Lowering the external pressure lowers the boiling point. For most substances, boiling point and vapour pressure are related by an equation of the form,

$$\log p = A + B/(t + 273),$$

where p is the pressure, t is in $^{\circ}C$, and A and B are constants. Hence, if the boiling points at two different pressures are known the boiling point at another pressure can be calculated from a simple plot of log p versus $1/(t + 273)$. For organic molecules that are not strongly associated, this equation can be written in the form,

$$\log p = 8.586 - 5.703 (T + 273)/(t + 273)$$

where T is the b.pt. in $^{\circ}C$ at 760 mm Hg. Table 1 gives computed b.pts over a range of pressures. Some examples illustrate its application. Ethyl acetoacetate, b.pt 180° (with decomposition) at 760 mmHg has a predicted b.pt. of 79° at 8 mm; the experimental value is 78°. Similarly, 2,4-diamino-toluene, b.pt. 292° at 760 mm Hg, has a predicted b.pt. of 147° at 8 mm.; the experimental value is 148-50°. For self-associated molecules the predicted b.pts. are lower than the experimental values. Thus, glycerol, b.pt. 290° at 760 mm Hg, has a predicted b.pt. of 168° at 20 mm Hg: the experimental value is 182°.

For pressures near to 760 mm, the change in boiling point is given approximately by (Crafts, Ber. 20, 709 (1887))

$$\Delta \underline{t} = \underline{a}(760 - \underline{p})(\underline{t} + 273)$$

where $\underline{a}$ = 0.00012 for most substances, but $\underline{a}$ = 0.00010 for water, alcohols, carboxylic acids and other associated ligands, and $\underline{a}$ = 0.00014 for very low-boiling substances such as nitrogen or ammonia.

When all the impurities are non-volatile, simple distillation is an adequate purification. The observed boiling point remains almost constant and approximately equal to that of the pure material. Usually, however, some of the impurities are appreciably volatile, so that the boiling point progressively rises during the distillation because of the progressive enrichment of the higher-boiling components in the distilling flask. In such cases, separation is effected by fractional distillation using an efficient column.

The principle involved in fractional distillation can be seen by considering a system which approximately obeys Raoult's law. (This law states that the vapour pressure of a solution at any given temperature is the sum of the vapour pressures of each substance multiplied by its mole fraction in the solution.) If two substances, A and B, having vapour pressures of 600 mm Hg and 360 mm Hg, respectively, were mixed in a mole ratio of 2:1, the mixture would have (ideally) a vapour pressure of 520 mm Hg and the vapour phase would contain 77% of A and 23% of B. If this phase was now condensed, the new liquid phase would, therefore, be richer in the volatile component A. Similarly, the vapour in equilibrium with this phase is still further enriched in A. Each such liquid-vapour equilibrium constitutes a "theoretical plate". The efficiency of a fractionating column is commonly expressed as the number of such plates to which it corresponds in operation. (Alternatively, this information may be given in the form of the height equivalent to a theoretical plate, or HETP.)

In most cases, systems deviate to a greater or less extent from Raoult's law, and vapour pressures may be greater or less than those calculated from it. In extreme cases, vapour pressure-composition curves pass through maxima or minima, so that attempts at fractional distillation lead finally to the separation of a constant-boiling (azeotropic) mixture and one (but not both) of the pure species if either of the latter is present in excess.

Technique

Distillation apparatus consists basically of a distilling flask, usually fitted with a vertical fractionating column (which may be empty or packed with suitable materials such as glass helices or stainless-steel wool) to which is attached a condenser leading to a receiving flask. The bulb of a thermometer projects into the vapour phase just below the region where the condenser joins the column. The distilling flask is heated so that its contents are steadily vaporized by boiling. The vapour passes up into the column where, initially, it condenses and runs back. The resulting heat transfer gradually warms the column so that there is a progressive movement of the vapour phase-liquid boundary up the column, with increasing

enrichment of the more volatile component. Because of this fractionation, the vapour finally passing into the condenser (where it condenses and flows into the receiver) is commonly that of the lowest-
boiling material in the system. The conditions apply until all of the
low-boiling material has been distilled, whereupon distillation ceases
until the column temperature is high enough to permit the next component to distil. This usually results in a temporary fall in the
temperature indicated by the thermometer.

The efficiency of a distillation apparatus used for purification of
liquids depends on the difference in boiling points of the pure
material and its impurities. For example, if two components of an
ideal mixture have vapour pressures in the ratio 2:1, it would be
necessary to have a still with an efficiency of at least seven plates
(giving an enrichment of $2^7 = 128$) if the concentration of the higher-
boiling component in the distillate was to be reduced to less than 1%
of its initial value. For a vapour pressure ratio of 5:1, three
plates would achieve as much separation.

In a fractional distillation, it is usual to reject the initial and
final fractions, which are likely to be richer in lower-boiling and
higher-boiling impurities. The centre fraction can be further
purified by repeated fractional distillation.

To achieve maximum separation by fractional distillation:

1. The column must be flooded initially to wet the packing.
   For this reason it is customary to operate a still at
   reflux for some time before beginning the distillation.

2. The reflux ratio should be high, so that the distillation
   proceeds slowly and with minimum disturbance of the
   equilibria in the column.

3. The hold-up of the column should not exceed one-tenth of
   the volume of any one component to be separated.

4. Heat loss from the column should be prevented but, if
   the column is heated to offset this, its temperature
   must not exceed that of the distillate in the column.

5. Heat input to the still-pot should remain constant.

6. For distillations under reduced pressure there must be
   careful control of the pressure to avoid flooding or
   cessation of reflux.

Distillation at Atmospheric Pressure

The distilling flask. To minimize superheating of the liquid (due
to the absence of minute air bubbles or other suitable nuclei for
forming bubbles of vapour), and to prevent bumping, one or more of
the following precautions should be taken:

(a) The flask is heated uniformly over a large part of its surface,
    either by using an electrical heating mantle or, much better,
    by partial immersion in a bath heated somewhat above the
    boiling point of the liquid to be distilled.

(b) Before heating begins, small pieces of unglazed fireclay
or porcelain ("porous pot", "boiling chips"), pumice,
carborundum, Teflon, diatomaceous earth, or platinum wire
are added to the flask. (They act as sources of air bubbles.)

(c) The flask may contain glass siphons or boiling tubes. The
former are inverted J-shaped tubes, the end of the shorter
arm being just above the surface of the liquid. The latter
comprise long capillary tubes sealed above the open lower end.

(d) A steady slow stream of gas is passed through the liquid.

(e) In some cases zinc dust can also be used. It reacts
chemically with acidic or strongly alkaline solutions
to liberate fine bubbles of hydrogen.

(f) The liquid in the flask is stirred mechanically. This is
especially necessary when suspended insoluble material is
present.

For simple distillations a Claisen flask (see, for example, Quickfit
and Quartz Ltd. catalogue of interchangeable laboratory glassware, or
Kontes Glass Co., Vineland, New Jersey, cat.no. TG-15) is often used.
This flask is, essentially, a round-bottomed flask to the neck of
which is joined another neck carrying a side arm. This second neck is
sometimes extended so as to form a Vigreux column.

For heating baths, see Table 2. For distillation on a semi-micro
scale, see Linstead, Elvidge and Whalley, 1955.

   Types of columns and packings. A slow distillation rate is necessary
to ensure that equilibrium conditions operate and also that the vapour
does not become superheated so that the temperature rises above the
boiling point. Efficiency is improved if the column is heat insulated
(either by vacuum jacketing or by lagging) and, if necessary, heated
to just below the boiling point of the most volatile component.
(Electrical heating tape is convenient for this purpose.) Efficiency
of separation also improves with increase in the heat of vaporization
of the liquids concerned (because fractionation depends on heat
equilibration at multiple liquid-gas boundaries). Water and alcohols
are more easily purified by distillation for this reason.

Columns used in distillation vary in their shapes and types of
packing. Packed columns are intended to give efficient separations by
maintaining a large surface of contact between liquid and vapour.
Efficiency of separation is further increased by operating under conditions approaching total reflux. (That is, under a high reflux ratio.
The reflux ratio is the volume of the condensate that is returned
(as reflux) to the distilling flask divided by the volume of liquid
that is allowed to distil over.) Better control of reflux ratio is
achieved by fitting a total condensation, variable take-off, still-
head (see, for example, catalogues by Quickfit and Quartz, or Kontes)
to the top of the fractionating column. However, great care must be
taken to avoid flooding of the column during distillation. The
minimum number of theoretical plates for satisfactory separation of
two liquids differing in boiling point by Δt is approximately $(273 + t)$
$/3\Delta t$, where t is the average boiling point in $^{\circ}C$.

Some of the more commonly used columns are:

Bruun column. A type of all-glass bubble-cap column.

Bubble-cap column. A type of plate column in which inverted cups
(bubble caps) deflect ascending vapour through reflux liquid lying
on each plate. Excess liquid from any plate overflows to the plate
lying below it and ultimately returns to the flask. (For further
details, see Bruun and Faulconer, Ind.Eng.Chem.(Anal.Ed.) 9, 192
(1937), Bruun and West, ibid. 9, 247 (1937). Like most plate
columns, it has a high through-put, but a relatively low number of
theoretical plates for a given height.

Dufton column. A plain tube, into which fits closely (preferably
ground to fit) a solid glass spiral wound round a central rod. It
tends to choke at temperatures above 100° unless it is well lagged.
(Dufton, J.Soc.Chem.Ind. (London) 38, 45T (1919).)

Hempel column. A plain tube (fitted near the top with a side arm)
which is almost filled with a suitable packing, which may be of rings
or helices.

Oldershaw column. An all-glas perforated-plate column. The plates
are sealed into a tube, each plate being equipped with a baffle to
direct the flow of reflux liquid, and a raised outlet which maintains
a definite liquid level on the plate and also serves as a drain on to
the next lower plate. (See Oldershaw, Ind.Eng.Chem.(Anal.Ed.)
13, 265 (1941).

Podbielniak column. A plain tube containing "Heli-Grid" Nichrome
or Inconel wire packing. This packing provides a number of passage-
ways for the reflux liquid, while the capillary spaces ensure very
even spreading of the liquid, so that there is a very large area of
contact between liquid and vapour while, at the same time, channell-
ing and flooding are minimized. A column 1 m high has been stated to
have an efficiency of 200-400 theoretical plates. (For further
details, see Podbielniak, Ind.Eng.Chem. (Anal.Ed.) 13, 639 (1941);
Mitchell and O'Gorman, Anal.Chem. 20, 315 (1948).)

Stedman column. A plain tube containing a series of wire-gauze
discs stamped into flat, truncated cones and welded together, alternatively base-to-base and edge-to-edge, with a flat disc across each
base. Each cone has a hole, alternately arranged, near its base,
vapour and liquid being brought into intimate contact on the gauze
surfaces. (See Stedman, Canad.J.Res. B 15, 383 (1937).)

Todd column. A column (which may be a Dufton type, fitted with a
Monel metal rod and spiral, or a Hempel type, filled with glass
helices) which is surrounded by an open heating jacket so that the
temperature can be adjusted to be close to the distillation temperature.(Todd, Ind.Eng.Chem.(Anal.Ed.) 17, 175 (1945).

Vigreux column. A glass tube in which have been made a number of
pairs of indentations which almost touch each other and which slope
slightly downwards. The pairs of indentations are arranged to form a
spiral of glass inside the tube.

Widmer column. A Dufton column, modified by enclosing within two
concentric tubes the portion containing the glass spiral. Vapour
passes up the outer tube and down the inner tube before entering the
centre portion. In this way flooding of the column, especially at
high temperatures, is greatly reduced. (Widmer, Helv.Chim.Acta,
7, 59 (1924).

The packing of a column greatly increases the surface of liquid film
in contact with the vapour phase, thereby increasing the efficiency
of the column but reducing its capacity (the quantities of vapour
and liquid able to flow in opposite directions in a column without
causing flooding). Material for packing should be of uniform size,
symmetrical shape, and have a unit diameter less than one-eighth
that of the column. (Rectification efficiency increases sharply as
the size of the packing is reduced but so, also, does the hold-up in
the column.) It should also be capable of uniform, reproducible
packing. The usual packings are:

(a) Rings. These may be of hollow glass or porcelain
    (Raschig rings), of stainless-steel gauze (Dixon
    rings), or hollow rings with a central partition
    (Lessing rings) which may be of porcelain,
    aluminium, copper or nickel.

(b) Helices. These may be of metal or glass (Fenske rings),
    the latter being used where resistance to chemical
    attack is important (e.g. in distilling acids, organic
    halides, some sulphur compounds, and phenols). Metal
    single-turn helices are available in aluminium, nickel,
    and stainless steel. Glass helices are less efficient,
    because they cannot be tamped to ensure more uniform
    packing.

(c) Balls. These are usually glass.

(d) Wire packing. For use of "Heli-Grid" and "Heli-Pak"
    packings see references given for Podbielniak column.
    For Stedman packing, see entry under Stedman column.

Condensers. Some of the more commonly used condensers are as
follows:

Air condenser. A glass tube such as the inner part of a Liebig
condenser. Used for liquids with boiling points above 90°.   Can be
of any length.

Allihn condenser. The inner tube of a Liebig condenser is
modified by having a series of bulbs to increase the condensing
surface. Further modifications of the bubble shapes give the Julian
and Allihn-Kronbitter condensers.

Bailey-Walker condenser. A type of all-metal condenser fitting
into the neck of extraction apparatus and being supported by the rim.
Used for high-boiling liquids.

Coil condenser. An open tube, into which is sealed a glass coil or
spiral through which water circulates. The tube is sometimes also
surrounded by an outer cooling jacket.

Double-surface condenser. A tube in which the vapour is condensed between an outer and an inner water-cooled jacket after impinging on the latter. Particularly useful for liquids boiling below 40°.

Friedrichs condenser. A "cold-finger" type of condenser sealed into a glass jacket open at the bottom and near the top. The cold finger is formed into glass screw threads.

Graham condenser. A type of coil condenser.

Hopkins condenser. A cold-finger type of condenser, resembling that of Friedrichs.

Liebig condenser. An inner glass tube surrounded by a glass jacket through which water is circulated.

Othmer condenser. A large-capacity condenser which has two coils of relatively large-bore glass tubing inside it, through which the water flows. The two coils join at their top and bottom.

West condenser. A Liebig condenser with a light-walled inner tube and a heavy-walled outer tube, with only a narrow space between them.

Wiley condenser. A condenser resembling the Bailey-Walker type.

VACUUM DISTILLATION

This expression is commonly used to denote a distillation under a pressure lower than that of the normal atmosphere. Because the boiling point of a substance depends on the pressure, it is often possible by sufficiently lowering the pressure to distil materials at a temperature low enough to avoid partial or complete decomposition, even if they are unstable when boiled under atmospheric pressure.

Sensitive or high-boiling liquids should invariably be distilled or fractionally distilled under reduced pressure. The apparatus is essentially as described for distillation except that ground joints connecting different parts of the apparatus should be greased with the appropriate vacuum grease. For low, moderately high, and very high temperatures Apiezon L, M and T, respectively, are very satisfactory. Alternatively, it is often preferable to avoid grease and to use thin Teflon sleeves in the joints. The distilling flask must be supplied with a capillary bleed (which allows a fine stream of air or nitrogen into the flask), and the receiver should be of the fraction collector type (e.g. a Perkin triangle, see Quickfit and Quartz Ltd. interchangeable glassware catalogue, or Kontes Glass Co., Vineland, New Jersey, cat.no. TG-15). When distilling under vacuum it is very important to place a loose packing of glass wool above the liquid to buffer sudden boiling of the liquid. The flask should be not more than two-thirds full of liquid. The vacuum must have attained a steady state before the heat source is applied, and the temperature of the heat source must be raised very slowly until boiling is achieved.

If the pump is a filter pump run off a high-pressure water supply, its performance will be limited by the temperature of the water because the vapour pressure of water at 10°, 15°, 20°, and 25° is

9.2, 12.8, 17.5, and 23.8 mm Hg, respectively. The pressure can be measured with an ordinary manometer. For vacuums in the range 10^{-2} mm Hg (10μ) to 10 mm Hg, rotary mechanical pumps (oil pumps) are used and the pressure can be measured with a Vacustat McLeod type gauge. If still higher vacuums are required, for example for high-vacuum sublimations, a mercury diffusion pump is suitable. In principle, this pump resembles an ordinary water pump. It has a single, double or triple jet through which the mercury vapour and condensate pass. Such a pump can provide a vacuum up to 10^{-6} mm Hg. Two pumps can be used in series. For better efficiency these pumps are backed by a mechanical pump. The pressure is measured with a Pirani gauge. Where there is fear of contamination with mercury vapour, the mercury can be replaced with vacuum oils, e.g. Apiezon type G or Silicone fluid (Dow Corning no. 702 or 703), which produce a vacuum range of 10^{-4} -10^{-7} mm Hg depending on pump design and system used. These fluids are resistant to oxidation, are non-corrosive and are non-toxic.

In all cases, the pump is connected to the still through several traps to remove vapours. These traps may operate by chemical action, for example the use of sodium hydroxide pellets to react with acids, or by condensation, in which case empty tubes cooled in solid carbon dioxide-ethanol or in liquid nitrogen (contained in wide-mouthed Dewar flasks) are used.

Special oil or mercury traps are available commercially and a liquid-nitrogen trap is the most satisfactory one to use between these and the apparatus. It has an advantage over liquid air or oxygen in that it is non-explosive if it becomes contaminated with organic matter. Air should not be sucked through the apparatus before starting a distillation or sublimation because this will cause liquid air to condense in the liquid-nitrogen trap and a good vacuum can never be achieved. Hence, it is advisable to degas the system for a short period before the trap is immersed into the liquid nitrogen (which is kept in a Dewar flask).

   Spinning-band columns. Factors which limit the performance of distillation columns include the tendency to flood (which occurs when the returning liquid blocks the pathways taken by the vapour through the column) and the increased hold-up (which decreases the attainable efficiency) in the columns that should, theoretically, be highly efficient. To overcome these difficulties, especially for distillations under high vacuum of heat-sensitive or high-boiling highly viscous fluids, spinning-band columns have become commercially available.

In such units, the distillation columns contain a rapidly rotating, motor-driven, spiral band, which may be of polymer-coated metal, stainless steel or platinum. The rapid rotation of the band in contact with the wall of the still gives intimate mixing of descending liquid and ascending vapour while the screw-like motion of the band drives the liquid towards the still-pot, helping to reduce hold-up. There is very little pressure drop in such a system, and high throughputs are possible, at high efficiency. For example, a 30-in. 10-mm diameter commercial column is reported to have an efficiency of 28 plates and a pressure drop of 0.2 mm Hg for a throughput of 330 ml/hr. The columns may be either vacuum jacketed or heated externally. The stills can be operated down to 10^{-5} mm Hg. The principle, which was first used commercially in the Podbielniak Centrifugal Superfractionator, has also been embodied in descending-film molecular distillation apparatus.

STEAM DISTILLATION

When two immiscible liquids distil, the sums of their (independent) partial pressures is equal to the atmospheric pressure. Hence, in steam distillation, the distillate has the composition

$$\frac{\text{Moles of substance}}{\text{Moles of water}} = \frac{p_{subst}}{p_{water}} = \frac{760 - p_{water}}{p_{water}} ,$$

where the p's are vapour pressures (in mm Hg) in the boiling mixture. One of the advantages of using water in this way lies in its low molecular weight.

The customary technique consists of heating the substance and water in a flask (to boiling), usually with the passage of steam, followed by condensation and separation of the aqueous and non-aqueous phases. Its advantages are those of selectivity (because only some water-insoluble substances, such as naphthalene, nitrobenzene, phenol and aniline are volatile in steam) and of ability to distil certain high-boiling substances well below their boiling point. It also facilitates the recovery of a non-steam-volatile solid at a relatively low temperature from a high-boiling solvent such as nitrobenzene. The efficiency of steam distillation is increased if superheated steam is used (because the vapour pressure of the organic component is increased relative to water). In this case the flask containing the material is heated (without water) in an oil bath and the steam passing through it is superheated by prior passage through a suitable heating device (such as a copper coil over a bunsen burner or in an oil bath.) (For further details, see Krell, 1963.)

AZEOTROPIC DISTILLATION

In some cases two or more liquids form constant-boiling mixtures, or azeotropes. Azeotropic mixtures are most likely to be found with components which readily form hydrogen bonds or are otherwise highly associated, especially when the components are dissimilar, for example an alcohol and an aromatic hydrocarbon, but have similar boiling points. (Many systems are summarized in Azeotropic Data - III, L.H. Horsley, Advances in Chemistry Series 116, American Chemical Society, Washington, 1973.)

Examples where the boiling point of the distillate is a minimum (less than either pure component) include:

water with ethanol, isopropyl alcohol, n-propyl alcohol, tert.-butyl alcohol, propionic acid, butyric acid, pyridine, methanol with methyl iodide, methyl acetate, chloroform, ethanol with ethyl iodide, ethyl acetate, chloroform, benzene, toluene, methyl ethyl ketone, benzene with cyclohexane, acetic acid with toluene.

Although less common, azeotropic mixtures are known which have higher boiling points than their components. These include water with most of the mineral acids (hydrofluoric, hydrochloric, hydrobromic, perchloric, nitric and sulphuric) and formic acid. Other examples are acetic acid—pyridine, acetone—chloroform, aniline—phenol, and chloroform—methyl acetate.

The following azeotropes are important commercially for drying ethanol:

     ethanol 95.5%(by weight)—4.5% water   b.p. 78.1°
     ethanol 32.4%—benzene 67.6%   b.p. 68.2°
     ethanol 18.5%—benzene 74.1%—water 7.4%   b.p. 64.9° .

Materials are sometimes added so as to form an azeotropic mixture with the substance to be purified. Because the azeotrope boils at a different temperature, this facilitates separation from substances distilling in the same range as the pure material. (Conversely, the impurity might form the azeotrope and be removed in this way.) This method is often convenient, especially where the impurities are isomers or are otherwise closely related to the desired substance. Formation of low-boiling azeotropes also facilitates distillation.

One or more of the following methods can generally be used for separating the components of an azeotropic mixture:

1. By using a chemical method to remove most of one species prior to distillation. (For example, water can be removed by suitable drying agents; aromatic and unsaturated hydrocarbons can be removed by sulphonation.)

2. By redistillation with an additional substance which can form a ternary azeotropic mixture (as in the ethanol-water-benzene example given above.)

3. By selective adsorption of one of the components. (For example, of water on to silica gel or molecular sieve, or of unsaturated hydrocarbons on to alumina.)

4. By fractional crystallization of the mixture, either by direct freezing or after solution in a suitable solvent.

SUBLIMATION

Sublimation differs from ordinary distillations because the vapour phase condenses to a solid instead of liquid. Usually, the pressure in the heated system is diminished by pumping, and the vapour is condensed (after travelling a relatively short distance) on to a cold finger or some other cooled surface. This technique, which is applicable to many organic solids, can also be used with aluminium chloride, ammonium chloride, arsenious oxide, iodine, and several other inorganic materials. In some cases, passage of a stream of inert gas over the heated substance secures adequate vaporization.

RECRYSTALLIZATION

Technique

The most commonly used procedure for the purification of a solid material by crystallization from a solution involves the following steps:

(a) The impure material is dissolved in a suitable solvent, by shaking or vigorous stirring, at or near the boiling point, to form a near-saturated solution.

(b) The hot solution is filtered to remove any insoluble particles. (To prevent crystallization during this filtration, a heated (jacketted) filter funnel can be used or the solution can be somewhat diluted with more of the solvent.

(c) It is then allowed to cool so that the dissolved substance crystallizes out.

(d) The crystals are separated from the mother liquor, either by centrifuging or by filtering, under suction, through a sintered glass, a Hirsch, or a Büchner, funnel. Usually, centrifuging is much to be preferred because of the much greater ease and efficiency of separating crystals and mother liquor, and also because of the saving of time and effort, particularly when very small crystals are formed or when there is entrainment of solvent.

(e) They are washed free of mother liquor with a little fresh cold solvent, then dried.

If the solution contains extraneous coloured material likely to contaminate the crystals, this can often be removed by adding some activated charcoal (decolorizing carbon) to the hot, but not boiling, solution which is then shaken frequently for several minutes before being filtered. (The large active surface of the carbon makes it a good adsorbent for this purpose.) In general, the cooling and crystallization step should be rapid so as to give small crystals which occlude less of the mother liquor. This is usually satisfactory with inorganic materials, so that commonly the filtrate is cooled in an ice-water bath while being vigorously stirred. In many cases, however, organic molecules crystallize much more slowly, so that the filtrate must be set aside to cool to room temperature or left in the refrigerator. It is often desirable to subject material that is very impure to preliminary purification, such as steam distillation, Soxhlet extraction, or sublimation, before recrystallizing it. A greater degree of purity is also to be expected if the crystallization process is repeated several times, especially if different solvents are used. The advantage of successive crystallization from different solvents lies in the fact that the material sought, and its impurities, are unlikely to have similar solubilities as solvents and temperatures are varied.

For the final separation of solid material, sintered-glass discs are preferable to filter paper. Sintered glass is unaffected by strongly acid solutions or by oxidizing agents. Also, with filter paper,

cellulose fibres are likely to become included in the sample. The
sintered-glass discs or funnels can be readily cleaned by washing in
freshly prepared "chromic acid cleaning mixture". This mixture is
made by adding 100 ml of concentrated sulphuric acid slowly and with
stirring to a solution of 5 g of sodium dichromate in 5 ml of water.
(The mixture warms to about 70°.)

For materials with melting points below 70° it is sometimes convenient
to use dilute solutions in acetone, methanol, pentane, ethyl ether or
$CHCl_3$-CCl_4. The solutions are cooled to -78° in Dry Ice, to give a
filterable slurry which is filtered off through a precooled
Büchner funnel. Experimental details, as applied to the purification
of nitromethane, are given by Parrett and Sun, J.Chem.Educ., 54, 448
(1977).

Where substances vary little in solubility with temperature, isothermal crystallization may sometimes be employed. This usually takes the
form of a partial evaporation of a saturated solution at room temperature by leaving it under reduced presure in a desiccator.

However, in rare cases, crystallization is not a satisfactory method
of purification, especially if the impurity forms crystals that are
isomorphous with the material being purified. In fact, the impurity
content may even be greater in such recrystallized material. For this
reason, it still remains necessary to test for impurities and to
remove or adequately lower their concentrations by suitable chemical
manipulation prior to recrystallization.

Filtration

Filtration removes particulate impurities from liquids and is also
used to collect insoluble or crystalline solids which separate or
crystallize from solution. The usual technique is to pass the solution, cold or hot, through a fluted filter paper in a conical glass
funnel (see Vogel's Textbook of Practical Organic Chemistry).

If a solution is hot and needs to be filtered rapidly a Büchner
funnel and flask are used and filtration is performed under a slight
vacuum (water pump), the filter medium being a circular cellulose
filter paper wet with solvent. If filtration is slow, even under high
vacuum, a pile of about twenty filter papers, wet as before, are
placed in the Büchner funnel and, as the flow of solution slows down
the upper layers of filter paper are progressively removed. Alternatively, a filter aid, e.g. Celite, Florisil or Hyflo-supercel, is
placed on top of a filter paper in the funnel. When the flow of the
solution (under suction) slows down the upper surface of the filter
aid is scratched gently. Filter papers with various pore sizes are
available covering a range of filtration rates. Hardened filter
papers are slow filtering but they can withstand acidic and alkaline
solutions without appreciable hydrolysis of the cellulose (see
Table 3). When using strong acids it is preferable to use glass micro
fibre filter papers which are commercially available (see Table 3).

Freeing a solution from extremely small particles (e.g. for ORD or CD
measurements) requires filters with very small pore size. Commercially
available (Millipore, Gelman, Nuclepore) fibres other than cellulose
or glass include nylon, Teflon, and polyvinyl chloride, and the pore

diameter may be as small as 0.01 micron (see Table 4). Special containers are used to hold the filters, through which the solution is
pressed by applying pressure, e.g. from a syringe. Some of these
filters can be used to clear strong solutions of sulphuric acid.

As an alternative to a Büchner funnel for collecting crystalline
solids, a funnel with a sintered glass plate under suction may be
used. Sintered glass funnels with various porosities are commercially
available and can be easily cleaned with warm chromic or nitric acid.

When the solid particles are too fine to be collected on a filter
funnel because filtration is extremely slow, separation by
centrifugation should be used. Bench type centrifuges are most convenient for this purpose. The solution is placed in the centrifuge
tube, the tubes containing the solution on opposite sides of the
rotor should be balanced accurately (at least to within 0.05 to 0.1g),
and the solution is spun at maximum speed for as long as it takes to
settle the solid (usually ca 3-5 minutes). The solid is washed with
cold solvent by centrifugation, and finally twice with a pure
volatile solvent in which the solid is insoluble, also by centrifugation. After decanting the supernatant the residue is dried in a
vacuum, at elevated temperatures if necessary. In order to avoid
"spitting" while the solid in the centrifuge tube is dried, the mouth
of the tube is covered with silver paper and held fast with a tight
rubber band near the lip. The flat surface of the silver paper is
then perforated in several places with a pin.

Choice of Solvents

The best solvents for recrystallization have these properties:

(a) The material is much more soluble at higher temperatures
    than it is at room temperature or below.

(b) Well-formed (but not large) crystals are produced.

(c) Impurities are either very soluble or only sparingly soluble.

(d) The solvent must be readily removed from the purified material.

(e) There must be no reaction between the solvent and the
    substance being purified.

(f) The solvent must not be inconveniently volatile or too
    highly flammable. (These are reasons why ethyl ether and
    carbon disulphide are not commonly used in this way.)

The following generalizations provide a rough guide to the selection
of a suitable solvent:

(a) Substances usually dissolve best in solvents to which they
    are most closely related in chemical and physical character-
    istics. Thus, hydroxylic compounds are likely to be most
    soluble in water, methanol, ethanol, acetic acid or acetone.
    Similarly, petroleum ether might be used with water-insoluble
    substances. However, if the resemblance is too close,
    solubilities may become excessive.

(b) Higher members of homologous series approximate more and more closely to their parent hydrocarbon.

(c) Polar substances are more soluble in polar, than in non-polar, solvents.

Although chapters 3 and 4 provide details of the solvents used for recrystallizing a large portion of commercially available laboratory chemicals, they cannot hope to be exhaustive, nor need they necessarily be the best choice. In other cases where it is desirable to use this process, it is necessary to establish whether a given solvent is suitable. This is usually done by taking only a small amount of material in a small test-tube and adding enough solvent to cover it. If it dissolves readily in the cold or on gentle warming, the solvent is unsuitable. Conversely, if it remains insoluble when the solvent is heated to boiling (adding more solvent if necessary), the solvent is again unsuitable. If the material dissolves in the hot solvent but does not crystallize readily within several minutes of cooling in an ice-salt mixture, another solvent should be tried.

Solvents commonly used for recrystallization, and their boiling points, are given in Table 5.

Mixed Solvents

Where a substance is too soluble in one solvent and too insoluble in another, for either to be used for recrystallization, it is often possible (provided they are miscible) to use them as a mixed solvent. (In general, however, it is preferable to use a single solvent if this is practicable.) Table 6 comprises many of the common pairs of miscible solvents.

The technique of recrystallization from a mixed solvent is as follows:

The material is dissolved in the solvent in which it is the more soluble, then the other solvent (heated to near boiling) is added cautiously to the hot solution until a slight turbidity persists or crystallization begins. This is cleared by adding several drops of the first solvent, and the solution is allowed to cool and crystallize in the usual way.

A variation of this procedure is simply to precipitate the material in a microcrystalline form from solution in one solvent at room temperature, by adding a little of a second solvent, filtering this off, adding a little more of the second solvent and repeating the process. This ensures, at least in the first or last precipitation, a material which contains as little as possible of the impurities which may also be precipitated in this way. With salts the first solvent is commonly water, and the second solvent is alcohol or acetone.

Solidification from the Melt

A crystalline solid melts when its temperature is raised sufficiently for the thermal agitation of its molecules or ions to overcome the restraints imposed by the crystal lattice. Usually, impurities weaken crystal structures, and hence lower the melting points of solids (or the freezing points of liquids). If an impure material is melted and cooled slowly (with the addition, if necessary, of a trace of solid material near the freezing point to avoid supercooling), the first crystals that form will usually contain less of the impurity, so that fractional solidification by partial freezing can be used as a purification process for solids with melting points lying in a convenient temperature range (or for more readily frozen liquids). In some cases, impurities form higher melting eutectics with substances to be purified, so that the first material to solidify is less pure than the melt. For this reason, it is often desirable to discard the first crystals and also the final portions of the melt. Substances having similar boiling points often differ much more in melting points, so that fractional solidification can offer real advantages, especially where ultrapurity is sought.

The technique of recrystallization from the melt as a means of purification dates from its use by Schwab and Wichers (J.Res.Nat.Bur.Stds., 25, 747 (1940)) to purify benzoic acid. It works best if material is already nearly pure, and hence tends to be a final purification step. A simple apparatus for purifying organic compounds by progressive freezing is described by Matthias and Coggeshall (Anal.Chem., 31, 1124 (1959). In principle, the molten substance is cooled slowly by progressive lowering of the tube containing it into a suitable bath. For temperatures between 0° and 100°, waterbaths are convenient. Where lower temperatures are required, the cooling baths given in Table 7 can be used. Cooling is stopped when part of the melt has solidified, and the liquid phase is drained off. Column crystallization has been used to purify stearyl alcohol, cetyl alcohol, myristic acid; fluorene, phenanthrene, biphenyl, terphenyls, dibenzyl; phenol, 2-naphthol; benzophenone and 2,4-dinitrotoluene; and many other organic (and inorganic) compounds. (See, for example, "Developments in Melt Crystallization", by G.R. Atwood, in Recent Developments in Separation Science, N.N. Li (ed.), CRC Press, Cleveland, Ohio, 1972.)

Thus, an increase in purity from 99.80 to 99.98 mole% was obtained when acetamide was slowly crystallized in an insulated round bottom flask until half the material had solidified, and the solid phase was then crystallized from benzene (Schwab and Wichers, J.Res.Nat.Bur.Stand., 32, 253 (1944)).

Fractional solidification and its applications to obtaining ultrapure chemical substances, has been treated in detail in Fractional Solidification, ed. M. Zief and W.R. Wilcox, Edward Arnold, Inc., London, 1967, and Purification of Inorganic and Organic Materials, ed. M. Zief, Marcel Dekker, Inc., New York, 1969. These monographs should be consulted for discussion of the basic principles of solid-liquid processes such as zone melting, progressive freezing and column crystallization, laboratory apparatus and industrial scale equipment, and examples of applications. These include the removal of cyclohexane from benzene, and the purification of aromatic amines, dienes and naphthalene, and inorganic species such as the alkali iodides, potassium chloride, indium antimonide and gallium trichloride. The authors also discuss analytical methods for assessing the purity of the final material.

Zone Refining

Zone refining (or zone melting) is a particular development of fractional solidification and is applicable to all crystalline substances
that show differences in soluble impurity concentration in liquid and
solid states at solidification. The apparatus used in this technique
consists, essentially, of a device by which a narrow molten zone
moves slowly down a long tube filled with the material to be purified.
The machine can be set to cycle repeatedly. At its advancing side, the
zone has a melting interface with the impure material whereas on the
upper surface of the zone there is a constantly growing face of
higher-melting, resolidified material. This leads to a progressive
concentration of the impurities in the liquid phase which, at the end
of the run, is discarded. Also, because of the progressive increase
in impurity in the liquid phase, the resolidified material becomes
correspondingly less further purified. For this reason, it is
usually necessary to make several zone-melting runs before a sample
is satisfactorily purified. This is also why the method works most
successfully if the material is already fairly pure. In all these
operations the zone must travel slowly enough to enable impurities
to diffuse or be convected away from the area where resolidification
is occurring.

The technique finds commercial application in the production of metals
of extremely high purity (impurities down to 10^{-9} p.p.m.), in purifying refractory oxides, and in purifying organic compounds, using
commercially available equipment. Criteria for indicating that
definite purification is achieved include elevation of melting point,
removal of colour, fluorescence or smell, and a lowering of
electrical conductivity.

Difficulties likely to be met with in organic compounds, especially
those of low melting points and low rates of crystallization, are
supercooling and, because of surface tension and contraction, the
tendency of the molten zone to seep back into the recrystallized
areas. The method is likely to be useful in cases where fractional
distillation is not practicable, either because of unfavourable
vapour pressures or ease of decomposition, or where super-pure
materials are required. It has been used for the latter purpose with
anthracene, benzoic acid,chrysene, morphine and pyrene.

For a description of an apparatus for purifying organic compounds by
zone refining, see Herington, Handley and Cook, Chem. & Ind. 292
(1956); and for a semi-micro version see Handley and Herington,
Chem. & Ind. 304 (1956).

DRYING

Removal of Solvents

Where substances are sufficiently stable, removal of solvents from
recrystallized materials presents no problems. The crystals, after
filtering at the pump (and perhaps air-drying by suction), are heated
in an oven above the boiling point of the solvent (but below their
melting point), followed by cooling in a desiccator. Where this
treatment is inadvisable, it is still often possible to heat to a
lower temperature under reduced pressure, for example in an

Abderhalden pistol. This device consists of a small chamber which is heated externally by the vapour of a boiling solvent. Inside this chamber, which can be evacuated by a water pump or some other vacuum pump, is placed a small boat containing the sample to be dried and also a receptacle with a suitable drying agent. Convenient liquids for use as boiling liquids in an Abderhalden pistol, and their temperatures, are given in Table 8. In cases where heating above room temperature cannot be used, drying must be carried out in a vacuum desiccator containing suitable absorbents. For example, hydrocarbons, such as benzene, cyclohexane and petroleum ether, can be removed by using shredded paraffin wax, and acetic acid and other acids are absorbed by pellets of sodium hydroxide or potassium hydroxide. However, in general, solvent removal is less of a problem than ensuring that the water content of solids and liquids is reduced below an acceptable level.

Removal of Water

Methods for removing water from solids depend on the thermal stability of the solids or the time available. The safest method is to dry in a vacuum desiccator over concentrated sulphuric acid, phosphorus pentoxide, silica gel, calcium chloride, or some other desiccant. Where substances are stable in air and melt above 100°, drying in an air oven may be adequate. In other cases, use of an Abderhalden pistol may be satisfactory.

Often, in drying inorganic salts, the final material that is required is a hydrate. In such cases, the purified substance is left in a desiccator to equilibrate above an aqueous solution having a suitable water-vapour pressure. A convenient range of solutions for use in this way is given in Table 9.

The choice of desiccants for drying liquids is more restricted because of the need to avoid all substances likely to react with the liquids themselves. In some cases, direct distillation of an organic liquid is a suitable method of drying both solids and liquids, especially if low-boiling azeotropes are formed. Examples include acetone, aniline, benzene, chloroform, carbon tetrachloride, ethylene dichloride, heptane, hexane, methanol, nitrobenzene, petroleum ether, toluene and xylene. Addition of benzene can be used for drying ethanol by distillation. In carrying out distillations intended to yield anhydrous products, the apparatus should be fitted with guard-tubes containing calcium chloride or silica gel to prevent entry of moist air into the system. (Many anhydrous organic liquids are appreciably hygroscopic.)

Removal of water from gases may be by physical or chemical means, and is commonly by adsorption on to a drying agent in a low-temperature trap. The effectiveness of drying agents depends on the vapour pressure of the hydrated compound - the lower the vapour pressure the less the remaining moisture in the gas.

The most usually applicable of the specific methods for detecting and determining water in organic liquids is due to Karl Fischer. (See J. Mitchell and D.M. Smith, Aquametry, Interscience, New York, 1948.) Other techniques include electrical conductivity measurements and observation of the temperature at which the first cloudiness

appears as the liquid is cooled (applicable to liquids in which water
is only slightly soluble.) Addition of anhydrous cobalt(II) iodide
(blue) provides a convenient method (colour change to pink on hydration)for detecting water in alcohols, ketones, nitriles and some
esters. Infrared absorption measurements of the broad band for water
near 3500 cm^{-1} can also sometimes be used for detecting water in
non-hydroxylic substances.

Intensity and Capacity of Common Desiccants

Drying agents can conveniently be grouped into three classes, depending on whether they combine with water reversibly, they react chemically (irreversibly) with water, or they are molecular sieves. The
first group vary in their drying intensity with the temperature at
which they are used, depending on the vapour pressure of the hydrate
that is formed. This is why, for example, drying agents such as
anhydrous sodium sulphate, magnesium sulphate or calcium chloride
should be filtered off from the liquids before the latter are heated.
The intensities of drying agents belonging to this group fall in the
sequence

$P_2O_5 \gg BaO > Mg(ClO_4)_2$, CaO, MgO, KOH (fused), conc. H_2SO_4,

$CaSO_4$, $Al_2O_3 >$ KOH (sticks), silica gel, $Mg(ClO_4)_2.3H_2O$

$>$ NaOH (fused), 95% H_2SO_4, $CaBr_2$, $CaCl_2$ (fused) $>$ NaOH (sticks),

$Ba(ClO_4)_2$, $ZnCl_2$ (sticks), $ZnBr_2 > CaCl_2$ (technical) $>$ $CuSO_4$

$>$ Na_2SO_4, K_2CO_3.

Where large amounts of water have to be removed, a preliminary drying of liquids is often possible by shaking with concentrated solutions of calcium chloride or potassium carbonate, or by adding sodium
chloride to salt out the organic phase (for example, in the drying
of lower alcohols).

Drying agents that combine irreversibly with water include the
alkali metals, the metal hydrides (discussed in Chapter 2), and
calcium carbide.

Suitability of Individual Desiccants

Alumina. (Preheated to 175° for about 7 hr). Mainly as a drying
agent in a desiccator or as a column through which liquid is
percolated.

Aluminium amalgam. Mainly used for removing traces of water from
alcohols, which are distilled from it after refluxing.

Barium oxide. Suitable for drying organic bases.

Barium perchlorate. Expensive. Used in desiccators. Unsuitable
for drying solvents or any organic material where contact is necessary, because of the danger of explosion.

Boric anhydride. (Prepared by melting boric acid in an air oven at a high temperature, cooling in a desiccator, and powdering.) Mainly used for drying formic acid.

Calcium chloride (anhydrous). Cheap. Large capacity for absorption of water, giving the hexahydrate below 30°, but is fairly slow in action and not very efficient. Its main use is for preliminary drying of alkyl and aryl halides, most esters, saturated and aromatic hydro-carbons, and ethers. Unsuitable for drying alcohols and amines (which form addition compounds), fatty acids, amides, aminoacids, ketones, phenols, or some aldehydes and esters. Calcium chloride is suitable for drying the following gases—hydrogen, hydrogen chloride, carbon monoxide, carbon dioxide, sulphur dioxide, nitrogen, methane, oxygen, paraffins, ethers, olefins and alkyl chlorides.

Calcium hydride. See Chapter 2.

Calcium oxide. (Preheated to 700-900° before use.) Suitable for alcohols and amines (but does not dry them completely). Need not be removed before distillation, but in that case the head of the distillation column should be packed with glass wool to trap any calcium oxide powder that might be carried over. Unsuitable for acidic compounds or esters. Suitable for drying gaseous amines and ammonia.

Calcium sulphate (anhydrous). (Prepared by heating the dihydrate or the hemihydrate in an oven at 235° for 2-3 hr; it can be regener-ated.) Available commercially as Drierite. It forms the hemihydrate, $2CaSO_4.H_2O$, so that its capacity is fairly low (6.6% of its weight of water), and hence is best used on partially dried substances. It is very rapid and efficient (being comparable with phosphorus pent-oxide and concentrated sulphuric acid). Suitable for most organic compounds. Solvents boiling below 100° can be dried by direct distillation from calcium sulphate.

Copper(II) sulphate (anhydrous). Suitable for esters and alcohol. Preferable to sodium sulphate in cases where solvents are sparingly soluble in water (for example, benzene or toluene).

Lithium aluminium hydride. See Chapter 2.

Magnesium amalgam. Mainly used for removing traces of water from alcohols, which are distilled from it after refluxing.

Magnesium perchlorate (anhydrous). (Available commercially as Dehydrite. Expensive.) Used in desiccators. Unsuitable for drying solvents or any organic material where contact is necessary, because of the danger of explosion.

Magnesium sulphate (anhydrous). (Prepared from the heptahydrate by drying at 300° under reduced pressure.) More rapid and effective than sodium sulphate. It has a large capacity, forming $MgSO_4.7H_2O$ below 48°. Suitable for the preliminary drying of most organic compounds.

Molecular sieves. See page 40.

Phosphorus pentoxide. Very rapid and efficient, but difficult to handle and should only be used after the organic material has been partially dried, for example with magnesium sulphate. Suitable for acid anhydrides, alkyl and aryl halides, esters, ethers, hydrocarbons and nitriles, and for use in desiccators. Not suitable with acids, alcohols, amines or ketones, or with organic molecules from which a molecule of water can fairly readily be abstracted by an elimination reaction. Suitable for drying the following gases—hydrogen, oxygen, carbon dioxide, carbon monoxide, sulphur dioxide, nitrogen, methane, ethylene and paraffins.

Potassium (metal). Properties and applications are similar to those for sodium, and it is a correspondingly hazardous substance.

Potassium carbonate (anhydrous). Has a moderate efficiency and capacity, forming the dihydrate. Suitable for an initial drying of alcohols, bases, esters, ketones and nitriles by shaking with them, then filtering off. Also suitable for salting out water-soluble alcohols, amines and ketones. Unsuitable for acids, phenols and other acidic substances.

Potassium hydroxide. Solid potassium hydroxide is very rapid and efficient. Its use is limited almost entirely to the initial drying of organic bases. Alternatively, sometimes the base is shaken first with a concentrated solution of potassium hydroxide to remove most of the water present. Unsuitable for acids, aldehydes, amides, esters, ketones or phenols. Also used for drying gaseous amines and ammonia.

Silica gel. Granulated silica gel is a commercially available drying agent for use with gases, in desiccators, and (because of its chemical inertness) in physical instruments (pH meters, spectrophotometers, balances). Its drying action depends on physical adsorption, so that silica gel must be used at room temperature or below. By incorporating cobalt chloride into the material it can be made self-indicating, redrying in an oven at 110° being necessary when the colour changes from blue to pink.

Sodium (metal). Used as fine wire or as chips, for more completely drying ethers, saturated hydrocarbons and aromatic hydrocarbons which have been partially dried (for example with calcium chloride or magnesium sulphate). Unsuitable for acids, alcohols, aldehydes, amines, esters, organic halides or ketones. Reacts violently if much water is present.

Sodium hydroxide. Properties and applications are similar to those for potassium hydroxide.

Sodium-potassium alloy. Used as lumps. Lower melting than sodium, so that its surface is readily renewed by shaking. Properties and applications are similar to those for sodium.

Sodium sulphate (anhydrous). Has a large capacity for the absorption of water, forming the decahydrate below 33°, but drying is slow and inefficient, especially for solvents that are sparingly soluble in water. It is suitable for the preliminary drying of most types of organic compounds.

Sulphuric acid (concentrated). Widely used in desiccators. Suitable for drying bromine, saturated hydrocarbons, alkyl and aryl halides. Also suitable for drying the following gases — hydrogen, nitrogen, carbon dioxide, carbon monoxide, chlorine, methane and paraffins. Unsuitable for alcohols, bases, ketones or phenols.

For convenience, the above drying agents are listed in Table 10 under the classes of organic compound for which they are commonly used.

CHROMATOGRAPHY

Chromatography is often used with advantage for the purification of small amounts of complex organic mixtures, either as liquid chromatography or as vapour phase (gas) chromatography.

The mobile phase in liquid chromatography is a liquid and the stationary phase is of four main types. These are for adsorption, partition, ion-exchange, and gel filtration. The technique of chromatography which applies to all liquid chromatography at atmospheric pressure comprises the following distinct steps. The material is adsorbed as a level bed onto the column of stationary phase. (It is important that this bed is as narrow as possible because the bands of components in the mixture that is applied widen as they move with the mobile phase down the column.) The column is washed (developed) with a quantity of pure solvent or solvent mixture. The column may be pushed out of the tube so that it can be divided into zones. The desired components are then extracted from the appropriate zones using a suitable solvent. Alternatively, and more commonly, the column is left intact and the bands are progressively eluted by passing more solvent through the column.

The mobile phase in vapour phase chromatography is a gas (e.g. hydrogen, nitrogen, helium or argon) and the stationary phase is a nonvolatile liquid impregnated onto a porous material. The mixture to be purified is injected into a heated inlet whereby it is vaporized and taken into the column by the carrier gas. It is separated into its components by partition between the liquid on the porous support and the gas. For this reason vapour-phase chromatography is sometimes referred to as gas-liquid chromatography.

Adsorption Chromatography

Adsorption chromatography is based on differences in the extent to which substances in solution are adsorbed onto a suitable surface. The substances to be purified are usually placed on the top of the column and the solvent is run down the column. In a more common variation of this method, the column containing the adsorbent is full of solvent before applying the mixture at the top of the column. In another application the mixture is adsorbed onto a small amount of stationary phase and placed at the bottom of the column with the dry stationary phase above it. By applying a slight vacuum at the top of the column, the eluting solvent can be sucked slowly upwards from the bottom of the column. When the solvent has reached the top of the column the separation is complete and the vacuum is released. The packing is pushed gently out of the tube and cut into strips as above.   Alternatively the vacuum is kept and the effluent from the top of the column is collected in fractions. The fractions are monitored by U.V. or visible spectra, colour reactions or other means for identifying the components.

Graded adsorbents and solvents. Materials used in columns for adsorption chromatography are grouped in Table 11 in an approximate order of decreasing effectiveness. Other adsorbents sometimes used include barium carbonate, calcium sulphate, charcoal (usually mixed with kieselguhr or other form of diatomaceous earth, for example, the filter aid Celite),cellulose, glucose and lactose. The alumina can be prepared in several grades of activity (see below).

In most cases, adsorption takes place most readily from non-polar solvents, such as petroleum ether or benzene, and least readily from polar solvents such as alcohols, esters, and acetic acid. Common solvents, arranged in approximate order of increasing eluting ability are also given in Table 11.

Eluting power roughly parallels the dielectric constants of the solvents. The series also reflects the extent to which the solvent binds to the column material, thereby displacing the substances that are already adsorbed. This preference of alumina and silica gel for polar molecules explains, for example, the use of percolation through a column of silica gel for the following purposes—drying of ethyl-benzene, removal of aromatics from paraffin hydrocarbons, of halogen-containing impurities from 2,4-dimethylpentane and of ultraviolet-absorbing substances from cyclohexane.

Mixed solvents are intermediate in strength, and so provide a finely graded series. In choosing a solvent for use as an eluent it is necessary to consider the solubility of the substance in it, and the ease with which it can subsequently be removed.

Preparation and standardization of alumina. The activity of alumina depends inversely on its water content, and a sample of poorly active material can be rendered more active by leaving for some time in a round-bottom flask heated up to about 200° in an oil bath or a heating mantle while a slow stream of a dry inert gas is passed through it. Alternatively, it is heated to red heat (380-400°) in an open vessel for 4-6 hr. with occasional stirring and then cooled in a vacuum desiccator: this material is then of grade I activity. Conversely, alumina can be rendered less active by adding small amounts of water and thoroughly mixing for several hours. Addition of about 3% (w/w) of water converts grade I alumina to grade II.

Used alumina can be regenerated by repeated extraction, first with boiling methanol, then with boiling water, followed by drying and heating. The degree of activity of the material can be expressed conveniently in terms of the scale due to Brockmann and Schodder (Ber. B 74, 73 (1941)).

This system is based on the extent of adsorption of five pairs of azo dyestuffs, being adjacent members of the set—azobenzene, p-methoxyazo-benzene, Sudan yellow, Sudan red, aminoazobenzene, hydroxyazobenzene. In testing the alumina, a tube 10 cm long by 1.5 cm internal diameter is packed with alumina to a depth of 5 cm and covered with a disc of filter paper. The dyestuff solutions are prepared by dissolving 2 mg of each azo dye of the pair in 2 ml of purified benzene (distilled

from potassium hydroxide) and 8 ml of petroleum ether. The solution
is applied to the column and developed with 20 ml of benzene—
petroleum ether mixture (1:4 v/v) at a flow rate of about 20-30 drops
per min. The following behaviour is observed:

| Grade | Position of zones | | |
	(a)	(b)	(c)
I	p-Methoxyazobenzene	Azobenzene	
II	p-Methoxyazobenzene (d)		Azobenzene
II	Sudan yellow	p-Methoxyazobenzene	
III		Sudan yellow	p-Methoxyazobenzene
III	Sudan red	Sudan yellow	
IV	Sudan red (d)		Sudan yellow
IV	Aminoazobenzene	Sudan red	
V	Hydroxyazobenzene	Aminoazobenzene	

(a) Near top of column. (b) Near bottom of column. (c) In effluent.
(d) 1 to 2 cm from top. Grade I is the most active, grade V the
least active.

Alumina is normally slightly alkaline. A (less strongly adsorbing)
neutral alumina can be prepared by making a slurry in water and
adding 2 M hydrochloric acid until the solution is acid to Congo red.
The alumina is then filtered off, washed with distilled water until
the wash water gives only a weak violet colour with Congo red paper,
and dried.

   Preparation of Other Adsorbents. Silica gel can be prepared from
commercial water-glass by diluting it with water to a density of 1.19
and, while keeping it cooled to 5°, adding concentrated hydrochloric
acid with stirring until the solution is acid to thymol blue. After
standing for 3 hr, the precipitate is filtered off, washed on a
Büchner funnel with distilled water, then suspended in 0.2M hydrochloric acid. The suspension is stood for 2-3 days, with occasional
stirring, then filtered, washed well with water and dried at 110°.
It can be activated by heating up to about 200° as described for
alumina.

Powdered commercial silica gel can be purified by suspending and
standing overnight in concentrated hydrochloric acid (6 ml/g),
decanting the supernatant and repeating with fresh acid until the
latter remains colourless. After filtering with suction on a
sintered-glass funnel, the residue is suspended in water and washed
by decantation until free of chloride ion. It is then filtered,
suspended in 95% ethanol, filtered again and washed on the filter
with 95% ethanol. The process is repeated with anhydrous ethyl ether
before the gel is heated for 24 hr at 100° and stored for another
24 hr in a vacuum desiccator over phosphorus pentoxide.

   Diatomaceous earth (Celite 535 or 545, Hyflo Super-Cel, Dicalite,
kieselguhr) is purified before use by washing with 3 M hydrochloric
acid, then with water, or it is made into a slurry with hot water,
filtered at the pump and washed with water at 50° until the filtrate

is no longer alkaline to litmus. Organic materials can be removed by
repeated extraction at 50° with methanol, benzene or chloroform,
followed by washing with methanol, filtering and drying at 90-100°.

Activation of charcoal is generally achieved satisfactorily by heating gently to red heat in a crucible or quartz beaker in a muffle
furnace, finally allowing to cool under an inert atmosphere in a
desiccator. To improve their porosity, charcoal columns are usually
prepared in admixture with diatomaceous earth.

Purification of cellulose for chromatography is by sequential washing
with chloroform, ethanol, water, ethanol, chloroform and acetone.
More extensive purification uses aqueous ammonia, water, hydrochloric
acid, water, acetone and ethyl ether, followed by drying in a vacuum.
Trace metals can be removed from filter paper by washing for several
hours with 0.1 M oxalic or citric acid, followed by repeated washing
with distilled water.

Partition Chromatography

Partition chromatography is concerned with the distribution of
substances between a mobile phase and a non-volatile liquid which is
itself adsorbed onto an inert supporting stationary phase. The mobile
phase may be a gas (see vapour phase chromatography) or a liquid.
Paper chromatography, and reverse-phase thin layer chromatography
are other applications of partition chromatography. Yet another
application is paired-ion chromatography which is used for the
separation of substances by virtue of their ionic properties. In
principle, the separation of components of a mixture depends on the
differences in their distribution ratios between the mobile phase and
the liquid stationary phase. The more the distribution of a substance
favours the stationary phase, the more slowly it progresses through
the column.

When cellulose is used as a stationary phase, with water or aqueous
organic solvents as eluents, the separation of substances is by
partition between the eluting mixture and the water adsorbed on the
column. This is similar to the cellulose in paper chromatography.

For chromatography on dextran gels see page 32.

Paired-ion Chromatography

Mixtures containing ionic compounds (e.g. acids and/or bases), ionic
and non-ionizable compounds, and zwitterions, can be separated
successfully by paired-ion chromatography (PIC). It utilizes the
'reverse-phase' technique (Eksberg and Schill, Anal.Chem. 45, 2092
(1973)). The stationary phase is lipophilic, such as μ-BONDAPAK C_{18}
(Waters Assoc.) or any other adsorbent that is compatible with
water. The mobile phase is water or aqueous methanol containing the
acidic or basic counter ions. Thus the mobile phase consists of
dilute solutions of strong acids (e.g. 5 mM 1-heptanesulphonic acid)
or strong bases (e.g. 5 mM tetrabutylammonium phosphate) that are
completely ionized at the operating pH values which are usually
between 2 and 8. An equilibrium is set up between the neutral species
of a mixture in the stationary phase and the respective ionized
(anion or cation) species which dissolve in the mobile phase

containing the counter ions. The extent of the equilibrium will depend on the ionization constants of the respective components of the mixture, and the solubility of the unionized species in the stationary phase. Since the ionization constants and the solubility in the stationary phase will vary with the water-methanol ratio of the mobile phase, the separation may be improved by altering this ratio gradually (gradient elution) or stepwise. If the compounds are eluted too rapidly the water content of the mobile phase should be increased, e.g. by steps of 10%. Conversely, if components do not move, or move slowly, the methanol content of the mobile phase should be increased by steps of 10%.

The application of pressure to the liquid phase in liquid chromato- graphy generally increases the separation (see HPLC). Also in PIC improved efficiency of the column is observed if pressure is applied to the mobile phase (Wittmer, Nuessle and Haney, Anal.Chem., 47, 1422 (1975)).

Ion-exchange Chromatography

Ion-exchange chromatography involves an electrostatic process which depends on the relative affinities of various types of ions for an immobilized assembly of ions of opposite charge.The stationary phases are weak or strong anion or cation supports. The mobile phase is an aqueous buffer with a fixed pH or an aqueous mixture of buffers in which the pH is continuously increased or decreased as the separation may require. This form of liquid chromatography can be performed at high inlet pressures of liquid with increased column performance.

   Ion-exchange resins. An ion-exchange resin is made up of particles of an insoluble elastic hydrocarbon network to which is attached a large number of ionizable groups. Materials commonly used comprise synthetic ion-exchange resins made, for example, by crosslinking polystyrene to which has been attached nondiffusible ionized or ionizable groups. These groups are further distinguished as strong ($-SO_2OH$, $-NR_3^+$) or weak ($-OH$, $-COOH$, $-PO(OH)_2$, $-NH_2$). (For an extensive collection listing most of the commercially available materials and their characteristics, see Ion Exchangers in Organic and Biochemistry, C. Calmon and T.R.E. Kressman, Interscience, New York, 1957, pp.116-129.) Their charges are counterbalanced by diffusible ions, and the operation of a column depends on its ability and selectivity to replace these ions. The exchange that takes place is primarily an electrostatic process but adsorptive forces and hydrogen bonding can also be important. A typical sequence for the relative affinities of some common anions (and hence the inverse order in which they pass through such a column), is the following, obtained using a quaternary ammonium (strong-base) anion-exchange column:

   fluoride < acetate < bicarbonate < hydroxide < formate < chloride < bromate < nitrate < cyanide < bromide < chromate < nitrate < iodide < thiocyanate < oxalate < sulphate < citrate.

For an amine (weak-base) anion-exchange column in its chloride form, the following order has been observed:

fluoride < chloride < bromide = iodide = acetate < molybdate
< phosphate < arsenate < nitrate < tartrate < citrate < chromate
< sulphate < hydroxide.

With strong cation-exchangers, the usual sequence is that polyvalent
ions bind more firmly than di- or mono-valent ones, a typical series
being as follows:

$$Th^{4+} > Fe^{3+} > Al^{3+} > Ba^{2+} > Pb^{2+} > Sr^{2+} > Ca^{2+} > Co^{2+} > Ni^{2+}$$

$$= Cu^{2+} > Zn^{2+} = Mg^{2+} > UO_2^+ = Mn^{2+} > Ag^+ > Tl^+ > Cs^+ > Rb^+$$

$$> NH_4^+ = K^+ > Na^+ > H^+ > Li^+.$$

Thus, if an aqueous solution of a sodium salt contaminated with heavy
metals is passed through the sodium form of such a column, the heavy
metal ions will be removed from the solution and will be replaced by
sodium ions from the column. This effect is greatest in dilute solution. Passage of sufficiently strong solutions of alkali metal salts
or mineral acids readily displaces all other cations from ion-
exchange columns. (The regeneration of columns depends on this
property.) However, when the cations lie well to the left in the
above series it is often advantageous to use complex-forming species
to facilitate removal. For example, iron can be displaced from ion
exchange columns by passage of sodium citrate or sodium ethylenediaminetetraacetate.

Some of the more common commercially available resins are listed in
Table 12.

Ion-exchange resins swell in water to an extent which depends on the
amount of crosslinking in the polymer, so that columns should be
prepared from the wet material by adding it as a suspension in water
to a tube already partly filled with water. (This also avoids trapping air bubbles.) The exchange capacity of a resin is commonly
expressed as mg equiv./ml of wet resin. This quantity is pH-dependent
for weak-acid or weak-base resins but is constant at about 0.6-2 for
most of the strong-acid or strong-base types.

Apart from their obvious applications to inorganic species, sulphonic
acid resins have been used in purifying amino acids, aminosugars,
organic bases, peptides, purines and pyrimidines. Thus, organic bases
can be applied to the H^+ form of such resins by adsorbing them from
neutral solution and, after washing with water, they are eluted
sequentially with suitable buffer solutions or dilute acids. Alternatively, by passing alkali solution through the column, the bases
will be displaced in an order that is governed by their pK values.
Similarly, strong-base anion exchangers have been used for aldehydes
and ketones (as bisulphite addition compounds), carbohydrates (as
their borate complexes), nucleosides, nucleotides, organic acids,
phosphate esters and uronic acids. Weakly acidic and weakly basic
exchange resins have also found extensive applications, mainly in
resolving weakly basic and acidic species. For demineralization of
solutions without large changes in pH, mixed-bed resins can be
prepared by mixing a cation-exchange resin in its H^+ form with an
anion-exchange resin in its OH form. Commercial examples include
Amberlite MB-1 (IR-120 + IRA-400) and Bio-Deminrolit (Zeo-Karb 225 and

Zerolit FF). The latter is also available in a self-indicating form.

   Ion-exchange celluloses and Sephadex. A different type of ion-exchange column that is finding extensive application in biochemistry for the purification of proteins, nucleic acids and acidic poly-saccharides derives from cellulose by incorporating acidic and basic groups to give ion-exchange resins of controlled acid and basic strengths. Commercially available cellulose-type resins are given in Tables 13 and 14.

Ion-exchange celluloses are available in different particle sizes. It is important that the amounts of 'fines' are kept to a minimum otherwise the flow of liquid through the column can be extremely slow and almost stop. Celluloses with a large range of particle sizes should be freed from 'fines' before use. This is done by suspending the powder in the required buffer and allowing it to settle for one hour and then decanting the 'fines'. This separation appears to be wasteful but it is necessary for reasonable flow rates without applying high pressures at the top of the column. Several ion-exchange celluloses require recycling before use, a process which must be performed with recovered celluloses. Recycling is done by stirring the cellulose with 0.1N aqueous sodium hydroxide, washing with water until neutral, then suspending in 0.1N hydro-chloric acid and finally washing with water until neutral. When regenerating a column it is advisable to wash with a salt solution (containing the required counter ions) of increasing ionic strength up to 2. The cellulose is then washed with water and recycled if necessary. Recycling can be carried out more than once if there are doubts about the purity of the cellulose. The basic matrix of these ion-exchangers is cellulose and it is important not to subject them to strongly acid (> 1$\underline{N}$) and strongly basic (> 1$\underline{N}$) solutions.

When storing ion-exchange celluloses, or during prolonged usage, it is important to avoid growth of microorganisms or moulds which slowly destroy the cellulose. Good inhibitors of microorganisms are phenyl mercuric salts (0.001%, effective in weakly alkaline sol-utions) and chlorohexidine (Hibitane at 0.002% for anion exchangers), 0.02% aqueous sodium azide or 0.005% of ethyl mercuric thiosali-cylate (Merthiolate) are most effective in weakly acidic solutions for cation exchangers. Trichlorobutanol (Chloretone, at 0.05% is only effective in weakly acidic solutions) can be used for both anion and cation exchangers. Most organic solvents are effective antimicrobial agents but only at high concentrations.

Sephadex ion-exchangers, unlike celluloses, are available in narrow ranges of particle size. These are of two medium types, the G-25 and G-50, and their dry bead diameter sizes are 50 to 150 microns. They are available as cation and as anion exchange Sephadex. With these it is not necessary to remove 'fines' and the flow rates are always satisfactory.

In preparing any of the above for use in columns, the dry powder is evacuated, then mixed under reduced pressure with water or the appropriate solution. Alternatively it is stirred gently with the solution until all air bubbles are removed. Because the wet powders change volumes reversibly with alteration of pH or ionic strength, it is imperative to make allowances when packing columns in order to avoid overflowing of packing when the pH or ionic strength are altered.

Gel Filtration

The corresponding CM, DEAE and SE resins based on dextran are sold in bead form as Sephadex ion exchangers. The gel-like nature of wet Sephadex (a modified dextran) enables small molecules such as inorganic salts to diffuse freely into it while, at the same time, protein molecules are unable to do so. Hence, passage through a Sephadex column can be used for complete removal of salts from protein solutions. Polysaccharides can be freed from monosaccharides and other small molecules because of their differential retardation. Similarly, amino acids can be separated from proteins and large peptides.

Gel filtration using Sephadex G-types (50 to 200, from Pharmacia, Uppsala, Sweden) is essentially useful for fractionation of large molecules with molecular weights above 1000. Fractionation of lower molecular weight solutes (e.g. ethylene glycols, benzyl alcohols) can now be achieved with Sephadex G-10 (up to Mol. Wt 700) and G-25 (up to Mol. Wt 1500). These dextrans are used only in aqueous solutions. More recently, however, Sephadex LH-20 and LH-60 (prepared by hydroxypropylation of Sephadex) have become available and are used for the separation of small molecules (Mol. Wt less than 500) using most of the common organic solvents as well as in water. Sephasorb HP (ultrafine, prepared by hydroxypropylation of cross-linked dextran) can also be used for the separation of small molecules in organic solvents and water, and in addition it can withstand pressures up to 1400 p.s.i. making it useful in HPLC. Because solutions with high and low pH values slowly decompose, these gels are best operated at pH values between 2 and 12.

High Performance Liquid Chromatography (HPLC)

When pressure is applied at the inlet of a liquid chromatographic column the performance of the column can be increased by several orders of magnitude. This is partly because of the increased speed at which the liquid flows through the column and partly because fine column packings can be used which have larger surface areas. Because of the improved efficiency of the columns this technique has been referred to as high performance, high pressure, or high speed liquid chromatography.

The basic equipment consists of a hydraulic system to provide the pressure at the inlet of the column, a column, a detector and a recorder. The pressures used in HPLC vary from a few p.s.i. to 4000-5000 p.s.i. The most convenient working pressures are, however, between 600 and 1800 p.s.i. The plumbing is made of stainless steel or non-corrosive metal tubing to withstand high pressures. Since increase of temperature has a very small effect on the performance of a column in liquid chromatography, the column is placed in an oven at ambient temperature in order to have a constant temperature along the column so as not to upset the equilibria. The packing (stationary phase) is specially prepared for withstanding high pressures. It may be an adsorbent (for adsorption or solid-liquid HPLC), a material impregnated with a high boiling insoluble liquid (reverse-phase or

liquid-liquid HPLC, or paired-ion HPLC), an ion-exchange material
(for ion-exchange HPLC), or a highly porous non-ionic gel (for high
performance gel filtration). The mobile phase is water, aqueous
buffers, salt solutions, aqueous organic solvents or organic solvents.
The more commonly used detectors have UV, visible, or fluorescence
monitoring for light absorbing substances, and refractive index
monitoring for transparent compounds. The sensitivity of the refractive index monitoring is usually lower than the light absorbing monitoring by a factor of ten or more. The cells of the monitoring
devices are very small (ca 5µl) and the detection is very good. The
volumes of the analytical columns are quite small (ca 2 ml for a 1 m
column) hence the result of an analysis is achieved very quickly.
Larger columns have been used for preparative work and can be used
with the same equipment. Most modern machines have solvent mixing
chambers for gradient or ion gradient elution. The solvent gradient
(for two solvents) or pH or ion gradient can be adjusted in a linear,
increasing or decreasing exponential manner. Some of the more common
columns are given in Table 15.

Other Types of Liquid Column Chromatography

New stationary phases for specific purposes in chromatographic separations are being continuously proposed. Charge transfer adsorption
chromatography makes use of a stationary phase which contains immobilized aromatic compounds and permits the separation of aromatic
compounds by virtue of their ability to form charge transfer complexes (sometimes coloured) with the stationary phase. The separation
is caused by the differences in stability of these complexes (Porath
and Dahlgren-Caldwell, J.Chromatog., 133, 180 (1977)).

In metal chelate adsorption chromatography a metal is immobilized by
partial chelation on a column which contains bi- or tri- dentate
ligands. Its application is in the separation of substances which can
complex with the bound metals and depends on the stability constants
of the various ligands (Porath, Carlsson, Olsson, and Belfrage,
Nature, 258, 598 (1975); Lönnerdal, Carlsson, and Porath,
Febs.Letts., 75, 89 (1977)).

An application of chromatography which has found extensive use in
biochemistry and has brought a new dimension in the purification of
enzymes is affinity chromatography. A specific enzyme inhibitor is
attached by covalent bonding to a stationary phase (e.g.
AH-Sepharose 4B for acidic inhibitors and CH-Sepharose 4B for basic
inhibitors), and will strongly adsorb only the specific enzyme which
is inhibited, allowing all other proteins to flow through the column.
The enzyme is then eluted with a solution of high ionic strength
(e.g. 1 M sodium chloride) or a solution containing a substrate or
reversible inhibitor of the specific enzyme. Considerable purification
can be achieved by one passage through the column and the column can
be reused several times.

Hydrophobic adsorption chromatography takes advantage of the hydrophobic properties of substances to be separated and has also found
use in biochemistry (Hoftsee, Biochem. Biophys. Res. Comm., 50, 751
(1973); Jennissen and Heilmeyer Jr., Biochemistry, 14, 754 (1975)).
Specific covalent binding with the stationary phase, a procedure that
was called covalent chromatography, has been used for the separation

of compounds and for immobilizing enzymes on a support: the column
was then used to carry out specific bioorganic reactions (Mosbach,
Methods in Enzymology, 1976, vol.44).

Vapour Phase Chromatography

Although this technique was first used for analytical purposes in
1952, its application to the purification of chemicals at a preparative level is much more recent and commercial instruments for this
purpose are currently in a state of rapid development. This type of
partition chromatography uses a tubular column packed with an inert
material which is impregnated with a liquid. This liquid separates
components of gases or vapours as they flow through the column. On a
preparative scale, use of a large column heated slightly above the
boiling point of the material to be processed makes it possible to
purify in this way small quantities of many volatile organic substances. For example, if the impurities have a greater affinity for the
liquid in the column than the desired component has, the latter will
emerge first and in a substantially pure form.

In operation, the organic material is carried as a vapour in a gas
such as hydrogen, helium, carbon dioxide, nitrogen or argon (in a
manner analogous to a solution in a suitable solvent in liquid-phase
chromatography). The technique that is almost invariably used is to
inject the substance (for example, by means of a hypodermic syringe)
over a relatively short time on to the surface of the column through
which is maintained a slow continuous passage of the chemically inert
carrier gas. This leads to the progressive elution of individual
components from the column in a manner analogous to the movement of
bands in conventional chromatography. As substances emerge from the
column they can be condensed in suitable traps. The carrier gas blows
the vapour through these traps hence these traps have to be very
efficient. Improved collection of the effluent vaporized fractions in
preparative work is attained by strong cooling, increasing the
surface of the traps by packing them with glass wool, and by applying
an electrical potential which neutralizes the charged vapour and
causes it to condense.

The choice of the gas for use as a carrier is largely determined by
the type of detection system that is available (see below). Column
efficiency is greater in argon, nitrogen or carbon dioxide than it is
in helium or hydrogen, but the latter are less impeded in flowing
through packed columns so that lower pressure differentials exist
between inlet and outlet. The packing in the column is usually an
inert supporting material such as powdered firebrick, or a firebrick
—Celite mixture coated with a high-boiling organic liquid as the
stationary phase. These liquids include Apiezon oils and greases,
di-esters (such as dibutylphthalate or di-(2-ethylhexyl)sebacate),
polyesters (such as diethyleneglycol sebacate), polyethylene glycols,
hydrocarbons (such as Nujol or squalane), silicone oils and
tricresyl phosphate.

The coating material (about 75 ml per 100 ml of column packing) is
applied as a solution in a suitable solvent such as methylene
dichloride, acetone, methanol or pentane, which is then allowed to
evaporate in air, over a steam-bath, or in a vacuum oven (provided
the adsorbed substance is sufficiently non-volatile). The order in

which a mixture of substances travels through such columns depends on
their relative solubilities in the materials making up the stationary
phases. Some examples of liquids used in gas chromatography for
particular types of mixtures are given in Table 16 and below.

The three main requirements of a liquid for use in a gas chromatograph column are that it must have a high boiling point, a low vapour
pressure, and at the same time permit adequate separation of components fairly rapidly. As a rough guide, the boiling point should be at
least 250° above the temperature of the column, and, at column
temperature, the liquid should not be too viscous, nor should it
react chemically with the sample. Liquids suitable for use as
stationary phases in gas chromatography are given in Table 16 and below.

Stationary phase	Mixture
Benzyl diphenyl	Aromatic molecules
Benzyl ether	Saturated hydrocarbons from olefins
Bis(2-n-butoxyethyl)phthalate	Saturated hydrocarbons from olefins
Diethylene glycol adipate	Methyl esters of fatty acids up to C_{24}
Dimethyl sulpholane(below 40°)	Saturated and unsaturated hydrocarbons
Dinonyl phthalate	Paraffins, olefins, low molecular weight aromatics, alcohols (up to amyl alcohol), lower ethers, esters and carbonyl compounds
Hexadecane	Low-boiling hydrocarbons
Mineral oil	Aliphatic and aromatic amines
β,β'-Oxydipropionitrile	Paraffins, cycloalkanes, olefins, ethers, alkylbenzenes, acetates, aldehydes, alcohols, acetals and ketones
Polyethylene glycols	Aromatic molecules from paraffins
Silicone oil	Aromatic hydrocarbons, alcohols, esters
Silicone-stearic acid	Fatty acids
Squalane	Saturated hydrocarbons
Tricresyl phosphate	Hexanes, heptanes, aromatics, organic sulphur compounds and aliphatic chlorides

Where the stationary phase is chemically similar to the material to be
separated, the main factors governing the separation will be the
molecular weight and the shape. Otherwise, polar interactions must
also be considered, for example hydroxylated compounds used for
stationary phases are likely to retard the movement through the
column of substances with hydrogen-accepting groups. A useful guide to
the selection of a suitable stationary phase is to compare, on the
basis of polarity, possible materials with the components to be separated. This means that, in general, solute and solvent will be members
of the same, or of adjacent, classes in the following groupings:

    A. Water, polyhydric alcohols, aminoalcohols, oxyacids,
       polyphenols, di- and tricarboxylic acids.

B. Alcohols, fatty acids, phenols, primary and secondary amines, oximes, nitro compounds, nitriles with α-H atoms.

C. Ethers, ketones, aldehydes, esters, tertiary amines, nitriles without α-H atoms.

D. Chlorinated, aromatic or olefinic hydrocarbons.

E. Saturated hydrocarbons, carbon disulphide, tetrachloromethane.

Material emerging from the column is detected by a thermal-conductivity cell, an ionization method, or a gas-density balance.

The first of these methods, which is applicable when hydrogen or helium is used as carrier gas, depends on the differences in heat conductivities between these gases and most others, including organic substances. The resistance of a tungsten or platinum wire heated by a constant electric current will vary with its temperature which, in turn, is a function of the thermal conductivity through the gas. These devices, also known as catharometers, can detect about 10^{-8} moles of substance. When argon is used as carrier gas, an ionization method is practicable. It is based on measurement of the current between two electrodes at different voltages in the presence of a suitable emitter of β-radiation. The gas-density balance method depends on measurement of the difference in thermal e.m.f. between two equally warmed copper-constantan thermocouples located in the cross-channel of what constitutes a mechanical equivalent to the Wheatstone bridge. Any increase in density of the effluent gas relative to the reference gas will cause movement of gas along the cross-channel, and hence cool one of the thermocouples relative to the other. The technique is comparable in sensitivity with the thermal-conductivity method.

Paper Chromatography

Paper chromatography is basically a type of partition chromatography between water adsorbed onto the cellulose fibre of the paper and a liquid mobile phase in a closed tank. The most common application is the ascending solvent technique. The paper is hung by means of clips or string and the lower end is made to dip into the eluting solvent. The material under test is applied as a spot 2.5 cm or so above the lower end of the paper and marked with a pencil. It is important that the spots are above the eluting solvent before it begins to rise up the paper by capillarity. Eluents are normally aqueous mixtures of organic solvents, acids or bases. (For solvent systems see Lederer and Lederer.) The descending technique has also been used, and in this case the top of the paper dips into a trough containing the eluent which travels downwards, also by capillarity. The spots are applied at the top of the paper close to the solvent trough. A closed tank is necessary for these operations because better reproducibility is achieved if the solvent and vapour in the tank are in equilibrium. The tanks have to be kept away from draughts. Elution times vary from several hours to a day depending on the solvent system and paper. For more efficient separations the dried paper is eluted with a different solvent along a direction which is 90° from that of the first elution. This is referred as two dimensional paper chromatography. In a third application ordinary

circular filter papers are used. The filter paper is placed between
two glass plates. The upper plate has a hole in the centre which is
coincident with the centre of the paper. A strong solution of the
mixture is placed in this hole followed by the eluting solvent. The
substances in the mixture are then separated radially by the
eluting solvent. After the solvents have travelled the required
distances in the above separations, the papers are air dried and the
spots are revealed by their natural colours or, by spraying with a
reagent that forms a coloured product with the spots. In many cases,
the positions of the spots can be seen as light fluorescing or
absorbing spots when viewed under UV light.

The use of thick filter paper such as Whatman nos. 3 or 31 (0.3-0.5 mm)
increases the amounts that can be handled (up to about 100 mg per
sheet). Larger quantities require multiple sheets or cardboard, e.g.
Schleicher and Schüll nos. 2071 (0.65 mm), 2230 (0.9 mm) or 2181
(4 mm). For even larger amounts recourse may be had to chromatopack
or chromatopile procedures. The latter use a large number (200-500)
of identical filter papers stacked and compressed in a column, the
material to be purified being adsorbed onto a small number of these
discs which, after drying, are placed almost at the top of the
column. The column is then subjected to descending development, and
bands are separated mechanically by disassembling the filter papers.
Instead of filter papers, cellulose powder may be suitable, the
column being packed by first suspending the powder in the solvent to
be used for development. Yet another variation employs tightly wound
paper roll columns contained in thin polythene skins. (These are
unsuitable for such solvents as benzene, chloroform, collidine,
ethyl ether, pyridine and toluene.)

The technique of paper chromatography has been almost entirely
superseded by thin- or thick-layer chromatography.

Thin or Thick Layer Chromatography (TLC)

Thin layer chromatography is in principle similar to paper chromatography when used in the ascending method, i.e. the solvent creeps up
the stationary phase by capillarity. The adsorbent (e.g.
silica, alumina, cellulose) is spread on a rectangular glass plate
(or solid inert plastic sheet). Some adsorbents (e.g. silica) are
mixed with a setting material (e.g. $CaSO_4$) by the manufacturers
which cause the film to set on drying. The adsorbent can be
activated by heating at 100-110° for a few hours. Other adsorbents
(e.g. celluloses) adhere on glass plates without a setting agent.
The materials to be purified are spotted in a solvent about 2.5 cm
from the bottom of the plate and allowed to dry. The plate is then
placed upright in a tank containing the eluting solvent which should
be a few millimetres below the spots. Elution is carried out in a
closed tank as in paper chromatography. It requires less than three
hours for the solvent to reach the top of the plate (200 cm). Better
separations are achieved with square plates if a second elution is
performed at right angles to the first as in two dimensional paper
chromatography. For rapid work microscope slides are used as plates
and the elution time is usually less than fifteen minutes.

The thickness of the plates could be between 0.2 mm to 2 mm or more.
The thicker plates are used for preparative work in which hundreds of

milligrams of mixtures can be purified conveniently and quickly. The spots or areas are easily scraped off the plates and eluted with the required solvent. These can be revealed on the plates by UV light if they are UV absorbing or fluorescing substances, by spraying with a reagent that gives a coloured product with the spot (e.g. iodine solution or vapour gives brown colours with amines), or with dilute sulphuric acid (organic compounds become coloured or black when the plates are then heated at 100°) if the plates are of alumina or silica, but not cellulose. Some alumina and silica powders are available with fluorescent materials in them, in which case the whole plate fluoresces under UV light. Non-fluorescing spots are thus clearly visible, and fluorescent spots invariably fluoresce with a different colour. The colour of the spots can be different under UV light of different wavelengths that is why it is better to view the plates under UV light at 254 nm and at 365 nm. Another useful way of showing up non-UV absorbing spots is to spray the plate with a 1-2% solution of Rhodamine 6G in acetone. Under UV light the dye fluoresces and reveals the non-fluorescing spot. If the material in the spot is soluble in ether, benzene or light petroleum, the spots can be extraced from the powder with these solvents which leave the water soluble dye behind.

Thin and thick layer chromatography has been used successfully with ion-exchange celluloses as stationary phases and various aqueous buffers as mobile phases. Also, gels (e.g. Sephadex G-50 to G-200 superfine) have been adsorbed on glass plates and are very good for fractionating substances of high molecular weight (1500 to 250,000). With this technique, which is called thin layer gel filtration (TLG), molecular weights of proteins can be determined when suitable markers of known molecular weights are run alongside.

Commercially available precoated plates with a variety of adsorbents are generally very good for quantitative work because they are of a standard quality. More recently plates of a standardized silica gel 60 (a medium porosity silica gel with a mean porosity of 6 mm) were released by Merck. These have a specific surface of 500 m^2/g and a specific pore volume of 0.75 ml/g. They are so efficient that they have been called high performance thin layer chromatography (HPTLC) plates (Ripphahn and Halpap, J. Chromatog., 112, 81 (1975)).

SOLVENT EXTRACTION AND DISTRIBUTION

Extraction of a substance from suspension or solution in another solvent can sometimes be used as a purification process. Thus, organic substances can often be separated from inorganic impurities by shaking an aqueous solution or suspension with suitable immiscible solvents such as benzene, carbon tetrachloride, chloroform, ethyl ether, isopropyl ether or petroleum ether. After several such extractions the combined organic phase is dried and the solvent is evaporated. Immiscible solvents suitable for extractions are given in Table 17. Addition of electrolytes (such as ammonium sulphate, calcium chloride or sodium chloride) to the aqueous phase helps to ensure that the organic layer separates cleanly and also increases the extent of extraction into the latter. Emulsions can also be broken up by filtration (with suction) through Celite, or by adding a little octyl alcohol or some other paraffinic alcohol. The main factor in selecting a suitable immiscible solvent is to find one in

which the material to be extracted is readily soluble, whereas the
substance from which it is being extracted is not. The same con-
siderations apply irrespective of whether it is the substance being
purified, or one of its contaminants, that is taken into the new
phase. (The second of these processes is described as "washing".)

Common examples of washing with aqueous solutions include the
following:

   Removal of acids from water-immiscible solvents by washing
      with aqueous alkali, sodium carbonate or sodium bicarbonate.
   Removal of phenols from similar solutions by washing with
      aqueous alkali.
   Removal of organic bases by washing with dilute hydrochloric
      or sulphuric acids.
   Removal of unsaturated hydrocarbons, of alcohols and of
      ethers from saturated hydrocarbons or alkyl halides by
      washing with cold concentrated sulphuric acid.

This process can also be applied to purification of the substance if
it is an acid, a phenol or a base, by extracting into the appropriate
aqueous solution to form the salt which, after washing with pure
solvent, is again converted to the free species and re-extracted.
Paraffin hydrocarbons can be purified by extracting them with phenol
(in which aromatic hydrocarbons are highly soluble) prior to
fractional distillation.

For extraction of solid materials with a solvent, a Soxhlet extractor
is commonly used. This technique is applied, for example, in the
alcohol extraction of dyes to free them from insoluble contaminants
such as sodium chloride or sodium sulphate.

Acids, bases and amphoteric substances can be purified by taking
advantage of their ionization constants. Thus an acid can be separated from other acidic impurities which have different pK_a values
and from basic and neutral impurities, by extracting a solution of
the acid in an organic solvent (e.g. benzene or amyl alcohol) with a
set of inorganic aqueous buffers of increasing pH. (See Table 18.)
The acid will dissolve to form its salt in a set of buffers of pH
greater than its pK_a value. It can then be isolated by adding excess
mineral acid to the buffer and extracting with an organic solvent.
On a large scale, a countercurrent distribution machine (e.g. Craig
type, see Quickfit and Quartz catalogue) can be used. In this way a
very large number of liquid-liquid extractions can be carried out
automatically. The closer the ionization constants of the impurities
are to those of the required material, the larger should be the
number of extractions to effect a good separation. A detailed discussion of the technique and theory of methods of liquid-liquid
extraction is available in review articles such as that by L.L.Craig,
D.Craig, and E.G.Scheibel. Applications are summarized in
C.G. Casinovi's review, "A Comprehensive Bibliography of Separations
of Organic Substances by Countercurrent Distribution",
Chromatographic Reviews, 5, 161 (1963). This technique, however,
appears to have been displaced almost completely by chromatographic
methods.

MOLECULAR SIEVES

Molecular sieves are types of adsorbents composed of crystalline
zeolites (sodium and calcium aluminosilicates). By heating them,
water of hydration is removed, leaving holes of molecular dimensions
in the crystal lattices. These holes are of uniform size and allow
the passage into the crystals of small molecules, especially water,
but not of large ones. This "sieving" action explains their use as
very efficient drying agents for gases and liquids. The pore size of
these sieves can be modified (within limits) by varying the cations
built into the latices. The three types of Linde (Union Carbide)
molecular sieves currently available are:

Type 4A, a crystalline sodium aluminosilicate,
Type 5A, a crystalline calcium aluminosilicate.
Type 13X, a crystalline sodium aluminosilicate.

They are unsuitable for use with strong acids but are stable over the
pH range, 5-11.

Type 4A sieves. The pore size is about 4 Å, so that, besides
water, the ethane molecule (but not butane) can be adsorbed. Other
molecules removed from mixtures include carbon dioxide, hydrogen
sulphide, sulphur dioxide, ammonia, methanol, ethanol, ethylene,
acetylene, propylene, n-propyl alcohol, ethylene oxide and
(below -30°) nitrogen, oxygen and methane. The material is supplied
as beads, pellets or powder.

Type 5A sieves. Because the pore size is about 5 Å, these sieves
adsorb larger molecules than type 4A. For example, as well as the
substances listed above, propane, butane, hexane, butene, higher
n-olefins, n-butyl alcohol and higher n-alcohols, and cyclopropane
can be adsorbed, but not branched-chain C_6 hydrocarbons, cyclic
hydrocarbons such as benzene and cyclohexane, or secondary and
tertiary alcohols, carbon tetrachloride or boron trifluoride. This
is the type generally used for drying gases.

Type 13X sieves. Their pore size of about 10 Å enables many
branched-chain and cyclic materials to be adsorbed, in addition to
all the substances taken out by type 5A sieves.

Because of their selectivity, molecular sieves offer advantages over
silica gel, alumina or activated charcoal, especially in their very
high affinity for water, polar molecules and unsaturated organic
compounds. Their relative efficiency is greatest when the impurity
to be removed is present at low concentration. Thus, at 25° and a
relative humidity of 2%, type 5A molecular sieve adsorbs 18% by
weight of water, whereas for silica gel and alumina the figures are
3.5% and 2.5% respectively. Even at 100° and a relative humidity of
1.3% molecular sieves adsorb about 15% by weight of water.

The much greater preference of molecular sieve for combining with
water molecules explains why this material can be used for drying
ethanol and why molecular sieves are probably the most universely
useful and efficient drying agents. Percolation of ethanol with an
initial water content of 0.5% through a 57-in long column of type 4A
molecular sieve reduced the water content to 10 p.p.m. Similar

results have been obtained with pyridine.

The main applications of molecular sieves to purifications comprise:

1. Drying of gases and liquids containing traces of water.
2. Drying of gases at elevated temperatures.
3. Selective removal of impurities (including water) from gas streams.

(For example, carbon dioxide from air or ethylene; nitrogen oxides from nitrogen; methanol from ethyl ether. In general, carbon dioxide, carbon monoxide, ammonia, hydrogen sulphide, mercaptans, ethane, ethylene, acetylene, propane and propylene are readily removed at 25°. In mixtures of gases, the more polar ones are preferentially adsorbed.)

Special applications include the removal of straight-chain from branched-chain or cyclic molecules. For example, type 5A sieve will adsorb n-butyl alcohol but not its branched-chain isomers. Similarly, it separates n-tetradecane from benzene, or n-heptane from methyl-cyclohexane. A logical development is the use of molecular sieves as chromatographic columns for particular purifications.

The following liquids have been dried with molecular sieves-- acetone, acetonitrile, acrylonitrile, allyl chloride, amyl acetate, benzene, butadiene, n-butane, butene, butyl acetate, n-butylamine, n-butyl chloride, carbon tetrachloride, chloroethane, 1-chloro-2-ethylhexane, cyclohexane, dibromoethane, dichloroethane, 1,2-dichloropropane, 1,2-dimethoxyethane, dimethyl ether, dimethylformamide, dimethyl sulphoxide, ethanol, ethyl ether, 2-ethylhexanol, 2-ethylhexylamine, n-heptane, n-hexane, isoprene, isopropyl alcohol, isopropyl ether, methanol, methyl ethyl ketone, oxygen, n-pentane, phenol, propane, n-propyl alcohol, propylene, pyridine, styrene, tetrachloroethylene, toluene, trichloroethylene and xylene. In addition, the following gases have been dried—acetylene, air, argon, carbon dioxide, chlorine, ethylene, helium, hydrogen, hydrogen chloride, hydrogen sulphide, nitrogen, oxygen and sulphur hexafluoride.

After use, molecular sieves can be regenerated by heating at between 150° and 300° for several hours, preferably in a stream of dry air, then cooling in a desiccator.

However, care must be exercised in using molecular sieves for drying organic liquids. Appreciable amounts of impurities were _formed_ when samples of acetone, 1,1,1-trichloroethane and methyl-t-butyl ether were dried in the liquid phase by contact with molecular sieve 4A (Connett, Lab. Practice, 21, 545 (1972)). Other, less reactive types of sieves may be more suitable but, in general, it seems desirable to make a preliminary test to establish that no unwanted reaction takes place.

TABLE 1. PREDICTED EFFECT OF PRESSURE ON BOILING POINT

P in mm Hg   T in Deg Cent

760	0	20	40	60	80	100	120	140	160	180	200	220
0.1	-111	-99	-87	-75	-63	-51	-39	-27	-15	-4	8	20
0.2	-105	-93	-81	-69	-56	-44	-32	-19	-7	5	17	30
0.4	-100	-87	-74	-62	-49	-36	-24	-11	2	15	27	40
0.6	-96	-83	-70	-57	-44	-32	-19	-6	7	20	33	46
0.8	-94	-81	-67	-54	-41	-28	-15	-2	11	24	38	51
1.0	-92	-78	-65	-52	-39	-25	-12	1	15	28	41	54
2.0	-85	-71	-58	-44	-30	-16	-3	11	25	39	53	66
4.0	-78	-64	-49	-35	-21	-7	8	22	36	51	65	79
6.0	-74	-59	-44	-30	-15	-1	14	29	43	58	72	87
8.0	-70	-56	-41	-26	-11	4	19	34	48	63	78	93
10.0	-68	-53	-38	-23	-8	7	22	37	53	68	83	98
14.0	-64	-48	-33	-18	-2	13	28	44	59	74	90	105
18.0	-61	-45	-29	-14	2	17	33	48	64	79	95	111
20.0	-59	-44	-28	-12	3	19	35	50	66	82	97	113
30.0	-54	-38	-22	-6	10	26	42	58	74	90	106	123
40.0	-50	-34	-17	-1	15	32	48	64	81	97	113	130
50.0	-47	-30	-14	3	19	36	52	69	86	102	119	135
60.0	-44	-28	-11	6	23	40	56	73	90	107	123	140
80.0	-40	-23	-6	11	28	45	62	79	97	114	131	148
100.0	-37	-19	-2	15	33	50	67	85	102	119	137	154
150.0	-30	-12	6	23	41	59	77	95	112	130	148	166
200.0	-25	-7	11	29	47	66	84	102	120	138	156	174
300.0	-18	1	19	38	57	75	94	113	131	150	169	187
400.0	-13	6	25	44	64	83	102	121	140	159	178	197
500.0	-8	11	30	50	69	88	108	127	147	166	185	205
600.0	-5	15	34	54	74	93	113	133	152	172	192	211
700.0	-2	18	38	58	78	98	118	137	157	177	197	217
750.0	0	20	40	60	80	100	120	140	160	180	200	220
770.0	0	20	40	60	80	100	120	140	160	180	200	220
800.0	1	21	41	61	81	101	122	142	162	182	202	222

P in mm Hg   T in Deg Cent

760	240	260	280	300	320	340	360	380	400	420	440	460
0.1	32	44	56	68	80	92	104	115	127	139	151	163
0.2	42	54	67	79	91	103	116	128	140	153	165	177
0.4	53	65	78	91	103	116	129	141	154	167	180	192
0.6	59	72	85	98	111	124	137	150	163	176	189	202
0.8	64	77	90	103	116	130	143	156	169	182	195	208
1.0	68	81	94	108	121	134	147	161	174	187	201	214
2.0	80	94	108	121	135	149	163	176	190	204	218	232
4.0	93	108	122	136	151	165	179	193	208	222	236	251
6.0	102	116	131	146	160	175	189	204	219	233	248	262
8.0	108	123	137	152	167	182	197	212	227	241	256	271
10.0	113	128	143	158	173	188	203	218	233	248	263	278
14.0	120	136	151	166	182	197	212	228	243	258	274	289
18.0	126	142	157	173	188	204	219	235	251	266	282	297

TABLE 1 continued

760	240	260	280	300	320	340	360	380	400	420	440	460
20.0	129	144	160	176	191	207	223	238	254	270	285	301
30.0	139	155	171	187	203	219	235	251	267	283	299	315
40.0	146	162	179	195	211	228	244	260	277	293	309	326
50.0	152	168	185	202	218	235	251	268	284	301	317	334
60.0	157	174	190	207	224	241	257	274	291	308	324	341
80.0	165	182	199	216	233	250	267	284	301	318	336	353
100.0	171	189	206	223	241	258	275	293	310	327	345	362
150.0	184	201	219	237	255	273	290	308	326	344	362	379
200.0	193	211	229	247	265	283	302	320	338	356	374	392
300.0	206	225	243	262	281	299	318	337	355	374	393	411
400.0	216	235	254	273	292	311	330	350	369	388	407	426
500.0	224	244	263	282	302	321	340	360	379	399	418	437
600.0	231	251	270	290	310	329	349	368	388	408	427	447
700.0	237	257	277	296	316	336	356	376	396	416	436	455
750.0	239	259	279	299	319	339	359	379	399	419	439	459
770.0	241	261	281	301	321	341	361	381	401	421	441	461
800.0	242	262	282	302	322	342	362	383	403	423	443	463

TABLE 2. HEATING BATHS

Up to 100°	Water baths
-20 to 200°	Glycerol or di-n-butyl phthalate
Up to about 220°	Medicinal paraffin
Up to about 250°	Hard hydrogenated cotton-seed oil (m.p. 40-60°), or a 1:1 mixture of cotton-seed oil and castor oil containing about 1% of hydroquinone
-40 to 250° (to 440° under nitrogen)	D.C. 550 silicone fluid
Up to about 260°	A mixture of 85% orthophosphoric acid (4 parts) and metaphosphoric acid (1 part)
Up to 340°	A mixture of 85% orthophosphoric acid (2 parts) and metaphosphoric acid (1 part)
150 to 500°	A mixture of $NaNO_2$ (40%), $NaNO_3$ (7%) and KNO_3 (53%)
73 to 350°	Wood's Metal[†]
250 to 800°	Solder[†]
350 to 800°	Lead[†]

† In using metal baths, the beaker or flask should be removed while the metal is still liquid.

TABLE 3. WHATMAN FILTERS

Grade No.	1	2	3	4	5	6	113
Particle size retained in microns	11	8	5	12	2.4	2.8	28
Filtration speed* sec./100ml.	40	55	155	20	<300	125	9

Whatman routine ashless filters

Grade No.	40	41	42	43	44
Particle size retained in microns	7.5	12	3	12	4
Filtration speed* sec./100ml.	68	19	200	38	125

Whatman

	← hardened →			← hardened ashless →		
Grade No.	50	52	54	540	541	542
Particle size retained in microns	3	8	20	9	20	3
Filtration speed* sec./100ml.	250	55	10	55	12	250

Whatman Glass Micro Filters

Grade	GF/A	GF/B	GF/C	GF/D	GF/F
Particle size retained in microns	1.6	1	1.1	2.2	0.8
Filtration speed* sec./100ml	8.3	20	8.7	5.5	17.2

* Filtration speeds are rough estimates of initial flowrates and should be considered on a relative basis.

TABLE 4. MICRO FILTERS

Nuclepore (polycarbonate) Filters

Mean Pore Size (microns)	8.0	2.0	1.0	0.1	0.03	0.015
Av. pores/cm^2	10^5	2×10^6	2×10^7	3×10^8	6×10^8	$1-6 \times 10^9$
Water flowrate ml/min/cm^2	2000	2000	300	8	0.03	0.1-0.5

Millipore Filters

Type	Cellulose ester MF/SC	MF/VF	Teflon LC	LS	Microweb[†] WS	WH
Mean Pore Size (microns)	8	0.01	10	5	3	0.45
Water flowrate ml/min/cm^2	850	0.2	170	70	155	55

Gelman Membranes

Type	Cellulose ester GA-1	TCM-450	VM-1	DM-800	Copolymer AN-200	Tuffryn-450
Mean Pore Size (microns)	5	0.45	5	0.8	0.2	0.45
Water flowrate ml/min/cm^2	320	50	700	200	17	50

Sartorius Membrane Filters (SM)

Application	Gravimetric	Biological clarification	Sterilization	Particle count in H_2O	For acids and alkali
Type No.	11003	11044	11006	11011	12801
Mean Pore Size (microns)	1.2	0.6	0.45	0.01	8
Water flowrate ml/min/cm^2	300	150	65	0.6	1100

* Only a few representative filters are tabulated (available ranges are more extensive).

† Reinforced nylon.

TABLE 5. COMMON SOLVENTS USED IN RECRYSTALLIZATION
(and their boiling points)

Acetic acid ($118°$)   Chloroform ($61°$)   Methyl isobutyl ketone
  (glacial)   ($116°$)

* Acetone ($56°$)   *Cyclohexane ($81°$)   Nitrobenzene ($210°$)

* Benzene ($80°$)   Dimethylformamide ($153°$) *Petroleum ether (various)

n-Butanol ($118°$)   *Dioxane ($101°$)   Pyridine ($115.5°$)
                      *Diethyl ether ($34.5°$)

γ-Butyrolactone   *Ethanol ($78°$)   Pyridine trihydrate ($93°$)
  ($206°$)

Carbon tetrachloride *Ethyl acetate ($78°$)   *Tetrahydrofuran ($64-66°$)
  ($77°$)

Cellosolve ($135°$)   *Methanol ($64.5°$)   *Toluene ($110°$)

Chlorobenzene ($132°$) *Methyl ethyl ketone   Water ($100°$)
                       ($80°$)

      * Highly flammable, should be heated or evaporated
      on steam baths or electrically heated water bath.

TABLE 6. PAIRS OF MISCIBLE SOLVENTS

Acetic acid with chloroform, ethanol, ethyl acetate, petroleum ether,
    or water
Acetone with benzene, butyl acetate, butyl alcohol, carbon tetra-
    chloride, chloroform, cyclohexane, ethanol, ethyl acetate, ethyl
    ether, petroleum ether or water
Ammonia with pyridine
Aniline with acetone, benzene, carbon tetrachloride, ethyl ether,
    n-heptane, methanol or nitrobenzene
Benzene with acetone, butyl alcohol, carbon tetrachloride, chloroform,
    cyclohexane, ethanol, petroleum ether or pyridine
Butyl alcohol with acetone, ethyl acetate
Carbon disulphide with petroleum ether
Carbon tetrachloride with cyclohexane
Chloroform with acetic acid, acetone, benzene, ethanol, ethyl acetate,
    hexane, methanol or pyridine
Cyclohexane with acetone, benzene, carbon tetrachloride, ethanol or
    ethyl ether
Dimethylformamide with benzene, ethanol or water
Dimethyl sulphoxide with acetone, benzene, chloroform, ethanol, ethyl
    ether or water
Dioxane with benzene, carbon tetrachloride, chloroform, ethanol, ethyl
    ether, petroleum ether, pyridine or water
Ethanol with acetic acid, acetone, benzene, chloroform, cyclohexane,
    dioxane ethyl ether, pentane, toluene, water or xylene
Ethyl acetate with acetic acid, acetone, butyl alcohol, chloroform
    or methanol

TABLE 6 continued

Ethyl ether with acetone, cyclohexane, ethanol, methanol, methylal,
  pentane or petroleum ether
Glycerol with ethanol, methanol or water
Hexane with benzene, chloroform or ethanol
Methanol with chloroform, ethyl ether, glycerol or water
Methylal with ethyl ether
Methyl ethyl ketone with acetic acid, benzene, ethanol or methanol
Nitrobenzene with aniline
Pentane with ethanol or ethyl ether
Petroleum ether with acetic acid, acetone, benzene, carbon disulphide
  or ethyl ether
Phenol with carbon tetrachloride, ethanol, ethyl ether or xylene
Pyridine with acetone, ammonia, benzene, chloroform, dioxane,
  petroleum ether, toluene or water
Toluene with ethanol or pyridine
Water with acetic acid, acetone, ethanol, methanol or pyridine
Xylene with ethanol or phenol

TABLE 7. MATERIALS FOR COOLING BATHS

Temperature	Composition
$0°$	Crushed ice
-5 to $-20°$	Ice-salt mixtures
$-33°$	Liquid ammonia
-40 to $-50°$	Ice (3.5-4 parts)
	$-CaCl_2.6H_2O$ (5 parts)
$-72°$	Solid CO_2 with ethanol
$-77°$	Solid CO_2 with chloroform or acetone
$-78°$	Solid CO_2 (powdered)
$-100°$	Solid CO_2 with ethyl ether
$-192°$	Liquid air
$-196°$	Liquid nitrogen

Alternatively, the following liquids can be used, partially frozen,
as cryostats, by adding solid CO_2 from time to time to the material
in a Dewar-type container and stirring to make a slush:

$13°$	p-Xylene
12	Dioxane
6	Cyclohexane
5	Benzene
2	Formamide
-8.6	Methyl salicylate
-9	Hexane-2,5-dione

TABLE 7 continued

-10.5°	Ethylene glycol
-11.9	tert-Amyl alcohol
-12	Cycloheptane
-12	Methyl benzoate
-15	Benzyl alcohol
-16.3	n-Octanol
-18	1,2-Dichlorobenzene
-22	Tetrachloroethylene
-22.4	Butyl benzoate
-22.8	Carbon tetrachloride
-24.5	Diethyl sulphate
-25	1,3-Dichlorobenzene
-29	o-Xylene
-29	Pentachloroethane
-30	Bromobenzene
-32	m-Toluidine
-32.6	Dipropyl ketone
-38	Thiophen
-41	Acetonitrile
-42	Pyridine
-42	Diethyl ketone
-44	Cyclohexyl chloride
-45	Chlorobenzene
-47	m-Xylene
-50	Ethyl malonate
-50	n-Butylamine
-52	Benzyl acetate
-52	Diethylcarbitol
-55	Diacetone
-56	n-Octane
-60	Isopropyl ether
-73	Trichloroethylene
-73	Isopropyl acetate
-74	o-Cymene
-74	p-Cymene
-77	Butyl acetate
-79	Isoamyl acetate
-83	Propylamine

By using liquid nitrogen* instead of solid CO_2, this range can be extended.

-83.6	Ethyl acetate
-86	Methyl ethyl ketone
-89	n-Butanol
-90	Nitroethane
-91	Heptane
-92	n-Propyl acetate
-93	2-Nitropropane
-93	Cyclopentane
-94	Ethyl benzene
-94	Hexane
-94.6	Acetone
-95.1	Toluene
-97	Cumene

TABLE 7 continued

-98°	Methanol
-98	Methyl acetate
-99	Isobutyl acetate
-104	Cyclohexene
-107	Isooctane
-108	1-Nitropropane
-116	Ethanol
-116	Ethyl ether
-117	Isoamyl alcohol
-126	Methylcyclohexane
-131	n-Pentane
-160	Isopentane

*Do not use liquid air instead because it will explode in contact with organic matter.

TABLE 8.   LIQUIDS FOR DRYING PISTOLS

	b.p.
Ethyl chloride	12.2°
Methylene dichloride	39.8
Acetone	56.1
Chloroform	62
Methanol	64.5
Carbon tetrachloride	76.5
Ethanol	78.3
Benzene	79.8
Trichloroethylene	86
Water	100
Toluene	110.5
Tetrachloroethylene	121.2
Chlorobenzene	132-133
m-Xylene	139.3
Isoamyl acetate	142.5
Tetrachloroethane	146.3
Bromobenzene	155
p-Cymene	176
Tetralin	207

TABLE 9. VAPOUR PRESSURES (in mm Hg) OF SATURATED AQUEOUS SOLUTIONS IN EQUILIBRIUM WITH THE SOLID SALTS

Salt	Temperature					% humidity at 20°
	10°	15°	20°	25°	30°	
$LiCl.H_2O$			2.6			15
$CaBr_2.6H_2O$	2.1	2.7	3.3	4.0	4.8	19
Potassium acetate			3.5			20
$CaCl_2.6H_2O$	3.5	4.5	5.6	6.9	8.3	32
CrO_3			6.1			35
$Zn(NO_3)_2.6H_2O$			7.4			42
$K_2CO_3.2H_2O$			7.7	10.7		44
KCNS			8.2			47
$Na_2Cr_2O_7.2H_2O$			9.1			52
$Ca(NO_3)_2.4H_2O$	6.0	7.7	9.6	11.9	14.3	55
$Mg(NO_3)_2.6H_2O$			9.8			56
$NaBr.2H_2O$	5.8	7.8	10.3	13.5	17.5	58
$NaNO_2$			11.6			66
$NaClO_3$			13.1			75
NaCl	6.9	9.6	13.2	17.8	21.4	75
Sodium acetate			13.3			76
NH_4Cl			13.8			79
$(NH_4)_2SO_4$			14.2			81
KBr			14.7			84
$KHSO_4$			15.1			86
KCl			15.1	20.2	27.0	86
K_2CrO_4			15.4			88
$ZnSO_4.7H_2O$			15.8			90
$NH_4.H_2PO_4$			16.3			93
$Na_2HPO_4.12H_2O$			16.7			95
KNO_3			16.7	22.3	29.8	95
$Pb(NO_3)_2$			17.2			98
(Water)	9.21	12.79	17.53	23.76	31.82	100

TABLE 10. DRYING AGENTS FOR CLASSES OF COMPOUNDS

Class	Dried with
Acetals	Potassium carbonate
Acids (organic)	Calcium sulphate, magnesium sulphate, sodium sulphate
Acyl halides	Magnesium sulphate, sodium sulphate
Alcohols	Calcium oxide, calcium sulphate, magnesium sulphate, potassium carbonate, followed by magnesium and iodine
Aldehydes	Calcium sulphate, magnesium sulphate, sodium sulphate
Alkyl halides	Calcium chloride, calcium sulphate, magnesium sulphate, phosphorus pentoxide, sodium sulphate
Amines	Barium oxide, calcium oxide, potassium hydroxide, sodium carbonate, sodium hydroxide
Aryl halides	Calcium chloride, calcium sulphate, magnesium sulphate, phosphorus pentoxide, sodium sulphate
Esters	Magnesium sulphate, potassium carbonate, sodium sulphate
Ethers	Calcium chloride, calcium sulphate, magnesium sulphate, sodium
Heterocyclic bases	Magnesium sulphate, potassium carbonate, sodium hydroxide
Hydrocarbons (arom., also satd.)	Calcium chloride, calcium sulphate, magnesium sulphate, phosphorus pentoxide, sodium (not for olefins)
Ketones	Calcium sulphate, magnesium sulphate, potassium carbonate, sodium sulphate
Mercaptans	Magnesium sulphate, sodium sulphate
Nitro compounds and nitriles	Calcium chloride, magnesium sulphate, sodium sulphate
Sulphides	Calcium chloride, calcium sulphate

TABLE 11. GRADED ADSORBENTS AND SOLVENTS

Adsorbents (in decreasing effectiveness)	Solvents (in increasing eluting ability)
Fullers' earth (hydrated aluminosilicate)	Petroleum ether, b.p. 40-60°
Magnesium oxide	Petroleum ether, b.p. 60-80°
Charcoal	Carbon tetrachloride
Alumina	Cyclohexane
Magnesium trisilicate	Benzene
Silica gel	Ethyl ether
Calcium hydroxide	Chloroform
Magnesium carbonate	Ethyl acetate
Calcium phosphate	Acetone
Calcium carbonate	Ethanol

TABLE 11 continued

Adsorbents   Solvents

Sodium carbonate   Methanol
Talc   Pyridine
Inulin   Acetic acid
Sucrose = Starch

TABLE 12. COMMON ION-EXCHANGE RESINS

Sulphonated polystyrene bead resins. (Strong-acid) cation exchangers:

    Amberlite IR-120   (Rohm and Haas)
    Dowex 50   (Dow Chemical Co.)
    Permutit RS   (Permutit A.G., Germany)
    Permutite C 50D   (Phillips and Pain-Vermorel, Paris)
    Zerolit 225   (Zerolit, Ltd., U.K.)

Carboxylic acid-type resins. (Weak-acid) cation exchangers:

    Amberlite IRC-50   (Rohm and Haas)
    Permutit C   (Permutit A.G.)
    Permutits H and H-70   (Permutit Co., U.S.A.)
    Zerolit 236   (Zerolit, Ltd., U.K.)

Aliphatic amine-type resins. (Weak-base) anion exchangers:

    Amberlites IR-4B and -45   (Rohm and Haas)
    Dowex 3   (Dow Chemical Co.)
    Permutit E   (Permutit A.G.)
    Permutit A 240A   (Phillips and Pain-Vermorel)

(Strong-base) anion exchangers:

    Amberlite IRA-400   (Rohm and Haas)
    Zerolit FF   (Zerolit, Ltd., U.K.)
    Dowex 1   (Dow Chemical Co.)
    Permutit ESB   (Permutit A.G.)
    Permutite A 330D   (Phillips and Pain-Vermorel)

TABLE 13. MODIFIED FIBROUS CELLULOSES FOR ION-EXCHANGE

Cation exchange	Anion exchange
CM cellulose (carboxy methyl)	DEAE cellulose (diethylamino-ethyl)
CM 22, 23 cellulose	DE 22, 23 cellulose
P cellulose (phosphate)	ECTEOLA cellulose
SE cellulose (sulphoethyl)	PAB cellulose (p-aminobenzyl)
SM cellulose (sulphomethyl)	TEAE cellulose (triethylamino-ethyl)

SE and SM are much stronger acids than CM, whereas P has two ionizable groups (pK 2-3, 6-7), one of which is stronger, the other weaker, than for CM (3.5-4.5).

For basic strengths, the sequence is:

TEAE ≫ DEAE (pK 8-9.5) > ECTEOLA (pK 5.5-7) > PAB.

Their exchange capacities lie in the range 0.3-1.0 mg equiv./g.

TABLE 14. BEAD FORM ION-EXCHANGE PACKINGS.[1]

Cation exchange	capacity (meq/g)	Anion exchange	capacity (meq/g)
CM-Sephadex C-25,C-50.[2] (weakly acidic)	4.5±0.5	DEAE-Sephadex A-25,A-50.[6] (weakly basic)	3.5±0.5
SP-Sephadex C-25,C-50.[3] (strongly acidic)	2.3±0.3	QAE-Sephadex A-25,A-50.[7] (strongly basic)	3.0±0.4
CM-Sepharose CL-6B.[4]	0.12±0.02	DEAE-Sepharose CL-6B.[4]	0.13±0.02
		DEAE-Sephacel.[8]	1.4±0.1
CM-32 cellulose.		DE-32 cellulose.	
CM-52 cellulose.[5]		DE-52 cellulose.[5]	

1. May be sterilized by autoclaving at pH 7 and below 120°.
2. Carboxymethyl.
3. Sulphopropyl.
4. Crosslinked agarose gel, no precycling required, pH range 3-10.
5. Microgranular, pre-swollen, does not require precycling.
6. Diethylaminoethyl.
7. Diethyl(2-hydroxypropyl)aminoethyl.
8. Bead form cellulose, pH range 2-12, no precycling required.

TABLE 15. COLUMNS FOR HPLC.[1,2]

Column	Mobile phase[3]	Application
Dupont		
ODS "permaphase" (octadecyl silane)	Most solvents except strong acids and bases, useful for gradient elution.	Aromatic compounds, sterols, drugs, natural products.
HCP (hydrocarbon polymer)	Aqueous alcohols up to 50% isopropanol.	Aromatic compounds, quinones.
CWT (carbowax 4000)	Hydrocarbons only.	Steroids and polar organic compounds.
TMG (trimethylene glycol)	Hydrocarbons, chloroform, dioxane and tetrahydrofuran. Not alcohols, mobile phase must be saturated with trimethylene glycol.	Hydroxy and amino compounds, pesticides, polymer intermediates.
BOP (β,β'-oxydipropionitrile)	Hydrocarbons, butyl ether, with up to 15% of THF. Mobile phase must be saturated with β,β'-oxydipropionitrile.	Alkaloids, pesticides, polymer additives, steroids.
WAX (weak anion exchange)	Water only, retention and resolution is modified by change of pH and ionic strength.	Ionic compounds.
SAX (strong anion exchange)	Water only, as above	As above.
SCX (strong cation exchange)	Water only, as above	As above.
Merck		
Silica Gel 60-Kieselgel 60	Ethanol, $CHCl_3$, CH_2Cl_2, $\underline{n}$-C_7H_{16}, EtOAC, AcOH.[2]	Vitamins, alkaloids, esters, sterols, drugs, aromatic compounds.
LiChrosorb SI60, SI 100 and Alex T.	Hydrocarbons, ether, aliphatic acids, Me_2SO, $CHCl_3$, CH_2Cl_2, $\underline{t}$-BuOH.	As above, phthalimido-acids, antioxidants.

TABLE 15 continued

Merck

Perisorb A.	Hexane, acetic acid, Isooctane, EtOAc.	Acids, esters, aromatic amines and hydrocarbons.
Perisorb PA6	Methanol, water, AcOH.	As above.
Perisorb KAT	Aqueous buffers to pH 11.	Heterocyclic compounds, nucleo-sides, acids, bases.

1. Only a few representative columns are tabulated.
2. Waters Assoc. and Altex also have a wide range of columns.
3. Not to be used above 50°, halide acids and salts are corrosive and must be avoided.

TABLE 16. LIQUIDS FOR STATIONARY PHASES IN GAS CHROMATOGRAPHY

Material	Temp.	Retards
Dimethylsulpholane	0-40°	Olefins and aromatic hydrocarbons
Di-n-butyl maleate	0-40	(General purpose)
Squalane	0-150	Volatile hydrocarbons and polar molecules
Silicone oil or grease	0-250	(General purpose)
Diglycerol	20-120	Water, alcohols, amines, esters and aromatic hydrocarbons
Dinonyl phthalate	20-130	(General purpose)
Polydiethyleneglycol succinate	50-200	Aromatic hydrocarbons, alcohols, ketones and esters
Polyethyleneglycol	50-200	Water, alcohols, amines, esters and aromatic hydrocarbons
Apiezon grease	50-250	Volatile hydrocarbons and polar molecules
Tricresyl phosphate		(General purpose)

TABLE 17. SOME COMMON IMMISCIBLE OR SLIGHTLY MISCIBLE PAIRS OF SOLVENTS

Carbon tetrachloride with ethanolamine, ethylene glycol, formamide
    or water
Dimethyl formamide   )
                   with cyclohexane or petroleum ether
Dimethyl sulphoxide)
Ethyl ether with ethanolamine, ethylene glycol or water
Methanol with carbon disulphide, cyclohexane or petroleum ether
Petroleum ether with aniline, benzyl alcohol, dimethylformamide,
    dimethyl sulphoxide, formamide, furfuryl alcohol, phenol or water
Water with aniline, benzene, benzyl alcohol, carbon disulphide,
    carbon tetrachloride, chloroform, cyclohexane, cyclohexanol,
    cyclohexanone, ethyl acetate, isoamyl alcohol, methyl ethyl
    ketone, nitromethane, petroleum ether, phenol, tributyl phosphate
    or toluene

TABLE 18. AQUEOUS BUFFERS

Approx. pH	Composition
0	2N sulphuric acid or N hydrochloric acid
1	0.1 N hydrochloric acid or 0.18 N sulphuric acid
2	Either 0.01 N hydrochloric acid or 0.013 N sulphuric acid or 50 ml of 0.1 M glycine (also 0.1 M in sodium chloride) + 50 ml of 0.1 N hydrochloric acid
3	Either 20 ml of 0.2 M Na_2HPO_4 + 80 ml of 0.1 M citric acid or 50 ml of 0.1 M glycine + 22.8 ml of 0.1 N hydrochloric acid in 100 ml
4	Either 38.5 ml of 0.2 M Na_2HPO_4 + 61.5 ml of 0.1 M citric acid or 18 ml of 0.2 M sodium acetate + 82 ml of 0.2 M acetic acid
5	Either 70 ml of 0.2 M sodium acetate + 30 ml of 0.2 M acetic acid or 51.5 ml of 0.2 M Na_2HPO_4 + 48.5 ml of 0.1 M citric acid
6	63 ml of 0.2 M Na_2HPO_4 + 37 ml of 0.1 M citric acid
7	82 ml of 0.2 M Na_2HPO_4 + 18 ml of 0.1 M citric acid
8	Either 50 ml of 0.1 M TRIS buffer + 29 ml of 0.1 N hydrochloric acid, in 100 ml or 30 ml of 0.05 M borax + 70 ml of 0.2 M boric acid
9	80 ml of 0.05 M borax + 20 ml of 0.2 M boric acid
10	Either 25 ml of 0.05 M borax + 43 ml of 0.1 N sodium hydroxide, in 100 ml or 50 ml of 0.1 M glycine + 32 ml of 0.1 N sodium hydroxide, in 100 ml
11	50 ml of 0.15 M Na_2HPO_4 + 15 ml of 0.1 N sodium hydroxide
12	50 ml of 0.15 M Na_2HPO_4 + 75 ml of 0.1 N sodium hydroxide
13	0.1 N sodium or potassium hydroxides
14	N sodium or potassium hydroxides

These buffers are suitable for use in obtaining ultraviolet spectra.
Alternatively, for a set of accurate buffers of low but constant
ionic strength (I=0.01), covering the pH range 2.2 to 11.6 at 20°,
see Perrin, Aust.J.Chem., 16, 572 (1963).

BIBLIOGRAPHY

The following books and reviews provide fuller details of the topics indicated.

Safety in the Chemical Laboratory

G. M. Muir (Ed.), Hazards in the Chemical Industry, The Chemical Society, London, 2nd edn., 1977.

N. V. Steere (Ed.), Handbook of Laboratory Safety, Chemical Rubber Co., Cleveland, Ohio, 1967.

N. I. Sax, Dangerous Properties of Industrial Materials, Van Nostrand Reinhold, London, 3rd edn., 1968.

College Safety Committee, Safety in the Chemical Laboratory and in the Use of Chemicals, Imperial College, London, 3rd edn., 1971.

College Safety Committee, Code of Practice against Radiation Hazards, Imperial College, London, 6th edn., 1973.

Laboratory Technique and Theoretical Discussion

E. W. Berg, Physical and Chemical Methods of Separation, McGraw-Hill, New York, 1963.

N. D. Cheronis and J. B. Entrikin, Identification of Organic Compounds, Interscience, New York, 1963.

L. F. Fieser, Experiments in Organic Chemistry, Heath, Boston, 3rd edn., 1957.

B. Keil, Laboratoriumstechnik der organischen Chemie, Akademie Verlag, Berlin, 1961. (German edn.,translated and revised by H.Fürst.)

R. P. Linstead, J. A. Elvidge and M. Whalley, A Course in Modern Techniques of Organic Chemistry, Butterworths, London, 1955.

F. G. Mann and B. C. Saunders, Practical Organic Chemistry, Longmans, London, 4th edn., 1960.

A. A. Morton, Laboratory Technique in Organic Chemistry, McGraw-Hill, New York, 1938.

B. S. Furniss et al., Vogel's Textbook of Practical Organic Chemistry, Longmans, London, 4th edn., 1978.

P. S. Diamond and R. F. Denman, Laboratory Techniques in Chemistry and Biochemistry, 2nd edn., Butterworths, London, 1973.

F. J. Wolf, Separation Methods in Organic Chemistry and Biochemistry, Academic Press, New York, 1969.

P. A. Ongley (Ed.), Organicum, Practical Handbook of Organic Chemistry, Pergamon Press, Oxford (English edn.), 1973.

Organic Solvents

J. A. Riddick and W. B. Bunger, Organic Solvents: Physical Properties and Methods of Purification, Techniques of Chemistry, vol. II, Wiley-Interscience, New York, 1970.

Solvent Extraction and Distribution

L. C. Craig, D. Craig and E. G. Scheibel, in A. Weissberger's (Ed.) Technique of Organic Chemistry, vol. III, pt I, Interscience, New York, 2nd edn., 1956.

F. A. von Metzsch, in W. G. Berl's (Ed.) Physical Methods in Chemical Analysis, vol. IV, Academic Press, New York, 1961.

Distillation

T. P. Carney, Laboratory Fractional Distillation, Macmillan, New York, 1949.

E. Krell, Handbook of Laboratory Distillation, Elsevier, Amsterdam, 1963.

A. Weissberger (Ed.), Technique of Organic Chemistry, vol. IV, Distillation, Interscience, New York, 1951.

Crystallization

R. S. Tipson, in A. Weissberger's (Ed.) Technique of Organic Chemistry, vol. III, pt.I, Interscience, New York, 2nd edn., 1956.

Zone Refining

M. Zief and W. R. Wilcox (Eds.), Fractional Solidification, vol. 1, Marcel Dekker, New York, 1967.

E. F. G. Herington, Zone Melting of Organic Compounds, Wiley, New York, 1963.

W. Pfann, Zone Melting, 2nd edn., Wiley, New York, 1966.

H. Schildknecht, Zonenschmelzen, Verlag Chemie, Weiheim, 1964.

W. R. Wilcox, R. Friedenberg et al., Chem. Rev., 64, 187 (1964).

Drying

G. Broughton, in A. Weissberger's (Ed.) Technique of Organic Chemistry, vol. III, pt. I, Interscience, New York, 2nd edn., 1956.

Chromatography

J. N. Balston, B. G. Talbot and T. S. G. Jones, A Guide to Filter Paper and Cellulose Powder Chromatography, Reeve Angel, London, and Balston, Maidstone, 1952.

R. J. Block, E. L. Durrum and G. Zweig, A Manual of Paper Chromatography and Paper Electrophoresis, Academic Press, New York, 1955.

H. G. Cassidy, Fundamentals of chromatography, Technique of Organic Chemistry, vol. X. Interscience, New York, 1957.

E. Heftman (Ed.) Chromatography, Reinhold, New York, 1961.

E. Lederer and M. Lederer, Chromatography, A Review of Principles and Applications, Elsevier, Amsterdam, 2nd edn., 1957.

T. I. Williams, An Introduction to Chromatography, Blackie, London, 1947.

R. Stock and C. B. F. Rice, Chromatographic Methods, 3rd edn., Chapman and Hall, London, 1974.

Molecular Sieves

D. W. Breck, Zeolite Molecular Sieves. Structure, Chemistry and Use, Wiley, New York, 1974.

P. Andrews, Molecular Sieve Chromatography, Brit. Med. Bull., 22, 109 (1966).

Union Carbide Molecular Sieves for Selective Adsorption, British Drug Houses, Poole, England, 2nd edn., 1961.

C. K. Hersh, Molecular Sieves, Reinhold, New York, 1961.

G. R. Landolt and G. T. Kerr, Sepn. Purif. Meth., 2, 283 (1973).

Ion Exchange

Dowex: Ion Exchange, Dow Chemical Co., Midland, Michigan, 1959.

Ion Exchange Resins, British Drug Houses, Poole, England, 5th edn., 1977.

C. Calmon and T. R. E. Kressman (Eds.) Ion Exchangers in Organic and Biochemistry, Interscience, New York, 1957.

Ion Exchange continued

R. Kunin, Ion Exchange Resins, Wiley, New York, 2nd edn., 1958.

C. G. Horvath, Ion Exchange, Dekker, New York, vol. 3, 1972.

E. A. Peterson in Laboratory Techniques in Biochemistry and Molecular Biology, vol. 2, pt II, T. S. Work and E. Work (Eds.), North Holland Publ. Co., Amsterdam, 1970.

H. F. Walton (Ed.) Ion-exchange Chromatography, Dowden, Hutchinson and Ross Inc., Stroudsburg, Pa., distributed by Halsted Press, 1976.

E. A. Peterson and H. A. Sober, J. Amer. Chem. Soc., 78, 751 (1956).

J. Porath, Ark. Kemi Min. Geol., 11, 97 (1957).

H. A. Sober et al., J. Amer. Chem. Soc., 78, 756 (1956).

Ionic Equilibria

D. D. Perrin and B. Dempsey, Buffers for pH and Metal Ion Control, Chapman and Hall, London, 1974.

G. Kortüm, W. Vogel and K. Andrussow, Dissociation Constants of Organic Acids in Aqueous Solution, Butterworths, 1961.

E. P. Serjeant and B. Dempsey, Ionization Constants of Organic Acids in Aqueous Solution, Pergamon Press, Oxford, 1979.

D. D. Perrin, Dissociation Constants of Organic Bases in Aqueous Solution, Butterworths, London, 1965; and Supplement 1972.

D. D. Perrin, Dissociation Constants of Inorganic Acids and Bases in Aqueous Solution, Butterworths, London, 1969.

Gas Chromatography

V. J. Coates, H. J. Noebels and I. S. Fagerson (Eds.) Gas Chromatography, Academic Press, New York, 1958.

D. H. Desty (Ed.) Vapour Phase Chromatography, Butterworths, London, 1957.

D. H. Desty (Ed.) Gas Chromatography, Butterworths, London, 1958.

R. Kaiser, Gas Chromatography, vol. 3, Butterworths, London, 1963 (practical details of liquid phases, adsorbents, etc., with references.)

A. I. M. Keulemans, Gas Chromatography, Reinhold, New York, 2nd edn., 1959. Materials for Gas Chromatography, May & Baker, 1958.

H. J. Noebels, R. F. Wall and N. Brenner (Eds.) Gas Chromatography, Academic Press, New York, 1961.

Gas Chromatography continued

R. L. Pecsok, Principles and Practice of Gas Chromatography, Wiley, New York, 1959.

A. B. Littlewood, Gas Chromatography: Principles, Techniques and Applications, Academic Press, New York, 2nd edn., 1970.

R. A. Jones, An Introduction to Gas-Liquid Chromatography, Academic Press, London, 1970.

D. Ambrose, Gas Chromatography, 2nd edn., Butterworths, London, 1971.

High Performance Liquid Chromatography

P. R. Brown, High Pressure Liquid Chromatography, Academic Press, New York, 1973.

N. A. Parris, Instrumental Liquid Chromatography. A Manual of High Performance Liquid Chromatography Methods, Elsevier, Amsterdam, 1976.

H. Engelhardt, High Performance Liquid Chromatography (translated from the German by G. Gutnikov), Springer-Verlag, Berlin, 1979.

Z. Deyl, K. Macek and J. Janak (Eds.), Liquid Column Chromatography. A survey of Modern Techniques and Applications, Elsevier, Amsterdam 1975.

P. F. Dixon, C. H. Gray, C. K. Lim and M. S. Stoll (Eds.), High Pressure Liquid Chromatography in Clinical Chemistry (Symposium), Academic Press, New York, 1976.

J. J. Kirkland, Columns for HPLC, Anal. Chem., $\underline{43}$ (12), 36A, 1971.

A. Zlatkis and R. E. Kaiser (Eds.), High Performance thin layer Chromatography, Elsevier, Amsterdam, 1977.

Affinity Chromatography

C. R. Lowe and P. D. G. Dean, Affinity Chromatography, Wiley, London, 1974.

W. B. Jakoby and M. Wilchek, Affinity Chromatography in Methods in Enzymology, vol. 34, 1975.

J. Turkova, Affinity Chromatography, Elsevier, Amsterdam, 1978.

Gel Chromatography

L. Fischer, <u>Introduction to Gel Chromatography</u>, North-Holland, Amsterdam, 1968.

H. Determann, <u>Gel Chromatography</u>, Springer-Verlag, Berlin, 1967.

P. Flodin, <u>Dextran Gels and their Applications in Gel Filtration</u>, Pharmacia, Uppsala, Sweden, 1962.

CHAPTER 2

CHEMICAL METHODS USED IN PURIFICATION

GENERAL REMARKS

Greater selectivity in purification can often be achieved by making use of differences in chemical properties between the substance to be purified and its contaminants. Unwanted metal ions may be removed by precipitation in the presence of a collector. Sodium borohydride and other metal hydrides transform organic peroxides and carbonyl-containing impurities such as aldehydes and ketones in alcohols and ethers. Many classes of organic chemicals can be purified by conversion to suitable derivatives, followed by regeneration. This Chapter describes relevant procedures.

REMOVAL OF TRACES OF METALS FROM REAGENTS

It is necessary to purify the reagents used for determinations of the more common heavy metals. Also, there should be very little if any metallic contamination of many of the materials required for biochemical studies. The main methods for removing impurities of this type are as follows.

Distillation. Reagents such as water, ammonia, hydrochloric acid, nitric acid, perchloric acid (under reduced pressure), and sulphuric acid can be purified in this way using an all-glass still. Isothermal distillation is convenient for ammonia: a beaker containing concentrated ammonia is left alongside a beaker of distilled water for several days in an empty desiccator so that some of the ammonia distils over into the water. Hydrochloric acid can be purified in the same way. The redistilled ammonia should be kept in a polyethylene or paraffin-waxed bottle. In some cases, instead of attempting to purify a salt it is simpler to synthesize it from distilled components. Ammonium acetate is an example.

Use of an ion-exchange resin. Application of ion-exchange columns has greatly facilitated the removal of heavy metal ions such as Cu^{2+}, Zn^{2+} and Pb^{2+} from aqueous solutions of many reagents. Thus, sodium salts and sodium hydroxide can be purified by passage through a column of a cation-exchange resin in its sodium form. Similarly, for acids, a column in its H^+ form is used. In some cases, where metals

form anionic complexes, they can be removed by passage through an
anion-exchange resin. Iron in hydrochloric acid solution is an
example.

Water, very low in ionic impurities (and approaching conductivity
standards), is readily obtained by percolation through alternate
columns of cation- and anion-exchange resins, or through a mixed-bed
resin, and many commercial devices are available for this purpose.
For some applications, this method is unsatisfactory because the
final water contains traces of organic material after passage through
the columns.

Precipitation. In removing trace impurities by precipitation it is
necessary to include a material to act as a "collector" of the
precipitated substance so as to facilitate its removal by filtration
or decantation. Aqueous hydrofluoric acid can be freed from lead by
adding 1 ml of 10% strontium chloride per 100 ml of acid, lead being
co-precipitated as lead fluoride with the strontium fluoride. If the
acid is decanted from the precipitate and the process is repeated,
the final lead content in the acid is less than 0.003 p.p.m. Similarly, lead can be precipitated from a nearly saturated sodium
carbonate solution by adding 10% strontium chloride dropwise (1-2 ml
per 100 ml), then filtering. (If the sodium carbonate is required
solid, the solution can be evaporated to dryness in a platinum dish.)
Removal of lead from potassium chloride uses precipitation as lead
sulphide, followed, after filtration, by evaporation and recrystallization of the potassium chloride.

Several precipitation methods are available for iron. It has been
removed from potassium thiocyanate solutions by adding a few milligrams of an aluminium salt, then precipitating aluminium and iron as
their hydroxides by adding a few drops of ammonia. Iron is also
carried down on the hydrated manganese dioxide precipitate formed in
cadmium chloride or cadmium sulphate solutions by adding 0.5%
aqueous potassium permanganate (0.5 ml per 100 ml of solution),
sufficient ammonia to give a slight precipitate, and 1 ml of ethanol.
The solution is heated to boiling to coagulate the precipitate, then
filtered. For the removal of iron from sodium potassium tartrate, a
small amount of cadmium chloride solution and a slight excess of
ammonium sulphide are added, the solution is stood for 1 hr, and the
sulphide precipitate is filtered off. Ferrous ion can be removed from
copper solutions by adding some hydrogen peroxide to the solution to
oxidize the iron, followed by precipitation of ferric hydroxide by
adding a little sodium hydroxide.

Traces of calcium can be removed from solutions of sodium salts by
precipitation at pH 9.5-10 as its 8-hydroxyquinolinate. The excess
8-hydroxyquinoline acts as collector. The magnesium content of
calcium chloride solutions can be reduced by making them about 0.1 M
in sodium hydroxide and filtering.

Extraction. In some cases, a simple solvent extraction is sufficient to remove a particular impurity. For example, traces of gallium
can be removed from titanous chloride in hydrochloric acid by extracting with isopropyl ether. Similarly, ferric chloride can be removed
from aluminium chloride solutions containing hydrochloric acid by
extraction with ethyl ether. Usually, however, it is necessary to

extract with an organic solvent in the presence of a suitable complexing agent such as dithizone or sodium diethyl dithiocarbamate.
When the former is used, weakly alkaline solutions are extracted with
dithizone in chloroform (at about 25 mg/l. of chloroform) or carbon
tetrachloride until the colour of some fresh dithizone solution
remains unchanged after shaking. Excess dithizone is taken out by
extracting with the pure solvent, the last traces of which, in turn,
are removed by aeration. This method has been used with aqueous
solutions of diammonium hydrogen citrate, potassium bromide, potassium cyanide, sodium acetate and sodium citrate. The advantage of
dithizone for such a purpose lies in the wide range of metals with
which it combines under these conditions. 8-Hydroxyquinoline (oxine)
can also be used in this way. Sodium diethyl dithiocarbamate has been
used to purify aqueous hydroxylamine hydrochloride (made just alkaline to thymol blue by adding ammonia) from copper and other heavy
metals by repeated extraction with chloroform until no more diethyl
dithiocarbamate remained in the solution (which was then acidified
to thymol blue by adding hydrochloric acid).

Complexation. Although not strictly a removal of an impurity,
addition of a suitable complexing agent such as ethylenediaminetetraacetic acid often overcomes the undesirable effects of contaminating
metal ions by reducing the concentrations of the free species to very
low levels. For a detailed discussion of this "masking", see
Masking and Demasking of Chemical Reactions, D. D. Perrin, Wiley-
Interscience, New York, 1970.

USE OF METAL HYDRIDES

This group of reagents has become commercially available in large
quantities; some of its members - notably lithium aluminium hydride
($LiAlH_4$), calcium hydride (CaH_2), sodium borohydride ($NaBH_4$) and
potassium borohydride (KBH_4) - have found widespread use in the
purification of chemicals.

Lithium aluminium hydride. Solid lithium aluminium hydride is
stable at room temperature, and is soluble in ether-type solvents.
It reacts violently with water, and is a powerful drying and reducing
agent for organic compounds. It reduces aldehydes, ketones, esters,
carboxylic acids, peroxides, acid anhydrides and acid chlorides to
alcohols. Similarly, amides, nitriles, aldimines and aliphatic nitro
compounds yield amines, while aromatic nitro compounds are converted
to azo compounds. For this reason it finds extensive application in
purifying organic chemical substances by the removal of water and
carbonyl-containing impurities. Reactions can generally be carried
out at room temperature, or in refluxing ethyl ether, at atmospheric
pressure.

Calcium hydride. This powerful drying agent is suitable for use
with hydrogen, argon, helium, nitrogen, hydrocarbons, chlorinated
hydrocarbons, esters and higher alcohols.

Sodium borohydride. Sodium borohydride is a solid which is stable
in dry air up to 300°. Like potassium borohydride, it is a less
powerful reducing agent than lithium aluminium hydride, from which
they differ also by being insoluble in ether-type solvents. Sodium
borohydride forms a dihydrate melting at $36-37^\circ$, and its aqueous

solutions decompose slowly unless stabilized to above pH 9 by alkali. (For example, a useful solution is one nearly saturated at 30-40° and containing 0.2% sodium hydroxide.) Its solubility in water is 25, 55 and 88.5 per 100 ml of water at 0°, 25° and 60°, respectively. Its aqueous solutions are rapidly decomposed by boiling or by acidification. This reagent, available either as a hygroscopic solid or as an aqueous sodium hydroxide solution, is useful as a water-soluble reducing agent for aldehydes, ketones and organic peroxides. This explains its use for the removal of carbonyl-containing impurities and peroxides from alcohols, polyols, esters, polyesters, amino-alcohols, olefins, chlorinated hydrocarbons, ethers, polyethers, amines (including aniline), polyamines and aliphatic sulphonates.

Purifications can be carried out conveniently using alkaline aqueous or methanolic solution, allowing the reaction mixture to stand at room temperature for several hours. Other solvents that can be used with this reagent include isopropyl alcohol (without alkali), amines (including liquid ammonia, in which its solubility is 104 g per 100 g of ammonia at 25°, and ethylenediamine), diglyme, formamide, dimethylformamide and tetrahydrofurfuryl alcohol. Alternatively, the material to be purified can be percolated through a column of the borohydride. In the absence of water, sodium borohydride solutions in organic solvents such as dioxane or amines decompose only very slowly at room temperature. Treatment of ethers with sodium boro-hydride appears to inhibit peroxide formation.

Potassium borohydride. Potassium borohydride is similar in properties and reactions to sodium borohydride, and, like it, is used as a reducing agent for removing aldehydes, ketones and organic peroxides. It is nonhygroscopic and can be used in water, methanol or water-methanol mixtures, provided some alkali is added to minimize decomposition, but it is somewhat less soluble than sodium borohydride in most solvents. For example, its solubility in water at 25° is 19 g per 100 ml of water (compare sodium borohydride, 55 g).

PURIFICATION via DERIVATIVES

Relatively few derivatives of organic substances are suitable for use as aids to purification. This is because of the difficulty in regenerating the starting material. For this reason, we list, below the common methods of preparation of derivatives that can be used in this way.

Whether or not any of these derivatives is likely to be satisfactory for use in any particular case will depend on the degree of difference in properties, such as solubility, volatility or melting point, between the starting material, its derivative and likely impurities, as well as on the ease with which the substance can be recovered. Purification via a derivative is likely to be of most use when the quantity of pure material that is required is not too large. Where large quantities (for example, more than 50 g) are available, it is usually more economical to purify the material directly and discard larger fractions (for example, in distillations and crystallizations).

The most generally useful purifications via derivatives are as follows:

Alcohols. Aliphatic or aromatic alcohols are converted to solid esters. p-Nitrobenzoates are the most convenient esters to form because of their sharp melting points, and the ease with which they can be recrystallized and the alcohols recovered. The p-nitrobenzoyl chloride used in the esterification is prepared by refluxing dry p-nitrobenzoic acid with a 3 mole excess of thionyl chloride for 30 min on a steam bath (in a fume cupboard). The solution is cooled slightly and the excess thionyl chloride is distilled off under (water-pump) vacuum, keeping the temperature below 40°. Dry toluene is added to the residue in the flask, then distilled off under vacuum, the process being repeated two or three times to ensure complete removal of thionyl chloride, hydrogen chloride and sulphur dioxide. (This freshly prepared p-nitrobenzoyl chloride cannot be stored without decomposition; ⁻it should be used directly.) A solution of the acid chloride (1 mol) in dry toluene or alcohol-free chloroform (distilled from phosphorus pentoxide or by passage through an activated Al_2O_3 column) under a reflux condenser is cooled in an ice bath while the alcohol (1 mol), with or without a solvent (preferably miscible with toluene or chloroform), is added dropwise to it. When addition is over and the reaction subsides, the mixture is refluxed for 30 min and the solvent is removed under reduced pressure. The solid ester is then recrystallized to constant melting point from toluene, acetone, light petroleum or mixtures of these, but not from alcohols.

Hydrolysis of the ester is achieved by refluxing in aqueous N or 2 N NaOH solution until the insoluble ester dissolves. The solution is then cooled, and the alcohol is extracted into a suitable solvent, e.g. ether, toluene or alcohol free chloroform. The extract is dried (calcium sulphate or magnesium sulphate) and distilled, then fractionally distilled. (The nitro acid can be recovered by acidification of the aqueous layer.) In most cases where the alcohol to be purified is readily freed from ethanol, the hydrolysis of the ester is best achieved with N or 2 N ethanolic NaOH or 85% aqueous ethanolic N NaOH. The former is prepared by dissolving the necessary alkali in a minimum volume of water and diluting with absolute alcohol. The alcoholic solution is refluxed for one to two hours and hydrolysis is complete when an aliquot gives a clear solution on dilution with four to five times its volume of water. The bulk of the ethanol is distilled off and the residue is extracted as above. Alternatively, use can be made of ester formation with benzoic acid, toluic acid or 3,5-dinitro-benzoic acid, by the above method.

Other derivatives can be prepared by reaction of the alcohol with an acid anhydride. For example, phthalic or 3-nitrophthalic anhydride (1 mol) and the alcohol (1 mol) are refluxed for ½-1 hr in a non-hydroxylic solvent, e.g. toluene or chloroform, and then cooled. The phthalate ester crystallizes out, is precipitated by the addition of light petroleum or is isolated by evaporation of the solvent. It is recrystallized from water, 50% ethanol, toluene or light petroleum. Such an ester has a characteristic melting point and the alcohol can be recovered by acid or alkaline hydrolysis.

Aldehydes. The best derivative from which an aldehyde can be recovered readily is its bisulphite addition compound, the main disadvantage being the lack of a sharp melting point. The aldehyde (sometimes in ethanol) is shaken with a saturated cold solution of sodium bisulphite until no more solid adduct separates. The adduct is filtered off, washed with a little water, then alcohol. A better reagent is freshly prepared saturated aqueous sodium bisulphite solution to which 75% ethanol is added to near-saturation. (Water may have to be added dropwise to render this solution clear.) With this reagent the aldehyde need not be dissolved separately in alcohol and the adduct is finally washed with alcohol. The aldehyde is recovered by dissolving the adduct in the least volume of water and adding an equivalent quantity of sodium carbonate (not sodium hydroxide) or concentrated hydrochloric acid to react with the bisulphite, followed by steam distillation or solvent extraction.

Other derivatives that can be prepared are the Schiff bases and semi-carbazones. Condensation of the aldehyde with an equivalent of prim-ary aromatic amine yields the Schiff base, for example aniline at 100° for 10-30 min.

Semicarbazones are prepared by dissolving semicarbazide hydrochloride (ca 1 g) and sodium acetate (ca 1.5 g) in water (8-10 ml) and adding the aldehyde or ketone (0.5 - 1 g) and stirring. The semicarbazide crystallizes out and is recrystallized from ethanol or aqueous ethanol. These are hydrolyzed by steam distillation in the presence of oxalic acid or better by exchange with pyruvic acid (Hershberg, J.Org.Chem., 13, 542 (1948)).

Amines. (a) Picrates: The most versatile derivative from which the free base can be readily recovered is the picrate. This is very satis-factory for primary and secondary aliphatic and aromatic amines and is particularly so for heterocyclic bases. The amine, dissolved in water, alcohol or benzene, is treated with excess of a saturated solution of picric acid in water, alcohol or benzene, respectively, until separ-ation of the picrate is complete. If separation does not occur, the solution is stirred vigorously and warmed for a few minutes, or diluted with a solvent in which the picrate is insoluble. Thus, a solution of the amine and picric acid in ethanol or benzene can be treated with benzene or light petroleum, respectively, to precipitate the picrate. Alternatively, the amine can be dissolved in alcohol and aqueous picric acid added. The picrate is filtered off, washed with water, ethanol, or benzene, and recrystallized from boiling water, ethanol, methanol, aqueous ethanol or methanol, chloroform or benzene. The solubility of picric acid in water, ethanol, and benzene, is 1.4, 6.23, and 5.27%, respectively at 20°.

It is not advisable to store large quantities of picrates for long periods. The free base should be recovered as soon as possible. The picrate is suspended in an excess of 2 M aqueous sodium hydroxide and warmed a little. Because of the limited solubility of sodium picrate, excess hot water must be added. Alternatively, because of the greater solubility of lithium picrate, aqueous 10% lithium hydroxide solution can be used. The solution is cooled, the amine is extracted with a suitable solvent such as ethyl ether or toluene, washed with 5 M sodium hydroxide until the alkaline solution remains colourless, then with water, and the extract is dried with anhydrous sodium carbonate. The solvent is distilled off and the amine is fractionally distilled (under reduced pressure if necessary) or recrystallized.

If the amines are required as their hydrochlorides, picrates can
often be decomposed by suspending them in much acetone and adding
two equivalents of 10 M hydrochloric acid. The hydrochloride of the
base is filtered off, leaving the picric acid in the acetone. Dowex
No.1 anion ion-exchange resin in the chloride form is useful for
changing solutions of the more soluble picrates (for example, of
adenosine) into solutions of their hydrochlorides, from which sodium
hydroxide precipitates the free base (Davoll and Lowy, J. Amer. Chem.
Soc., 73, 1650 (1951)).

   (b) Salts: Amines can also be purified via their salts, e.g.
hydrochlorides. A solution of the amine in dry toluene, ether or
chloroform is saturated with dry hydrogen chloride (generated by
addition of concentrated sulphuric acid to dry sodium chloride) and
the insoluble hydrochloride is filtered off and dissolved in water.
The solution is made alkaline and the amine is extracted, as above.
Hydrochlorides can also be prepared by dissolving the amine in
ethanolic HCl and adding ether or light petroleum. Where hydrochlorides are too hygroscopic or too soluble for satisfactory isolation,
other salts, e.g. nitrate, sulphate, bisulphate or oxalate, can be
used.

   (c) N-Acetyl derivatives: Purification as their N-acetyl derivatives is satisfactory for primary, and to a limited extent secondary,
amines. The base is refluxed with slightly more than one equivalent
of acetic anhydride for ½-2 hr, cooled and poured into ice-cold
water. The insoluble derivative is filtered off, dried, and
recrystallized from water, ethanol, aqueous ethanol, benzene or
benzene-light petroleum. The derivative is then hydrolyzed by
refluxing with 70% sulphuric acid for ½-1 hr. The solution is cooled,
poured onto ice, and made alkaline. The amine is steam distilled or
extracted as above. Alkaline hydrolysis is very slow.

   (d) N-Tosyl derivatives: Primary and secondary amines are converted into their N-tosyl derivatives by mixing equimolar amounts of
amine and toluene-p-sulphonyl chloride in dry pyridine (ca 5-10 moles)
and allowing to stand at room temperature overnight. The solution is
poured into ice water and the pH adjusted to 2 with hydrochloric
acid. The solid derivative is filtered off, washed with water, dried
(in a vac. des.) and recrystallized from an alcohol or aqueous
alcohol solution to a sharp melting point. The derivative is
decomposed by dissolving in liquid ammonia (fume cupboard) and adding
sodium metal (in small pieces with stirring) until the blue colour
persists for 10-15 min. Ammonia is allowed to evaporate (fume
cupboard), the residue treated with water and the solution checked
that the pH is above 10. If the pH is below 10 then the solution has
to be basified with 2 N NaOH. The mixture is extracted with ether or
toluene, the extract is dried (K_2CO_3), evaporated and the residual
amine recrystallized if solid or distilled if liquid.

   (e) As double salts: The amine (1 mole) is added to a solution of
anhydrous zinc chloride (1 mole) in concentrated hydrochloric acid
(42 ml) in ethanol (200 ml, or less depending on the solubility of
the double salt). The solution is stirred for 1 hr and the precipitated salt is filtered off and recrystallized from ethanol. The free
base is recovered by adding excess of 5-10 N sodium hydroxide (to
dissolve the zinc hydroxide that separates) and steam distilled.

Mercuric chloride in hot water can be used instead of zinc chloride and the salt is crystallized from 1% hydrochloric acid. Other double salts have been used; for example cuprous salts, but are not as convenient as the above salts.

Aromatic hydrocarbons. (a) Adducts: Aromatic hydrocarbons can be purified as their picrates using the procedures described for amines. Instead of picric acid, 1,3,5-trinitrobenzene or 2,4,7-trinitro-fluorenone can also be used. In all these cases, following recryst-allization, the hydrocarbon can be isolated either as described for amines or by passing a solution of the adduct through an activated alumina column and eluting with toluene or light petroleum. The picric acid and nitro compounds are more strongly adsorbed on the column.

(b) By sulphonation: Naphthalene, xylenes and alkyl benzenes can be purified by sulphonation with conc. sulphuric acid and crystall-ization of the sodium sulphonate. The latter is then decomposed by heating with dilute aqueous sulphuric acid. The hydrocarbon is distilled out of the mixture with superheated steam.

Carboxylic acids. (a) p-Bromophenacyl esters: (See Lee Judefind and Reid, J. Amer. Chem. Soc., 42, 1043 (1920).) A solution of the sodium salt of the acid is prepared. If the salt is not available, the acid is dissolved in an equivalent of aqueous sodium hydroxide and the pH is adjusted to 8-9. A solution of one equivalent of p-bromophenacyl bromide (for a monobasic acid , two equivalents for a dibasic acid, etc.) in ten times its volume of ethanol is then added. The mixture is heated to boiling, and, if necessary, enough ethanol is added to clarify the solution which is refluxed for ½-3 hr depending on the number of carboxyl groups that have to be esterified. (One hour is generally sufficient for monocarboxylic acids.) On cooling, the ester should crystallize out. If it does not do so, the solution is heated to boiling, and enough water is added to produce a slight turbidity. The solution is again cooled. The ester is collected, and recrystallized from ethanol, aqueous ethanol or toluene until it has a sharp melting point.

The ester is hydrolyzed by refluxing for 1-2 hr with 1-5% of barium carbonate suspended in water or with aqueous sodium carbonate solution. The solution is cooled and extracted with ether, toluene or chloroform. It is then acidified and the acid is collected by filtration or extraction, and recrystallized or fractionally distilled.

p-Nitrobenzyl esters can be prepared in an analogous manner using the sodium salt of the acid and p-nitrobenzyl bromide. They are readily hydrolyzed.

(b) Alkyl esters: Of the alkyl esters, methyl esters are the most useful because of their more rapid hydrolysis. The acid is refluxed with one or two equivalents of methanol in excess alcohol-free chloroform containing about 0.1 g of toluene-p-sulphonic acid (as catalyst), using a Dean and Stark trap. (The water formed by the esterification is carried away into the trap.) When the theoretical amount of water is collected in the trap, esterification is complete. The chloroform solution in the flask is washed with 5% aqueous sodium

carbonate solution, then water, and dried over sodium sulphate or
magnesium sulphate. The chloroform is distilled off and the ester is
fractionally distilled through an efficient column. The ester is
hydrolyzed by refluxing with 5-10% aqueous sodium hydroxide solution
until the insoluble ester has completely dissolved. The aqueous
solution is concentrated a little by distillation to remove all of
the methanol. It is then cooled and acidified. The acid is either
extracted with ether, toluene or chloroform, or filtered off and
isolated as above.

(c) Salts: The most convenient salt derivatives for carboxylic
acids are the isothiouronium salts. These are prepared by mixing
almost saturated solutions containing the acid (carefully neutralized with N-NaOH using phenolphthalein indicator) then add
two drops of N HCl and an equimolecular amount of S-benzyl-
isothiouronium chloride in ethanol and filtering off the salt that
crystallizes out. After recrystallization from water, alcohol or
aqueous alcohol the salt is decomposed by suspending or dissolving
in 2N-HCl and extracting the carboxylic acid in ether, chloroform
or toluene.

Hydroperoxides. These can be converted to their sodium salts by
precipitation below 30° with aqueous 25% sodium hydroxide. The salt
is then decomposed by addition of solid (powdered) carbon dioxide
and extracted with low-boiling petroleum ether. The solvent should
be removed under reduced pressure below 20°. The apparatus should be
adequately shielded for the safety of the operator.

Ketones. (a) Bisulphite adduct: The adduct can be prepared and
decomposed as described for aldehydes. Alternatively, because no
Cannizzaro reaction is possible, it can also be decomposed with
0.5N sodium hydroxide.

(b) Semicarbazone: A powdered mixture of semicarbazide hydrochloride (1 mol) and anhydrous sodium acetate (1.3 mol) is dissolved
in water by gentle warming. A solution of the ketone (1 mol) in the
least volume of ethanol needed to dissolve it is then added. The
mixture is warmed on a water bath until separation of the semicarbazone is complete. The solution is cooled, and the solid is filtered
off. After washing with a little ethanol followed by water, it is
recrystallized from ethanol or dilute aqueous ethanol. The derivative should have a characteristic melting point. The semicarbazone is
decomposed by refluxing with excess aqueous oxalic acid or with
aqueous sodium carbonate solution. The ketone (which steam distils)
is distilled off. It is extracted or separated from the distillate
(after saturating with sodium chloride), dried with calcium sulphate
or magnesium sulphate and fractionally distilled using an efficient
column (under vacuum if necessary).

Phenols. The most satisfactory derivatives for phenols that are
of low molecular weight or monohydric are the benzoate esters.
(Their acetate esters are generally liquids or low-melting solids.)
Acetates are more useful for high molecular weight and polyhydric
phenols.

(a) Benzoates: The phenol (1 mol) in 5% aqueous sodium hydroxide
is treated (while cooling) with benzoyl chloride (1 mol) and the

mixture is stirred in an ice bath until the separation of the solid
benzoyl derivative is complete. This derivative is filtered off,
washed with alkali, then water, and dried (in a vacuum desiccator
over solid hydroxide). It is recrystallized from ethanol or dilute
aqueous ethanol. The benzoylation can also be carried out in dry
pyridine at low temperature ($\sim0°$) instead of in sodium hydroxide
solution, finally pouring the mixture into water and collecting the
solid as above. The ester is hydrolyzed by refluxing in an alcohol
(for example, ethanol, n-butanol) containing two to three equivalents
of the alkoxide of the corresponding alcohol (for example sodium
ethoxide or sodium n-butoxide) and a few (~5-10) millilitres of
water, for ½-3 hr. When hydrolysis is complete, an aliquot will
remain clear on dilution with four to five times its volume of water.
Most of the solvent is distilled off. The residue is diluted with
cold water and acidified and the phenol is steam distilled. The
latter is collected from the distillate, dried and either fractionally distilled or recrystallized.

(b) Acetates: These can be prepared as for the benzoates using
either acetic anhydride with 3N sodium hydroxide or acetyl chloride
in pyridine. They are hydrolyzed as described for the benzoates.
This hydrolysis can also be carried out with aqueous 10% sodium
hydroxide solution, completion of hydrolysis being indicated by the
complete dissolution of the acetate in the aqueous alkaline solution.
On steam distillation, acetic acid also distils off but in these
cases the phenols (see above) are invariably solids which can be
filtered off and recrystallized.

Phosphate and phosphonate esters. These can be converted to their
uranyl nitrate addition compounds. The crude or partially purified
ester is saturated with uranyl nitrate solution and the adduct
filtered off. It is recrystallized from n-hexane, benzene or
ethanol. For the more soluble members crystallization from hexane
using lower temperatures (-40°) has been successful. The adduct is
decomposed by shaking with sodium carbonate solution and water, the
solvent is steam distilled (if hexane or benzene is used) and the
ester is collected by filtration. Alternatively, after decomposition,
the organic layer is separated, dried with calcium chloride or
barium oxide, filtered, and fractionally distilled at high vacuum.

Alternatively, impurities can sometimes be removed by conversion to
derivatives under conditions where the major component does not
react. For example, normal (straight-chain) paraffins can be freed
from unsaturated and branched-chain components by taking advantage
of the greater reactivity of the latter with chlorosulphonic acid
or bromine. Similarly, the preferential nitration of benzene can be
used to remove benzene from cyclohexane by shaking for some hours
with a mixture of concentrated nitric acid (25%), sulphuric acid
(58%), and water (17%).

BIBLIOGRAPHY

Trace Metal Analysis

N. Zief and J. W. Mitchell, Contamination Control in Trace Analysis, Wiley, New York, 1976.

F. Feigl and V. Anger, Spot Tests in Inorganic Analysis, Elsevier, Amsterdam, 1972.

Metal Hydrides

Sodium Borohydride and Potassium Borohydride: A Manual of Techniques, Metal Hydrides, Beverly, Massachusetts, 1958.

N. G. Gaylord, Reductions with Complex Metal Hydrides, Interscience, New York, 1956.

Characterization of Organic and Inorganic Compounds

N. D. Cheronis and J. B. Entrikin, Identification of Organic Compounds, Interscience, New York, 1963.

W. J. Hickinbottom, Reactions of Organic Compounds, Longmans, London, 3rd edn., 1957.

S. M. McElvain, The Characterization of Organic Compounds, Macmillan, New York, 2nd edn., 1958.

R. L. Shriner and R. C. Fuson, The Systematic Identification of Organic Compounds, Wiley, New York, 3rd edn., 1948.

B. S. Furniss et al., Vogel's Textbook of Practical Organic Chemistry, Longmans, London, 4th edn., 1978.

M. Fieser and L. F. Fieser, Reagents for Organic Synthesis, vols 1-6, J. Wiley, 1967-1977.

"Physical Constants of Inorganic Compounds" and "Physical Constants of Organic Compounds" in R. C. Weast, CRC Handbook of Chemistry and Physics, CRC Press, Cleveland, Ohio, 58th edn., 1978.

CHAPTER 3

PURIFICATION OF INDIVIDUAL
ORGANIC CHEMICALS

Most organic liquids, and a number of solids, can readily be purified
by fractional distillation, usually at atmospheric pressure: sometimes,
especially with higher-boiling materials, distillation at reduced
pressure is desirable. The general principles and techniques are
described in Chapter 1. For this reason, and to save space, the present
chapter omits those substances for which published purification methods involve only distillation. Where boiling points are given, purification by distillation is another means of separating impurities.
Similarly, references are omitted for methods such as simple recrystallization from solution. Otherwise, substances are listed alphabetically, usually with some physical criteria of purity, giving brief
details of how they can be purified, and relevant literature references.

In Chapters 3 and 4, "drying with" and "distillation from" are to be
taken as implying physical contact between the substances concerned,
whereas "drying over (or above)" indicates remoteness, for example as
in the drying of a solid in a desiccator containing sulphuric acid.

To save space, the following abbreviations are used: aq. (aqueous),
cryst. (crystallized), crystn. (crystallization), dil. (dilute),
distd. (distilled), distn. (distillation), pet. ether (petroleum
ether), pptd. (precipitated), ppte. (precipitate), pptn. (precipitation), satd. (saturated), soln. (solution) and vac. (vacuum).

In addition, the following journals are designated by their initials:

Anal. Chem.	AC
Biochem. J.	BJ
Ind. Eng. Chem. (Anal. Ed.)	IECAE
J. Amer. Chem. Soc.	JACS
J. Biol. Chem.	JBC
J. Chem. Phys.	JCP
J. Chem. Soc.	JCS
J. Org. Chem.	JOC
J. Phys. Chem.	JPC
Pure Appl. Chem.	PAC
Trans. Faraday Soc.	TFS

As a good general rule all low boiling (<110°) organic liquids
should be treated as highly flammable and the necessary precautions
should be taken.

Abietic acid, m.p. 172-175°, $[\alpha]_D^{25}$ -116° (c = 1 in ethanol)
    Cryst. from aq. ethanol.

Abscisic acid, m.p. 160-161° (with sublimation) $[\alpha]_{287}$ +24,000,
    $[\alpha]_{245}$ -69,000 (c = 1-50 µg/ml in acidified methanol or ethanol)
    Cryst. from chloroform-light petroleum.

Acenaphthene, m.p. 94°.
    Cryst. from ethanol.
    Purified by chromatography of CCl_4 solns. on alumina, with benzene
as eluant. [McLaughlin and Zainal, JCS, 2485 (1960).]

Acenaphthoquinone, m.p. 260-261°.
    Extracted with, then recryst. twice from, benzene. [Le Fevre,
Sundaram and Sundaram, JCS, 974 (1963).]

Acenaphthylene, m.p. 92-93°.
    Dissolved in warm redist. methanol, filtered through a sintered-
glass crucible and cooled to -78° to ppte. the material as yellow
plates. [Dainton, Ivin and Walmsley, TFS, 56, 1784 (1960).]

Acetal, b.p. 103.7-104°, n_D^{20} 1.38054, n_D^{25} 1.3682.
    Dried over sodium to remove alcohols and water and to polymerize
aldehydes, then fractionally distd.
    Treated with alkaline H_2O_2 soln. at 40-45° to remove aldehydes,
then the soln. was satd. with NaCl, separated, dried with anhydrous
K_2CO_3 and dist. from sodium. [Vogel, JCS, 616 (1948).]

Acetaldehyde, b.p. 20.2°, n_D^{20} 1.33113.
    Usually purified by fractional distn. in a glass helices-packed
column under dry nitrogen, discarding the first portion of distillate.
    Shaken for 30 min with $NaHCO_3$, dried with $CaSO_4$ and fractionally
distd. at atmospheric pressure through a 70-cm Vigreux column. The
middle fraction was taken and further purified by standing for 2 hr

Acetaldehyde (continued)
at 0° with a small amount of hydroquinone, followed by distn.
[Longfield and Walters, JACS, 77, 810 (1955).]

Acetamide, m.p. 81°.

Cryst. by soln. in hot methanol (0.8 ml/g), dilution with ethyl
ether and allowing to stand. [Wagner, J.Chem.Ed., 7, 1135 (1930).]
Alternative crystns. are from acetone, benzene, chloroform, dioxane,
methyl acetate or from benzene-ethyl acetate mixture (3:1 and 1:1).
It has also been cryst. from hot water after treating with HCl-washed
activated charcoal (which had been washed repeatedly with water until
free of chloride ions), then cryst. again from hot 50% aq. ethanol
and finally twice from hot 95% ethanol. [Christoffers and Kegeles,
JACS, 85, 2562 (1963).] Final drying is in a vac. desiccator over
$Mg(ClO_4)_2$ or P_2O_5.

Acetamide is also purified by distn. (b.p. $221-223^{\circ}$) or by
sublimation at reduced pressure.

Acetamidine hydrochloride, m.p. 174°.

Cryst. from ethanol.

N-(2-Acetamido)-2-aminoethanolsulphonic acid (ACES).

Recryst. from hot aq. ethanol.

4-Acetamidobenzaldehyde, m.p. 156°.

Cryst. from water.

N-(2-Acetamido)iminodiacetic acid (ADA), m.p. 200° (decomp.).

Dissolved in water by adding one equivalent of NaOH soln. (to
final pH of 8-9), then acidified with HCl to ppte. the free acid.
Filtered and washed with water.

2-Acetamido-5-nitrothiazole, m.p. $264-265^{\circ}$.

Cryst. from ethanol or glacial acetic acid.

2-Acetamidophenol, m.p. 209°.

Cryst. from water or aq. ethanol.

3-Acetamidophenol, m.p. 148-149°.
    Cryst. from water.

4-Acetamidophenol, m.p. 169-170.5°.
    Cryst. from water or ethanol.

5-Acetamido-1,3,4-thiadiazole-2-sulphonamide, m.p.258-259° (decomp.).
    Cryst. from water.

Acetanilide, m.p. 114°.
    Cryst. from water, aq. ethanol, benzene or toluene.

Acetic acid (glacial), m.p. 16.6°, b.p. 118°, n_D^{20} 1.37171,
    n_D^{25} 1.36995.
    Usual impurities are traces of acetaldehyde and other oxidizable
substances and water. (Glacial acetic acid is very hygroscopic. The
presence of 0.1% water in acetic acid lowers its m.p. by 0.2°.)
Purified by adding some acetic anhydride to react with the water
present, heating for 1 hr to just below boiling in the presence of
2 g CrO_3 per 100 ml and then fractionally distilling. [Orton and
Bradfield, JCS, 960 (1924); 983 (1927).] Instead of CrO_3, 2-5% (w/w)
of $KMnO_4$, with refluxing for 2-6 hr, has been used.
    Traces of water have been removed by refluxing with tetraacetyl
diborate (prepared by warming 1 part of boric acid with 5 parts (w/w)
of acetic anhydride at 60°, cooling, and filtering off), followed by
distn. [Eichelberger and La Mer, JACS, 55, 3633 (1933).]
    Refluxing with acetic anhydride in the presence of 0.2 g %
β-naphthalene-sulphonic acid as catalyst has also been used. [Orton
and Bradfield, JCS, 983 (1927).] Other suitable drying agents include
$Mg(ClO_4)_2$, $CuSO_4$ and chromium triacetate: P_2O_5 converts some of the
acid to the anhydride. Azeotropic removal of water by distn. with
thiophene-free benzene or with butyl acetate has been used. [Bird-
whistell and Griswold, JACS, 77, 873 (1955).] An alternative purification uses fractional freezing.

Acetic anhydride, b.p. 138.0°, n_D^{15} 1.39299, n_D^{20} 1.3904.
    Adequate purification can usually be obtained by fractional distn.
through an efficient column. Acetic acid can be removed by prior

Acetic anhydride (continued)

refluxing with CaC_2 or with coarse magnesium filings at 80-90° for 5
days, or by distn. from synthetic quinoline (1% of total charge) at
75 mm pressure. Acetic anhydride can also be dried by standing with
sodium wire for up to a week, and distilling from it under vac.
(Sodium reacts vigorously with acetic anhydride at 65-70°.)

    Dippy and Evans (JOC, 15, 451 (1950)] let acetic anhydride (500 g)
stand with P_2O_5 (50 g) for 3 hr, then decanted it and stood it with
ignited K_2CO_3 for a further 3 hr. The supernatant liquid was distd.
and the fraction, b.p. 136-138°, was further dried with P_2O_5 for 12
hr, followed by shaking with ignited K_2CO_3, before two further distns.
through a five-section Young and Thomas fractionating column. The
final material distd. at 137.8-138.0°. Can also be purified by azeotropic distn. with toluene: the acetic acid-toluene azeotrope boils
at 100.6°. After removal of the remaining toluene, the acetic
anhydride is distd. (Sample had a specific conductivity of 5×10^{-9}
$ohm^{-1} cm^{-1}$.)

Acetin blue.
    Cryst. from 3:1 benzene-methanol.

Acetoacetanilide, m.p. 86°.
    Cryst. from water, aq. ethanol or pet. ether (b.p. 60-80°).

Acetoacetopiperidide, b.p. 88.9°/0.1 mm, n_D^{25} 1.4983.
    Dissolved in benzene, extracted with 0.5 M HCl to remove basic
impurities, washed with water, dried, and distd. at 0.1 mm Hg.
[Wilson, JOC, 28, 314 (1963).]

α-Acetobromglucose, m.p. 88-89°, $[α]_D^{19}$+ 199.3° (c = 3 in $CHCl_3$).
    Cryst. from di-isopropyl ether, or pet. ether (b.p. 40-60°).

Acetoin, see 3-Hydroxy-2-butanone.

β-Acetonaphthalene, m.p. 55-56°.
    Cryst. from pet. ether, ethanol or glacial acetic acid.

2-Acetonaphthone, see β-Acetonaphthalene.

Acetone, b.p. 56.2°, d_{25}^{25} 0.7880, n_D^{20} 1.35880, n_D^{25} 1.35609.

Analytical reagent quality generally contains less than 0.1% organic impurities but may have up to about 1% water. It can be dried with anhydrous $CaSO_4$, K_2CO_3 or type 4A Linde molecular sieve, and then distd. Silica gel and alumina cause acetone to undergo the aldol condensation, so that its water content is increased by passage through these reagents. This also occurs to some extent when P_2O_5 or sodium amalgam is used. Anhydrous $MgSO_4$ is an inefficient drying agent, and $CaCl_2$ forms an addition compound.

Organic impurities have been removed from acetone by adding 4 g of silver nitrate in 30 ml of water to 1 l. of acetone, followed by 30 ml 1 M NaOH, shaking for 10 min, filtering, drying with anhydrous $CaSO_4$ and distilling. [Werner, Analyst, 58, 335 (1933).] Alternatively, successive small portions of $KMnO_4$ have been added to acetone at reflux, until the violet colour persists, followed by drying and distn. Refluxing with chromic anhydride has also been used. Methanol has been removed from acetone by azeotropic distn. (at 35°) with methyl bromide, and by treatment with acetyl chloride.

Small amounts of acetone can be purified as the sodium iodide addition compound, by dissolving 100 g of finely powdered NaI in 440 g of boiling acetone, then cooling in ice and salt to -8°. Crystals of $NaI.3C_3H_6O$ are filtered off and, on warming in a flask, acetone distils readily. (This method is more convenient than the one using the bisulphite addition compound.) Also purified by gas chromatography on a 20% free fatty acid phthalate (Chromsorb P) column at 100°.

Acetonedicarboxylic acid, m.p. 138° (decomp.).

Cryst. from ethyl acetate, and stored over P_2O_5.

Acetone semicarbazone, m.p. 187°.

Cryst. from water or from aq. ethanol.

Acetonitrile, b.p. 81.6°, d_4^{25} 0.77683, n_D^{20} 1.34411, n_D^{25} 1.34163.

Usual contaminants include water, acetamide, ammonium acetate and ammonia. Anhydrous $CaSO_4$ and $CaCl_2$ are inefficient drying agents. Preliminary treatment of acetonitrile with cold, satd. aq. KOH is undesirable because of base-catalyzed hydrolysis and the introduction

<u>Acetonitrile</u> (continued)

of water. Drying by shaking with silica gel or Linde 4A molecular
sieves removes most of the water in acetonitrile. Subsequently stirring with calcium hydride until no further hydrogen is evolved leaves
only traces of water and removes acetic acid. The acetonitrile is
then fractionally distd. at high reflux, taking precaution to exclude
moisture by refluxing over CaH_2 [Coetzee, <u>PAC</u>, <u>13</u>, 429 (1966)].
Alternatively, 0.5-1% (w/v) P_2O_5 is often added to the distilling
flask to remove most of the remaining water. Excess P_2O_5 should be
avoided because it leads to the formation of an orange polymer.
Traces of P_2O_5 can be removed by distilling from anhydrous K_2CO_3.

Kolthoff, Bruckenstein and Chantooni [<u>JACS</u>, <u>83</u>, 3297 (1961)]
removed acetic acid from 3 l. of acetonitrile by shaking for 24 hr
with 200 g of freshly activated alumina (which was then reactivated
by heating at 250° for 4 hr). The decanted solvent was again shaken
with activated alumina, followed by five batches of 100-150 g of
anhydrous $CaCl_2$. (The water content of the solvent was then less than
0.2%.) It was shaken for 1 hr with 10 g of P_2O_5, twice, and distd. in
a column 1 m x 2 cm, packed with stainless-steel wool and protected
from atmospheric moisture by $CaCl_2$ tubes. The middle fraction had a
water content between 0.7-2 mM.

Traces of unsaturated nitriles can be removed by an initial refluxing with a small amount of aq. KOH (1 ml of 1% solution per
litre). Acetonitrile can be dried by azeotropic distn. with dichloromethane, benzene or trichloroethylene. Isonitrile impurities can be
removed by treatment with conc. HCl until the smell of isonitrile has
gone, followed by drying with K_2CO_3 and distn.

Acetonitrile was refluxed with, and distd. from, alk. $KMnO_4$ and
$KHSO_4$, followed finally by fractional distn. from CaH_2. (This was
better than fractionation from molecular sieves or passage through a
type H activated alumina column, refluxing with KBH_4 for 24 hr and
fractional distn.) [Bell, Rodgers and Burrows, <u>JCS Faraday Trans I</u>,
<u>73</u>, 315 (1977).]

Material suitable for polarography was obtained by refluxing over
anhydrous $AlCl_3$ (15g/l) for 1 hr, distg., refluxing over Li_2CO_3
(10g/l) for 1 hr, and redistg. It was then refluxed over CaH_2 (2g/l)
for 1 hr and fractionally distd, retaining the middle portion. [The
product was not suitable for uv spectroscopic use. A better purific-

Acetonitrile (continued again)
ation used refluxing over anhydrous $AlCl_3$ (15 g/l) for 1 hr., distn.,
refluxing over alk. $KMnO_4$ (10 g $KMnO_4$, 10 g Li_2CO_3/l.) for 15
minutes, and distn. A further reflux for 1 hr. over $KHSO_4$ (15 g/l),
then distn., was followed by refluxing over CaH_2 (2 g/l) for 1 hr.,
and fractional distn. The product was protected from atmospheric
moisture and stored under nitrogen.] [Walter and Ramalay, AC, 45,
165 (1973).]

p-Acetophenetidine, m.p. 136°.
   Cryst. from water or purified by soln. in cold dil. alkali
and repptd. by addn. of acid to neutralization point. Air-dried.

Acetophenone, m.p. 19.6^O, b.p. $54^O/2.5$ mm, $202^O/760$ mm,
d_4^{25} 1.0238, n_D^{15} 1.53631, n_D^{25} 1.5322.

   Dried by fractional distn. or by standing with anhydrous $CaSO_4$
or $CaCl_2$ for several days, followed by fractional distn. under
reduced pressure (from P_2O_5, optional), and careful, slow and
repeated partial crystns. from the liquid at 0° excluding light and
moisture. It can also be cryst. at low temps. from isopentane.
Distn. can be followed by purification using gas-liquid chromatography. [Earls and Jones, JCS Faraday Trans I, 71, 2186 (1975).]

Acetoxime, m.p. 60°.
   Cryst. from pet. ether (b.p. $40-60^O$). Can be sublimed.

21-Acetoxypregnenolone, m.p. $184-185^O$.
   Cryst. from acetone.

Acet-o-toluidide, m.p. 110°, b.p. $296^O/760$ mm.
Acet-m-toluidide, m.p. 65.5^O, b.p. $307^O/760$ mm; $182-183^O/14$ mm.
   Cryst. from water.

Acet-p-toluidide, m.p. 146°, b.p. $307^O/760$ mm.
   Cryst. from aq. ethanol.

Acetylacetone, b.p. 45°/30 mm, 138.9°/750 mm, $d_4^{30.2}$ 0.9630, $n_D^{18.5}$ 1.45178.

Small amounts of acetic acid were removed by shaking with small portions of 2 M NaOH until the aq. phase remained faintly alkaline. The sample, after washing with water, was dried with anhydrous Na_2SO_4, distd. through a modified vigreux column. [Cartledge, JACS, 73, 4416 (1951).] An additional purification step is fractional crystn. from the liquid. Alternatively, there is less loss of acetylacetone if it is dissolved in four volumes of benzene and the soln. is shaken three times with an equal volume of distd. water (to extract acetic acid): the benzene is then removed by distn. Acetylacetone was refluxed over $NaHCO_3$ (20g/l) for ½ hr, then distd. at $43-53^\circ$ under reduced pressure (20-30 mm Hg) through a helices-packed column. Then refluxed over P_2O_5 (10g/l) and fractionally distd. under reduced pressure. The distillate (specific conductivity, 4×10^{-8} $ohm^{-1}cm^{-1}$) was suitable for polarography. [Fujinaga and Lee, Talanta, 24, 395 (1977).] To recover used acetylacetone, metal ions were stripped from the soln. at pH 1 (using 100 ml 0.1 M H_2SO_4/l. of acetylacetone). The acetyl-acetone was washed with dil. (1:10) ammonia soln. (100 ml/l) and then with distd. water (100 ml/l, twice), and treated as above.

N-Acetyl-β-alanine, m.p. $78.3-80.3^\circ$.
    Cryst. from acetone.

Acetyl-α-amino-n-butyric acid.
    Cryst. twice from water (charcoal) and air dried. [King and King, JACS, 78, 1089 (1956).]

2-Acetylaminofluorene, see N-2-Fluorenylacetamide.

N-Acetylanthranilic acid, m.p. 185°.
    Cryst. from water, ethanol or acetic acid.

4-Acetylbiphenyl, m.p. $120-121^\circ$, b.p. $325-327^\circ$/760 mm.
    Cryst. from ethanol or acetone.

Acetyl-5-bromosalicylic acid, m.p. $168-169^\circ$.
    Cryst. from ethanol.

Acetyl chloride, b.p. 52°.

Refluxed with PCl_5 for several hours to remove traces of acetic acid, then dist. Redistd. from one-tenth volume of dimethylaniline or quinoline to remove free HCl. A.R. quality is freed from HCl by pumping it for 1 hour at -78° and distg. into a trap at -196°.

Acetylcholine bromide, m.p. 146°.
    Cryst. from ethanol.

Acetyldigitoxin-α, m.p. 217-221°.
    Cryst. from methanol.

Acetylene.

Purified by successive passage through spiral wash bottles containing, in order, satd. aq. $NaHSO_3$, water, 0.2 M iodine in aq. KI (two bottles), sodium thiosulphate soln. (two bottles), alkaline sodium hydrosulphite with sodium anthraquinone-β-sulphonate as indicator (two bottles), and 10% aq. KOH soln. (two bottles). The gas was then passed through a Dry-Ice trap and two drying tubes, the first containing $CaCl_2$ and the second, Dehydrite. [Conn, Kistiakowsky and Smith, JACS, 61, 1868 (1939).] Acetone vapour can be removed from acetylene by passage through two traps at -65°.

Acetylenedicarboxamide, m.p. 294° (decomp.).
    Cryst. from aq. methanol.

Acetylenedicarboxylic acid, m.p. 179° (anhydrous).
    Cryst. from aq. ether as dipicrate.

2-Acetylfluorene, m.p. 132°.
    Cryst. from ethanol.

Acetyl fluoride, b.p. 20.5°/760 mm.
    Purified by fractional distn.

N-Acetyl-D-galactosamine, m.p. 160-161°, $[\alpha]_{546}^{20}$ +102° (c = 1 in H_2O).

N-Acetyl-D-glucosamine, m.p. ∿215°, $[\alpha]_{546}^{20}$ +49° after 2 hr. (c = 2 in H_2O).
    Cryst. from methanol-ether.

N-Acetylglutamic acid, m.p. 185° (RS), 201° (S), $[\alpha]_D^{25}$ -16.6°
(in water), (S).
   Likely impurity: glutamic acid. Cryst. from boiling water.

N-Acetylglycine, m.p. 206-208°.
   Treated with washed charcoal and recryst. three times from water.
[King and King, JACS, 78, 1089 (1956).]

N-Acetylhistidine, monohydrate, m.p. 148° (RS), 169° (S),
$[\alpha]_D^{25}$ +46.2° (water), (S).
   Likely impurity: histidine. Cryst. from water, then from 4:1
acetone: water.

N-Acetylimidazole, m.p. 101.5-102.5°.
   Cryst. from isopropenyl acetate. Dried in vac. over P_2O_5.

Acetyl iodide, b.p. 108°/760 mm.
   Purified by fractional distn.

N-Acetylmethionine, m.p. 104°, $[\alpha]_{546}^{20}$ +24.5° (c = 1 in water).
   Cryst. from water or ethyl acetate.

Acetylmethionine nitrile, m.p. 44-46°.
   Cryst. from ethyl ether.

3-Acetyl-6-methoxybenzaldehyde, m.p. 144°.
   Cryst. from ethanol or ethyl ether.

Acetylmethylcarbinol, see 3-Hydroxy-2-butanone.

N-Acetyl-D-penicillamine, m.p. 189-190° (decomp.) $[\alpha]_D^{20}$ +18°
(c = 1 in 50% EtOH).
   Cryst. from water.

N-Acetylphenylalanine, (L) m.p. 170-171°, $[\alpha]_D$ +49.3°,
(DL) m.p. 152.5-153°.
   Cryst. from $CHCl_3$ and stored in a desiccator at 4°. (DL)-isomer
cryst. from water or acetone.

N-Acetyl-L-phenylalanine ethyl ester.
   Cryst. from water.

1-Acetyl-2-phenylhydrazine, m.p. 128.5°.
   Cryst. from aq. ethanol.

Acetylsalicylic acid, m.p. 133.5-135°.
   Cryst. twice from benzene, washed with cyclohexane and dried at 60°
under vac. for several hours. [Davis and Hetzer, J.Res.Nat.Bur.Stand.
60, 569 (1958).] Has also been recryst. from isopropyl alcohol and
from ethyl ether-pet. ether (b.p. 40-60°).

Acetylsalicylsalicylic acid, m.p. 159°.
   Cryst. from dil. acetic acid.

N(4)-Acetylsulphanilamide, m.p. 216°.
   Cryst. from aq. ethanol.

N-Acetylsulphanilyl chloride, m.p. 149°.
   Cryst. from $CHCl_3$ or ethylene dichloride.

Acetyl-o-toluidine, m.p. 110°.
   Cryst. from ethanol.

Acetyl-p-toluidine, m.p. 153°.
   Cryst. from aq. ethanol.

N-Acetyltryptophan, m.p. 206° (RS), 188° (S), $[\alpha]_D^{25}$ +30.1°.
   (in aq. NaOH), (S).
   Likely impurity: tryptophan. Cryst. from ethanol by adding water.

Aconitic acid.
   Cryst. from water, by cooling (sol. 1 g in 2 ml of water at 25°).
Dried in vac. desiccator.

Aconitide, m.p. 204°, $[\alpha]_{546}^{20}$ +20° (c = 1 in $CHCl_3$).
   Cryst. from ethanol, chloroform or benzene.

Aconitine hydrobromide, m.p. 207°.
  Cryst. from water or ethanol-ether.

Acraldehyde, see Acrolein.

Acridine, m.p. 111°.
  Cryst. twice from benzene-cyclohexane, or from aq. ethanol, then
sublimed, removing and discarding the first 25% of the sublimate.
The remainder was again cryst. and sublimed, discarding the first
10-15%. [Wolf and Anderson, JACS, 77, 1608 (1955).]
  Acridine can also be purified by crystn. from n-heptane and then
from ethanol-water, or by chromatography on alumina with pet. ether
in a darkened room. Alternatively, acridine can be pptd. as the
hydrochloride from benzene soln. by adding HCl, after which the base
is regenerated, dried at 110°/50 mm, and cryst. to constant m.p. from
pet. ether. [Cumper, Ginman and Vogel, JCS, 4518 (1962).] The regenerated free base may be recryst., chromatographed on basic alumina,
then vac.-sublimed and zone-refined. [Williams and Clarke,
JCS Faraday Trans I, 73, 514 (1977).]

Acridine orange, m.p. 181-182° (free base).
  The double salt with $ZnCl_2$ (6 g) was dissolved in water (200 ml)
and stirred with four successive portions (12 g each) of Dowex-50
ion-exchange resin (K^+ form) to remove the zinc. The soln. was then
concentrated in vac. to 20 ml, and 100 ml of ethanol was added to
ppte. KCl which was removed. Ether (160 ml) was added to the soln.
from which, on chilling, the dye cryst. as its chloride. It was
separated by centrifuging, washed with chilled ethanol and ether, and
dried under vac. before being recryst. from ethanol (100 ml) by adding ether (50 ml), and chilling. Yield 1 g. [Pal and Schubert, JACS,
84, 4384 (1962).]
  It was recryst. twice as the free base from ethanol or methanol-
water by dropwise addition of NaOH (less than 0.1 M). The ppte. was
washed with water and dried under vac. It was dissolved in $CHCl_3$ and
chromatographed on alumina: the main sharp band was collected, concentrated and cooled to -20°. The ppte. was filtered, dried in air,
then dried for 2 hr under vac. at 70°. [Stone and Bradley, JACS,
83, 3627 (1961); Blauer and Linschitz, JPC, 66, 453 (1962).]

Acridine yellow.

Cryst. from 1:1 benzene-methanol.

Acridinol, see 4-Hydroxyacridine.

Acriflavine.

Treated twice with freshly pptd. AgOH to remove proflavine, then recryst. from absolute methanol. [Wen and Hsu, JPC, 66, 1353 (1962).]

Acrolein, b.p. 52.1°, n_D^{20} 1.3992.

Purified by fractional distn. under nitrogen, drying with anhydrous $CaSO_4$ and then distilling under vac. Blacet, Young and Roof [JACS, 59, 608 (1937)] distd. under nitrogen through a 90 cm column packed with glass rings. To avoid formation of diacryl, the vapour was passed through an ice-cooled condenser into a receiver cooled in an ice-salt mixture and containing 0.5 g catechol. The acrolein was then distd. twice from anhydrous $CuSO_4$ at low pressure, catechol being placed in the distilling flask and the receiver to avoid polymerization. (Alternatively, hydroquinone [1% of final soln.] can be used.)

Acrolein semicarbazone, m.p. 171°.

Cryst. from water.

Acrylamide, m.p. 84°.

Cryst. from acetone, chloroform, ethyl acetate, methanol or benzene-chloroform mixture, then vac. dried and kept in the dark under vac.

Acrylic acid, m.p. 11°, b.p. 30°/3 mm.

Can be purified by steam distn. or vac. distn. through a column packed with copper gauge to inhibit polymerization. (This treatment also removes inhibitors such as methylene blue that may be present.) Azeotropic distn. of the water with benzene converts aq. acrylic acid to the anhydrous material.

Acrylonitrile, b.p. 78°, n_D^{25} 1.3886.

Washed with dil. H_2SO_4 or dil. H_3PO_4, then with dil. Na_2CO_3 and water. Dried with Na_2SO_4, $CaCl_2$ or (better) by shaking with molecular sieve. Fractionally distd. under nitrogen. Can be stabilized by

Acrylonitrile (continued)
adding 10 p.p.m. tert.-butyl catechol. Immediately before use, the
stabilizer can be removed by passage through a column of activated
alumina (or by washing with 1% NaOH soln. if traces of water are
permissible in the final material), followed by distn.

Actidione, see Cycloheximide.

Actinomycin.
    Cryst. from ethyl acetate or from methanol.

Adamantane, 269.6-270.8°.
    Cryst. from acetone or cyclohexane, sublimes in vac. below m.p.

Adamantane-1-carboxylic acid, m.p. 177°.
    Cryst. from absolute ethanol and dried under vac. at 100°.

Adenine, m.p. 360-365° (decomp., rapid heating).
    Cryst. from distd. water.

Adenosine, m.p. 229°.
    Cryst. from distd. water.

Adenosine-3'-phosphoric acid, m.p. 194°.
    Cryst. from large volume of distd. water, as monohydrate.

Adenosine-5'-phosphoric acid, m.p. 196-200°.
    Cryst. from water by addition of acetone.
    Purified by chromatography on Dowex 1 (in formate form), eluting
with 0.25 M formic acid. It was then adsorbed onto charcoal (which
had been boiled for 15 min with M HCl, washed free of chloride and
dried at 100°), and recovered by stirring three times with isoamyl
alcohol-water (1:9 v/v). The aq. layer from the combined extracts
was evaporated to dryness under reduced pressure, and the product
was cryst. twice from hot water. [Morrison and Doherty, BJ, 79, 433
(1961).]

Adenosine triphosphate.

Pptd. as its barium salt when excess barium acetate soln. was added
to a 5% soln. of ATP in water. After filtering off, the ppte. was
washed with distd. water, redissolved in 0.2 M HNO_3, and again pptd.
with barium acetate. The ppte., after several washings with distd.
water, was dissolved in 0.2 M HNO_3 and slightly more 0.2 M H_2SO_4 than
was needed to ppte. all the barium as $BaSO_4$ was added. After filtering off the $BaSO_4$, the ATP was pptd. by addition of a large excess of
95% ethanol, filtered off, washed several times with 100% ethanol and
finally with dry ethyl ether. [Kashiwagi and Rabinovitch, JPC, 59,
498 (1955).]

Adenylic acid, see Adenosinephosphoric acid.

Adipic acid, m.p. 154°.

For use as a volumetric standard, adipic acid was cryst. once from
hot water with the addition of a little animal charcoal, dried at
120° for 2 hr, then recryst. from acetone and again dried at 120°
for 2 hr.

Other purification procedures include crystn. from ethyl acetate
and from acetone-petroleum ether, fusion followed by filtration and
crystn. from the melt, and preliminary distn. under vac.

Adonitol, m.p. 102°.

Cryst. from ethanol by addition of ethyl ether.

Adrenalin, see Epinephrine.

Adrenochrome, m.p. 125-130°.

Cryst. from methanol-formic acid, as hemihydrate, and stored in a
vac. desiccator.

Adrenosterone, m.p. 220-224°.

Cryst. from ethanol. Can be sublimed under high vac.

Agaric acid, m.p. 142° (decomp.), $[\alpha]_D$ -19° (in NaOH).

Cryst. from ethanol.

Agmatine sulphate, m.p. 231°.
   Cryst. from aq. methanol.

Agroclavin, m.p. 198-203° (decomp.), $[\alpha]_D^{30}$ -242°.
   Cryst. from ethyl ether.

Ajmalicine, m.p. 250-252°.
   Cryst. from methanol.

Ajmalicine hydrochloride, m.p. 290° (decomp.), $[\alpha]_D^{20}$ -17°
   (c = 0.5 in methanol)
   Cryst. from ethanol.

Ajmaline, m.p. 160°.
   Cryst. from methanol.

Ajmaline hydrochloride, m.p. 140°.
   Cryst. from water.

Alanine, (RS), m.p. 295-296°, (S), m.p. 297° (decomp.),
   $[\alpha]_D^{15}$ +14.7° (in 1 M. HCl).
   Cryst. from water or aq. ethanol, e.g. cryst. from 25% ethanol in
water, recryst. from 62.5% ethanol, washed with ethanol and dried to
constant weight in a vac. desiccator over P_2O_5. [Gutter and Kegeles,
JACS, 75, 3893 (1953).] 2,2'-Iminodipropionic acid is a likely
impurity.

β-Alanine, m.p. 200°.
   Cryst. from filtered hot satd. aq. soln. by adding four volumes of
absolute ethanol and cooling in an ice-bath. Recryst. in same way
and then finally, cryst. from a warm satd. soln. in 50% ethanol by
adding four volumes of absolute ethanol cooled in an ice-bath.
Crystals were dried in a vac. desiccator over P_2O_5. [Donoian and
Kegeles, JACS, 83, 255 (1961).]

Albumin (bovine serum).
   Purified by soln. in conductivity water and passage at 2-4°
through two ion-exchange columns, each containing a 2:1 mixture of

Albumin (continued)

anionic and cationic resins (Amberlite IR-120, H-form; Amberlite
IRA-400, OH-form). This treatment removed ions and lipoid impurities.
Care was taken to exclude CO_2, and the soln. was stored at $-15°$.
[Möller, van Os and Overbeek, TFS, 57, 312 (1961).]

More complete lipid removal was achieved by lyophilizing the de-
ionized soln., covering the dried albumin (human serum) with a
mixture of 5% glacial acetic acid (v/v) in iso-octane (previously
dried with Na_2SO_4) and allowing to stand at $0°$ (without agitation)
for upwards of 6 hr before decanting and discarding the extraction
mixture, washing with iso-octane, re-extracting, and finally washing
twice with iso-octane. The purified albumin was dried under vac. for
several hours, then dialyzed against water for 12-24 hr at room temp.,
lyophilized, and stored at $-10°C$. [Goodman, Science, 125, 1296 (1957).]

Aldol, b.p. $80-81°/20$ mm.

An ethereal soln. was washed with a satd. aq. soln. of $NaHCO_3$, then
with water. The non-aqueous layer was dried with anhydrous $CaCl_2$ and
distd. immediately before use. The fraction, b.p. $80-81°/20$ mm, was
taken. [Mason, Wade and Pouncy, JACS, 76, 2255 (1954).]

Aldosterone, m.p. $108-112°$ (hydrate), $164°$ (anhydr.).

Cryst. from aq. acetone.

Aldrin, m.p. $103-104.5°$.

Cryst. from methanol.

Aleuritic acid, m.p. $100-101°$.

Cryst. from aq. ethanol.

Alginic acid.

To 5 g in 550 ml water containing 2.8 g $KHCO_3$, were added 0.3 ml
acetic acid and 5 g potassium acetate. Ethanol to make the soln. 25%
(v/v) in ethanol was added and any insoluble material was discarded.
Further addition of ethanol, to 37% (v/v), pptd. alginic acid. [Pal
and Schubert, JACS, 84, 4384 (1962).]

Aliquat 336.
   A 30% (v/v) soln. in benzene was washed twice with an equal volume
of 1.5 M HBr. [Petrow and Allen, AC, 33, 1303 (1961).]

Alizarin, m.p. 289°.
   Cryst. from glacial acetic acid or 95% ethanol. Can also be sublimed.

Alizarin orange, see 3-Nitroalizarin.

Allantoin, m.p. 238°.
   Cryst. from water or ethanol.

Allene.
   Frozen in liquid nitrogen, pumped on, then thawed out. This cycle
was repeated several times, then the allene was frozen in a methyl
cyclohexane-liquid nitrogen bath and pumped for some time.

Allopregnane-3α, 20α-diol, m.p. 248-248.5°.
   Cryst. from ethanol.

Alloxan, m.p. ∿170° (decomp.).
   Cryst. from water gives the tetrahydrate. Anhydrous crystals are
obtained by cryst. from acetone, glacial acetic acid or by sublimation in vacuo.

Alloxantin, m.p. 253-255° (decomp.)(yellow at 225°).
   Cryst. from water or ethanol and kept under nitrogen. Turns red in
air.

Allyl acetate, b.p. 103°, n_4^{20} 1.40488, n_D^{27} 1.4004.
   Freed from peroxides by leaving with crystalline ferrous ammonium
sulphate, then washed with 5% $NaHCO_3$, followed by satd. $CaCl_2$ soln.
Dried with Na_2SO_4 and fractionally distd. in an all-glass apparatus.

Allyl alcohol, b.p. 98°, d_4^{20} 0.857, n_D^{20} 1.4134.
   Can be dried with K_2CO_3 or $CaSO_4$, or by azeotropic distn. with
benzene followed by distn. under nitrogen. It is difficult to obtain
peroxide-free.

Allylamine, b.p. 52.9°, n_D^{20} 1.42051.

   Purified by fractional distn. from calcium hydride.

1-Allyl-6-amino-3-ethyluracil, m.p. 143-144° (anhydr.).

   Cryst. from water (as monohydrate).

Allyl bromide, b.p. 70°, n_D^{20} 1.46924.

   Washed with $NaHCO_3$ soln. then distd. water. Dried with $CaCl_2$ or $MgSO_4$, and fractionally distd. Protect from strong light.

Allyl chloride, b.p. 45.1°, n_D^{20} 1.4151, n_D^{25} 1.4130.

   Likely impurities include 2-chloropropene, propyl chloride, iso-propyl chloride, 3,3-dichloropropane, 1,2-dichloropropane and 1,3-dichloropropane. Purified by washing with conc. HCl, then with Na_2CO_3 soln., drying with $CaCl_2$, and distn. through an efficient column. [Oae and Vanderwerf, JACS, 75, 2724 (1953).]

1-N-Allyl-3-hydroxymorphinan, m.p. 180-182°.

   Cryst. from aq. ethanol.

Allyl iodide, b.p. 103°, d^{12} 1.848.

   Purified in a dark room by washing with aq. Na_2SO_3 to remove free iodine, then drying with $MgSO_4$ and distilling at 21 mm pressure, to give a very pale yellow liquid.(This material, dissolved in hexane, was stored in a light-tight container at -5° for up to three months before free iodine could be detected, by its colour, in the soln.) [Sibbett and Noyes, JACS, 75, 761 (1953).]

Allyl thiourea, m.p. 78°.

   Cryst. from acetone, ethanol or ethyl acetate, after decolorizing with charcoal.

N-Allylurea, m.p. 85°.

   Cryst. from ethanol, ethanol-ether, ethanol-chloroform or ethanol-toluene.

D-Altrose, m.p. 103-105°.

   Cryst. from aq. ethanol.

Amethopterin, m.p. 185-204° (decomp.).
  Cryst. from water.

p-Aminoacetanilide, m.p. 162-163°.
  Cryst. from water.

Aminoacetic acid, m.p. 262° (decomp., goes brown at 226°).
  Cryst. from distd. water by dissolving at 90-95°, filtering, cooling to about -5°, and draining the crystals centrifugally. Alternatively, cryst. from distd. water by addition of methanol or ethanol
(e.g. 50 g dissolved in 100 ml of warm water, and 400 ml of methanol
added). The crystals can be washed with methanol or ethanol, then
with ethyl ether. Likely impurities: ammonium glycinate, iminodiacetic
acid, nitrilotriacetic acid, ammonium chloride.

Aminoacetonitrile bisulphate, m.p. 101°.
  Cryst. from aq. ethanol.

α-Aminoacetophenone hydrochloride, m.p. 188°.
  Cryst. from acetone-ethanol.

m-Aminoacetophenone, m.p. 98-99°.
  Cryst. from ethanol.

p-Aminoacetophenone, m.p. 106-107°.
  Cryst. from water.

9-Aminoacridine, m.p. 241°.
  Cryst. from ethanol or acetone.

α-Aminoadipic acid, m.p. 206° (anhyd.).
  Cryst. from water.

2-Amino-4-anilino-s-triazine, m.p. 235-236°.
  Cryst. from dioxane or 50% aq. ethanol.

1-Aminoanthraquinone-2-carboxylic acid, m.p. 295-296°.
  Cryst. from nitrobenzene.

4-Aminoantipyrine, m.p. 109°.
  Cryst. from ethanol or ethanol-ether.

p-Aminoazobenzene, m.p. 126°.
  Cryst. from ethanol, CCl$_4$, pet. ether-benzene, or a methanol-water
mixture.

o-Amino-5-azotoluene, m.p. 101.4-102.6°.
  Cryst. twice from ethanol, once from benzene, then dried in an
Abderhalden drying apparatus. [Cilento, JACS, 74, 968 (1952).]
Carcinogenic.

5-Aminobarbituric acid, see Uramil.

2-Aminobenzaldehyde, m.p. 39-40°.
  Distd. in steam, and cryst. from water or ethanol-ether.

3-Aminobenzaldehyde, m.p. 28-30°.
  Cryst. from ethyl acetate.

p-Aminobenzeneazodimethylaniline, m.p. 182-183°.
  Cryst. from aq. ethanol.

p-Aminobenzenesulphonamide, see Sulphanilamide.

m-Aminobenzenesulphonic acid, see Metanilic acid.

p-Aminobenzenesulphonic acid, see Sulphanilic acid.

o-Aminobenzoic acid, m.p. 145°.
  Cryst. from water (charcoal). Has also been cryst. from 50% aq.
acetic acid. Can be vac. sublimed.

m-Aminobenzoic acid, m.p. 174°.
  Cryst. from water.

p-Aminobenzoic acid, m.p. 187-188°.
  Purified by dissolving in 4-5% aq. HCl at 50-60°, decolourising

p-Aminobenzoic acid (continued)
with charcoal and carefully pptg. with 30% Na_2CO_3 to pH 3.5-4 in
the presence of ascorbic acid. It can be cryst. from water, ethanol
or ethanol-water mixtures.

p-Aminobenzonitrile, m.p. 86-86.5°.
    Cryst. from water or 5% ethanol and dried over P_2O_5.

4-Aminobenzophenone, m.p. 119-122°.
    Cryst. from ethanol-water.

2-Aminobenzothiazole, m.p. 132°.
6-Aminobenzothiazole, m.p. 87°.
    Cryst. from aq. ethanol.

N-(p-Aminobenzoyl)glutamic acid, m.p. 173° (L-form), 197° (DL).
    Cryst. from water.

3-o-Aminobenzyl-4-methylthiazolium chloride hydrochloride,
m.p. 213° (decomp.).
    Cryst. from aq. ethanol.

o-Aminobiphenyl, m.p. 49.0°.
    Cryst. from aq. ethanol (charcoal).

p-Aminobiphenyl, m.p. 53°.
    Cryst. from water or ethanol. Carcinogenic.

2-Amino-5-bromotoluene, m.p. 59°.
    Steam distd., and cryst. from ethanol.

S-α-Aminobutyric acid, m.p. 292° (decomp.).
    Cryst. from aq. ethanol.

RS-α-Aminobutyric acid, m.p. 304°.
    Cryst. from water.

β-Aminobutyric acid, m.p. 193-194°.

γ-Aminobutyric acid, m.p. 202° (decomp.).
   Cryst. from aq. ethanol.

RS-α-Amino-n-caproic acid, see RS-Norleucine.

2-Amino-5-chlorobenzoic acid, m.p. 100°.
   Cryst. from water, ethanol or chloroform.

3-Amino-4-chlorobenzoic acid, m.p. 216-217°.
   Cryst. from water.

4-Amino-4'-chlorobiphenyl, m.p. 134°.
   Cryst. from pet. ether.

2-Amino-4-chloro-6-methylpyrimidine.
   Cryst. from ethanol.

2-Amino-5-chloropyridine, m.p. 135-136°.
   Cryst. from pet. ether, sublimes at 50°/0.5 mm.

2-Amino-3,5-dibromopyridine, m.p. 103-104°.
   Steam distd., and cryst. from aq. ethanol or pet. ether.

2-Amino-4,6-dichlorophenol, m.p. 95-96°.
   Cryst. from CS_2 or benzene.

3-Amino-2,6-dichloropyridine, m.p. 119°, b.p. 110°/0.3 mm.
   Cryst. from water.

p-Amino-N,N-diethylaniline hydrochloride.
   Cryst. from ethanol.

4-Amino-3,5-diiodobenzoic acid, m.p. >350°.
   Purified by soln. in dil. NaOH and pptn. with dil. HCl. Air-dried.

2-Amino-4,6-dimethylpyridine, m.p. 69-70.5°.
   Cryst. from hexane, ether-pet. ether or benzene. Residual benzene
was removed from crystals over paraffin-wax chips in an evacuated
desiccator.

2-Amino-4,6-dimethylpyridine, m.p. 69-70.5°.
  Cryst. from ether-pet. ether.

2-Amino-4,6-dimethylpyrimidine.
  Cryst. from water gives m.p. 197°, and cryst. from acetone gives
m.p. 153°.

2-Aminodiphenylamine, m.p. 79-80°.
  Cryst. from water.

4-Aminodiphenylamine, b.p. 155°/0.026 mm.
  Cryst. from ethanol gives m.p. 66°, and cryst. from ligroin gives
m.p. 75°.

2-Amino-1,2-diphenylethanol, m.p. 165°.
  Cryst. from ethanol.

2-Aminodiphenylmethane, m.p. 52°, b.p. 190°/22 mm and 172°/12 mm.
  Cryst. from ether.

2-Aminoethanethiol, m.p. 97-98.5°.
  Sublimed under vac.

2-Aminoethanol, f.p. 10.5°, b.p. 171.1°, n_D^{20} 1.4539.
  Decomposes slightly when distd. at atmospheric pressure, with the
formation of conducting impurities. Fractional distn. at about 5 mm
pressure is satisfactory.
  After dist., 2-aminoethanol was further purified by repeated washing with ether and crystn. from ethanol. After fractional distn. in
the absence of CO_2, it was twice cryst. by cooling, followed by
distn. [Reitmeier, Silvertz and Tartar, JACS, 62, 1943 (1940).] Can
be dried by azeotropic distn. with dry benzene.

2-Aminoethanol hydrochloride, m.p. 75-77°.
  Cryst. from ethanol. Is deliquescent.

2-Aminoethanol hydrogen sulphate.
  Cryst. from water.

2-(2-Aminoethylamino)ethanol, see Hydroxyethyl-ethylenediamine.

S-(2-Aminoethyl)isothiouronium bromide hydrobromide, m.p. 194-195°.
   Cryst. from absolute ethanol + ethyl acetate. Is hygroscopic.

2-Amino-4-(ethylthio)butyric acid, see Ethionine.

(2-Aminoethyl)trimethylammonium chloride hydrochloride, (Cholamine chloride hydrochloride).
   Cryst. from ethanol. (Material very soluble in water.)

RS-α-Aminohexanoic acid, see RS-Norleucine.

4-Aminohippuric acid, m.p. 198-199°.
   Cryst. from water.

1-Amino-4-hydroxyanthraquinone, m.p. 207-208°.
   Purified by thin layer chromatography on SiO_2 gel plates using toluene/acetone (9:1) as eluant. The main band was scraped off and extracted with methanol. The solvent was evapd. off and the dye was dried in a drying pistol. [Land, McAlpine, Sinclair and Truscott, JCS Faraday Trans I, 72, 2091 (1976).] Cryst. from aq. ethanol.

2-Amino-4-hydroxybutyric acid, see Homoserine.

dl-4-Amino-3-hydroxybutyric acid, m.p. 225° (decomp.).
   Cryst. from water or aq. ethanol.

2-Amino-2-hydroxymethyl-1,3-propanediol,
   see Tris(hydroxymethyl)aminomethane.

3-Amino-4-hydroxytoluene, m.p. 137-138°.
   Cryst. from water or benzene.

4-Amino-5-hydroxytoluene, m.p. 179°.
   Cryst. from 50% ethanol.

6-Amino-3-hydroxytoluene, m.p. 162° (decomp.).
   Cryst. from 50% aq. ethanol.

5-Aminoindane, m.p. 37-38°, b.p. 247-249°/745 mm, 146-147°/25 mm, 131°/15 mm.
  Cryst. from pet. ether.

6-Aminoindazole, m.p. 210°.
  Cryst. from water or ethanol and sublimes _in vacuo_.

2-Amino-5-iodotoluene, m.p. 87°.
  Cryst. from 50% ethanol.

α-Aminoisobutyric acid, sublimes 280°.
  Cryst. from aq. ethanol and dried at 110°.

D-4-Amino-3-isoxazolidone, m.p. 154-155° (decomp.), $[\alpha]_{546}^{20}$ +139° (c = 2 in H$_2$O).
  Cryst. from aq. ammoniacal soln. at pH 10.5 (100 mg/ml) by diluting with 5 vols. of isopropyl alcohol and then adjusting to pH 6 with acetic acid.

8-Amino-6-methoxyquinoline, m.p. 41-42°.
  Distd. under nitrogen at about 50 microns, then recryst. several times from methanol (0.4 ml/g).

1-Amino-4-methylaminoanthraquinone.
  Purified by thin-layer chromatography on silica gel plates using toluene/acetone (3:1) as eluant. The main band was scraped off and extracted with methanol. The solvent was evaptd. and the dye was dried in a drying pistol. [Land, McAlpine, Sinclair and Truscott, JCS Faraday Trans I, 72, 2091 (1976).]

4-Amino-2-methyl-1-naphthol hydrochloride, m.p. 283° (decomp.).
  Cryst. from dil. HCl.

2-Amino-2-methyl-1,3-propanediol, m.p. 111°., b.p. 151-152°/10 mm.
  Cryst. three times from methanol, dried in a stream of dry nitrogen at room temp., then in a vacuum oven at 55°. Stored over CaCl$_2$.
[Hetzer and Bates, JPC, 66, 308 (1962).]

2-Amino-2-methyl-1-propanol, m.p. 31°, b.p. 164-166°/760 mm.
    Purified by distn. and fractional freezing.

2-Amino-3-methylpyridine, m.p.,33.2°
    Cryst. three times from benzene, most of the residual benzene being
removed from the crystals over paraffin-wax chips in an evacuated
desiccator. The amine, transferred to a separating funnel under nitrogen, was left in contact with NaOH pellets for 3 hr with occasional
shaking. It was then placed in a vac. distilling flask where it was
refluxed gently in a stream of dry nitrogen before being fractionally
distd. [Mod, Magne and Skau, JPC, 60, 1651 (1956).]

2-Amino-4-methylpyridine, m.p. 99.2°.
    Cryst. from ethanol or a 2:1 benzene-acetone mixture, and dried
under vac.

2-Amino-5-methylpyridine, m.p. 76.5°.
    Cryst. from acetone.

2-Amino-6-methylpyridine, m.p. 44.2°
    Cryst. three times from acetone, drying under vac. at about 45°.
After leaving in contact with NaOH pellets for 3 hr., with occasional
shaking, it was decanted and fractionally distd. [Mod, Magne and
Skau, JPC, 60, 1651 (1956).]

2-Amino-5-methylpyridimine, m.p. 193.5°.
    Cryst. from water and benzene. Sublimes at 50°/0.5 mm.

4-Amino-2-methylquinoline, m.p. 168°, b.p. 333°/760 mm.
    Cryst. from benzene-pet. ether.

2-Amino-4-(methylsulphonyl)butyric acid, see Methionine sulphoxide.

Aminonaphthalenesulphonic acid, see Naphthylaminesulphonic acid.

3-Amino-2-naphthoic acid, m.p. 214°.
    Cryst. from aq. ethanol.

1-Amino-8-naphthol-3,6-disulphonic acid.

Enough Na_2CO_3 (about 22 g) to make the soln. slightly alkaline to litmus was added to a soln. of 100 g of the dry acid in 750 ml of hot distd. water, followed by 5 g of activated charcoal and 5 g of Celite. The suspension was stirred for 10 min and filtered by suction. The acid was pptd. by adding approx. 40 ml of conc. HCl (soln. blue to Congo red), then filtered by suction through sharkskin filter paper and washed with 100 ml of distd. water. The purification process was repeated. The acid was dried overnight in an oven at 60° and stored in a dark bottle. [Post and Moore, AC, 31, 1872 (1959).]

1-Amino-2-naphthol hydrochloride.

Cryst. from the minimum volume of hot water containing a few drops of stannous chloride in an equal weight of hydrochloric acid (to reduce atmospheric oxidation).

1-Amino-2-naphthol-4-sulphonic acid.

Purified by warming 15 g of the acid, 150 g of $NaHSO_3$ and 5 g of Na_2SO_3 (anhydrous) with 1 l. of water to approx. 90°, shaking until most of the solid had dissolved, then filtering hot. The ppte. obtained by adding 10 ml of conc. HCl to the cooled filtrate was collected, washed with 95% ethanol until the washings were colourless, and dried under vac. over $CaCl_2$. It was stored in a dark coloured bottle, in the cold. [Chanley, Gindler and Sobotka, JACS, 74, 4347 (1952).]

6-Aminonicotinic acid, m.p. 312° (decomp.).

Cryst. from aq. acetic acid.

2-Amino-4-nitrobenzoic acid, m.p. 269° (decomp.).

Cryst. from water or aq. ethanol.

5-Amino-2-nitrobenzoic acid, m.p. 235° (decomp.).

Cryst. from water.

1-Amino-4-nitronaphthalene, m.p. 195°.

Cryst. from ethanol or ethyl acetate.

2-Amino-4-nitrophenol, m.p. 80-90° (hydrate), 142-143° (anhydrous).

2-Amino-5-nitrophenol, m.p. 207-208°.

6-Aminopenicillanic acid, m.p. 208-209°, $[\alpha]_{546}^{20}$ +327° (in 0.1 N HCl).

Cryst. from water.

o-Aminophenol, m.p. 175-176°.

Purified by soln. in hot water, decolorization with activated charcoal, filtration and cooling to induce crystn. It was necessary to maintain an atmosphere of nitrogen over the hot phenol soln. to prevent its oxidation. [Charles and Freiser, JACS, 74, 1385 (1952).] Can also be cryst. from ethanol.

m-Aminophenol, m.p. 122-123°.

Cryst. from water or from toluene.

p-Aminophenol, m.p. 190° (under N$_2$).

Cryst. from ethanol, then water, excluding oxygen. Can be sublimed at 110°/0.3 mm. Has been purified by chromatography on alumina with a 1:4 (v/v) mixture of absolute ethanol-benzene as eluant.

p-Aminophenol hydrochloride, m.p. 306° (decomp.).

Purified by treating an aq. soln. with satd. Na$_2$S$_2$O$_3$, filtering under an inert atmosphere, then recryst. from 50% ethanol twice and from absolute ethanol once. [Livingston and Ke, JACS, 72, 909 (1950).]

p-Aminophenylacetic acid, m.p. 199-200°.

Cryst. from hot water (60-70 ml/g).

p-Aminopropiophenone, m.p. 140°.

Cryst. from water or ethanol.

α-(α-Aminopropyl)benzyl alcohol, m.p. 79-80°.

Cryst. from benzene-pet.ether.

p-(2-Aminopropyl)phenol, m.p. 125-126°.

Cryst. from benzene.

1-Aminopyrene, m.p. 117-118°.
   Cryst. from hexane.

2-Aminopyridine, m.p. 58°.
   Cryst. from benzene-pet. ether (b.p. 40-60°) or CHCl$_3$-pet. ether.

3-Aminopyridine, m.p. 64°.
   Cryst. from benzene, CHCl$_3$-pet. ether (b.p. 60-70°), or
benzene-pet. ether (4:1).

4-Aminopyridine, m.p.160°, b.p. 180°/12-13 mm.
   Cryst. from benzene-ethanol, then recryst. twice from water,
crushed and dried for 4 hr at 105°. [Bates and Hetzer, J.Res.Nat.
Bur.Stand. 64A, 427 (1960).] Has also been cryst. from ethanol,
benzene, benzene-pet. ether, and sublimes in vacuo.

2-Aminopyrimidine, m.p. 126-127.5°.
   Cryst. from benzene, ethanol or water.

Aminopyrine, m.p. 107-109°.
   Cryst. from pet. ether.

3-Aminoquinoline, m.p. 93.5°.
   Cryst. from benzene.

4-Aminoquinoline, m.p. 158°.
   Purified by zone melting.

5-Aminoquinoline, m.p. 110°, b.p. 310°/760 mm, 184°/10 mm.
   Cryst. from pentane, then from benzene or ethanol.

8-Aminoquinoline, m.p. 70° .
   Cryst. from ethanol or ligroin.

p-Aminosalicylic acid, m.p. 150-151° (decomp.).
2-Amino-5-sulphanilylthiazole, m.p. 219-221° (decomp.).
   Cryst. from ethanol.

4-Amino-2-sulphobenzoic acid.
    Cryst. from water.

2-Aminothiazole, m.p. 93°, b.p. 140°/11 mm.
    Cryst. from pet. ether (b.p. 100-120°), or ethanol.

3-Amino-1,2,4-triazole, m.p. 159°.
    Cryst. from ethanol (charcoal), then three times from dioxane.
[Williams, McEwan and Henry, JPC, 61, 261 (1957).]

DL-α-Amino-n-valeric acid, see Norvaline.

5-Amino-n-valeric acid, m.p. 157-158°.
    Cryst. by dissolving in water and adding ethanol.

5-Amino-n-valeric acid hydrochloride, m.p. 103-104°.
    Cryst. from chloroform.

Ammonium benzoate, m.p. 200° (decomp.).
    Cryst. from ethanol.

Ammonium d-α-bromocamphor-π-sulphonate, $[\alpha]_D^{25}$ +84.8°.
    Passage of a hot aq. soln. through an alumina column removed
water-soluble coloured impurities which remained on the column when
the ammonium salt was eluted by hot water. The salt was cryst. from
water and dried over CaCl$_2$. [Craddock and Jones, JACS, 84, 1098
(1962).]

Ammonium nitrosophenylhydroxylamine, see Cupferron.

Ammonium picrate. Explodes above 200°.
    Cryst. from alcohol and acetone.

Amodiaquin, m.p. 208°.
    Cryst. from 2-ethoxyethanol.

Amygdalin, m.p. 214-216°.
    Cryst. from water.

n-Amyl acetate, b.p. 149.2°, n_D^{20} 1.40228.

Shaken with satd. NaHCO$_3$ soln. until neutral, washed with water, dried with MgSO$_4$ and distd.

n-Amyl alcohol, b.p. 138.1°, n_D^{20} 1.4100, d^{15} 0.8184.

Dried with anhydrous K$_2$CO$_3$ or CaSO$_4$, filtered and fractionally distd. Has also been treated with 1-2% of sodium and heated at reflux for 15 hr to remove water and chlorides. Traces of water can be removed from the near-dry alcohol by refluxing with a small amount of sodium in the presence of 2-3% n-amyl phthalate or succinate, followed by distn. (See Ethanol.)

Small amounts of amyl alcohol have been purified by esterifying with p-hydroxybenzoic acid, recrystallizing the ester from CS$_2$, saponifying with ethanolic KOH, drying with CaSO$_4$ and fractionally distilling. [Olivier, Rec.Trav.chim.Pays-Bas, 55, 1027 (1936).]

tert.-Amyl alcohol, b.p. 102.3°, n_D^{20} 1.4058, d^{15} 0.8135.

Refluxed with anhydrous K$_2$CO$_3$, CaH$_2$, CaO or sodium, then fractionally distd. Near-dry alcohol can be further dried by refluxing with magnesium activated with iodine, as described for ethanol. Further purification is possible using fractional crystn., zone melting or preparative gas chromatography.

n-Amylamine, b.p. 105°.

Dried by prolonged shaking with NaOH pellets, then distd.

n-Amyl bromide, b.p. 129.7°.

Washed with conc. H$_2$SO$_4$, then water, 10% Na$_2$CO$_3$ soln., again with water, dried with CaCl$_2$ or K$_2$CO$_3$, and fractionally distd. just before use.

n-Amyl chloride, b.p. 107.8°, n_D^{20} 1.41177, n_D^{25} 1.41026.
sec.-Amyl chloride, b.p. 96-97°.

Purified by stirring vigorously with 95% H$_2$SO$_4$, replacing the acid when it became coloured, until the layer remained colourless after 12 hr stirring. The amyl chloride was then washed with satd. Na$_2$CO$_3$ soln., then distd. water, and dried with anhydrous MgSO$_4$, followed by filtration, and distn. through a 10-in Vigreux column.

sec.-Amyl chloride (continued)

Alternatively a stream of oxygen containing 5% ozone was passed
through the amyl chloride for three times as long as it took to
cause the first colouration of starch iodide paper by the exit gas.
Washing the liquid with $NaHCO_3$ soln. hydrolyzed ozonides and
removed organic acids prior to drying and fractional distn. [Chien
and Willard, JACS, 75, 6160 (1953).]

tert.-Amyl chloride, b.p. 86°.

Methods of purification commonly used for other alkyl chlorides
lead to decomposition. Unsatd. materials were removed by chlorinat-
ion with a small amount of chlorine in bright light, followed by
distn. [Chien and Willard, JACS, 75, 6160 (1953).]

Amyl ether, b.p. 186.8°, n_D^{20} 1.41195, n_D^{25} 1.40985.

Repeatedly refluxed over sodium, then distd.

n-Amyl mercaptan, see 1-Pentanethiol.

Amylose (for use in iodine complex formation).

Amylopectin was removed from impure amylose by dispersing in aq.
15% pyridine at 80-90° (concn. 0.6-0.7%) and leaving the soln. stand
at 44-45° for 7 days. The ppte. was redispersed and recryst. during
5 days. After a further dispersion in 15% pyridine, it was cooled to
45°, allowed to stand at this temperature for 12 hr, then cooled to
25° and left for a further 10 hr. The combined ppte. was dispersed
in warm water, pptd. with ethanol, washed with absolute ethanol, and
vac. dried. [Foster and Paschall, JACS, 75, 1181 (1953).]

p-tert.-Amylphenol, m.p. 93.5-94.2°.

Purified via its benzoate, as for phenol. After evaporating the
solvent from its soln. in ether, the material was cryst. (from the
melt) to constant m.p. [Berliner, Berliner and Nelidow, JACS, 76,
507 (1954).]

2-n-Amylpyridine, b.p. 63.0°/2 mm, n_D^{26} 1.4861.

Purification as for 4-n-amylpyridine.

<u>4-n-Amylpyridine</u>, b.p. 78.0°/2.5 mm, n_D^{20} 1.4908, $n_D^{25.4}$ 1.4892.
    Dried with NaOH for several days, then distd. from CaO under
reduced pressure, taking the middle fraction and redistilling it.

<u>α-Amyrin</u>, m.p. 186°.
    Cryst. from ethanol.

<u>β-Amyrin</u>, m.p. 197-197.5°.
    Cryst. from pet. ether or ethanol.

<u>Androstane</u>, m.p. 50-50.5°.
    Cryst. from acetone-methanol.

<u>epi-Androsterone</u>, m.p. 172-173°, $[α]_{546}^{20}$ +115° (c = 1 in MeOH).
    Cryst. from aq. ethanol.

<u>cis-Androsterone</u>, m.p. 185-185.5°.
    Cryst. from acetone-ethyl ether.

<u>Angelic acid</u>, m. p. 45°.
    Steam distd., then cryst. from water.

<u>Aniline</u>, f.p. -6.0°, b.p. 184.4°, d_4^{20} 1.0220, n_D^{20} 1.5855, n_D^{25} 1.5832.
    Aniline is hygroscopic. It can be dried with KOH or CaH_2, and
distd. at reduced pressure (b.p. 65°/8 mm). Treatment with stannous
chloride removes sulphur-containing impurities, reducing the tendency to become coloured by aerial oxidation. Can be recryst. from
ethyl ether at low temperatures. More extensive purifications involve preparation of derivatives, such as the double salt of
aniline hydrochloride and cuprous chloride or zinc chloride, or
N-acetylaniline (m. p. 114°) which can be recryst. from water.
    Redistd. aniline was dropped slowly into an aq. soln. of recryst.
oxalic acid. Aniline oxalate was filtered off, washed several times
with water and recryst. three times from 95% ethanol. Treatment with
satd. Na_2CO_3 soln. regenerated aniline which was distd. from the
soln., dried, and redistd. under reduced pressure. [Knowles,
<u>Ind. Eng. Chem.</u>, <u>12</u>, 881 (1920).]
    After refluxing with 10% acetone f or 10 hr, aniline was acidified

Aniline (continued)

with HCl (Congo red as indicator) and extracted with ethyl ether
until colourless. The hydrochloride was purified by repeated crystn.
before aniline was liberated by addition of alkali, then dried with
solid KOH, and distd. The product was sulphur-free and remained
colourless in air. [Hantzsch and Freese, Ber. 27, 2529, 2966 (1894).]

Non-basic materials, including nitro compounds, were removed from
aniline dissolved in 40% H_2SO_4 by passing steam through the soln. for
1 hr. Pellets of NaOH were added to liberate the aniline which was
steam distd., dried with KOH, distd. twice from zinc dust at 20 mm,
dried with freshly prepared BaO, and finally distd. from BaO in an
all-glass apparatus. [Few and Smith, JCS, 753 (1949).]

Aniline hydrochloride, m.p. 200.5-201°.

Cryst. from water or ethanol and dried at 5 mm over P_2O_5.

Cryst. four times from methanol containing a few drops of conc.
HCl by addition of pet. ether (b.p. 60-70°), then dried to constant
weight over paraffin chips, under vac. [Gutbezahl and Grunwald,
JACS, 75, 559 (1953).]

p-Anilinophenol, see p-Hydroxydiphenylamine.

Anisaldehyde, b.p. 117°/11 mm.

Washed with $NaHCO_3$, then water, dried with anhydrous $MgSO_4$ and
distd. under reduced pressure under nitrogen. Stored under nitrogen
in sealed glass ampoules.

Anisic acid, see p-Methoxybenzoic acid.

p-Anisidine, m.p. 57°.

Cryst. from water or aq. ethanol. Dried in vac. oven and stored
in a dry box.

Anisole, f.p. -37.5°, b.p. 43°/11 mm, 153.8°/760 mm, d_{15}^{15} 0.9988,
n_D^{25} 1.5143.

Shaken with half volume of 2M NaOH, and emulsion allowed to separate. Repeated 3 times, then washed twice with water, dried over $CaCl_2$,
filtered, dried over sodium wire and finally distd. from fresh sodium

Anisole (continued)

under nitrogen, using a Dean-Stark trap, samples in the trap being rejected until free from turbidity. [Caldin, Parbov, Walker and Wilson, JCS Faraday Trans I, 72, 1856 (1976).]

Dried with $CaSO_4$ or $CaCl_2$, or by refluxing with sodium or BaO prior to fractional distn. Peroxides have been removed by standing with crystalline $FeSO_4$ or by passage through an alumina column. Traces of phenols have been removed by prior shaking with 2 M NaOH, following by washing with water. Can be purified by zone-refining.

2-p-Anisyl-1,3-indandione, m.p. 156-157°.

Cryst. from acetic acid or ethanol.

Anserine.

Cryst. from aq. ethanol. Hygroscopic.

S-Anserine nitrate, m.p. 225°, $[\alpha]_D^{30}$ +12.2°.

Likely impurities: 1-methylimidazole-5-alanine, histidine. Cryst. from aq. methanol.

Anthanthrone, m.p. 300°.

Cryst. from chlorobenzene or nitrobenzene.

Antheraxanthin, m.p. 205°, λ_{max} 460.5, 490.5 nm, in $CHCl_3$.

Likely impurities: violaxanthin and mutatoxanthin. Purified by chromatography on columns of $Ca(OH)_2$ and of $ZnCO_3$. Cryst. from benzene/methanol as needles or thin plates. Stored in the dark, in an inert atmosphere, at -20°.

Anthracene, m.p. 218°.

Likely impurities are anthraquinone, anthrone, carbazole, fluorene, 9,10-dihydroanthracene, tetracene and bianthryl. Carbazole is removed by continuous-absorption chromatography [see Sangster and Irvine JCP, 24, 670 (1956)] using a neutral alumina column and passing n-hexane. (Sherwood, in "Purification of Inorganic and Organic Materials", ed. Zief, Marcel Dekker, New York, 1969). The solvent is evapd. and anthracene is sublimed under vac., then purified by zone refining, under nitrogen in darkness or non-actinic light.

Anthracene (continued)

Has been purified by co-distillation with ethylene glycol (boils at 197.5°), from which it can be recovered by addition of water, followed by crystn. from 95% ethanol, benzene, toluene, a 4:1 benzene-xylene mixture, or ethyl ether. It has also been chromatographed on alumina with pet. ether in a darkened room (to avoid photo-oxidation of adsorbed anthracene to anthraquinone). Other purification methods include sublimation in a nitrogen atmosphere (in some cases after refluxing with sodium).

Anthracene-9-carboxylic acid, m.p. 207° (decomp.).
    Cryst. from ethanol.

9-Anthraldehyde, m.p. 104°.
    Cryst. from acetic acid.

Anthranilic acid, see o-Aminobenzoic acid.

Anthranol, m.p. 160-170° (decomp.).
    Cryst. from glacial acetic acid or aq. ethanol.

Anthraquinone, m.p. 286°.
    Cryst. from CHCl$_3$ (38 ml/g), benzene, or boiling acetic acid, washing with a little alcohol and drying under vac. over P$_2$O$_5$.

Anthraquinone blue B.
Anthraquinone blue RXO.
Anthraquinone green G.
    Purified by salting out three times with sodium acetate, followed by repeated extraction with ethanol. [McGrew and Schneider, JACS, 72, 2547 (1950).]

Anthrarufin, m.p. 280°.
    Cryst. from acetic acid.

1,8,9-Anthratriol, m.p. 176-181°.
    Cryst. from pet. ether.

Anthrimide.

 Cryst. from chlorobenzene or nitrobenzene.

Anthrone, m.p. 155°.

 Cryst. from a 3:1 mixture of benzene and pet. ether (b.p. 60-80°) (10-12 ml/g), or successively from benzene then ethanol. Dried under vac.

Antipyrine, m.p. 114°, b.p. 319°.

 Cryst. from ethanol-water mixture and dried under vac.

β-Apocarotenal, m.p. 139°, $E_{1cm}^{1\%}$ = 2640 at 461 nm.

β-Apocarotenoic acid ethyl ester, m.p. 134-138°, $E_{1cm}^{1\%}$ 2550 = 449 nm, 2140 = 475 nm, in cyclohexane.

β-Apocarotenoic acid methyl ester, m.p. 136-137°, $E_{1cm}^{1\%}$ 2575 = 446 nm, 2160 = 471 nm, in pet.ether.

 Cryst. from pet. ether or pet. ether-ethyl acetate. Stored in the dark, in inert atmosphere at -20°.

Apocodeine, m.p. 124°.

 Cryst. from methanol and dried at 80°/2 mm.

Apomorphine, m.p. 195° (decomp.).

 Cryst. from $CHCl_3$ and pet. ether.

β-L-Arabinose, m.p. 158°, $[\alpha]_D^{20}$ +104° (c = 4 in water).

 Cryst. slowly twice from 80% aq. ethanol, then dried under vac. over P_2O_5.

D-Arabinose, m.p. 155.5-156.5°.

 Cryst. three times from ethanol, vac. dried at 60° for 4 hr and stored in a vac. desiccator.

L-Arabitol, m.p. 102°.

DL-Arabitol, m.p. 105-106°.

 Cryst. from 90% ethanol.

Araboascorbic acid, see Isoascorbic acid.

Arachidic acid, m.p. 77°.   Cryst. from absolute ethanol.

Arachidic alcohol, m.p. 65.5° (71°), b.p. 200°/3 mm.
    Cryst. from benzene or benzene-pet. ether.

Arbutin, m.p. 163-164°
    Cryst. from water.

S-Arginine, m.p. 207° (decomp.), $[\alpha]_D^{20}$ +26.5° (c = 5 in 5N HCl)
$[\alpha]_{546}^{20}$ +32 (c = 5N HCl).
    Cryst. from 66% ethanol.

S-Arginine hydrochloride, m.p. 217° (decomp.), $[\alpha]_D^{25}$ +26.9°
(c = 6M HCl).
    Likely impurity: ornithine. Cryst. from water at pH 5-7, by
adding ethanol to 80% v/v.

S-Argininosuccinic acid, $[\alpha]_D^{24}$ +16.4° (in water).
    Likely impurity: fumaric acid. In neutral or alk. soln., readily
undergoes ring closure. Cryst. from water by adding 1½ volumes of
ethanol. Barium salt is stable at 0-5° if dry.

S-Argininosuccinic anhydride, $[\alpha]_D^{23}$ -10.0° (in water), for anhydride
formed at neutral pH.
    Cryst. from water by adding two volumes of ethanol. An isomeric
anhydride is formed if argininosuccinic acid is allowed to stand at
acid pH. In soln., mixtures of anhydride and free acid are formed.

Ascorbic acid, m.p. 192°.
    Cryst. from methanol-ethyl ether-pet.ether [Herbert et al.,
JCS, 1270 (1933).]

S-Asparagine, m.p. 234-235° (monohydrate), $[\alpha]_D^{20}$ +32.6° (in 0.1M HCl).
    Likely impurities: aspartic acid, tyrosine. Cryst. from water or
water and ethanol. Slowly effloresces in dry air.

Aspartic acid, m.p. 338-339° (RS), m.p. 271° (S-isomer, required heating in a sealed tube), $[\alpha]_D^{25}$ +25.4^{o} (in 3M HCl).

Likely impurities: glutamic acid, cystine and asparagine. Cryst. from water by adding 4 volumes of ethanol and dried at 110°.

Aspergillic acid, m.p. 97-99°.

Sublimed at 80°/10^{-3} mm. Cryst. from methanol.

Astacin, $E_{1cm}^{1\%}$ =10$^{5.5}$ (498 nm), in pyridine.

Probable impurity: astaxanthin. Purified by chromatography on alumina/fibrous clay (1:4) or sucrose, or by partition between pet. ether and methanol (alkaline). Cryst. from pyridine/water. Stored in the dark, in an inert atmosphere, at -20°.

Atrolactic acid, m.p. 94.5^{o} (anhydr.).

Cryst. from water and dried at 55°/0.5 mm.

Atropine, m.p. 114-116°.

Cryst. from acetone or hot water.

Aureomycin, m.p. 172-174° (decomp.).

Dehydrated by azeotropic distn. of its soln. in toluene. On cooling, cryst. anhydrous material separated and was recryst. from benzene, then dried under vac. at 100° over paraffin wax. (If it is cryst. from methanol, it contains methanol which is not removed on drying.) [Stephens et al., JACS, 76, 3568 (1954).]

Aureomycin hydrochloride, m.p. 234-236° (decomp.), $[\alpha]_D^{25}$ -235° (in water).

Purified by dissolving 1 g rapidly in 20 ml of hot water, cooling rapidly to 40°, treating with 0.1 ml of 2M HCl, and chilling in an ice-bath. The process was repeated twice. [Stephens et al., JACS, 76, 3568 (1954).]

8-Azaadenine, m.p. 345° (decomp.).

Cryst. from water.

8-Azaguanine.

Dissolved in hot 1 M NH_4OH, filtered, recryst. in the cold and washed with water.

1-Azaindolizine, b.p. $72-73°/1.0$ mm.
   Purified by gas-liquid chromatography.

Azaserine, m.p. $146-162°$ (decomp.), $[\alpha]_D^{27.5}$ $-0.5°$ (c = 8.5, H_2O, pH 5.2).
   Cryst. from 90% ethanol.

Azelaic acid, m.p. $105-106°$.
   Cryst. from water (charcoal) or thiophene-free benzene. The material cryst. from water was dried by azeotropic distn. in toluene: the residual toluene soln. was cooled and filtered, the ppte. being dried in a vac. oven.
   Purified by zone melting or by sublimation onto a cold finger at 10^{-3} torr.

Azobenzene, m.p. $68°$.
   Ordinary azobenzene is nearly all in the trans-form. It is partly converted into the cis-form on exposure to light. [For isolation, see Hartley, JCS, 633 (1938), and for spectra of cis- and trans-azo-benzenes, see Winkel and Siebert, Ber. 74B 670 (1941).] trans-Azo-benzene is obtained by chromatography on alumina using 1:4 benzene-heptane or pet.ether, and cryst. from ethanol (after refluxing for several hours) or hexane. All operations should be carried out in diffuse red light or in the dark.

α,α'-Azobisisobutyronitrile, m.p. $103°$ (decomp.).
   Cryst. from ethyl ether, $CHCl_3$ or methanol. Has also been cryst. from absolute ethanol below $40°$ in subdued light, with storage under vac. in the dark at about $-10°$ until used, and from $CHCl_3$ soln. by addition of pet.ether (b.p. $<40°$).

1.1'-Azobis(cyclohexane carbonitrile), m.p. $114-114.5°$, ε_{3500A} = 16.
   Cryst. from aq. ethanol.

Azobisisobutyramidinium chloride.
    Cryst. from water.

Azobisisobutyronitrile, m.p. 101-101.5°.
    Cryst. from acetone, $CHCl_3$ or methanol. Dried under vac. at room temp. and stored in the dark, under vac. at -18°.

2,2'-Azobis(2-methylpropionitrile), m.p. 101-101.5°.
    Cryst. from anhydrous ethyl ether, acetone or methanol.

Azolitmin.
    Cryst. from water.

Azomethane, m. p. -78°, b.p. 1.5°.
    Purified by vac. distn. and stored in the dark at -80°.

p,p'-Azoxyanisole, transition temps., 118.1-118.8°,135.6-136.0°.
    Cryst. from absolute ethanol or acetone, and dried by heating under vac.

Azoxybenzene, m.p. 36°.
    Cryst. from ethanol or methanol, and dried for 4 hours at 25° and 10^{-3} torr. Sublimed before use.

p,p'-Azoxyphenetole, m.p. 137-138° (turbid liquid clarifies at 167°).
    Cryst. from toluene or ethanol.

Azulene, m.p. 98.5-99°.
    Cryst. from ethanol.

B.A.L., see 1,2-Dimercapto-3-propanol.

Barbituric acid, m.p. 250°.
    Cryst. twice from water, then dried for 2 days at 100°.

Batyl alcohol, m.p. 70.5-71°.
    Cryst. from aq. acetone.

Behenic acid, see Docosanoic acid.

Behenyl alcohol, see 1-Docosanol.

Benzalacetone, m.p. 42°.
   Cryst. from b.p. 40-60° pet.ether, or distd. (b.p. 137-142°/16 mm).

Benzalacetophenone, m.p. 56-58°.
   Cryst. from ethanol warmed to 50° (about 5 ml/g), iso-octane, or
benzene-pet.ether.

Benzaldehyde, f.p. -26°, b.p. 62°/10 mm, 179.0°/760 mm, n_D^{20} 1.5455,
n_D^{25} 1.5428.
   To diminish its rate of oxidation, benzaldehyde usually contains
additives such as hydroquinone or catechol. It can be purified via
its bisulphite addition compound but usually distn. (under nitrogen,
at reduced pressure) is sufficient. Prior to distn. it is washed
with NaOH or 10% Na_2CO_3 (until no more CO_2 is evolved), then with
satd. Na_2SO_3 and water, followed by drying with $CaSO_4$, $MgSO_4$ or $CaCl_2$.

anti-Benzaldoxime, m.p. 130°.
   Cryst. from ethyl ether by adding pet.ether (b.p. 60-80°).

Benzamide, m.p. 129.5°.
   Cryst. from hot water (about 5 ml/g), ethanol or 1,2-dichloro-
ethane, and air dried.
   Cryst. from dil. aq. ammonia, water, acetone and then benzene
(using a Soxhlet extractor). Dried in an oven at 110° for 8 hr and
stored in a desiccator over 99% H_2SO_4. [Bates and Hobbs, JACS, 73,
2151 (1951).]

Benzanilide, m.p. 164°.
   Cryst. from pet.ether (b.p. 70-90°) using a Soxhlet extractor, and
dried overnight at 120°. Also cryst. from ethanol.

Benz[α]anthracene, m.p. 159-160°.
   Cryst. from methanol, ethanol or benzene (charcoal), then chromatographed on alumina from sodium-dried benzene (twice), using vac.distn.
to remove the benzene. Final purification was by vac. sublimation.

Benz[α]anthracene-7,12-dione, m.p. 169.5-170.5°.
  Cryst. from methanol (charcoal).

Benzanthrone, m.p. 170°.
  Cryst. from ethanol or xylene.

Benzene, f.p. 5.5°, b.p. 80.1°, d^{30} 0.86800, n_D^{20} 1.50110, n_D^{25} 1.49790.
  For most purposes, benzene can be purified sufficiently by shaking
with conc. H_2SO_4 until free from thiophen, then with water, dil. NaOH,
and water, followed by drying (with P_2O_5, sodium, $LiAlH_4$, CaH_2, 4X
Linde molecular sieve, or $CaSO_4$, or by passage through a column of
silica gel. For a preliminary drying, $CaCl_2$ is suitable), and distn.
A further purification step to remove thiophen, acetic acid and
propionic acid, is crystn. by partial freezing. The usual contaminants
in dry thiophen-free benzene are non-benzenoid hydrocarbons such as
cyclohexane, methylcyclohexane, and heptanes, together with naphthenic
hydrocarbons and traces of toluene. Carbonyl-containing impurities
can be removed by percolation through a Celite column impregnated
with 2,4-dinitrophenylhydrazine, phosphoric acid and water. (Prepared
by dissolving 0.5 g DNPH in 6 ml of 85% H_3PO_4 by grinding together,
then adding and mixing 4 ml of distd. water and 10 g of Celite.)
[Schwartz and Parks, AC, 33, 1396 (1961)]. Benzene has been freed
from thiophen by refluxing with 10% (w/v) of Raney nickel for 15 min,
after which the nickel was removed by filtration or centrifugation.
  Mair et al. [J. Res. Nat. Bur. Stand., 37, 229 (1946)] cooled a
mixture of 200 ml of benzene and 50 ml of ethanol in a cylindrical
brass container (5 cm diam. x 20 cm) in an ice-salt cooling bath at
about -10°. The slurry which was produced on vigorous stirring was
transferred to a centrifuge cooled to near -10°. After 5 min the
benzene crystals were removed from the basket of the centrifuge,
allowed to melt, washed three times with distd. water, and filtered
through silica gel to remove any alcohol and water before distn.
  Very dry benzene was obtained by doubly distilling high purity
benzene form a soln. containing the blue ketyl formed by the reaction
of sodium-potassium alloy with a small amount of benzophenone.
[Bercaw and Garrett, JACS, 78, 1841 (1956).]
  Thiophen has been removed from benzene (absence of bluish-green
coloration when 3 ml of benzene is shaken with a soln. of 10 mg of

Benzene (continued)

isatin in 10 ml conc. H_2SO_4) by refluxing the benzene (1 kg) for
several hours with 40 g HgO (freshly pptd.) dissolved in 40 ml
glacial acetic acid and 300 ml of water. The ppte. was filtered off,
the aq. phase was removed and the benzene was washed twice with
water, dried and distd. Alternatively, benzene dried with $CaCl_2$ has
been shaken vigorously for half an hour with anhydrous $AlCl_3$
(12 g/l. at 25-35°, then decanted, washed with 10% NaOH, and water,
dried and distd. The process was repeated, giving thiophen-free
benzene. [Holmes and Beeman, Ind. Eng. Chem., 26, 172 (1934).]

   After shaking successively for about an hour with conc. H_2SO_4,
distd. water (twice), 6M NaOH, and distd. water (twice), benzene was
distd. through a 3-ft glass column to remove most of the water.
Absolute ethanol was added and the benzene-alcohol azeotrope was
distd. (This low-boiling distn. leaves any non-azeotrope-forming
impurities behind.) The middle fraction was shaken with distd. water
to remove alcohol, and again redistd. Final slow and very careful
fractional distn. from sodium, then $LiAlH_4$ under nitrogen, removed
traces of water and peroxides. [Peebles, Clarke and Stockmayer,
JACS, 82, 2780 (1960).]

[$^2H_6$]Benzene, b.p. 80°/773.6 mm, 70°/562 mm, 60°/399.4 mm, 40°/186.3 mm,
20°/77.1 mm, 10°/49.9 mm, 0°/27.5 mm; d_4^{20} 0.9488, d_4^{40} 0.9257,
n_D^{20} 1.4991, n_D^{40} 1.4865.

   Hexadeuterobenzene of 99.5% purity is refluxed over and distd.
from CaH_2 onto Linde type 5A sieves under nitrogen.

Benzeneazodiphenylamine, m.p. 82°.

   Purified by chromatography on neutral alumina using anhydrous
benzene with 1% anhydrous methanol. The major component, which gave
a stationary band, was cut and eluted. [Högfeldt and Bigeleisen,
JACS, 82, 15 (1960).] Cryst. from pet.ether or ethanol.

1-Benzeneazo-2-naphthol, m.p. 134°.

   Cryst. from ethanol.

1-Benzeneazo-2-naphthylamine, m.p. 102-104°.

   Cryst. from acetic acid-water.

P.L.C.—E

m-Benzenedisulphonic acid.

Freed from sulphuric acid by conversion to the calcium or barium salts (using $Ca(OH)_2$ or $Ba(OH)_2$, and filtering). The calcium salt was then converted to the potassium salt, using K_2CO_3. Both the potassium and the barium salts were recryst. from water, and the acid was regenerated by passage through the H^+ form of a cation exchange resin. The acid was recryst. twice from conductivity water and dried over $CaCl_2$ at 25°. [Atkinson, Yokoi and Hallada, JACS, 83, 1570 (1961).] It has also been cryst. from ethyl ether and dried in a vac. oven.

m-Benzenedisulphonyl chloride, m.p. 63°.

Cryst. from $CHCl_3$ and dried at 20 mm pressure.

Benzenephosphinic acid, m.p. about 70°.

Purified by allowing to stand for several days under ethyl ether, with intermittent shaking and several changes of solvent. After filtration, the excess ether was removed under vac.

Benzenesulphonic anhydride, m.p. $88-91^O$.

Cryst. from ethyl ether.

Benzenesulphonyl chloride, m.p. 14.5^O, b.p. $120^O/10$ mm, $251.2^O/760$ mm.

Distd., then treated with 3 mole% each of toluene and $AlCl_3$, and allowed to stand overnight. The free benzenesulphonyl chloride was distd. off at 1 mm pressure, and then carefully fractionally distd. at 10 mm in an all-glass column. [Jensen and Brown, JACS, 80, 4042 (1958).]

Benzene-1,2,4,5-tetracarboxylic acid, m.p. 275°.

Cryst. from water.

Benzenethiol, f.p. -14.9^O, b.p. 168.0^O, n_D^{20} 1.58973.

Dried with $CaCl_2$ or $CaSO_4$, and distd. at 15 mm pressure or at 100 mm (b.p. 103.5^O) in a stream of nitrogen.

Benzene-1,2,3-tricarboxylic acid, m.p. 190°.
Benzene-1,3,5-tricarboxylic acid, m.p. 380°.

Cryst. from water.

1,2,4-Benzenetriol, m.p. 141°.
    Cryst. from ethyl ether.

Benzethonium chloride, m.p. 164-166°.
    Cryst. from 1:9 ethanol-ethyl ether mixture.

Benzhydrol, m.p. 69°, b.p. 297°/748 mm, 180°/20 mm.
    Cryst. from hot water or b.p. 60-70° pet.ether, pet.ether contain-
ing a little benzene, from CCl_4, or ethanol (1 ml/g). An additional
purification step is passage of a benzene soln. through an activated
alumina column. Sublimes in vacuo.

Benzidine, m.p. 128-129°.
    Its soln. in benzene was decolorized by percolation through two
2-cm columns of activated alumina, then concentrated until benzidine
cryst. on cooling. Recryst. alternately from ethanol and benzene to
constant absorption spectrum. [Carlin, Nelb and Odioso, JACS, 73,
1002 (1951).] Has also been cryst. from hot water (charcoal) and from
ethyl ether. Dried under vac. in an Abderhalden pistol. Stored in the
dark in a stoppered container. Carcinogenic.

Benzidine hydrochloride.
    Cryst. by soln. in hot water, with addition of conc. HCl to the
slightly cooled soln.

Benzil, m.p. 96-96.5°.
    Cryst. from benzene after washing with alkali. (Crystn. from
ethanol did not free benzil from material reacting with alkali.)
[Hine and Haworth, JACS, 80, 2274 (1958).] Has also been cryst. from
CCl_4 and diethyl ether.

Benzilic acid, m.p. 150°.
    Cryst. from benzene (about 6 ml/g), or hot water.

Benzil monohydrazone, m.p. 151°.
    Cryst. from ethanol.

α-Benzil monoxime, m.p. 140°.
    Cryst. from benzene. (Must not use animal charcoal.)

Benzimidazole, m.p. 172-173°.

Cryst. from water or aq. ethanol (charcoal), and dried at 100° for 12 hr.

2-Benzofurancarboxylic acid, m.p. 192-193°.

Cryst. from water.

Benzoic acid, m.p. 122.6-123.1°

For use as a volumetric standard, analytical reagent grade benzoic acid should be carefully fused at about 130° (to dry it) in a platinum crucible, and then powdered in an agate mortar. Benzoic acid has been cryst. from boiling water (charcoal), aq. acetic acid, glacial acetic acid, benzene, aq. ethanol, pet.ether (b.p. 60-80°), and from ethanol soln. by adding water. It is readily purified by fractional crystn. from its melt and by sublimation under vac. at 80°.

Benzoic anhydride, m.p. 42°.

Freed from benzoic acid by washing with aq. $NaHCO_3$, then water, and drying. Cryst. from benzene (0.5 ml/g) by adding just enough pet. ether (b.p. 40-60°), to cause a cloudiness, then cooling in ice. Can be distd. at 210-220°/20 mm.

Benzoin, m.p. 137°

Cryst. from CCl_4, hot ethanol (8 ml/g), or 50% acetic acid.

Cryst. from high purity benzene, then twice from high purity methanol, to remove fluorescent impurities. [Elliott and Radley, AC, 33, 1623 (1961).]

α-Benzoinoxime, m.p. 151°.

Cryst. from ethyl ether.

Benzonitrile, f.p. -12.9°, b.p. 191.1°, n_D^{20} 1.52823.

Dried with $CaSO_4$, $CaCl_2$, $MgSO_4$ or K_2CO_3, and distd. from P_2O_5 in an all-glass apparatus, under reduced pressure, collecting the middle portion. Distn. from CaH_2 causes some decomposition of solvent. Isonitriles can be removed by preliminary treatment with conc. HCl until the smell of isonitrile has gone, followed by a preliminary drying with K_2CO_3. (This treatment also removes amines.)

Benzonitrile (continued)

Steam distd. (to remove small quantities of carbylamine). The distillate was extracted into ether, washed with dil. Na_2CO_3, dried overnight with $CaCl_2$, and the ether removed by evaporation. The residue was distd. at 40 mm (b.p. 96°). [Kice, Parham and Simons, JACS, 82, 834 (1960).]

Conductivity-grade benzonitrile (specific conductance 2×10^{-8} mho) was obtained by treatment with anhydrous $AlCl_3$, followed by rapid distn. at $40-50^\circ$ under vac. After washing with alkali and drying with $CaCl_2$, the distillate was vac. distd. several times at 35° before being fractionally cryst. several times by partial freezing. It was dried over finely divided activated alumina from which it was withdrawn as required. [Van Dyke and Harrison, JACS, 73, 402 (1951).]

Benzophenone, m.p. $48.5-49.0^\circ$

Cryst. from methanol, ethanol or pet.ether, then dried in a current of warm air and stored over BaO or P_2O_5. Also purified by zone melting and by sublimation.

Benzophenone oxime, m.p. 142°.

Cryst. from methanol (4 ml/g).

Benzopinacol, m.p. $170-180^\circ$ (depends on heating rate).

Cryst. from ethanol.

Benzopurpurin 4B.

Cryst. from water.

Benzo[α]pyrene, m.p. $179.0-179.5^\circ$.

Chromatographed on activated alumina, eluted with a cyclohexane-benzene mixture containing up to 80% benzene, and the solvent evapd. under reduced pressure. [Cahnmann, AC, 27, 1235 (1955).]

2,3-Benzoquinoline, m.p. $108.5-109^\circ$.
3,4-Benzoquinoline, m.p. $102.5-103.5^\circ$.
5,6-Benzoquinoline, m.p. $85.5-86^\circ$.
7,8-Benzoquinoline, m.p. $52.0-52.5^\circ$.

Chromatographed on activated alumina from a benzene soln., with

Benzoquinoline(s), (continued)
ethyl ether as eluant. Evapn. of ether gave crystalline material
which was freed from residual solvent under vac., then further purified by fractional crystn. under nitrogen, from its melt. [Slough and
Ubbelodhe, JCS, 911 (1957).]

p-Benzoquinone, m.p. 115.7°.
   Usually purified in one or more of the following ways: steam distn.,
followed by filtration and drying (e.g. in a desiccator over CaCl$_2$);
crystn. from pet.ether (b.p. 80-100°), benzene (with, then without,
charcoal), water or 95% ethanol; sublimation under vac. (e.g. from
room temp. to liquid nitrogen). It slowly decomposes, and should be
stored, refrigerated, in an evacuated or sealed glass vessel in the
dark.

1,2,3-Benzothiadiazole, m.p. 35°.
2,1,3-Benzothiadiazole, m.p. 44°, b.p. 206°/760 mm.
   Cryst. from pet.ether.

1,2,3-Benzotriazole, m.p. 100.0°.
   Cryst. from benzene or CHCl$_3$ and dried at room temp. or in a vac.
oven at 65°.

Benzotrifluoride, see α,α,α-Trifluorotoluene.

Benzoylacetone, m.p. 58.5-59.0°.
   Cryst. from ethyl ether or methanol and dried under vac. at 40°.

Benzoyl chloride, b.p. 56°/4 mm, 196.8°/745 mm.
   A soln. of benzoyl chloride (300 ml) in benzene (200 ml) was
washed with two 100 ml portions of cold 5% NaHCO$_3$ soln., separated,
dried with CaCl$_2$ and distd. [Oakwood and Weisgerber, Org. Synth. III,
113 (1955).]
   Repeated fractional distn. at 4 mm through a glass helices-packed
column (avoiding porous porcelain or silicon-carbide boiling chips,
and hydrocarbon or silicone greases on the ground joints) gave
benzoyl chloride that did not darken on addition of AlCl$_3$. Further
purification was achieved by adding 3 mole % each of AlCl$_3$ and

Benzoyl chloride (continued)

toluene, standing overnight, and distilling off the benzoyl chloride at 1-2 mm. [Brown and Jensen, JACS, 80, 2291 (1958).] Refluxing for 2 hr with an equal weight of thionyl chloride before distn., has also been used.

Benzoyl disulphide, m.p. 131.2-132.3°.

About 300 ml of solvent was blown off from a filtered soln. of benzoyl disulphide (25 g) in acetone (350 ml). The remaining acetone was decanted from the solid which was recryst. first from 300 ml of 1:1 (v/v) ethanol-ethyl acetate, then from 300 ml of ethanol, and finally from 240 ml of 1:1 (v/v) ethanol-ethyl acetate. Yield about 40%. [Pryor and Pickering, JACS, 84, 2705 (1962).]

Benzoyl glycine, m.p. 188°.

Cryst. from boiling water.

Benzoyl peroxide.

Dissolved in $CHCl_3$ at room temp. and pptd. by adding an equal volume of methanol or pet.ether. Similarly, pptd. from acetone by adding two volumes of distd. water. Has also been cryst. from 50% methanol, and from ethyl ether. Dried under vac. Stored in a desiccator in the dark.

p-Benzoylphenol, m.p. 133.4-134.8°.

Dissolved in hot ethanol (charcoal), cryst. once from ethanol-water and twice from benzene. [Grunwald, JACS, 73, 4934 (1951).]

3,4-Benzphenanthrene, m.p. 68°.

Cryst. from ethanol, pet.ether, or ethanol-acetone.

3,4-Benzpyrene, m.p. 177.5-178°.

A soln. of 250 mg in 100 ml of benzene was diluted with an equal volume of hexane, then passed through a column of alumina, calcium hydroxide and Celite (3:1:1). The adsorbed material was developed with a 2:3 benzene-hexane mixture. (It showed as an intensely purple fluorescent zone.) The main zone was eluted with 3:1 acetone-ethanol, and was transferred into 1:1 benzene-hexane by adding water. The soln. was washed, dried with Na_2SO_4, evapd. and cryst. from benzene by the

3,4-Benzpyrene (continued)
addition of methanol. [Lijinsky and Zechmeister, JACS, 75, 5495
(1953).] Carcinogenic.

Benzylacetophenone, m.p. 73°.
  Cryst. from ethanol (about 1 ml/g).

Benzyl alcohol, f.p. -15.3°, b.p. 205.5°, n_D^{20} 1.54033, n_D^{25} 1.5371.
  Usually purified by careful fractional distn. at reduced pressure
in the absence of air. Benzaldehyde, if present, can be detected by
u.v. absorption at 283 mμ.
  Purified by shaking with aq. KOH and extracting with peroxide-free
ethyl ether. After washing with water, the extract was treated with
satd. NaHS soln., filtered, washed and dried with K_2CO_3. After
evaporating the ether, the alcohol was dried with CaO and distd. under
reduced pressure. [Mathews, JACS, 48, 562 (1926).]

Benzylamine, b.p. 178°/742 mm, 185°/768 mm, n_D^{25} 1.5392.
  Dried with NaOH or KOH, then distd. from sodium, under nitrogen,
through a column packed with glass helices, taking the middle
fraction. Has also been distd. from zinc dust at reduced pressure.

Benzylamine hydrochloride, m.p. 248° (rapid heating).
  Cryst. from water.

Benzylaniline, m.p. 36°.
  Cryst. from pet.ether (b.p. 60-80°)(about 0.5 ml/g).

Benzyl bromide, m.p. -4°, b.p. 85°/12 mm, 192°/760 mm.
  Washed with conc. H_2SO_4, water, 10% Na_2CO_3 or $NaHCO_3$ soln., and
again with water. Dried with $CaCl_2$, Na_2CO_3 or $MgSO_4$, and fractionally
distd. in the dark, under reduced pressure.

Benzyl chloride, b.p. 63°/8 mm, n_D^{23} 1.5375.
  Dried with $MgSO_4$ or $CaSO_4$, or refluxed with fresh calcium turnings,
then fractionally distd. under reduced pressure, collecting the
middle fraction and storing with CaH_2 or P_2O_5. Has also been purified
by passage through a column of alumina.

N-Benzyl-β-chloropropionamide, m.p. 94°.
   Cryst. from methanol.

Benzyl cyanide, b.p. 100°/8 mm, 233.5°/760 mm, n_D^{20} 1.52327,
n_D^{25} 1.52086.
   Benzyl isocyanide can be removed by shaking vigorously with an
equal volume of 50% H_2SO_4 at 60°, washing with satd. aq. $NaHCO_3$,
then half-satd. NaCl soln., drying, and fractionally distilling under
reduced pressure. Distn. from CaH_2 causes some decomposition of
solvent: it is better to use P_2O_5. Other purification procedures
include passage through a column of highly activated alumina, and
distn. from Raney nickel.

Benzyldimethyloctadecylammonium chloride, m.p. 63°.
   Cryst. from acetone.

Benzyl disulphide, see Dibenzyl disulphide.

Benzyl ether, b.p. 298°, n_D^{20} 1.54057, n_D^{25} 1.53851.
   Refluxed over sodium, then distd. under vac. Also purified by
fractional freezing.

Benzyl ethyl ether, b.p. 186°, n_D^{20} 1.4955.
   Dried with $CaCl_2$ or NaOH, then fractionally distd.

O-Benzylhydroxylamine, m.p. 234-238° (sublimes).
   Cryst. from water or ethanol.

Benzylideneacetone, see Benzalacetone.

Benzylideneacetophenone, see Benzalacetophenone.

N-Benzylideneaniline, m.p. 48° (54°), b.p. 300°/760 mm.
   Steam volatile and cryst. from benzene or 85% ethanol.

S-Benzyl-isothiuronium chloride, two forms, m.p. 150° and 175°.
   Cryst. from 0.2 M HCl (2 ml/g), and dried in air.

Benzylmalonic acid, m.p. 121°.
  Cryst. from benzene.

Benzyl mercaptan, b.p. 70.5-70.7°/9.5 mm, n_D^{20} 1.5761.
  Purified via the mercury salt [see Kern, JACS, 75, 1865 (1953)],
which was cryst. from benzene as needles (m.p. 121°), and then
dissolved in $CHCl_3$. Passage of H_2S gas regenerated the mercaptan.
The HgS ppte. was filtered off, and washed thoroughly with $CHCl_3$.
The filtrate and washings were evapd. to remove $CHCl_3$, then residue
was fractionally distd. under reduced pressure. [Mackle and McClean,
TFS, 58, 895 (1962).]

Benzyl methyl ketone, see Phenyl-2-propanone.

p-(Benzyloxy)phenol, m.p. 122.5°.
  Cryst. from water, and dried over P_2O_5 under vac.

2-Benzylphenol, b.p. 312°/760 mm, 175°/18 mm.
  Cryst. from ethanol, stable form has m.p. 52° and unstable form
has m.p. 21°.

4-Benzylphenol, m.p. 84°.
  Cryst. from water.

2-Benzylpyridine, b.p. 98.5°/4 mm, n_D^{26} 1.5771.
4-Benzylpyridine, b.p. 110.0°/6 mm, n_D^{25} 1.5814.
  Dried with NaOH for several days, then distd. from CaO under
reduced pressure, redistilling the middle fraction.

N^4-Benzylsulphanilamide, m.p. 175°.
  Cryst. from dioxane + water.

Benzyl sulphide, m.p. 50°.
  Cryst. from ethanol, then chromatographed on alumina using pentane
as eluent, and finally recryst. from ethanol. [Kice and Bowers,
JACS, 84, 2390 (1962).]

Benzylthiocyanate, m.p. 43°, b.p. 256° (decomp.).
  Cryst. from ethanol or aq. ethanol.

S-Benzylthiuronium chloride, m.p. 172-174°
   Cryst. from ethanol or dil. HCl.

Benzyl toluene-p-sulphonate, m.p. 58°.
   Cryst. from pet. ether (b.p. 40-60°).

Benzyltrimethylammonium chloride.
   A 60% aq. soln. was evapd. to dryness under vac. on a steam-bath,
and then left in a vac. desiccator containing a suitable dehydrating
agent. The solid residue was dissolved in a small amount of boiling
absolute ethanol and pptd. by adding an equal volume of ethyl ether
and cooling. After washing, the ppted. was dried under vac.
[Karusch, JACS, 73, 1246 (1951).]

Benzyltrimethylammonium hydroxide.
   A 38% aq. soln. (as supplied) was decolorized (charcoal), then
evapd. under reduced pressure to a syrup, with final drying at 75°
and 1 mm pressure.

Berbamine, m.p. 197-210°.
   Cryst. from pet. ether.

Berberine, m.p. 145°
   Cryst. from ethyl ether.

Berberine hydrochloride.
   Cryst. from water.

Betaine, m.p. 301-305° (decomp.)(anhydrous).
   Cryst. from aq. ethanol.

Biacetyl, b.p. 88°, $n_D^{18.5}$ 1.3933.
   Dried with anhydrous $CaSO_4$, $CaCl_2$ or $MgSO_4$, then vac. distd. under
nitrogen, taking the middle fraction and storing it at Dry-Ice
temperature in the dark (to prevent polymerization).

Bibenzyl, m.p. 52.5-53.5°.
   Cryst. from hexane, methanol or 95% ethanol. Has also been sublimed

Bibenzyl (continued)
under vac., and further purified by percolation through columns of
silica gel and activated alumina.

Bicuculline, m.p. 215° (196°, 177°), $[\alpha]_{546}^{20}$ +159° (c = 1 in CHCl$_3$).
    Cryst. by dissolving in chloroform and adding methanol or ethanol.

Bicyclohexyl, b.p. 238° (cis-cis), 217-219° (trans-trans).
    Shaken repeatedly with aq. KMnO$_4$ and with conc. H$_2$SO$_4$, washed with
water, dried, first from CaCl$_2$ then from sodium, and distd.
[MacKenzie, JACS, 77, 2214 (1955).]

Bicyclo(3.2.1)octane, m.p. 141°.
    Purified by zone melting.

Biguanide, m.p. 130°.
    Cryst. from ethanol.

Bilirubin, $\varepsilon_{450\ m\mu}$ = 55,600 in CHCl$_3$.
    Meso-type impurities eliminated by successive Soxhlet extraction
with ethyl ether and methanol. Then cryst. from CHCl$_3$, and dried to
constant weight at 80° under vac. [Gray et al., JCS, 2264 (1961).]

Biliverdin, m.p. >300°.
    Cryst. from methanol.

2,2'-Binaphthyl, m.p. 188°.
    Cryst. from benzene.

D-Biotin, m.p. 230.2° (decomp.).
    Cryst. from hot water.

Biphenyl, m.p. 70-71°.
    Cryst. from ethanol, methanol, b.p. 40-60° pet.ether or glacial
acetic acid. Freed from polar impurities by passage of its soln. in
benzene through an alumina column, followed by evapn. of the benzene.
Its soln. in CCl$_4$ has been chromatographed on alumina, with benzene
as eluant.Has also been purified by distn. under vac. and by zone refining.

p-Biphenylamine, m.p. 53°.
  Cryst. from water.

Biphenyl-2-carboxylic acid, m.p. 114°, b.p. 343-344°.
Biphenyl-4-carboxylic acid, m.p. 228°.
  Cryst. from benzene-pet.ether or aq. ethanol.

2,4'-Biphenyldiamine, m.p. 45°, b.p. 363°/760 mm.
  Cryst. from aq. ethanol.

α-(4-Biphenylyl)butyric acid, m.p. 175-177°.
γ-(4-Biphenylyl)butyric acid, m.p. 118°.
  Cryst. from methanol.

2-Biphenylyl diphenyl phosphate, n_D^{25} 1.5925.
  Vac. distd., then percolated through an alumina column. Passed
through a packed column maintained at 150° to remove residual traces
of volatile materials by a countercurrent stream of nitrogen at
reduced pressure. [Dobry and Keller, JPC, 61, 1448 (1957).]

2,2'-Bipyridyl, m. p. 70.5°.
  Cryst. from hexane, aq. ethanol, or (after charcoal treatment of a
$CHCl_3$ soln.) from pet.ether. Also pptd. from a conc. soln. in
ethanol by addition of water. Dried in vac. over P_2O_5. Further purified by chromatography on alumina or by sublimation.

4,4'-Bipyridyl, m.p. 73° (hydrate), 114° (171-172°)(anhydrous),
b.p. 305°/760 mm, 293°/743 mm.
  Cryst. from water, benzene-pet.ether, and sublimes in vacuo.

2,2'-Bipyridylamine, m.p. 95.1°.
  Cryst. from acetone.

2,2'-Biquinolyl, m.p. 196°.
  Decolorized in $CHCl_3$ soln. (charcoal), then cryst. to constant m.p.
from pet.ether [Cumper, Ginman and Vogel, JCS, 1188 (1962).]

Bis-(4-aminophenyl)methane, m.p. 92-93°, b.p. 232°/9 mm.
  Cryst. from 95% ethanol.

Bis-(p-bromophenyl) ether, m.p. 60.1-61.7°.
  Cryst. twice from ethanol, once from benzene and dried under vac.
[Purcell and Smith, JACS, 83, 1063 (1961).]

Bis-(p-tert.-butylphenyl)phenyl phosphate, n_D^{25} 1.5412, b.p. 281°/5 mm.
  Same as for 2-Biphenylyl diphenyl phosphate.

Bis-(β-chloroethyl)amine hydrochloride, m.p. 114-116°.
  Cryst. from acetone.

Bis-(2-chloroethyl)ether, b.p. 178.8°, n_D^{20} 1.45750, n_D^{25} 1.45534.
  Washed with conc. H_2SO_4, then Na_2CO_3 soln., drying with anhydrous
Na_2CO_3, and finally passing through a 50 cm column of activated
alumina. Alternatively, washed with 10% ferrous sulphate soln., then
water, and dried with $CaSO_4$. Distd. at 94°/33 mm.

Bis-(chloromethyl)durene, m.p. 197-198°.
  Cryst. three times from benzene, then dried under vac. in an
Abderhalden pistol.

N,N-Bis(2-chloroethyl)-2-naphthylamine, m.p. 54-56°, b.p. 210°/5 mm.
  Cryst. from pet.ether. Carcinogenic.

3,3'-Bis-(chloromethyl)oxacyclobutane, m.p. 18.9°.
  Shaken with aq. $NaHSO_3$ or $FeSO_4$ to remove peroxides. Septd., dried
with anhydrous Na_2SO_4, then distd. under reduced pressure from a
little CaH_2. [Dainton, Ivin and Walmsley, TFS, 56, 1784 (1960).]

Bis-(o-chlorophenyl) phenyl phosphate, n_D^{25} 1.5767, b.p. 254°/4 mm.
  Same as for 2-Biphenylyl diphenyl phosphate.

2,2-Bis-(p-chlorophenyl)-1,1,1-trichloroethane (D.D.T.), m.p. 108°.
  Cryst. from n-propyl alcohol (5 ml/g), then dried in air or in an
air oven at 50-60°.

2,2'-Bis-[di-(carboxymethyl)-amino]diethyl ether, $(HOOCCH_2)_2$-$NCH_2CH_2OCH_2CH_2N(CH_2COOH)_2$.

1,2-Bis-[2-di-(carboxymethyl)-aminoethoxy]ethane, $(HOOCCH_2)_2NCH_2CH_2OCH_2CH_2OCH_2CH_2N(CH_2COOH)_2$.

   Cryst. from ethanol.

4,4'-Bis-(dimethylamino)benzophenone, m.p. 175°.
   Cryst. from ethanol (25 ml/g) and dried under vac.

Bis(p-dimethylaminobenzylidene)benzidine, m.p. 318°.
   Cryst. from nitrobenzene.

1,8-Bis(dimethylamino)naphthalene, m.p. $47-48^{\circ}$.
   Cryst. from ethanol and dried in vac. oven. Stored in dark.

Bis-(dimethylthiocarbamyl)disulphide, m.p. $155-156^{\circ}$.
   Cryst. from $CHCl_3$, by addition of ethanol.

Bis-(2-ethoxyethyl) ether, see Diethylene glycol diethyl ether.

Bis-(2-ethylhexyl) 2-ethylhexyl phosphonate, n_D^{25} 1.4473.
   Purified by stirring an 0.4 M soln. in benzene with an equal volume of 6 M HCl at approx. 60° for 8 hr. The benzene layer was then
shaken successively with equal volumes of water (twice), aq. 5%
Na_2CO_3 (three times), and water (eight times), followed by evapn. of
the benzene and dissolved water under reduced pressure at room temp.
(Using a rotating evacuated flask.) Stored in dry, dark conditions.
[Peppard et al., J. Inorg. Nuclear Chem., 24, 1387 (1962).]
   Vac. distd., then percolated through an alumina column before
finally passed through a packed column maintained at 150° where
residual traces of volatile materials were removed by a countercurrent stream of nitrogen at reduced pressure. [Dobry and Keller,
JPC, 61, 1448 (1957).]

Bis-(2-ethylhexyl) phosphoric acid.
   See Peppard, Ferraro and Mason, J. Inorg. Nuclear Chem., 7, 231
(1958) or Stewart and Crandall, JACS, 73, 1377 (1951).

N,N-Bis(2-hydroxyethyl)-2-aminoethanesulphonic acid, (BES),
m.p. 150-155°.
   Cryst. from aq. ethanol.

N,N'-Bis(2-hydroxyethyl)glycine, (Bicine), m.p. 191-194° (decomp.).
   Cryst. from 80% methanol.

[Bis-(2-hydroxyethyl)imino]-tris-[(hydroxymethyl)methanol], m.p. 89°.
   Cryst. from hot 1-butanol. Dried in vac. at 25°.

3,4-Bis(4-hydroxyphenyl)hexane, m.p. 187°.
   Purified from diethylstilboestrol by zone melting.

2,2-Bis-(4-hydroxyphenyl)propane, m.p. 151-152°.
   Cryst. from chlorobenzene or dil. acetic acid. The 2,4'-isomer can
be removed by vac. sublimation around 95° at a pressure of 1 mm. The
2,2-bis-isomer can then be sublimed rather quickly at 156° from the
residue.

Bis-(2-methoxyethyl) ether.
   Passed through a 12 in. column of molecular sieve to remove water
and peroxides.

1,4-Bismethylaminoanthraquinone.
   Purified by thin-layer chromatography on silica gel plates, using
toluene/acetone (3:1) as eluent. The main band was scraped off and
extracted with methanol. The solvent was evaporated and the dye was
dried in a drying pistol. [Land, McAlpine, Sinclair and Truscott,
JCS Faraday Trans I, 72, 2091 (1976).]

Bis(1-naphthylmethyl)amine, m.p. 62°.
   Cryst. from pet.ether.

N,N'-Bis(nicotinic acid)hydrazide, m.p. 227-228°.
   Cryst. from water.

Bis-(p-nitrophenyl)ether, m.p. 142-143°.
Bis-(p-nitrophenyl)methane, m. p. 183°.
   Cryst. twice from benzene, and dried under vac.

Bisnorcholanic acid, m.p. 214° (α-form), 242° (β-form), 210-211° (γ-form), 184° (δ-form), 181° (ε-form).
Cryst. from ethanol (α-form), or acetic acid (all forms).

3,3'-Bis-(phenoxymethyl)oxacyclobutane, m.p. 67.5-68°.
Cryst. from methanol.

Biuret. (Sinters at 218° and chars at 270°.)
Cryst. from ethanol.

Bixin, m.p. 198°.
Cryst. from acetone.

Blue tetrazolium.
Cryst. from 95% ethanol-anhydrous ethyl ether, to constant absorbance at 254 mμ.

R-2-endo-Bornanol, m.p. 208°, $[α]_D$ +15.8° (in ethanol).
Cryst. from boiling ethanol (charcoal).

(±) Borneol, m.p. 208°.
Cryst. to constant m.p. from pet.ether (b.p. 60-80°).

Brazilin, m.p. 130° (decomp.).
Cryst. from ethanol.

Brilliant cresyl blue.
Cryst. from pet.ether.

Brilliant green.
Purified by pptn. as the perchlorate from aq. soln. (0.3%) after filtering, heating to 75° and adjustment to pH 1-2. Recryst. from ethanol-water (1:4).[Kerr and Gregory, Analyst, 94, 1036 (1969).]

N-Bromoacetamide, m.p. 102-105°.
Cryst. from $CHCl_3$-hexane.

p-Bromoacetanilide, m.p. 167°.
    Cryst. from aq. methanol or ethanol. Purified by zone melting.

Bromoacetic acid, m.p. 50°, b.p. 208°, 118°/15 mm.
    Cryst. from pet.ether (b.p. 40-60°). Ethyl ether soln. passed
through an alumina column, and the ether evapd. at room temp. under
vac.

Bromoacetone, b.p. 31.5°/8 mm.
    Stood with anhydrous $CaCO_3$, distd. under low vac., and stored
with $CaCO_3$.

p-Bromoacetophenone, m.p. 54°.
ω-Bromoacetophenone, m.p. 57-58°.
    Cryst. from ethanol, or from pet.ether (b.p. 80-100°).

p-Bromoaniline, m.p. 66°.
    Cryst. (with appreciable loss) from aq. ethanol.

2-Bromoanisole, f.p. 2.5°, b.p. 124°/40 mm, n_D^{25} 1.5717.
4-Bromoanisole, f.p. 13.4°, b.p. 124°/40 mm, n_D^{25} 1.5617.
    Cryst. by partial freezing (repeatedly), then distd. at reduced
pressure.

p-Bromobenzal diacetate, m.p. 95°.
    Cryst. from hot ethanol (3 ml/g).

Bromobenzene, b.p. 155.9°, n_D^{15} 1.56252, n_D^{20} 1.5588.
    Washed vigorously with conc. H_2SO_4, then 10% NaOH or $NaHCO_3$ solns.,
and water. Dried with $CaCl_2$ or Na_2SO_4, or passed through activated
alumina, before refluxing with, and distilling from, calcium
turnings or sodium, using a glass helix-packed column.

p-Bromobenzenesulphonyl chloride, m.p. 74.3-75.1°.
    Cryst. from pet.ether, or from ethyl ether cooled in powdered Dry
Ice after the ether soln. had been washed with 10% NaOH until
colourless, then dried with anhydrous Na_2SO_4.

o-Bromobenzoic acid, m.p. 148.9°.
   Cryst. from benzene or methanol.

m-Bromobenzoic acid, m.p. 155°.
   Cryst. from acetone-water, methanol or acetic acid.

p-Bromobenzoic acid, m.p. 251-252°.
   Cryst. from methanol, or methanol-water mixture.

p-Bromobenzophenone, m.p. 81°.
p-Bromobenzyl bromide, m.p. 61°.
p-Bromobenzyl chloride, m.p. 40-41°, b.p. 105-115°/12 mm.
   Cryst. from ethanol.

p-Bromobiphenyl, m.p. 88.8-89.2°.
   Cryst. from absolute ethanol and dried under vac.

1-Bromobutane, see n-Butyl bromide.

2-Bromobutane, b.p. 91.2°, n_D^{20} 1.4367, n_D^{25} 1.4341.
   Washed with conc. HCl, water, 10% aq. $NaHSO_3$, 10% Na_2CO_3, and then
water. Dried with $CaCl_2$, Na_2SO_4 or anhydrous K_2CO_3, and fractionally
distd. through a 1 m column packed with glass helices.

3-Bromocamphor-8-sulphonic acid, m.p. 195-196° (anhydrous),
$[\alpha]_D^{20}$ +88.3° (in H_2O).
3-Bromocamphor-10-sulphonic acid, m.p. 47.5°, $[\alpha]_D^{20}$ +98.3° (in H_2O).
3-Bromocamphor-8-sulphonic acid ammonium salt, m.p. 270° (decomp.),
$[\alpha]_D^{20}$ +84.8° (in H_2O).
   Cryst. from water.

Bromocresol green, m.p. 218-219°.
   Cryst. from glacial acetic acid or dissolved in aq. 5% $NaHCO_3$
soln. and pptd. from hot soln. by dropwise addition of aq. HCl.
Repeated until extinction coefficients did not increase.

Bromocresol purple, m.p. 241-242°.

Dissolved in aq. 5% NaHCO$_3$ soln and pptd from hot soln by dropwise addn of aqueous HCl. Repeated until extinction coefficients did
not increase. Can also be recryst. from benzene.

1-Bromodecane, see n-Decyl bromide.

p-Bromo-N,N-dimethylaniline, m.p. 55°, b.p. 264°.

Refluxed for 3 hr with two equivalents of acetic anhydride, then
fractionally distd. at reduced pressure.

1-Bromo-2,4-dinitrobenzene, m.p. 75°.

Cryst. from ethyl ether, isopropyl ether, 80% ethanol or
absolute ethanol.

Bromoethane, see Ethyl bromide.

Bromoform, f.p. 8.1°, b.p. 55-56°/35 mm, 149.6°/760 mm, d^{15} 2.90380,
d^{30} 2.86460, n$_D^{15}$ 1.60053, n$_D^{20}$ 1.5988.

Storage and stability of bromoform and chloroform are similar.
Ethanol, added as a stabilizer, is removed by washing with water or
with satd. CaCl$_2$ soln., and the CHBr$_3$, after drying with CaCl$_2$ or
K$_2$CO$_3$, is fractionally distd. Prior to distn., CHBr$_3$ has also been
washed with conc. H$_2$SO$_4$ until the acid layer no longer became
coloured, then with dil. NaOH or NaHCO$_3$, and water. A further
purification step is fractional crystn. by partial freezing.

1-Bromoheptane, see n-Heptyl bromide.

2-Bromohexadecanoic acid, m.p. 53°.

Cryst. from ligroin.

1-Bromohexane, see n-Hexyl bromide.

4-Bromo-1-isopropylaminopentane hydrobromide, m.p. 167-167.5°.

Cryst. from acetone-ethyl ether.

Bromomethane, see Methyl bromide.

1-Bromo-2-methylpropane, see Isobutyl bromide.

1-Bromonaphthalene, b.p. 118°/6 mm.
  Purified by passage through activated alumina, and three vac.
distns.

2-Bromonaphthalene, m.p. 59°.
  Purified by fractional elution from a chromatographic column.

1-Bromo-2-naphthol.
6-Bromo-2-naphthol.
  Cryst. from ethanol.

ω-Bromo-4-nitroacetophenone, m.p. 98°.
  Cryst. from benzene-pet. ether.

o-Bromonitrobenzene, m.p. 43°.
p-Bromonitrobenzene, m.p. 127°.
m-Bromonitrobenzene, m.p. 42°.
  Cryst. twice from pet. ether, using charcoal before the first
crystn.

α-Bromo-p-nitrotoluene, see p-Nitrobenzyl bromide.

1-Bromooctane, see n-Octyl bromide.

1-Bromopentane, see n-Amyl bromide.

p-Bromophenacyl bromide, m.p. 110-111°.
  Cryst. from ethanol (about 8 ml/g).

o-Bromophenol, b.p. 194°.
  Purified by at least two passes through a chromatograph column.

p-Bromophenol, m.p. 64°.
  Cryst. from $CHCl_3$, CCl_4, pet.ether (b.p. 40-60°), or water, and
dried at 70° under vac. for 2 hr.

Bromophenol blue, m.p. 270-271° (decomp.).
   Cryst. from benzene or acetone-glacial acetic acid, and air dried.

(p-Bromophenoxy)acetic acid, m.p. 158°.
β-(p-Bromophenoxy)propionic acid, m.p. 146°.
   Cryst. from ethanol.

2-Bromo-4'-phenylacetophenone, see p-Phenylphenacyl bromide.

p-Bromophenylhydrazine, m.p. 108-109°.
   Cryst. from water.

p-Bromophenyl isocyanate, m.p. 41-42°.
   Cryst. from pet. ether (b.p. 30-40°).

Bromopicrin, m.p. 10.2-10.3°, b.p. 85-87°/16 mm, d_4^{20} 2.7880, n_D^{20} 1.5790.
   Steam distd., dried with anhydrous Na_2SO_4 and vac. distd.

1-Bromopropane, see n-Propyl bromide.

2-Bromopropane, see Isopropyl bromide.

3-Bromopropene, see Allyl bromide.

β-Bromopropionic acid, m.p. 62.5°.
   Cryst. from CCl_4.

(3-Bromopropyl)benzene, see 3-Phenylpropyl bromide.

2-Bromopyridine, b.p. 49.0°/2.7 mm, n_D^{20} 1.5713.
   Dried over NaOH for several days, then distd. from CaO under
reduced pressure, taking the middle fraction.

Bromopyrogallol red.
   Cryst. from 50% ethanol.

5-Bromosalicyl hydroxamic acid, m. p. 232° (decomp.).
   Cryst. from ethanol.

N-Bromosuccinimide, m.p. 108°.

N-Bromosuccinimide (30 g) was dissolved rapidly in 300 ml of boiling water and filtered through a fluted filter into a flask immersed in ice bath, and left for 2 hr. The crystals were filtered, washed thoroughly with about 100 ml of ice-cold water and drained on a Büchner funnel before drying under vac. over P_2O_5 or $CaCl_2$. [Dauben and McCoy, JACS, 81, 4863 (1959).] Has also been cryst. from acetic acid.

Bromotetronic acid, m.p. 183° (decomp.).

Decolorized and free bromine removed by charcoal treatment of an ethyl acetate soln, then recryst. from ethyl acetate. [Schuler, Bhatia and Schuler, JPC, 78, 1063 (1974).]

Bromothymol blue.

Dissolved in aq. 5% $NaHCO_3$ soln. and pptd. from the hot soln. by dropwise addition of aqueous HCl. Repeated until extinction coefficients did not increase.

α-Bromotoluene, see Benzyl bromide.

p-Bromotoluene, m.p. 28°.

Cryst. from ethanol.

α-Bromo-4-toluic acid, m.p. 229-230°.

Cryst. from acetone.

Bromotrichloromethane, f.p. -5.6°, b.p. 104.1°, n_D^{20} 1.5061, n_D^{25} 1.5032.

Washed with aq. NaOH soln. or dil. Na_2CO_3, then with water, and dried with $CaCl_2$, BaO or $MgSO_4$ before distilling in diffuse light and storing in the dark. Has also been purified by treatment with charcoal and by fractional crystn. by partial freezing.

Purified by vigorous stirring with portions of conc. H_2SO_4 until the acid did not discolour during several hours stirring. Washed with Na_2CO_3 and water, dried with $CaCl_2$ and then P_2O_5. Illuminated with a 1000 W projection lamp at 6 in. for 10 hr after making 0.01 M in bromine. Passed through a 30 x 1.5 cm column of activated alumina

Bromotrichloromethane (continued)
before fractionally distilling through a 12 in. Vigreux column. The
middle fraction was passed through a fresh activated alumina column.
[Firestone and Willard, JACS, 83, 3551 (1961).]

Bromotrifluoromethane, b.p. -59°.
    Passed through a tube containing P_2O_5 on glass wool into a vac.
system where it was frozen out in a quartz sample tube and degassed
by a series of cycles of freezing, evacuating and thawing.

5-Bromovaleric acid, m.p. 40°.
    Cryst. from pet. ether.

α-Bromo-p-xylene, m.p. 35°, b.p. 218-220°/740 mm.
    Cryst. from ethanol.

Bromural, m.p. 154-155°.
    Cryst. from toluene, and air dried.

Brucine, m.p. 178-179°, $[\alpha]_{5461}^{20}$ -149.9° (c = 1 in $CHCl_3$).
    Cryst. once from water, as tetrahydrate, then suspended in $CHCl_3$
and shaken with anhydrous Na_2SO_4 (to dehydrate the brucine, which
then dissolves). Pptd. by pouring the soln. into a large bulk of dry
pet.ether (b.p. 40-60°), filtered and heated to 120° in high vac.
[Turner, JCS, 842 (1951).]

[α]-Brucine sulphate.
    Cryst. from water.

Bufotenine hydrogen oxalate, m.p. 96.5°.
    Cryst. from ether.

1,3-Butadiene, b.p. -2.6°.
    Dried by condensing with a soln. of aluminium triethyl in deca-
hydronaphthalene; then flash distd. Also dried by passage over
anhydrous $CaCl_2$ or distd. from $NaBH_4$.

n-Butane, m.p. -135°, b.p. -0.5°.

Dried by passage over anhydrous Mg(ClO$_4$)$_2$ and molecular sieve type 5A. Air removed by prolonged and frequent degassing at -107°.

1,4-Butanediol, f.p. 20.4°, b.p. 107-108°/4 mm, 127°/20 mm, n_D^{20} 1.4467.

Distd. and stored over Linde type 4A molecular sieve, or cryst. twice from anhydrous ethyl ether-acetone, and redistd.

meso-2,3-Butanediol, m.p. 25°.

Cryst. from isopropyl ether.

2,3-Butanedione, see Biacetyl.

1-Butanethiol, b.p. 98.4°, d_4^{25} 0.83679, n_D^{20} 1.44298, n_D^{25} 1.44034.

Dried with CaSO$_4$, then refluxed and distd. from magnesium; or dried with, and distd. from CaO, under nitrogen. Has been separated from hydrocarbons by extractive distn. with aniline.

Dissolved in 20% NaOH, extracted with a small amount of benzene, then steam distd., until clear. The soln. was then cooled and acidified slightly with 15% H$_2$SO$_4$. The thiol was distd. out, dried with CaSO$_4$ or CaCl$_2$, and fractionally distd. under nitrogen. [Mathias and Filho, JPC, 62, 1427 (1958).] Also purified by pptn. as lead mercaptide from alcoholic soln., with regeneration by adding dil. HCl to the residue after steam distn.

2-Butanethiol, b.p. 37.4°/134 mm, d_4^{25} 0.82456, n_D^{25} 1.43385.

Purified as for 1-butanethiol.

n-Butanol, b.p. 117.7°, d^{25} 0.80572, n_D^{15} 1.40118, n_D^{20} 1.39922.

Dried with MgSO$_4$, CaO, K$_2$CO$_3$, Ca or solid NaOH, followed by refluxing with, and distn. from, calcium, magnesium activated with iodine, aluminium amalgam or sodium. Can also dry with molecular sieves, or by refluxing with n-butyl phthalate or succinate. (For method, see Ethanol.) n-Butanol can also be dried by efficient fractional distn., water passing over in the first fraction as a binary azeotrope (contains about 37% water). An ultraviolet-transparent distillate has been obtained by drying with magnesium, and distilling

n-Butanol (continued)

from sulphanilic acid. To remove bases, aldehydes and ketones,
n-butyl alcohol has been washed with dil. H_2SO_4, then $NaHSO_3$ soln.;
esters were removed by boiling for 1½ hr with 20% NaOH.

Also purified by adding 2 g $NaBH_4$ to 1.5 l butanol, gently
bubbling with argon and refluxing for 1 day at $50°$. Then added 2 g
freshly cut sodium (washed with butanol) and refluxed for 1 day.
Distd. and the middle fraction taken. [Jou and Freeman, JPC, 81,
909 (1977).]

2-Butanol, see sec-Butyl alcohol.

2-Butanone, b.p. $79.6°$, d_4^{20} 0.8053, n_D^{20} 1.37850, n_D^{25} 1.37612.

In general, purification methods are the same as for acetone.
Aldehydes can be removed by refluxing with $KMnO_4$ + CaO, until the
Schiff aldehyde test is not positive, prior to distn. Shaking with
satd. K_2CO_3, or passage through a small column of activated alumina,
removes acidic impurities. The ketone can be dried by careful distn.
(an azeotrope containing 11% water boils at $73.4°$), or by $CaSO_4$,
P_2O_5, Na_2SO_4, or K_2CO_3, followed by fractional distn. Purification
as the bisulphite addition compound is achieved by shaking with
excess satd. $NaHSO_3$, cooled to $0°$, filtering off the ppte., washing
with a little ethyl ether and drying in air; this is followed by
decomposition with a slight excess of Na_2CO_3 soln. and steam distn.,
the distillate being satd. with K_2CO_3 so that the ketone can be
separated, dried with K_2CO_3, filtered, and distd. Purification as the
NaI addition compound (m.p. $73-74°$) is more convenient. (For details,
see Acetone.) Small quantities of 2-butanone can be purified by
conversion to the semicarbazone, recrystn. to constant m.p., drying
under vac. over $CaCl_2$ and paraffin wax, refluxing for 30 min with
excess oxalic acid, followed by steam distn., salting out, drying and
distn. [Cowan, Jeffery and Vogel, JCS, 171 (1940).]

cis-2-Butene, b.p. $2.95-3.05°$/746 mm.
trans-2-Butene, b.p. $0.3-0.4°$/744 mm.

Dried with CaH_2. Purified by gas chromatography.

2-Butene-1,4-dicarboxylic acid, m.p. 194-197°.
    Cryst. from boiling water, then dried at 50-60° in a vac. oven.

But-1-en-3-one, 79-80°.
    Dried with K_2CO_3, then Na_2SO_4, and fractionally distd.

2-Butoxyethanol, b.p. 171°/745 mm, n_D^{20} 1.4191.
    Peroxides can be removed by refluxing with anhydrous $SnCl_2$ or by
passage under slight pressure through a column of activated alumina.
Dried with anhydrous K_2CO_3 and $CaSO_4$, filtered and distd., or
refluxed with, and distd. from, NaOH.

2-(2-Butoxyethoxy)ethanol, see Diethylene glycol mono-n-butyl ether.

n-Butyl acetate, b.p. 126.1°, n_D^{20} 1.39406.
    Distd., refluxed with successive small portions of $KMnO_4$ until
the colour persisted, dried with anhydrous $CaSO_4$, filtered and
redistd.

tert.-Butyl acetate, b.p. 97-98°.
    Washed with 5% Na_2CO_3 soln., then satd. aq. $CaCl_2$, dried with
$CaSO_4$, and distd.

Butyl acrylate, b.p. 59°/25 mm, n_D^{12} 1.4254.
    Washed repeatedly with aq. NaOH to remove inhibitors such as
hydroquinone, then with distd. water. Dried with $CaCl_2$. Fractionally
distd. under reduced pressure in an all-glass apparatus. The middle
fraction was sealed under nitrogen and stored at 0° in the dark
until used. [Mallik and Das, JACS, 82, 4269 (1960).]

n-Butyl alcohol, see n-Butanol.

sec.-Butyl alcohol, b.p. 99.4°.
    Purification methods are the same as for n-Butanol. These include
drying with K_2CO_3 or $CaSO_4$, followed by filtration and fractional
distn.; refluxing with CaO, distn., then refluxing with magnesium
and redistn.; and refluxing with, then distn. from, CaH_2. Calcium
carbide has also been used as a drying agent. Anhydrous alcohol

sec.-Butyl alcohol (continued)

is obtained by refluxing with sec.-butyl phthalate or succinate. (For method, see Ethanol.) Small amounts of the alcohol can be purified by conversion to the alkyl hydrogen phthalate and recrystn. [Hargreaves, JCS, 3679, (1956).] For purification of optical isomers, see Timmermans and Martin, JCP., 25, 411 (1928).

tert.-Butyl alcohol, m.p. 25°, b.p. 82.5°, $n_D^{25.5}$ 1.38516.

Dried with CaO, K_2CO_3, $CaSO_4$ or $MgSO_4$, filtered and fractionally distd. Dried further by refluxing with, and distilling from, either magnesium activated with iodine, or small amounts of calcium, sodium or potassium, under nitrogen. Passage through a column of type 4A molecular sieve is another effective method of drying. So, also, is refluxing with tert.-butyl phthalate or succinate. (For method, see Ethanol.) Other methods include refluxing with excess aluminium tert.-butylate, or standing with CaH_2, distilling as needed. Further purification is achieved by fractional crystn. by partial freezing, taking care to exclude moisture. tert.-Butyl alcohol samples containing much water can be dried by adding benzene, so that the water distils off as a ternary azeotrope, b.p. 67.3°. Traces of isobutylene have been removed from dry tert.-butyl alcohol by bubbling dry "prepurified" nitrogen through for several hours at 40-50° before using.

n-Butylamine, b.p. 77.8°/ d_4^{30} 0.7302, n_D^{20} 1.4009, n_D^{25} 1.3992.

Dried with solid KOH, K_2CO_3, $LiAlH_4$, CaH_2 or $MgSO_4$, then refluxed with, and fractionally distd. from, P_2O_5, CaH_2, CaO or BaO.

Further purified by pptn. as the hydrochloride, m.p. 213-213.5°, from ether soln. by bubbling HCl gas into it. Repptd. three times from ethanol by adding ether, followed by liberation of the free amine using excess strong base. The amine was extracted into ether, which was separated, dried with solid KOH, the ether removed by evaporation and then the amine was distd. It was stored in a desiccator over solid NaOH. [Bunnett and Davis, JACS, 82, 665 (1960).]

tert.-Butylamine, b.p. 42°.

Dried with KOH or $LiAlH_4$. Distd. from CaH_2 or BaO.

n-Butyl p-aminobenzoate, m.p. 57-59°.
    Cryst. from ethanol.

n-Butylbenzene, b.p. 183.3°, n$_D^{20}$ 1.48979, n$_D^{25}$ 1.48742.
    Distd. from sodium. Washed with small portions of conc. H$_2$SO$_4$
until the acid was no longer coloured, then with water and aq.
Na$_2$CO$_3$. Dried with anhydrous MgSO$_4$, and distd. twice from sodium,
collecting the middle fraction. [Vogel, JCS, 607 (1948).]

tert.-Butylbenzene, b.p. 169.1°, n$_D^{20}$ 1.49266, n$_D^{25}$ 1.49024.
    Washed with cold conc. H$_2$SO$_4$ until a fresh portion of acid was no
longer coloured, then with 10% aq. NaOH, followed by distd. water
until neutral. Dried with CaSO$_4$ and distd. in a glass helices-packed
column, taking the middle fraction.

n-Butyl bromide, b.p. 101-102°, d$_4^{25}$ 1.2678, n$_D^{20}$ 1.4399, n$_D^{25}$ 1.4374.
    Washed with conc. H$_2$SO$_4$, water, 10% Na$_2$CO$_3$ and again with water.
Dried with CaCl$_2$, CaSO$_4$ or K$_2$CO$_3$, and distd. Redistd. after drying
with P$_2$O$_5$, or passed through two columns containing 5:1 silica
gel-Celite mixture and stored with freshly activated alumina.

sec.-Butyl bromide, see 2-Bromobutane.

Butyl carbitol, see Diethylene glycol mono-n-butyl ether.

4-tert.-Butylcatechol, m.p. 55-56°.
    Cryst. from pet. ether.

Butyl cellosolve, see 2-Butoxyethanol.

n-Butyl chloride, b.p. 78°, n$_D^{20}$ 1.4021.
    Shaken repeatedly with conc. H$_2$SO$_4$ (until no further colour dev-
eloped in the acid), then washed with water, aq. NaHCO$_3$ or Na$_2$CO$_3$,
and more water. Dried with CaCl$_2$, or MgSO$_4$ (then with P$_2$O$_5$ if
desired), decanted and fractionally distd.
    Alternatively, a stream of oxygen containing about 5% ozone was
passed through the material for about three times as long as was
necessary to obtain the first coloration of starch iodide paper by

n-Butyl chloride (continued)
the exit gas. After washing with $NaHCO_3$ soln. to hydrolyze ozonides
and to remove the resulting organic acids, the liquid was dried and
distd. [Chien and Willard, JACS, 75, 6160 (1953).]

sec.-Butyl chloride, see 2-Chlorobutane.

tert.-Butyl chloride, f.p. $-24.6°$, b.p. $50.4°$, n_D^{20} 1.38564.
    Purification methods commonly used for other alkyl halides lead to
decomposition. Some impurities can be removed by photochlorination
with a small amount of chlorine prior to use. The liquid can be
washed with ice water, dried with $CaCl_2$ or $CaCl_2$ + CaO, and fractionally distd. It has been further purified by repeated fractional
crystn. by partial freezing.

6-tert.-Butyl-1-chloro-2-naphthol, m.p. $76°$, b.p. $185°/15$ mm.
    Cryst. from pet. ether.

4-tert.-Butylcyclohexanone, m.p. $49-50°$.
    Cryst. from pentane.

n-Butyl disulphide, b.p. $110-113°/15$ mm, n_D^{22} 1.494.
    Shaken with lead peroxide, filtered, and dist. at 3 mm pressure
under nitrogen.

n-Butyl ether, b.p. $52-53°/26$ mm, $142.0°/760$ mm, n_D^{20} 1.39925,
n_D^{25} 1.39685.
    Peroxides (detected by liberation of iodine from weakly acid
(HCl) solns. of 2% KI) can be removed by shaking 1 1. of ether with
5-10 ml of a soln. comprising 6.0 g ferrous sulphate + 6 ml conc.
H_2SO_4 + 110 ml of water, with aq. Na_2SO_3, or with acidified NaI,
water, then $Na_2S_2O_3$. After washing with dil NaOH, KOH, or Na_2CO_3,
then water, the ether is dried with $CaCl_2$ and distd. It can be
further dried by distilling from CaH_2 or sodium (after drying with
P_2O_5), and stored in the dark with sodium or sodium hydride. The
ether can also be purified by treating with CS_2 and NaOH, expelling
the excess sulphide by heating. The ether is then washed with water,

n-Butyl ether (continued)

dried with NaOH and distd. [Kusama and Koike, JCS Jap. Pure Chem. Sect., 72, 229 (1951).]

Other purification procedures include passage through an activated alumina column to remove peroxides, or through a column of silica gel, and distn. after adding about 3% (v/v) of a 1 M soln. of CH_3MgI in n-butyl ether.

n-Butyl ethyl ether, b.p. 92.7°, n_D^{20} 1.38175, n_D^{25} 1.3800.

Purified by drying with $CaSO_4$, by passage through a column of activated alumina (to remove peroxides), followed by prolonged refluxing with sodium and then fractional distn.

tert.-Butyl ethyl ether, b.p. 71-72°.

Dried with $CaSO_4$, passed through an alumina column, and fractionally distd.

n-Butyl formate, b.p. 106.6°, n_D^{20} 1.38903.

Washed with satd. $NaHCO_3$ soln. in the presence of satd. NaCl, until no further reaction occurred, then with satd. NaCl soln., dried ($MgSO_4$) and fractionally distd.

tert.-Butyl hydroperoxide, f.p. 5.4°, b.p. 38°/18 mm, n_D^{20} 1.4013.

Alcoholic and volatile impurities can be removed by prolonged refluxing at 40° under reduced pressure, or by steam distn. For example, Bartlett, Benzing and Pincock [JACS, 82, 1762 (1960)] refluxed at 30 mm pressure in an azeotropic separation apparatus until two phases no longer separated, and then distilled through a 30 cm column, collecting the fraction distilling at 41° and 23 mm. Pure material is stored under nitrogen, in the dark, in a refrigerator.

Crude commercial material has been added to 25% NaOH below 30°, and the crystals of the sodium salt have been collected, washed twice with benzene and dissolved in distd. water. After adjusting the pH of the soln. to 7.5 by adding solid CO_2, the peroxide was extracted into pet.ether, from which, after drying with K_2CO_3, it was recovered by distilling off the solvent under reduced pressure at room temperature. [O'Brien, Beringer, and Mesrobian, JACS, 79, 6238 (1957).] Similarly, a soln. in pet. ether has been extracted with cold aq. NaOH, and the

tert.-Butyl hydroperoxide (continued)

hydroperoxide has been regenerated by adding, at 0°, $KHSO_4$ at a pH
not higher than 4.5, then extracted into ethyl ether, dried with
$MgSO_4$, filtered and the ether evapd. in a rotating evaporator under
reduced pressure. [Milac and Djokić, JACS, 84, 3098 (1962).]

n-Butyl iodide, b.p. 130.4°, n_D^{25} 1.4967.

Dried with $MgSO_4$ or P_2O_5, fractionally distd. through a column
packed with glass helices, taking the middle fraction and storing
with calcium or mercury in the dark. Also purified by prior passage
through activated alumina or by shaking with conc. H_2SO_4 then washing
with Na_2SO_3 soln. It has also been treated carefully with sodium to
remove free HI and water, before distilling in a column containing
copper turnings at the top. Another purification consisted of treatment with bromine, followed by extraction of free halogen with
$Na_2S_2O_3$, washing with water, drying and fractional distn.

sec.- Butyl iodide, see 2-Iodobutane.

tert.-Butyl iodide, b.p. 100° (decomp.).

Vac. distn. has been used to obtain a distillate which remained
colourless for several weeks at -5°. More extensive treatment has
been used by Boggs, Thompson and Crain [JPC, 61, 625 (1957)] who
washed with aq. $NaHSO_3$ soln. to remove free iodine, dried for 1 hr
with Na_2SO_4 at 0°, and purified by four or five successive partial
freezings of the liquid to obtain colourless material which was
stored at -78°.

n-Butyl mercaptan, see 1-Butanethiol.

sec.-Butyl mercaptan, see 2-Butanethiol.

tert.-Butyl mercaptan, see 2-Methyl-2-propanethiol.

n-Butyl methacrylate.
tert.-Butyl methacrylate.

Purified as for Butyl acrylate.

2-t-Butyl-4-methoxyphenol, m.p. 64.1°.

Fractionally distd. in vac., then passed as a soln. in CHCl₃ through alumina, and the solvent evapd. from the eluate.

n-Butyl methyl ether, b.p. 70°.

Dried with CaSO₄, passed through an alumina column to remove peroxides, and fractionally distd.

dl-sec.-Butyl methyl ether, b.p. 59°.

Distd. from, and stored with, potassium.

tert.-Butyl methyl ether, b.p. 54°.

Same as for n-Butyl methyl ether.

8-sec.-Butylmetrazole, m.p. 70°.

Cryst. from pet. ether, and dried for 2 days under vac. over P_2O_5.

p-tert.-Butylnitrobenzene, m.p. 28.4°.

Fractionally cryst. three times by partially freezing a mixture of the mono-nitro isomers, then recryst. from methanol twice and dried under vac. [Brown, JACS, 81, 3232 (1959).]

tert.-Butyl peracetate, b.p. 23-24°/0.5 mm, n_D^{25} 1.4030.

Washed with NaHCO₃, then redist. to remove benzene. [Kochi, JACS, 84, 774 (1962).]

tert.-Butyl perbenzoate.

Purified by fractional crystn. by partial freezing.

tert.-Butylperoxy isobutyrate, f.p. -45.6°.

After diluting 90 ml of the material with 120 ml of pet.ether, the mixture was cooled to 5° and shaken twice with 90 ml portions of 5% NaOH soln. (also at 5°). The non-aqueous layer, after washing once with cold water, was dried at 0° with a mixture of anhydrous MgSO₄ and MgCO₃ containing about 40% MgO. After filtering, this material was passed, twice, through a column of silica gel at 0° (to remove tert.-butyl hydroperoxide). The soln. was evapd. at 0°/0.5-1 mm to remove solvent, and the residue was recryst. several

tert.-Butylperoxy isobutyrate (continued)
times from pet. ether at -60°, then subjected to high vac. to remove
traces of solvent. [Milos and Golubović, JACS, 80, 5994 (1958).]

tert.-Butylperphthalic acid.
    Cryst. from ethyl ether and dried over H_2SO_4.

p-tert.-Butylphenol, m.p. 99°.
    Cryst. to constant m.p. from pet. ether (b.p. 60-80°). Also
purified via its benzoate, as for phenol.

p-tert.-Butylphenoxyacetic acid, m.p. 88-89°.
    Cryst. from a pet.ether-benzene mixture.

n-Butylphenyl n-butylphosphonate.
    Cryst. three times from hexane as its compound with uranyl nitrate.
See Tributyl phosphate.

p-tert.-Butylphenyl diphenyl phosphate, b.p.261°/6mm, n_D^{25} 1.5522.
    Purified by vac. distn., and percolation through an alumina
column, followed by passage through a packed column maintained at
150° to remove residual traces of volatile materials in a countercurrent stream of nitrogen at reduced pressure. [Dobry and Keller,
JPC, 61, 1448 (1957).]

n-Butyl phenyl ether, b.p. 210.5°.
    Dissolved in ethyl ether, washed first with 10% aq. NaOH to remove
traces of phenol, then repeatedly with distd. water, followed by
evapn. of the solvent and distn. under reduced pressure. [Arnett and
Wu, JACS, 82, 5660 (1960).]

Butyl phosphate, b.p. 289°/760 mm.
    Washed with water, then with 1% NaOH or 5% Na_2CO_3 for several
hours, then finally with water. Dried under reduced pressure and
vac. distd.

Butyl phthalate, f.p. -35°, b.p. 340°/760 mm.

Freed from alcohol by washing with water, or from acids and butyl hydrogen phthalate by washing with dil. NaOH. Distd. at 10 torr or less.

Butyl stearate, m.p. 26.3°.

Acidic impurities removed by shaking with 0.05 M NaOH or a 2% $NaHCO_3$ soln., followed by several water washes, then purified by fractional freezing of the melt and fractional crystn. from solvents with boiling points below 100°.

p-tert.-Butyltoluene, f.p. -53.2°, b.p. 91°/28 mm, n_D^{20} 1.4920.

A sample containing 5% of the meta isomer was purified by select-ive mercuration. Fractional distn. of the solid arylmercuric acetate, after removal from the residual hydrocarbon, gave pure p-tert.-butyltoluene. [Stock and Brown, JACS, 81, 5615 (1959).]

n-Butyl vinyl ether, b.p. 93.3°.

After five washings with equal volumes of water to remove alcoh-ols (made slightly alkaline with KOH), the ether was dried with sodium and dist. under vac., taking the middle fraction. [Coombes and Eley, JCS, 3700 (1957).] Stored over KOH.

2-Butyne, b.p. 0°/253 mm.

Stood with sodium for 24 hr, then fractionally distd. at reduced pressure.

2-Butyne-1,4-diol, m.p. 54-57°.

Cryst. from ethyl acetate.

n-Butyraldehyde, b.p. 74.8°, n_D^{15} 1.38164, n_D^{20} 1.37911.

Dried with $CaCl_2$ or $CaSO_4$, then fractionally distd. under nitrogen. Lin and Day [JACS, 74, 5133 (1952)] shook with batches of $CaSO_4$ for 10-min intervals until a 5 ml sample, on mixing with 2.5 ml of CCl_4 containing 0.5 g of aluminium isopropoxide, gave no ppte. and caused the soln. to boil within 2 min. Water can be removed from n-butyr-aldehyde by careful distn. as an azeotrope distilling at 68°. The aldehyde has also been purified through its bisulphite compound

n-Butyraldehyde (continued)

which, after decomposing with excess $NaHCO_3$ soln., was steam distd., extracted under nitrogen into ether and, after drying, the extract was fractionally distd. [Kyte, Jeffery and Vogel, JCS, 4454 (1960).]

n-Butyramide, m.p. 115°, b.p. 230°.

   Cryst. from acetone, benzene, CCl_4-pet.ether, 20% ethanol or water. Dried under vac. over P_2O_5, $CaCl_2$ or 99% H_2SO_4.

n-Butyric acid, f.p. -5.3°, b.p. 163.3°, n_D^{20} 1.39796, n_D^{25} 1.39581.

   Distd., mixed with $KMnO_4$ (20 g/l.), and fractionally redistd., discarding the first third. [Vogel, JCS, 1814 (1948).]

n-Butyric anhydride, b.p. 198°.

   Dried by shaking with P_2O_5, then distd.

γ-Butyrolactone, b.p. $83.8^{\circ}/12$ mm.

   Dried with anhydrous $CaSO_4$, then fractionally distd.

Butyronitrile, b.p. 117.9°, n_D^{15} 1.38600, n_D^{30} 1.37954.

   Treated with conc. HCl until the smell of isonitrile had gone, then dried with K_2CO_3 and fractionally distd. [Turner, JCS, 1681 (1956).]

Caffeic acid, m.p. 195°.

   Cryst. from water.

Caffeine, m.p. 237°.

   Cryst. from water or absolute ethanol.

RS-Camphene, m.p. $51-52^{\circ}$.

   Cryst. twice from ethanol, then repeatedly melted and frozen at 30 mm pressure. [Williams and Smyth, JACS, 84, 1808 (1962).]

R-Camphor, m.p. 178.8°, b.p. 204°, $[\alpha]_{5461}^{35}$ + 59.61° (in ethanol).

   Cryst. from ethanol, 50% ethanol-water, or pet.ether, of from glacial acetic acid by addition of water. Can be sublimed, and also fractionally cryst. from its own melt.

Camphoric acid, m.p. 186-188°.

  Cryst. from water.

R-Camphoric anhydride, transition temp. 135°, m.p. 223.5°.

  Cryst. from ethanol.

(+)Camphor-10-sulphonic acid, m.p. 193°.

  Cryst. from ethyl acetate and dried under vac.

S-Canavanine, m.p.184°, $[\alpha]_D^{20}$ +7.9° (c = 3.2 in H_2O), $[\alpha]_D^{17}$ + 19.4°(c = 2 in H_2O).
S-Canavanine sulphate, m.p. 172° (decomp.).

  Cryst. from aq. ethanol.

Cannabinol, m.p. 76-77°, b.p. 185°/0.05 mm.

  Cryst. from pet. ether. Sublimed.

Canthaxanthin, m.p. 211-212°, $E_{1cm}^{1\%}$ = 2200 (470 nm) in cyclohexane.

  Purified by chromatography on a column of deactivated alumina or
magnesia, or on a thin layer of silica gel G (Merck), using dichloro-
methane-ethyl ether (9:1) to develop the chromatogram. Stored in the
dark and in inert atmosphere at -20°.

Capric acid, n_D^{25} 1.4239.

  Purified by conversion to its methyl ester, b.p. 114.0°/15 mm
(using excess methanol, in the presence of H_2SO_4). After removal of
the H_2SO_4 and excess methanol, the ester was distd. under vac.
through a 3-ft column packed with glass helices. The acid was then
obtained from the ester by saponification. [Trachtman and Miller,
JACS, 84, 4828 (1962).]

n-Caproamide, m.p. 100°.

  Cryst. from hot water.

Caproic acid, b.p. 205.4°, n_D^{15} 1.4188, n_D^{25} 1.4149.

  Dried with $MgSO_4$, and fractionally distd. from $CaSO_4$.

ε-Caprolactam, m.p. 70°.

  Distd. at reduced pressure, cryst. from acetone or pet.ether, and
redistd.

ε-Caprolactam (continued)

Purified by zone melting. Very hygroscopic. Discolours in contact with air unless small amounts (0.2 g/l) of NaOH, Na_2CO_3 or $NaBO_2$ are present. Cryst. from a mixture of pet.ether (185 ml of b.p. 70°) and 30 ml of 2-methyl-2-propanol., from acetone, or pet.ether. Distd. under reduced pressure and stored under nitrogen.

Capronitrile, b.p. 163.7°, n_D^{20} 1.4069, n_D^{25} 1.4048.

Washed twice with half-volumes of conc. HCl, then with satd. aq. $NaHCO_3$, dried with $MgSO_4$, and distd.

Caprylic acid, see n-Octanoic acid.

Capsorubin, m.p. 218°, λ_{max} 443, 468, 503 nm, in hexane.

Probable impurities: zeaxanthin and capsanthin. Purified by chromatography on a column of $CaCO_3$ or MgO. Cryst. from benzene-pet. ether or CS_2.

Captan, m.p. 172-173°.

Cryst. from CCl_4.

p-(Carbamoylmethoxy)acetanilide, m.p. 208°.

Cryst. from water.

3-Carbamoyl-1-methylpyridinium chloride.

Cryst. from methanol.

Carbanilide, m . p. 242°.

Cryst. from ethanol or a large volume (40 ml/g) of hot water.

Carbazole, m.p. 240-243°.

Dissolved (60 g) in conc. H_2SO_4 (300 ml), extracted with three 200 ml portions of benzene, then stirred into 1600 ml of an ice-water mixture. The ppte. was filtered off, washed with a little water, dried, cryst. from benzene and then from pyridine-benzene. [Feldman, Pantages and Orchin, JACS, 73, 4341 (1951).] Has also been cryst. from ethanol or toluene, sublimed, and zone-refined.

9-Carbazoleacetic acid, m.p. 215°.
  Cryst. from ethyl acetate.

1-Carbethoxy-4-methylpiperazine hydrochloride, m.p. 168.5-169°.
  Cryst. from absolute ethanol.

Carbitol, see Diethylene glycol monoethyl ether.

γ-Carboline, m.p. 225°.
  Cryst. from water.

Carbon black.
  Leached for 24 hr with 1:1 HCl to remove oil contamination, then
washed repeatedly with distd. water. Dried in air, and eluted for
one day each with benzene and acetone. Again dried in air at room
temp., then heated in a vacuum for 24 hr at 600° to remove adsorbed
gases. [Tamamushi and Tamaki, TFS, 55, 1007 (1959).]

Carbon disulphide, b.p. 46.3°, d^o 1.29268, n_D^{15} 1.63189.
  Shaken for 3 hr with three portions of $KMnO_4$ soln. (5 g/l.), twice
for 6 hr with mercury (to remove sulphide impurities) until no
further darkening of the interface occurred, and finally with a soln.
of $HgSO_4$ (2.5 g/l.) or cold, satd. $HgCl_2$. Dried with $CaCl_2$, $MgSO_4$, or
CaH_2 (with further drying by refluxing with P_2O_5), followed by fractional distn. in diffuse light. Alkali metals cannot be used as
drying agents. Has also been purified by standing with bromine
(0.5 ml/l.) for 3-4 hr, shaking with KOH soln., then copper turnings
(to remove unreacted bromine), and drying with $CaCl_2$.
  Small quantities of carbon disulphide have been purified (including
removal of hydrocarbons) by mechanical agitation of a 45-50 g sample
with a soln. of 130 g of sodium sulphide in 150 ml of water for 24
hr at 35-40°. The aq. sodium thiocarbonate soln. was separated from
unreacted CS_2, then pptd. with 140 g of copper sulphate in 350 g of
water, with cooling. After filtering off the copper thiocarbonate, it
was decomposed by passing steam into it. The distillate was separated
from water and distd. from P_2O_5. [Ruff and Golla, Z. anorg. Chem.,
138, 17 (1924).]

Carbon tetrabromide, m.p. 92.5°.

Reactive bromide was removed by refluxing with dil. aq. Na_2CO_3, then steam distd., cryst. from ethanol, and dried in the dark under vac. [Sharpe and Walker, JCS, 157 (1962).] Can also be sublimed at 70° at low pressure.

Carbon tetrachloride, b.p. 76.8°, d^{15} 1.60370, d^{25} 1.5842, n_D^{15} 1.46305, n_D^{25} 1.45759.

For many purposes, careful fractional distn. gives adequate purification. Carbon disulphide can be removed by shaking vigorously for several hours with satd. KOH, separating, and washing with water: this treatment is repeated. The CCl_4 is shaken with conc. H_2SO_4 until there is no further coloration, then washed with water, dried with $CaCl_2$ or $MgSO_4$, and distd. (from P_2O_5 if desired). It must not be dried with sodium. An initial refluxing with mercury for 2 hr removes sulphides. Other purification steps include passage of dry CCl_4 through activated alumina, and distn. from $KMnO_4$. Carbonyl-containing impurities can be removed by percolation through a Celite column impregnated with 2,4-dinitrophenylhydrazine, H_3PO_4 and water. (Prepared by dissolving 0.5 g DNPH in 6 ml of 85% H_3PO_4 by grinding together, then mixing with 4 ml of distd. water and 10 g of Celite.) [Schwartz and Parks, AC, 33, 1396 (1961).] Photochlorination of CCl_4 has also been used: CCl_4 to which a small amount of chlorine has been added is illuminated in a glass bottle (e.g. for 24 hr with a 200 W tungsten lamp near it), and, after washing out the excess chlorine with 0.02 M Na_2SO_3, the CCl_4 is washed with distd. water and distd. from P_2O_5. It can be dried by passing through 4A molecular sieves and distd.

Carbon tetrafluoride, b.p. -15°.

Purified by repeated passage over activated charcoal at solid-CO_2 temperatures. Traces of air were removed by pumping while alternately freezing and melting. Alternatively, liquefied by cooling in liquid air and then fractionally distilled under vac. (The chief impurity originally present was probably CF_3Cl.)

N,N'-Carbonyldiimidazole, m.p. 115.5-116°.

Cryst. from benzene or tetrahydrofuran, in a dry-box.

Carbonyl sulphide, b.p. -47.5°.

  Passed through traps containing satd. aq. lead acetate and then through a column of anhydrous $CaSO_4$.

Carbostyril, see 2-Hydroxyquinoline.

(Carboxymethyl)trimethylammonium chloride hydrazide, see Girard reagent T.

o-Carboxyphenylacetonitrile, m.p. 114-115°.

  Cryst. (with considerable loss) from benzene or glacial acetic acid.

S-Carnosine, m.p. 246-250° (decomp.), $[\alpha]_D^{25}$ +20.5° (water).

  Likely impurities: histidine, β-alanine. Cryst. from water by adding ethanol in excess.

ξ-Carotene, m.p. 38-42°, λ_{max} 378, 400, 425 nm, $E_{1cm}^{1\%}$ = 2270 (400 nm), in pet. ether.

  Purified by chromatography on 50% magnesia-HyfloSupercel, developing with hexane and eluting with 10% ethanol in hexane. Stored in the dark, under inert atmosphere, at -20°.

α-Carotene, m.p. 184-188°, λ_{max} 422, 446, 474 nm, in hexane, $E_{1cm}^{1\%}$ = 2725 (at 446), 2490 (at 474 nm).

  Purified by chromatography on columns of calcium hydroxide, alumina or magnesia. Cryst. from CS_2-methanol, ethyl ether-pet. ether, or acetone-pet.ether. Stored in the dark, under inert atmosphere, at -20°.

β-Carotene, m.p. 178-180°, $E_{1cm}^{1\%}$ = 2590 (450 nm), 2280 (478 nm), in hexane.

  Purified by chromatography on magnesia column, thin layer of kieselguhr or magnesia. Cryst. from CS_2-methanol, ethyl ether-pet. ether, or acetone-pet.ether. Stored in the dark, under inert atmosphere, at -20°.

γ-Carotene, m.p. 150°, $E_{1cm}^{1\%}$ = 2055 (437 nm), 3100 (462 nm), 2720 (494 nm).

Purified by chromatography on alumina or magnesia columns. Cryst. from benzene-methanol (2:1). Stored in the dark, under inert atmosphere, at 0°.

λ-Carrageenan.

Pptd. from a soln. of 4 g in 600 ml of water containing 12 g of potassium acetate by addition of ethanol. The fraction taken pptd. between 30 and 45% (v/v) ethanol. [Pal and Schubert, JACS, 84, 4384 (1962).]

Catechin, m.p. 177° (anhydr.).
   Cryst. from hot water. Dried at 100°.

Catechol, m.p. 105°.
   Cryst. from benzene or toluene. Sublimed under vac.

Cation exchange resin.
   Conditioned before use by successive washing with water, ethanol and water, and taken through two H$^+$ - Na$^+$ - H$^+$ cycles by successive treatment with N-sodium hydroxide, water, and N-hydrochloric acid then washed with water until neutral. (Ion exchange resins, BDH Handbook, 5th edn., 1971).

β-Cellobiose, m.p. 228-229° (decomp.), $[\alpha]_D^{25}$ +33.3° (c = 2.0 in water).
   Cryst. from 75% aq. ethanol.

Cellosolve, see 2-Ethoxyethanol.

Cellulose triacetate.
   Extracted with cold ethanol, dried in air, washed with hot distd. water, again dried in air, then dried at 50° for 30 min. [Madorsky, Hart and Straus, J. Res. Nat. Bur. Stand., 60, 343 (1958).]

Cetane, see n-Hexadecane.

Cetyl acetate, m.p. 18.3°.

   Vac. distd. twice, then cryst. several times from ethyl ether-methanol.

Cetyl alcohol, m.p. 49.3°.

   Cryst. from aq. ethanol, or from cyclohexane. Purified by zone refining. Purity checked by gas chromatography.

Cetylamide.
Cetylamine, m.p. 78°.

   Cryst. from thiophen-free benzene and dried under vac. over P_2O_5.

Cetylammonium chloride.

   Cryst. from methanol.

Cetyl bromide, m.p. 15°, b.p. 193-196°/14 mm.

   Shaken with H_2SO_4, washed with water, dried with K_2CO_3 and fractionally distd.

Cetyl ether, m.p. 54°.

   Vac. distd., then cryst. several times from methanol-benzene.

Cetylpyridinium chloride, m.p. 80-83°.

   Cryst. from methanol.

Cetyltrimethylammonium bromide, m.p. 227-235° (decomp.).

   Cryst. from ethanol-benzene or from wet acetone after extracting twice with pet. ether.

   Shaken with anhydrous ethyl ether, filtered and dissolved in a little hot methanol. After cooling in the refrigerator, the ppte. was filtered at room temp. and redissolved in methanol. Anhydrous ether was added and, after warming to obtain a clear soln., it was cooled and crystalline material was filtered. [Duynstee and Grunwald, JACS, 81, 4540 (1959).]

Cetyltrimethylammonium chloride.

   Cryst. from acetone-ether mixture or from methanol.

Chalcone, see Benzalacetophenone.

Charcoal.

Charcoal (50 g) was added to 1 1. of 6 M HCl and boiled for 45 min.
The supernatant was discarded, and the charcoal was boiled with two
more lots of HCl, then with distilled water until the supernatant no
longer gave a test for chloride ion. The charcoal (which was now
phosphate-free) was filtered on a sintered-glass funnel and air dried
at 120° for 24 hr. [Lippin, Talbert and Cohn, JACS, 76, 2871 (1954).]
The purification can be carried out using a Soxhlet extractor
(without cartridge), allowing longer extraction times. Treatment with
conc. H_2SO_4 instead of HCl has been used to remove reducing substances.

Chaulmoogric acid, m.p. 68.5°, b.p. 247-248°/20 mm.
    Cryst. from pet. ether or ethanol.

Chelerythrine, m.p. 207°.
    Cryst. from $CHCl_3$ by addition of methanol.

Chelex 100.
    Washed successively with 2 M ammonia, water, 2 M nitric acid and
water.

Chelidamic acid, see 4-Hydroxypyridine-2,6-dicarboxylic acid.

Chelidonic acid, m.p. 262°.
    Cryst. from aq. ethanol.

Chenodesoxycholic acid, m.p. 143°.
    Cryst. from ethyl acetate.

Chimyl alcohol, m.p. 64°.
    Cryst. from hexane.

Chloral, b.p. 98°.
    Distd., then dried by distilling through a heated column of $CaSO_4$.

Chloralacetonechloroform, m.p. 65°.
    Cryst. from benzene.

α-Chloralose, m.p. 187°.
    Cryst. from ethanol or ethyl ether.

Chlorambucil, m.p. 64-66°.
    Cryst. from pet. ether.

Chloramphenicol, m.p. 150.5-151.5°.
    Cryst. from water or ethylene dichloride. Sublimed under high vac.

Chloramphenicol palmitate, m.p. 90°, $[\alpha]_D^{26}$ +24.6 (c = 5, in ethanol).
    Cryst. from benzene.

Chloranil, m.p. 294.2-294.6° (sealed tube).
    Cryst. from acetone, benzene, ethanol or toluene, drying under vac.
over P_2O_5, or from acetic acid, drying over NaOH in a vac. desiccator. Can be sublimed under vac.

Chloranilic acid, m.p. 283-284°.
    A soln. of 8 g in 1 l. of boiling water was filtered while hot,
then extracted twice at about 50° with 200 ml portions of benzene.
The aq. phase was cooled in ice-water. The crystals were filtered
off, washed with three 10 ml portions of water, and dried at 115°.
[Thamer and Voight, JPC, 56, 225 (1952).]

Chlorazol sky blue FF.
    Freed from other electrolytes by adding aq. 40% sodium acetate to
a boiling soln. of the dye in distd. water. After standing, the
salted-out dye was filtered on a Büchner funnel, the process being
repeated several times. Finally, the pptd. dye was boiled several
times with absolute ethanol to wash out any sodium acetate, then
dried (as the sodium salt) at 105°. [McGregor, Peters and
Petropolous, TFS, 58, 1045 (1962).]

α-Chloroacetamide, m.p. 121°, b.p. 224-225°/ 743 mm.
    Cryst. from acetone and dried under vac. over P_2O_5.

p-Chloroacetanilide, m.p. 179°.
    Cryst. from ethanol or aq. ethanol.

Chloroacetic acid, m.p. $62.8°$, b.p. $189°$.

Cryst. from $CHCl_3$, CCl_4, benzene or water. Dried over P_2O_5 or conc. H_2SO_4 in a vac. desiccator. Further purification by distn. from $MgSO_4$, and by fractional crystn. from the melt. Stored under vac. or under dry nitrogen.

Chloroacetic anhydride, m.p. $46°$, d_4^{20} 1.5494.

Cryst. from benzene.

Chloroacetone, b.p. $119°$.

Dissolved in water and shaken repeatedly with small amounts of ethyl ether which extracts, preferentially, 1,1-dichloroacetone present as an impurity. The chloroacetone was then extracted from the aq. phase using a larger amount of ethyl ether, and distd. at low pressure. It was dried with $CaCl_2$ and stored at Dry-Ice temperature. Alternatively, it was stood with $CaSO_4$, distd. and stored over $CaCO_3$.

Chloroacetonitrile, b.p. $125°$.

Refluxed with P_2O_5 for one day, then distd. through a helices-packed column.

o-Chloroaniline, m.p. $-1.9°$, b.p. $208.8°$, n_D^{20} 1.58807, n_D^{25} 1.58586.

Freed from small amounts of the p-isomer by dissolving in one equivalent of H_2SO_4 and steam distilling. The p-isomer remains behind as the sulphate. [Sidgwick and Rubie, JCS, 1013 (1921).] An alternative method is to dissolve in warm 10% HCl (11 ml/g of amine), and, on cooling, the hydrochloride of o-chloroaniline separates out. The latter can be recryst. until the acetyl derivative has a constant m.p. (In this way, yields are better than for the recrystn. of the picrate from ethanol or of the acetyl derivative from pet. ether.) [King and Orton, JCS, 1377 (1911).]

p-Chloroaniline, m.p. $70-71°$.

Cryst. from methanol, pet. ether (b.p. $30-60°$), or 50% aq. ethanol, then benzene-pet.ether (b.p. $60-70°$), then dried in a vac. desiccator. Can be distd. under vac. (b.p. $75-77°/33$ mm).

p-Chloroanisole, b.p. 79°/11.5 mm, 196.6°/760 mm, $n_D^{25.5}$ 1.5326.

Washed with 10% (vol.) aq. H_2SO_4 (three times), 10% aq. KOH (three times) and then with water until neutral. Dried with $MgSO_4$ and fractionally distd. from CaH_2 through a glass helices-packed column under reduced pressure.

o-Chlorobenzaldehyde, m.p. 11°, b.p. 213-214°.

Washed wtth 10% Na_2CO_3 soln., then fractionally distd. in the presence of a small amount of catechol.

m-Chlorobenzaldehyde, m.p. 18°, b.p. 213-214°.

Purified by low temp. crystn. from pet. ether (b.p. 40-60°).

p-Chlorobenzaldehyde, m.p. 47°.

Cryst. from ethanol-water (3:1), then sublimed twice at 2 mm pressure at a temp. slightly above the m.p.

Chlorobenzene, b.p. 131.7°, n_D^{15} 1.52748, n_D^{20} 1.52480.

The main impurities are likely to be chlorinated impurities orig-inally present in the benzene used in the synthesis of chlorobenzene, and also some of the unchlorinated hydrocarbons. A common purification procedure is to wash several times with conc. H_2SO_4 then with aq. $NaHCO_3$ or Na_2CO_3, and water, followed by drying with $CaCl_2$, K_2CO_3 or $CaSO_4$, then with P_2O_5, and distn. It can also be dried with Linde 4A molecular sieve. Passage through, and storage over, activated alumina has been used to obtain low conductance material. [Flaherty and Stern, JACS, 80, 1034 (1958).]

p-Chlorobenzenesulphonyl chloride, m.p. 53°, b.p. 141°/15 mm.

Cryst. from ether in powdered Dry Ice, after the soln. had been washed with 10% NaOH until colourless and dried with Na_2SO_4.

p-Chlorobenzhydrazide, m.p. 164°.

Cryst. from water.

o-Chlorobenzoic acid, m.p. 139-140°.

Cryst. successively from glacial acetic acid, aq. ethanol, and pet. ether (b.p. 60-80°). Other solvents include hot water or toluene (about 4 ml/g). Crude material can be given an initial purification by dissolving 30 g in 100 ml of hot water containing 10 g of Na_2CO_3, boiling with 5 g of charcoal for 15 min, then filtering and adding 31 ml of 1:1 aq. HCl: the ppte. is washed with a little water and dried at 100°.

m-Chlorobenzoic acid, m.p. 158°.

Cryst. successively from glacial acetic acid, aq. ethanol and pet. ether (b.p. 60-80°).

p-Chlorobenzoic acid, m.p. 238-239°.

Same as for m-Chlorobenzoic acid. Has also been cryst. from hot water, and from ethanol.

o-Chlorobenzonitrile, m.p. 45-46°.

Cryst. to constant m.p. from benzene-pet.ether (b.p. 40-60°).

o-Chlorobenzotrifluoride, b.p. 152.3°.
m-Chlorobenzotrifluoride, b.p. 137.6°.
p-Chlorobenzotrifluoride, b.p. 138.6°.

Dried with $CaSO_4$, and distd. at high reflux ratio through a silvered vacuum-jacketed glass column packed with one-eighth inch glass helices. [Potter and Saylor, JACS, 73, 90 (1951).]

p-Chlorobenzyl chloride, m.p. 28-29°, b.p. 96°/15 mm.

Dried with $CaSO_4$, then fractionally distd. under reduced pressure. Cryst. from heptane or dry ethyl ether.

p-Chlorobenzylpseudothiuronium chloride, m.p. 197°.

Cryst. from conc. HCl by addition of water.

1-Chlorobutane, see n-Butyl chloride.

2-Chlorobutane, b.p. 68.5°, n_D^{25} 1.3945.

Purified in the same way as n-Butyl chloride.

Chlorocresol see Chloromethylphenol.

Chlorocyclohexane, b.p. 142.5°, n_D^{25} 1.46265.
  Washed several times with dil. NaHCO$_3$, then repeatedly with distd.
water. Dried with CaCl$_2$ and fractionally distd.

2-Chloro-1,4-dihydroxybenzene, see Chloroquinol.

4-Chloro-3,5-dimethylphenol, m.p. 115.5°.
  Cryst. from benzene.

1-Chloro-2,4-dinitrobenzene, m.p. 51°.
  Usually cryst. from ethanol or methanol. Has also been cryst. from
ethyl ether, benzene or isopropyl alcohol. A preliminary purification
step has been to pass its soln. in benzene through an alumina column.
Also purified by zone melting.

4-Chloro-3,5-dinitrobenzoic acid, m.p. 159-161°.
  Cryst. from ethanol-water, ethanol or benzene.

Chloroethane, see Ethyl chloride.

2-Chloroethanol, b.p. 51.0°/31 mm, 128.6°/760 mm, n_D^{15} 1.44380.
  Dried with, then distd. from, CaSO$_4$ in the presence of a little
Na$_2$CO$_3$ to remove traces of acid.

β-Chloroethyl bromide, b.p. 106-108°.
  Washed with conc. H$_2$SO$_4$, water, 10% Na$_2$CO$_3$ soln., and again with
water, then dried with CaCl$_2$ and fractionally distd. just before use.

2-Chloroethyl vinyl ether.
  Washed repeatedly with equal volumes of water made slightly alkaline with KOH, dried with sodium, and distd. under vac.

Chloroform, b.p. 61.2°, d^{15} 1.49845, d^{30} 1.47060, n_D^{15} 1.44858.
  Reacts slowly with oxygen or oxidizing agents, when exposed to air
and light, giving, mainly, COCl$_2$, Cl$_2$ and HCl. Commercial CHCl$_3$ is
usually stabilized by addition of up to 1% ethanol or of dimethyl-

Chloroform (continued)

aminoazobenzene. Simplest purifications involve washing with water to remove the ethanol, drying with K_2CO_3 or $CaCl_2$, refluxing with P_2O_5, $CaCl_2$, $CaSO_4$ or Na_2SO_4, and distilling. It must not be dried with sodium. The distd. $CHCl_3$ should be stored in the dark to avoid photo-chemical formation of phosgene. As an alternative purification, $CHCl_3$ can be shaken with several small portions of conc. H_2SO_4, washed thoroughly with water, and dried with $CaCl_2$ or K_2CO_3 before filtering and distilling. Ethanol can be removed from $CHCl_3$ by passage through a column of activated alumina, or through a column of silica gel 4 ft long by $1^3/4$ in. diameter at a flow rate of 3 ml/min. (The column, which can hold about 8% of its weight of ethanol, is regenerated by air drying and then heating at $600°$ for 6 hr. It is pre-purified by washing with $CHCl_3$, then ethanol, leaving in conc. H_2SO_4 for about 8 hr, washing with water until the washings are neutral, then air drying, followed by activation at $600°$ for 6 hr. Just before use it is reheated for 2 hr to $154°$.) [McLaughlin, Kaniecki and Gray, AC, 30, 1517 (1958).]

Carbonyl-containing impurities can be removed from $CHCl_3$ by percolation through a Celite column impregnated with 2,4-dinitro-phenylhydrazine, phosphoric acid and water. (Prepared by dissolving 0.5 g DNPH in 6 ml of 85% H_3PO_4 by grinding together, then mixing with 4 ml of distd. water and 10 g of Celite.) [Schwartz and Parks, AC, 33, 1396 (1961).] Chloroform can be dried by distn. from powdered type 4A Linde molecular sieve. For use as a solvent in i.r. spectroscopy, chloroform is washed with water (to remove ethanol), then dried for several hours over anhydrous $CaCl_2$ and fractionally distd. This treatment removes material absorbing near 1600 cm^{-1}. (Percolation through activated alumina increases this impurity.) [Goodspeed and Millson, Chem. Ind., 1594 (1967).]

Chlorogenic acid, m.p. $208°$.
    Cryst. from water. Dried at $110°$.

p-Chloroiodobenzene, m.p. $53-54°$.
    Cryst. from ethanol.

5-Chloro-7-iodo-8-hydroxyquinoline, m.p. $178-179°$.
    Cryst. from absolute ethanol.

4-Chloro-2-methylphenol, m.p. 49°.
    Purified by zone melting.

4-Chloro-3-methylphenol, m.p. 66°.
    Cryst. from pet. ether.

1-Chloro-2-methylpropane, see Isobutyl chloride.

2-Chloro-2-methylpropane, see tert.-Butyl chloride.

Chloromycetin, m.p. 150.5-151.5°.
    Cryst. from water (sol. 2.5 mg/ml at 25°) or ethylene dichloride.

Chloromycetin palmitate, m.p. 90°.
    Cryst. from benzene.

1-Chloronaphthalene, f.p. -2.3°, b.p. 136.0-136.5°/20 mm,
259.3°/760 mm, n_D^{20} 1.6326.
    Washed with dil. $NaHCO_3$, then dried with Na_2SO_4 and fractionally
distd. under reduced pressure. Alternatively, before distn., passed
through a column of activated alumina, or dried with $CaCl_2$, then
distd. from sodium. It can be further purified by fractional crystn.
by partial freezing or by crystn. of its picrate to constant m.p.
(132-133°) from ethanol.

2-Chloronaphthalene, m.p. 61°, b.p. 264-266°.
    Cryst. from 25% alcohol-water and dried under vac.

1-Chloro-2-naphthol.
2-Chloro-1-naphthol.
4-Chloro-1-naphthol.
    Cryst. from ethanol.

4-Chloro-2-nitroaniline, m.p. 116.0-116.5°.
    Cryst. from hot water or ethanol-water and dried for 10 hr at 60°
under vac.

2-Chloro-4-nitrobenzamide, m.p. 172°.
    Cryst. from ethanol.

o-Chloronitrobenzene, m.p. 32.8-33.2°.
  Cryst. from ethanol, methanol or pentane (charcoal).

m-Chloronitrobenzene, m.p. 45.3-45.8°.
  Cryst. from methanol or 95% ethanol (charcoal), then pentane.

p-Chloronitrobenzene, m.p. 83.5-84°.
  Cryst. from 95% ethanol (charcoal).

1-Chloronitroethane, b.p. 37-38°/20 mm, n_D^{20} 1.4224, n_D^{25} 1.4235.
  Dissolved in alkali, extracted with ether (discarded), then the
aq. phase acidified with hydroxylamine hydrochloride, and the nitro
compound fractionally distd. under reduced pressure. [Pearson and
Dillon, JACS, 75, 2439 (1953).]

2-Chloro-5-nitropyridine, m.p. 108°.
  Cryst. from benzene or benzene-pet. ether.

α-Chloro-m-nitrotoluene, see m-Nitrobenzyl chloride.

1-Chloropentane, see n-Amyl chloride.

o-Chlorophenol, m.p. 8.8°, b.p. 176°.
  Passed at least twice through a gas chromatograph column.

m-Chlorophenol, m.p. 33°, b.p. 214°.
  Could not be obtained solid by crystn. from pet. ether. Purified
by distn. under reduced pressure.

p-Chlorophenol, m.p. 43.0°.
  Distd., then cryst. from pet. ether (b.p. 40-60°), and dried under
vac. over P_2O_5 at room temp.

Chlorophenol red.
  Cryst. from glacial acetic acid.

p-Chlorophenoxyacetic acid, m.p. 157°.
α-p-Chlorophenoxypropionic acid, m.p. 116°.
β-p-Chlorophenoxypropionic acid, m.p. 138°.
  Cryst. from ethanol.

m-Chlorophenylacetic acid, m.p. 74°.
p-Chlorophenylacetic acid, m.p. 106°.
  Cryst. from ethanol-water.

o-Chlorophenyl diphenyl phosphate, n_D^{25} 1.5707, b.p. 236°/4 mm.
  Purified by vac. distn., percolated through a column of alumina,
then passed through a packed column maintained at 150° to remove
residual traces of volatile materials by a countercurrent stream of
nitrogen at reduced pressure. [Dobry and Keller, JPC, 61, 1448
(1957).]

p-Chlorophenyl isocyanate, m.p. 29-30°.
  Cryst. from pet. ether (b.p. 30-40°).

p-Chlorophenyl o-nitrobenzyl ether, m.p. 69°.
p-Chlorophenyl p-nitrobenzyl ether, m.p. 102°.
  Cryst. from ethanol.

Chloropicrin, b.p. 112°.
  Dried with $MgSO_4$ and fractionally distd.

1-Chloropropane, see n-Propyl chloride.

2-Chloropropane, see Isopropyl chloride.

Chloro-2-propanone, see Chloroacetone.

3-Chloropropene, see Allyl chloride.

α-Chloropropionic acid, b.p. 98°/3 mm.
  Dried with P_2O_5 and fractionally distd. under vac.
β-Chloropropionic acid, m.p. 41°.
  Cryst. from pet. ether or benzene.

γ-Chloropropyl bromide, b.p. 142-145°, n_D^{25} 1.4732.

Washed with conc. H_2SO_4, water, 10% Na_2CO_3 soln., again with water, then dried with $CaCl_2$ and fractionally distd. just before use. [Akagi, Oae and Murakami, JACS, 78, 4034 (1956).]

6-Chloropurine, m.p. 179° (decomp.).

Cryst. from water.

2-Chloropyridine, b.p. 49.0°/7 mm, n_D^{20} 1.5322.

Dried with NaOH for several days, then distd. from CaO under reduced pressure.

Chloroquinol, m.p. 106°.

Cryst. from chloroform or toluene.

2-Chloroquinoline, m.p. 34°, b.p. 147-148°/15 mm, d_4^{35} 1.2351, n_D^{35} 1.62923.

Purified by crystn. of its picrate to constant m.p. (123-124°) from benzene, regenerating the base and distilling under vac. [Cumper, Redford and Vogel, JCS, 1183 (1962).] 2-Chloroquinoline can be cryst. from ethanol.

8-Chloroquinoline, b.p. 171-171.5°/24 mm, d_4^{20} 1.2780, n_D^{20} 1.64403.

Purified by crystn. of its $ZnCl_2$ complex (m.p. 228°) from aq. ethanol.

4-Chlororesorcinol, m.p. 105°.

Cryst. from boiling CCl_4 (10 g/l.), after charcoal, and air dried.

5-Chlorosalicyaldehyde, m.p. 98.5-99°.

Steam distd., then cryst. from aq. ethanol.

N-Chlorosuccinimide, m.p. 149-150°.

Rapidly cryst. from benzene.

8-Chlorotheophylline, m.p. 311° (decomp.).

Cryst. from water.

α-Chlorotoluene, see Benzyl chloride.

o-Chlorotoluene, b.p. 159°, n_D^{20} 1.5255.

Dried for several days with $CaCl_2$, then distd. from sodium using a glass helices-packed column.

p-Chlorotoluene, f.p. 7.2°, b.p. 162.4°, n^{20} 1.5208.

Dried with BaO, fractionally distd., then fractionally cryst. by partial freezing.

2-Chlorotriethylamine hydrochloride.

Cryst. from absolute methanol (to remove highly coloured impurities).

Chlorotrifluoroethylene, b.p. -26 to -24°.

Scrubbed with 10% KOH soln., then 10% H_2SO_4 soln. to remove inhibitors, and dried. Passed through silica gel.

Chlorotrifluoromethane, m.p. -180°, b.p. -81.5°.

Main impurities were CO_2, O_2, and N_2. The CO_2 was removed by passage through satd. aq. KOH, followed by conc. H_2SO_4. The O_2 was removed using a tower packed with activated copper on kieselguhr at 200°, and the gas was dried over P_2O_5.

Chlorotriphenylmethane, m.p. 112-113°.

Cryst. from benzene soln. (100 ml) containing a little acetyl chloride, by addition of 200 ml of pet. ether and cooling. Alternatively, a soln. in ethyl ether was satd. with dry HCl (by dripping conc. HCl into conc. H_2SO_4 and passing the gas through P_2O_5 towers) at 0°, then cooled in a Dry Ice-acetone bath. The crystals so obtained were recryst. from pet. ether (b.p. 30-60°) using Dry Ice-acetone baths. [Thomas and Rochow, JACS, 79, 1843 (1957).]

4-Chloro-3,5-xylenol, see 4-Chloro-3,5-dimethylphenol.

Cholamine chloride hydrochloride, see (2-Aminoethyl)trimethyl-ammonium chloride hydrochloride.

Cholanic acid, m.p. 163–164°.
    Cryst. from ethanol.

Cholanthrene, m.p. 173°.
    Cryst. from benzene-ethyl ether.

Cholestane, m.p. 80°.
    Cryst. from ethyl ether-ethanol.

Cholestanol, m.p. 141.5–142° (monohydrate).
    Cryst. from slightly aq. ethanol.

Cholesterol, m.p. 148.9–149.4°, $[\alpha]_D^{25}$ -35° (hexane).
    Cryst. from ethyl acetate, ethanol or isopropyl ether-methanol.
For extensive details of purification through the dibromide, see
Fieser [JACS, 75, 5421 (1953)] and Schwenk and Werthessen [Arch.
Biochem. Biophys., 40, 334 (1952)], and by repeated crystn. from
acetic acid, see Fieser [JACS, 75, 4395 (1953).]

Cholesteryl acetate, m.p. 112–115°.
    Cryst. from n-pentanol.

Cholesteryl myristate.
    Cryst. from n-pentanol. Purified by column chromatography with
methanol and evapd. to dryness. Dissolved in water and pptd with HCl
(spot 1) or passed through a cation-exchange column (spot 2).
Finally, dried in vac. over P_2O_5. [Malanik and Malat, Anal. Chim.
Acta, 76, 464 (1975).]

Cholesteryl oleate, m.p. 48.8–49.4°.
    Purified by chromatography on silica gel.

Cholic acid, m.p. 198–200°.
    Cryst. from ethanol. Dried under vac. at 94°.

Choline chloride.
    Extremely deliquescent. Purity checked by $AgNO_3$ titration or by
titration of free base after passage through an anion-exchange column.
Cryst. from absolute ethanol, or ethanol-ethyl ether, dried under vac.

Choline chloride (continued)
and stored in a vac. desiccator over P_2O_5 or $Mg(ClO_4)_2$.

Chromotropic acid.
   Cryst. from water by addition of ethanol.

Chrysene, m.p. 255-256°.
   Purified by chromatography on alumina from pet. ether in a darkened
room. Its soln. in benzene was passed through a column of decolorizing charcoal, then cryst. by concentration of the eluate. Also
purified by crystn. from benzene-pet.ether, and by zone melting.
   Dissolved in N,N-dimethylformamide and successive small portions
of alkali and iodomethane were added until the fluorescent colour of
the 5H-benzo(b)carbazole anion no longer appeared when alkali was
added. The chrysene (and the alkylated 5H-benzo(b)carbazole) separated on addition of water. Final purification was by crystn. from
ethylcyclohexane and from 2-methoxyethanol. [Bender, Sawicki and
Wilson, AC, 36, 1011 (1964).]

γ-Chymotrypsin.
   Cryst. twice from four-tenths satd. ammonium sulphate soln., then
dissolved in 0.001 M HCl and dialyzed against 0.001 M HCl. The soln.
was stored at 2°. [Lang, Frieden and Grunwald, JACS, 80, 4923 (1958).]

Cinchonidine, m.p. 210.5°, $[\alpha]_{546}^{20}$ -127.5° (c = 5 in EtOH).
   Cryst. from aq. ethanol.

Cinchonine, m.p. 265°.
   Cryst. from ethanol or ethyl ether.

Cincophen, see 2-Phenylquinoline-4-carboxylic acid.

1,8-Cineole, f.p. 1.3°, b.p. 176.0°, d_4^{20} 0.9251.
   Purified by dilution with an equal vol. of pet.ether, then satn.
with dry HBr. The ppte. was filtered off, washed with small portions
of pet.ether, and the cineole was regenerated by stirring the crystals in water. It can also be purified through its o-cresol or
resorcinol addition compounds. Stored with sodium until required.

trans-Cinnamic acid, m.p. 134.5-135°.
   Cryst. from benzene, CCl$_4$, hot water, water-ethanol (3:1), or 20%
ethanol-water. Dried at 60° under vac.

Cinnamyl alcohol, λ_{max} 2510 Å (ε = 18,180), m.p. 33°,
b.p. 143.5°/14 mm.
   Cryst. from ethyl ether-pentane.

Cinnamic anhydride, m.p. 136°.
   Cryst. from ethanol.

Cinnoline, m.p. 38°.
   Cryst. from pet. ether. Kept under nitrogen in sealed tubes.

Citraconic acid, m.p. 91°.
   Steam distilled and cryst. from ethanol-ligroin.

Citranaxanthin, m.p. 155-156°, $E_{1cm}^{1\%}$ = 410 (349 nm), 275 (466 nm),
in hexane.
   Purified by chromatography on a column of 1:1 magnesia-HyfloSupercel. Cryst. from pet. ether. Stored in the dark, under inert
atmosphere, at 0°.

Citric acid (monohydrate).
   Cryst. from hot water.

S-Citrulline, m.p. 222°, $[\alpha]_D^{25}$ +24.2° (in 5 M HCl).
   Likely impurities: arginine, ornithine. Cryst. from water by adding
5 volumes of ethanol. Also cryst. from water by addition of methanol.

Cocaine, m.p. 98°.
   Cryst. from ethanol.

Cocarboxylase.
   Cryst. from ethanol slightly acidified with HCl.

Codeine, m.p. 154-156°.
   Cryst. from water or aq. ethanol. Dried at 80°.

Colchicine, m.p. 155-157° (decomp.).

Commercial material contains up to 4% desmethylcolchicine. Purified by chromatography on alumina, eluting with $CHCl_3$. [Ashley and Harris, JCS, 677 (1944).] Alternatively acetone soln. on alkali-free alumina have been used, eluting with acetone. [Nicholls and Tarbell, JACS, 75, 1104 (1953).] Cryst. from ethyl acetate.

Colchicoside, m.p. 216-218°.
    Cryst. from ethanol.

2,4,6-Collidine, see 2,4,6-Trimethylpyridine.

Conessine, m.p. 125°.
    Cryst. from acetone.

Congo red.
    Cryst. from aq. ethanol (1:3). Dried in air.

Convallatoxin, m.p. 238-239°.
    Cryst. from ethyl acetate.

Coprostane, m.p. 72°.
    Cryst. from ethanol.

4,5-Coprosten-3-ol, m.p. 132°.
    Cryst. from ethyl ether-methanol.

Coprosterol, m.p. 101°.
    Cryst. from methanol.

Coronene, m.p. 438-440°.
    Cryst. from benzene.

Corticosterone, m.p. 180-182°.
    Cryst. from acetone.

Cortisone, m.p. 230-231°, $[\alpha]_{546}^{20}$ +255° (c = 1 in EtOH).
    Cryst. from 95% ethanol or acetone.

Cortisone-21-acetate, m.p. 242-243°, $[\alpha]_{546}^{20}$ +277° (c = 1 in $CHCl_3$).
  Cryst. from acetone.

Coumalic acid, m.p. 205-210° (decomp.).
  Cryst. from methanol.

Coumarilic acid, m.p. 192-193°.
  Cryst. from water.

Coumarin, m.p. 68-69°.
  Cryst. from aq. ethanol.

Coumarin-3-carboxylic acid, m.p. 188° (decomp.).
  Cryst. from water.

Creatine, m.p. 303°.
  Likely impurities: creatinine and other guanidino compounds.
Cryst. from water as monohydrates. Dried under vac. over P_2O_5 to
give anhydrous material.

Creatinine, m.p. 260° (decomp.).
  Likely impurities: creatine, ammonium chloride. Dissolved in dil.
HCl, then neutralized by adding ammonia. Recryst. from water by
adding acetone in excess.

o-Cresol, m.p. 30.9°, b.p. 191.0°, n_D^{41} 1.53610, n_D^{46} 1.53362.
  Can be freed from m- and p-isomers by repeated fractional distn.
Cryst. from benzene by addition of pet. ether. Fractionally cryst.
by partial freezing of its melt.

m-Cresol, f.p. 12.0°, b.p. 202.7°, n_D^{20} 1.5438.
  Separation of m and p-cresols requires chemical methods, such as
conversion to their sulphonates [Brückner, JAC.,75, 289 (1928)].
An equal volume of H_2SO_4 is added to m-cresol, stirring with a glass
rod until soln is complete. Heat for 3 hr at 103-105°. Dilute carefully with 1-1½ volumes of water, heat to b.p. and steam distil until
all unsulphonated cresol has been removed. Cool and extract residue
with ether. Evaporate the soln until the boiling point reaches 134°
and steam distil off the m-cresol.

m-Cresol (continued)

Another purification of m-cresol involves distn., fractional cryst. from its melt, then redistn. Freed from p-cresol by soln in glacial acetic acid and bromination by about half of an equivalent amount of bromine in glacial acetic acid. The acetic acid was distd. off, then fractional distn. under vac. gave bromocresols from which 4-bromo-m-cresol was obtained by crystn. from hexane. Addition of the bromocresol in glacial acetic acid slowly to a reaction mixture of HI and red phosphorus or (more smoothly) of HI and hypophosphorous acid, in glacial acetic acid, at reflux, removed the bromine. After an hour, the soln. was distd. at atmospheric pressure until layers formed. Then it was cooled and diluted with water. The cresol was extracted into ethyl ether, washed with water, $NaHCO_3$ soln. and again with water, dried with a little $CaCl_2$ and distd. [Baltzly, Ide and Phillips, JACS, 77, 2522 (1955).]

p-Cresol, m.p. 34.8°, b.p. 201.9°, n_D^{41} 1.53115, n_D^{46} 1.52870.

Can be separated from m-cresol by fractional crystn. of its melt. Purified by distn., by pptn. from benzene soln. with pet.ether, and via its benzoate, as for Phenol. Dried under vac. over P_2O_5. Has also been cryst. from pet. ether (b.p. 40-60°) and by conversion to sodium p-cresoxyacetate which, after crystn. from water was decomposed by heating with HCl in an autoclave. [Savard, Ann.Chim. (Paris), 11, 287 (1929).]

o-Cresolphthalein, see Cresol red.

o-Cresol red.

Cryst. from glacial acetic acid. Air dried. Dissolved in aq. 5% $NaHCO_3$ soln. and pptd from hot soln. by dropwise addition of aq.HCl. Repeated until extinction coefficients did not increase.

o-Cresotic acid, m.p. 163-164°.
m-Cresotic acid, m.p. 177°.
p-Cresotic acid, m.p. 151°.

Cryst. from water.

Crocetin diethyl ester, m.p. 218-219°, $E_{1cm}^{1\%}$ = 2340 (400 nm),
3820 (422 nm), 3850 (450 nm), in pet. ether.
    Purified by chromatography on a column of silica gel G. Cryst. from
benzene. Stored in the dark, under inert atmosphere, at 0°.

Crotonaldehyde, b.p. 104-105°.
    Fractionally distd. under nitrogen, through a short Vigreux
column. Stored in sealed ampoules.

trans-Crotonic acid, m.p. 72-72.5°.
    Distd. under reduced pressure. Cryst. from pet. ether (b.p. 60-80°)
or water, or by partial freezing of the melt.

Crotyl bromide, b.p. 103-105°/740 mm, n_D^{25} 1.4792.
    Dried with $MgSO_4$, $CaCO_3$ mixture. Fractionally distd. through an
all-glass Todd column.

18-Crown-6, m.p. 37-39°.
    Recryst. from acetonitrile and vac. dried.

Cryptopine, m.p. 220-221°.
    Cryst. from benzene.

Cryptoxanthin, $E_{1cm}^{1\%}$ = 2370 (452 nm), 2080 (480 nm), in pet. ether.
    Purified by chromatography on MgO, $CaCO_3$ or deactivated alumina,
using ethanol or ethyl ether to develop the column. Cryst. from
$CHCl_3$-ethanol. Stored in the dark, under inert atmosphere, at -20°.

Crystal violet.
    Cryst. from water (20 ml/g), the crystals being separated from the
chilled soln. by centrifuging, then washed with chilled ethanol and
ethyl ether and dried under vac.

Crystal violet chloride.
    The carbinol was pptd. from an aq. soln. of the dye, using excess
NaOH, then dissolved in HCl and recryst. from water as the chloride.
[Turgeon and La Mer, JACS, 74, 5988 (1952).]

Cumene, b.p. 69-70°/41 mm, 152.4°/760 mm, n_D^{20} 1.49146, n_D^{25} 1.48892.

Usual purification is by washing with several small portions of conc. H_2SO_4 (until the acid layer is no longer coloured), then with water, 10% aq. Na_2CO_3, again with water, and drying with $MgSO_4$, $MgCO_3$ or Na_2SO_4, followed by fractional distn. It can then be dried with, and distd. from, sodium, sodium hydride or CaH_2. Passage through columns of activated alumina or silica gel removes oxidation products. Has also been steam distd. from 3% NaOH, and azeotropically distd. with 2-ethoxyethanol (which was subsequently removed by washing out with water).

Cumene hydroperoxide, b.p. 60°/0.2 mm, n_D^{24} 1.5232.

Purified by adding 100 ml of 70% material slowly and with agitation to 300 ml of 25% NaOH in water, keeping the temperature below 30°. The resulting crystals of the sodium salt were filtered off, washed twice with 25 ml portions of benzene, then stirred with 100 ml of benzene for 20 min. After filtering off the crystals and repeating the washing, they were suspended in 100 ml of distd. water and the pH was taken to 7.5 by addition of 4 M HCl. The free hydroperoxide was extracted into two 20 ml portions of n-hexane, and the solvent was evapd. under vac. at room temp., last traces being removed at 40-50° and 1 mm. [Fordham and Williams, Canad. J. Res., 27 B, 943 (1949).] Petroleum ether, but not ethyl ether, can be used instead of benzene, and powdered solid CO_2 can replace the 4 M HCl.

Cupferron, m.p. 162.5-163.5°.

Cryst. from ethanol (charcoal), washed with ethyl ether and air dried.

Cupreine, m.p. 202°, $[\alpha]_D^7$ -176° (in methanol).

Cryst. from ethanol.

Cuproine, see 2,2'-Biquinolyl.

Curcumin, m.p. 183°.

Cryst. from ethanol or acetic acid.

Cyanamide, m.p. 41°.

Cryst. from ethyl ether, then vac.distd. at 80°. Hygroscopic.

2-Cyanoacetamide, m.p. 119.4°.
  Cryst. from methanol-dioxane (6:4), then water. Dried over P_2O_5 under vac.

Cyanoacetic acid, m.p. 70.9-71.1°.
  Cryst. to constant m.p. from benzene-acetone (2:3), and dried over silica gel.

Cyanoacetic acid hydrazide, m.p. 114.5-115°.
  Cryst. from ethanol.

p-Cyanoaniline, m.p. 85-87°.
p-Cyanobenzoic acid, m.p. 219°.
  Cryst. from water.

Cyanoguanidine, m.p. 209.5°.
  Cryst. from water or ethanol.

p-Cyanophenol, m.p. 113°.
  Cryst. from pet. ether or benzene and kept under vac. over P_2O_5.

3-Cyanopyridine, m.p. 50°.
  Cryst. to constant m.p. from o-xylene-hexane.

4-Cyanopyridine, m.p. 76-79°.
  Cryst. from mixture of dichloromethane and ethyl ether.

Cyanuric acid.
  Cryst. from water. Dried at room temp. in desiccator over conc. H_2SO_4 or P_2O_5.

Cyanuric chloride, m.p. 154°.
  Cryst. from CCl_4 or pet. ether (b.p. 90-100°), and dried under vac.

Cyclobutane-1,1-dicarboxylic acid, m.p. 158°.
  Cryst. from ethyl acetate.

trans-Cyclobutane-1,2-dicarboxylic acid, m.p. 131°.
  Cryst. from benzene.

Cyclobutanone, b.p. 96-97°, n_D^{25} 1.4189.

Treated with dil. aq. $KMnO_4$, dried with molecular sieves and fractionally distd. Purified via the semicarbazone, then regenerated,
dried with $CaSO_4$, and distd. in a spinning-band column. Alternatively,
purified by preparative gas chromatography using a Carbowax 20-M
column at 80°. (This treatment removes acetone.)

Cyclodecanone, b.p. 100-102°/12 mm.

Purified by sublimation.

Cycloheptanone, b.p. 105°/80 mm, 172.5°/760 mm, n_D^{24} 1.4607.

Shaken with aq. $KMnO_4$ to remove material absorbing around 230-240
nm, then dried with Linde type 13X molecular sieve and fractionally
distd.

Cycloheptatriene, b.p. 114-115°.

Washed with alkali, then fractionally distd.

Cycloheptane, b.p. 114.4°, n_D^{20} 1.4588.

Distd. from sodium, under nitrogen.

1,3-Cyclohexadiene, b.p. 83-84°, n_D^{20} 1.4707.

Distd. from $NaBH_4$.

Cyclohexane, f.p. 6.6°, b.p. 80.7°, d^{24} 0.77410, n_D^{20} 1.42623, n_D^{25}
1.42354.

Commonly, washed with conc. H_2SO_4 until the washings are colourless,
followed by water, aq. Na_2CO_3 or 5% NaOH, and again water until
neutral. It is next dried with P_2O_5, Linde type 4A molecular sieve,
$CaCl_2$, or $MgSO_4$ then sodium, and distd. Cyclohexane has been refluxed with, and distd. from sodium, CaH_2, $LiAlH_4$ (which also removes
peroxides), sodium-potassium alloy, or P_2O_5. Traces of benzene can
be removed by passage through a column of silica gel that has been
freshly heated: this gives material suitable for ultraviolet and
infrared spectroscopy. If there is much benzene in the cyclohexane,
most of it can be removed by a preliminary treatment with nitrating
acid (a cold mixture of 30 ml conc. HNO_3 and 70 ml conc. H_2SO_4)
which converts benzene to nitrobenzene. The impure cyclohexane and

<u>Cyclohexane</u> (continued)
the nitrating acid are placed in an ice-bath and stirred vigorously
for 15 min, after which the mixture is allowed to warm to 25° during
1 hr. The cyclohexane is decanted, washed several times with 25%
NaOH, then water, dried with $CaCl_2$, and distd. from sodium.

Carbonyl-containing impurities can be removed as described for
Chloroform. Other purification procedures include passage through
columns of activated alumina and repeated crystn. by partial freezing. Small quantities may be purified by chromatography on a Dowex
710-Chromosorb W gas-liquid chromatograph column.

<u>Cyclohexane-1,2-diaminetetraacetic acid.</u>
Dissolved in aq. NaOH as its disodium salt, then pptd. by adding
HCl. The free acid was filtered off and boiled with distd. water to
remove traces of HCl. [Bond and Jones, <u>TFS</u>, <u>55</u>, 1310 (1959).]
Recryst. from water and dried under vac.

<u>trans-1,2-Cyclohexanediol</u>, m.p. 104°.
Cryst. from acetone and dried at 50° for several days.

<u>cis-Cyclohexane-1,3-diol</u>, m.p. 86°.
Cryst. from ethyl acetate and acetone.

<u>trans-Cyclohexane-1,3-diol</u>, m.p. 117°.
Cryst. from ethyl acetate.

<u>cis-1,4-Cyclohexanediol</u>, m.p. 102.5°.
Cryst. from acetone (charcoal), then dried and sublimed under vac.

<u>Cyclohexane-1,3-dione</u>, m.p. $107-108^\circ$.
Cryst. from benzene.

<u>Cyclohexane-1,4-dione</u>, m.p. 78°.
Cryst. from water, then benzene.

<u>Cyclohexane-1,2-dione dioxime</u>, m.p. $189-190^\circ$.
Cryst. from alcohol-water and dried in vac. at 40°.

<u>Cyclohexanol</u>, m.p. 25.2°, b.p. 161.1°, d_4^{22} 0.9459,
n_D^{25} 1.4365, n_D^{30} 1.4629.

Refluxed with freshly ignited CaO, or dried with Na_2CO_3,
then fractionally distd. Redistd. from sodium. Further
purified by fractional crystn. from the melt in dry air.
Peroxides and aldehydes can be removed by prior washing with
ferrous sulphate and water, followed by distillation under
nitrogen from 2,4-dinitrophenylhydrazine, using a short
fractionating column: water distils as the azeotrope. Dry
cyclohexanol is very hygroscopic.

<u>Cyclohexanone</u>, f.p. -16.4°, b.p. 155.7°, n_D^{15} 1.45203,
n_D^{20} 1.45097.

Dried with $MgSO_4$, $CaSO_4$, Na_2SO_4 or Linde type 13X
molecular sieve, then distd. Cyclohexanol and other oxidizable
impurities can be removed by treatment with chromic acid or
dil. $KMnO_4$. More thorough purification is possible by
conversion to the bisulphite addition compound, or the
semicarbazone, followed by decomposition with Na_2CO_3 and
steam distn. (For example, equal weights of the bisulphite
adduct (cryst. from water) and Na_2CO_3 are dissolved in hot
water and, after steam distn., the distillate is saturated
with NaCl and extracted with benzene which is then dried and
the solvent evaporated prior to further distn.)

<u>Cyclohexanone oxime</u>, m.p. 90°.

Cryst. from water or pet. ether (b.p. 60-80°).

Cyclohexanone phenylhydrazone, m.p. 77°.

   Cryst. from aq. ethanol.

Cyclohexene, b.p. 83°, n_D^{20} 1.4464, n_D^{25} 1.4437.

   Freed from peroxides by washing with successive portions of dil.
acidified ferrous sulphate, or with $NaHSO_3$ soln. then with distd.
water, dried with $CaCl_2$ or $CaSO_4$, and distd. under nitrogen.
Alternative methods of removing peroxides include passage through
a column of alumina, refluxing with sodium wire or cupric stearate
(then distg. from sodium). Diene is removed by refluxing with
maleic anhydride before distg. under vac. Treatment by 0.1 moles
of CH_3MgI in 40 ml of ethyl ether removes traces of oxygenated
impurities. Other purification procedures include washing with aq.
NaOH, drying and distg. under nitrogen through a spinning-band
column; redistg. from CaH_2; storage with sodium wire; and passage
through an alumina column, under nitrogen, immediately before use.

Cycloheximide, m.p. 119.5-121°.

   Cryst. from water-methanol (4:1), amyl acetate, isopropyl
acetate-isopropyl ether or water.

Cyclohexylamine, b.p. 134.5°, d^{25} 0.8625, n_D^{20} 1.45926, n_D^{25} 1.4565.

   Dried with $CaCl_2$ or $LiAlH_4$, then distd. from BaO, KOH or sodium,
under nitrogen. Also purified by conversion to the hydrochloride,
with crystn. several times from water, then liberation of the amine
with alkali and fractional distn. under nitrogen.

Cyclohexylbenzene, f.p. 6.8°, b.p. 237-239°.

   Purified by fractional distn., and fractional freezing.

Cyclohexyl bromide, b.p. $72^{\circ}/29$ mm, n_D^{25} 1.4935.
 Shaken with 60% aq. HBr to remove the free alcohol. After separation from the excess HBr, the sample was dried and fractionally distd.

Cyclohexyl chloride, b.p. $142-142.5^{\circ}$.
 Dried with $CaCl_2$ and distd.

Cyclohexyl methacrylate.
 Purification as for Methyl methacrylate.

1-Cyclohexyl-5-methyltetrazole, m.p. $124-124.5^{\circ}$.
 Cryst. from absolute ethanol, then sublimed at $115^{\circ}/3$ mm.

Cyclononanone, m.p. $142.0-142.8^{\circ}$.
 Repeatedly sublimed at 0.05-0.1 mm pressure.

Cyclooctanone, m.p. 42°.
 Purified by sublimation after drying with Linde type 13X molecular sieve.

Cyclooctatetraene, b.p. $141-141.5^{\circ}/19$ mm, n_D^{25} 1.5350.
 Purified by shaking 3 ml with 20 ml of 10% aq. $AgNO_3$ for 15 min, then filtering off the silver nitrate complex as a ppte. The ppte. was dissolved in water and added to cold conc. ammonia to regenerate the cyclo-octatetraene which was fractionally distd. under vac.

Cyclooctene, b.p. $32-34^{\circ}/12$ mm.
 Cis isomer freed from trans isomer by fractional distn. through a spinning-band column, followed by preparative gas chromatography on a Dowex 710-Chromosorb W g.l.c. column.

Cyclopentadecanone, m.p. 63°.
 Sublimation is better than crystn. from aq. ethanol.

Cyclopentadiene, b.p. $41-42^{\circ}$.
 Dried with $Mg(ClO_4)_2$ and distd.

Cyclopentane, b.p. 49.3°, n^{20} 1.40645, n$_D^{25}$ 1.40363.

Freed from cyclopentene by two passages through a column of carefully dried and degassed activated silica gel.

Cyclopentanone, b.p. 130-130.5°, n$_D^{20}$ 1.4370, n$_D^{25}$ 1.4340.

Shaken with aq. KMnO$_4$ to remove materials absorbing around 230 to 240 nm. Dried with Linde type 13X molecular sieve and fractionally distd. Has also been purified by conversion to the NaHSO$_3$ adduct which, after crystallizing four times from ethanol-water (4:1), was decomposed by adding to an equal weight of Na$_2$CO$_3$ in hot water. The free cyclopentanone was steam distd. from the soln. The distillate was satd. with NaCl and extracted with benzene which was then dried and evapd.; the residue was distd. [Allen, Ellington and Meakins, JCS, 1909 (1960).]

Cyclopentene, b.p. 43-44°.

Freed from hydroperoxide by refluxing with cupric stearate. Fractionally distd. from sodium. Chromatographed on a Dowex 710-Chromosorb W g.l.c. column. Methods for Cyclohexene should be applicable.

Cyclopropane, b.p. -34°.

Washed with a soln. of HgSO$_4$, and dried with CaCl$_2$, then Mg(ClO$_4$)$_2$.

R-Cycloserine, see R-4-amino-3-isoxazolidone.

Cyclotrimethylenetrinitramine (RDX), m.p. 203.8° (decomp.).

Cryst. from acetone.

p-Cymene, b.p. 177.1°, d$_4^{20}$ 0.8569, n$_D^{20}$ 1.4909, n$_D^{25}$ 1.4885.

Washed with cold, conc. H$_2$SO$_4$ until there is no further colour change, then repeatedly with water, 10% aq. Na$_2$CO$_3$ and water again. Dried with Na$_2$SO$_4$, CaCl$_2$ or MgSO$_4$, and distd. Further purification steps include steam distn. from 3% NaOH, percolation through silica gel or activated alumina, and a preliminary refluxing for several days over powdered sulphur.

Cysteic acid, m.p. 260° (decomp.).

S-Cysteic acid, m.p. 289°, $[\alpha]_D^{20}$ +8.7° (water).

   Likely impurities: cystine, oxides of cysteine. Cryst. from water
by adding 2 volumes of ethanol.

S-Cysteine hydrochloride monohydrate, m.p. 175-178° (decomp.),

$[\alpha]_D^{25}$ +6.53 (5 M HCl).

   Likely impurities: cystine, tyrosine. Cryst. from methanol by
adding ether, or from hot 20% HCl. Dried under vac. over P_2O_5.
Hygroscopic.

Cysteine hydrochloride.

   Cryst. from hot 20% HCl; dried under vac. over P_2O_5.

S-Cystine, $[\alpha]_D^{18.5}$ -229° (c = 0.92 in M HCl).

   Cystine disulphoxide was removed by treating an aq. suspension with
H_2S. The cystine was filtered off, washed with distd. water and dried
at 100° under vac. over P_2O_5. Cryst. by dissolving in 1.5 M HCl, then
adjusting to neutral with ammonia. Likely impurities: D-cystine,
meso-cystine, tyrosine.

Cytidine, m.p. 230° (decomp.).

   Cryst. from 90% aq. ethanol.

Cytisine, m.p. 152-153°.

   Cryst. from acetone.

Cytosine, m.p. 320-325° (decomp.).

   Cryst. from water.

Daidzein, see 4',7-Dihydroxyisoflavone.

D.D.T., see 1,1,1-Trichloro-2,2-bis(p-chlorophenyl)ethane.

Decahydronaphthalene (mixed isomers), b.p. 191.7°.

   Stirred with conc. H_2SO_4 for some hours. Then the organic phase was
separated, washed with water, satd. aq. Na_2CO_3, again with water,
dried with $CaSO_4$ or CaH_2 (and perhaps dried further with sodium),

Decahydronaphthalene (continued)
filtered and distd. under reduced pressure. Also purified by repeated
passage through long columns of silica gel previously activated at
200-250°, followed by distn. from LiAlH$_4$ and storage under nitrogen.
Type 4A molecular sieve can be used as a drying agent. Storage over
silica gel removes water and other polar substances.

cis-Decahydronaphthalene, f.p. -43.2^{o}, b.p. 195.7^{o}, n$_D^{20}$ 1.48113.
trans-Decahydronaphthalene, f.p. -30.6^{o}, b.p. 187.3^{o}, n$_D^{20}$ 1.46968.
   Purification methods described for the mixed isomers are applicable.
The individual isomers can be separated by very efficient fractional
distn., followed by fractional crystn. by partial freezing. The cis
isomer reacts preferentially with AlCl$_3$ and can be removed from the
trans isomer by stirring the mixture with a limited amount of AlCl$_3$
for 48 hr at room temp., filtering and distilling.

Decalin, see Decahydronaphthalene.

Decamethylene glycol, see Decane-1,10-diol.

n-Decane, b.p. 174.1^{o}, n$_D^{20}$ 1.41189, n$_D^{25}$ 1.40967.
   Can be purified by shaking with conc. H$_2$SO$_4$, washing with water, aq.
NaHCO$_3$, and more water, then drying with MgSO$_4$, refluxing with sodium
and distilling. Passed through a column of silica gel or alumina. Can
also be purified by azeotropic distn., with 2-butoxyethanol, the
alcohol being washed out of the distillate, using water; the decane
is next dried and redistd. It can be stored with sodium hydride.
   Further purified by preparative gas chromatography on a column
packed with 30% SE-30 (General Electric methyl-silicone rubber) on
42/60 Chromsorb P at 150° and 40 psig, using helium. [Chu, JCP, 41,
226 (1964).] Also purified by zone melting.

Decan-1,10-diol, m.p. 72.5-74°.
   Cryst. from dry ethylene dichloride.

Decanoic acid, see Capric acid.

n-Decanol, f.p. 6.0°.
   Fractionally distd. in an all-glass unit at 10 mm pressure, then
fractionally cryst. by partial freezing. Also purified by preparative
g.l.c. and by passage through alumina before use.

Decyl alcohol, see Decanol.

n-Decyl bromide, b.p. 117-118°/15.5 mm.
   Shaken with H_2SO_4, washed with water, dried with K_2CO_3, and
fractionally distd.

Decyltrimethylammonium bromide.
   Cryst. from 50% (v/v) ethanol/ethyl ether, or from acetone. Dried
under vac.

Dehydroascorbic acid, m.p. 196° (decomp.), $[\alpha]_{546}^{20}$ +42.5° (c = 1 in H_2O).
7-Dehydrocholesterol, m.p. 142-143°.
   Cryst. from methanol.

Dehydrocholic acid, m.p. 237°.
   Cryst. from acetone.

Dehydroepiandrosterone, m.p. 140-141° and m.p. 152-153° (dimorphic).
   Cryst. from methanol and sublimed under vac.

Delphinine, m.p. 197.5-199°.
   Cryst. from ethanol.

3-Deoxy-D-allose, n_D^{18} 1.4735, $[\alpha]_D^{20}$ +8° (c = 0.25 in water).
   Cryst. from ethyl ether as a colourless syrup.

Deoxybenzoin, m.p. 60°, b.p. 320°/760 mm, 177°/12 mm.
   Cryst. from ethanol.

Deoxycholic acid, m.p. 176°.
   Refluxed with CCl_4 (50 ml/g), filtered, evapd. under vac. at 25°,
recryst. from acetone and dried under vac. at 155°. [Trenner et al.,
JACS, 76, 1196 (1954).] A soln. of (cholic acid-free) material
(100 ml) in 500 ml of hot ethanol was filtered, evapd. to less than

Deoxycholic acid (continued)

500 ml on a hotplate, and poured into 1500 ml of cold ethyl ether. The ppte., filtered by suction, was cryst. twice from 1-2 parts of absolute ethanol, to give an alcoholate, m.p. 118-120°, which was dissolved in ethanol (100 ml for 60 g) and poured into boiling water. After boiling for several hours the ppte. was filtered off, dried, ground and dried to constant weight. [Sobotka and Goldberg, BJ, 26, 555 (1932).]

11-Deoxycorticosterone, m.p. 141-142°, $[\alpha]_D^{20}$ +178° and $[\alpha]_{546}^{20}$ +223° (c = 1 in EtOH).

  Cryst. from ether.

2-Deoxy-β-galactose, m.p. 126-128°.

  Cryst. from ethyl ether.

2-Deoxy-α-D-glucose, m.p. 146°.

  Cryst. from methanol-acetone.

2-Deoxy-β-L-ribose, m.p. 77°.

2-Deoxy-β-D-ribose, m.p. 86-87°.

  Cryst. from ethyl ether.

Desoxycholic acid, see Deoxycholic acid.

Desthiobiotin, n.p. 156-158°.

Desyl bromide, m.p. 57.1-57.5°.

  Cryst. from 95% ethanol.

Dextrose, see D-Glucose.

p-Diacetobenzene, m.p. 113.5-114.2°.

  Cryst. from benzene and vac. dried over $CaCl_2$.

Diacetyl, see Biacetyl.

3,6-Diaminoacridine sulphate.

An aq. soln., after treatment with charcoal, was concentrated, chilled overnight, filtered and the ppte. was rinsed with a little ethyl ether. The ppte. was dried in air, then overnight in a vac. oven at 70°.

1,4-Diaminoanthraquinone, m.p. 268°.

Purified by thin-layer chromatography on silica gel plates using toluene-acetone (9:1) as eluant. The main band was scraped off and extracted with methanol. The solvent was evapd. and the dye was dried in a drying pistol. [Land, McAlpine, Sinclair and Truscott, JCS Faraday Trans. I, 72, 2091 (1976).] Cryst. from ethanol in dark violet crystals.

2,6-Diaminoanthraquinone, m.p. 310-320°.

Cryst. from pyridine.

3,4-Diaminobenzoic acid, m.p. 213° (decomp.).

3,5-Diaminobenzoic acid, m.p. 235-240° (decomp.).

Cryst. from water.

1,4-Diaminobutane dihydrochloride, m.p. >290°.

Cryst. from alcohol-water.

1,2-Diaminocyclohexanetetraacetic acid, see Cyclohexane-1,2-diamine-tetraacetic acid.

2,2'-Diaminodiethylamine, b.p. 208°.

Dried with sodium and distd. under reduced pressure.

4,4'-Diamino-3,3'-dinitrobiphenyl, m.p. 275°.

Cryst. from aq. ethanol.

p,p'-Diaminodiphenylamine, m.p. 158°.

Cryst. from water.

p,p'-Diaminodiphenylmethane, m.p. 91.6-92°.

Cryst. from water or benzene.

2,2'-Diaminodipropylamine.
    Dried with sodium and distd. under vac.

6,9-Diamino-2-ethoxyacridine, m.p. 226°.
    Cryst. from 50% ethanol.

2,7-Diaminofluorene, m.p. 165°.
2,4-Diamino-6-hydroxypyrimidine, m.p. 260-270° (decomp.).
1,5-Diaminonaphthalene, m.p. 190°.
    Cryst. from water.

1,8-Diaminonaphthalene, m.p. 66.5°.
    Cryst. from water or aq. ethanol, and sublimes in vacuo.

2,3-Diaminonaphthalene, m.p. 199°.
    Cryst. from water.

2,4-Diamino-5-phenylthiazole, m.p. 163-164° (decomp.).
    Cryst. from aq. ethanol or water. Stored in the dark under nitrogen.

d,1-2,6-Diaminopimelic acid, m.p. 313-315° (decomp.).
    Cryst. from water.

1,3-Diaminopropane dihydrochloride, m.p. 243°.
    Cryst. from ethanol-water.

1,3-Diaminopropan-2-ol, m.p. 38-40°.
    Dissolved in an equal amount of water, shaken with charcoal and
vac. distd. at 68°/0.1 mm. It is too viscous to be distd. through a
packed column.

2,3-Diaminopyridine, m.p. 116°.
2,6-Diaminopyridine, m.p. 121.5°.
    Cryst. from benzene and sublimes in vacuo.

3,4-Diaminopyridine, m.p. 218-219°.
    Cryst. from benzene and stored under H_2 because it is deliquescent
and absorbs CO_2.

meso-2,3-Diaminosuccinic acid, m.p. 305-306° (decomp. and sublimes).
  Cryst. from water.

Diaminotoluene, see Toluenediamine.

3,5-Diamino-1,2,4-triazole, m.p. 206°.
  Cryst. from water or ethanol.

Di-n-amyl n-amylphosphonate.
  Purified by three crystns. of its compound with uranyl nitrate,
from hexane. For method, see Tributyl phosphate.

2,5-Di-tert.-amylhydroquinone, m.p. 185.8-186.5°.
  Cryst. under nitrogen from boiling glacial acetic acid (7 ml/g)
plus boiling water (2.5 ml/g). [Stolow and Bonaventura, JACS, 85
3636 (1963).]

Di-n-amyl phthalate, b.p. 204-206°/11 mm, d_4^{25} 1.0230, n_D^{20} 1.4885.
  Washed with aq. Na_2CO_3, then distd. water. Dried with $CaCl_2$ and
distd. under reduced pressure. Stored in a vac. desiccator over P_2O_5.

1,4-Diazabicyclo(2,2,2)octane, see Triethylenediamine.

Diazoaminobenzene, m.p. 99°.
  Cryst. from pet. ether (b.p. 60-80°), 60% methanol-water or 50% aq.
ethanol (charcoal) containing a small amount of KOH. Also purified by
chromatography on alumina. Stored in the dark.

6-Diazo-5-oxo-L-norleucine, m.p. 145-155° (decomp.).
  Cryst. from aq. ethanol.

Dibenzalacetone, m.p. 112°.
  Cryst. from hot ethyl acetate (2.5 ml/g.) or ethanol.

Dibenz[a,h]anthracene, m.p. 266-267°.
  The yellow-green colour (due to other pentacyclic impurities) has
been removed by crystn. from benzene or by selective oxidation with
lead tetraacetate in acetic acid. [Moriconi et al., JACS, 82, 3441
(1960).]

Dibenzo-18-crown-6, m.p. 163-164°.
   Cryst. from toluene and dried under vac.

Dibenzofuran, m.p. 82.4°.
   Dissolved in ethyl ether, then shaken with two portions of aq. NaOH
(2 M), washed with water, separated and dried (MgSO$_4$). After evaporating the ether, dibenzofuran was cryst. from aq. 80% ethanol and
dried under vac. [Cass et al., JCS, 1406 (1958).]

Dibenzopyran, m.p. 100.5°.
   Cryst. from 95% ethanol.

Dibenzothiophen, m.p. 99°.
   Purified by chromatography on alumina with pet. ether, in a
darkened room.

Dibenzoylmethane, m.p. 80°.
   Cryst. from pet. ether or methanol.

Dibenzoyl peroxide, see Benzoyl peroxide.

Di-O-benzoyl-R-tartaric acid, $[\alpha]_{546}^{20}$ +136° (c = 2 in EtOH).
Di-O-benzoyl-S-tartaric acid, $[\alpha]_{546}^{20}$ -136° (c = 2 in EtOH).
   Cryst. from water as monohydrate, m.p. 88-89°, and cryst. from
xylene as anhydrous acid, m.p. 173°.

2,3,6,7-Dibenzphenanthrene, m.p. 257°.
   Cryst. from xylene.

Dibenzyl disulphide, m.p. 71-72°.
   Cryst. from ethanol.

Dibenzyl ketone, m.p. 34.0°.
   Fractionally cryst. from its melt, then cryst. from pet. ether.
Stored in the dark.

Dibenzyl sulphide, m.p. 48.5°.

Cryst. from ethanol-water (10:1), or repeatedly from purified hot ethyl ether. Vac. dried at 30° over P_2O_5, fused under nitrogen and redried.

2,4'-Dibromoacetophenone, see p-Bromophenacyl bromide.

2,4-Dibromoaniline, m.p. 79-80°.

Cryst. from aq. ethanol.

9,10-Dibromoanthracene, m.p. 226°.

Cryst. from xylene.

p-Dibromobenzene, m.p. 87.8°.

Steam distd., cryst. from ethanol or methanol and dried in the dark under vac. Purified by zone melting.

2,5-Dibromobenzoic acid, m.p. 157°.

Cryst. from water or ethanol.

4,4'-Dibromobiphenyl, m.p. 164°, b.p. 355-360°/760 mm.

Cryst. from methanol.

trans-1,4-Dibromobut-2-ene, m.p. 54°, b.p. 85°/10 mm.

Cryst. from ligroin.

Dibromodeoxybenzoin, m.p. 111.8-112.7°.

Cryst. from acetic acid.

Dibromodichloromethane, m.p. 22°.

Cryst. repeatedly from its melt, after washing with aq. $Na_2S_2O_3$ and drying with BaO.

1,2-Dibromoethane, f.p. 10.0°, b.p. 131.7°, n_D^{15} 1.54160.

Washed with conc. HCl or H_2SO_4, then water, aq. $NaHCO_3$ or Na_2CO_3, more water, and dried with $CaCl_2$. Fractionally distd. Alternatively, kept in daylight with excess bromine for 2 hr, then extracted with aq. Na_2SO_3, washed with water, dried with $CaCl_2$, filtered and distd. Can also be purified by fractional crystn. by partial freezing. Stored in the dark.

4',5'-Dibromofluorescein, m.p. 285°.
   Cryst. from aq. 30% ethanol.

5,7-Dibromo-8-hydroxyquinoline, m.p. 196°.
   Cryst. from acetone-ethanol. Can be sublimed.

2,5-Dibromonitrobenzene, m.p. 84°.
   Cryst. from acetone.

2,6-Dibromo-4-nitrophenol, m.p. 143-144°.
   Cryst. from aq. ethanol.

2,4-Dibromophenol, m.p. 37°.
   Cryst. from $CHCl_3$ at -40°.

2,6-Dibromophenol, m.p. 56-57°.
   Vac. distd. (at 18 mm), then cryst. from cold $CHCl_3$ or from
ethanol-water.

1,3-Dibromopropane, f.p. -34.4°, b.p. 165°.
   Washed with dil. aq. Na_2CO_3, then water. Dried and fractionally
distd. under reduced pressure.

5,7-Dibromo-8-quinolinol, see 5,7-Dibromo-8-hydroxyquinoline.

meso-2,3-Dibromosuccinic acid, m.p. 288-290° (sealed tube, decomp.).
   Cryst. from distd. water, keeping the temperature below 70°.

1,2-Dibromotetrafluoroethane, b.p. 47.3°/760 mm.
   Washed with water, then with weak alkali. Dried with $CaCl_2$ or H_2SO_4
and distd. [Locke et al., JACS, 56, 1726 (1934).] Also purified by
gas chromatography on a silicone DC-200 column.

α,α'-Dibromo-o-xylene, m.p. 95°, b.p. 128-130°/4.5 mm.
   Cryst. from chloroform.

α,α'-Dibromo-m-xylene, m.p. 77°, b.p. 156-160°/12 mm.
   Cryst. from acetone.

α,α'-Dibromo-p-xylene, m.p. 145-147°, b.p. 245°/760 mm, 155-158°/12-15 mm.

Cryst. from benzene or chloroform.

Di-n-butylamine, b.p. 159°, n_D^{20} 1.41766, n_D^{25} 1.4095.

Dried with LiAlH$_4$, CaH$_2$ or KOH pellets, filtered and distd. from BaO or CaH$_2$.

α-Dibutylamino-α-(p-methoxyphenyl)acetamide, m.p. 134°.

Cryst. from ethanol containing 10% ethyl ether.

p-Di-tert.-butylbenzene, m.p. 76-77.5°.

Cryst. from ethyl ether.

Di-n-butyl n-butylphosphonate, b.p. 150-151°/10 mm, 160-162°/20 mm, n_D^{25} 1.4302.

Purified by three crystns. of its compound with uranyl nitrate, from hexane. For method, see Tributyl phosphate.

Dibutylcarbitol.

Freed from peroxides by slow passage through a column of activated alumina. The eluate was shaken with Na$_2$CO$_3$ (to remove any remaining acidic impurities), washed with water, and stored with CaCl$_2$ in a dark bottle. [Tuck, JCS, 3202 (1957).]

2,6-Di-tert.-butyl-p-cresol, m.p. 69-70°.

Dissolved in n-hexane at room temp., then cooled, with rapid stirring, to -60°. The ppte. was separated, redissolved in hexane, and the process was repeated until the mother liquor was no longer coloured. The final product was stored under nitrogen at 0°. [Blanchard, JACS, 82, 2014 (1960).] Cryst. from ethanol.

Di-n-butyl cyclohexylphosphonate.

The compound with uranyl nitrate was cryst. three times from hexane. For method, see Tributyl phosphate.

2,6-Di-tert.-butyl-4-dimethylaminomethylphenol.

Cryst. from n-hexane.

Di-tert.-butyl diperphthalate.
   Cryst. from ethyl ether. Dried over H_2SO_4.

Di-n-butyl ether, see n-Butyl ether.

2,6-Di-tert.-butyl-4-ethylphenol, m.p. 42-44°.
   Cryst. from aq. ethanol or n-hexane.

2,5-Di-tert.-butylhydroquinone, m.p. 222-223°.
   Cryst. from benzene or glacial acetic acid.

2,6-Di-tert.-butyl-4-isopropylphenol, m.p. 39-41°.
   Cryst. from n-hexane or aq. ethanol.

2,6-Di-tert.-butyl-4-methylphenol, m.p. 71.5°.
   Cryst. from benzene, n-hexane or methylcyclohexane and dried under
vac.

Di-tert.-butyl peroxide, n_D^{20} 1.3889.
   Washed with water, dried. Freed from tert.-butyl hydroperoxide by
passage through an alumina column 24 x 0.5 in., and two high vac.
distns. from room temp. to liquid-air temp. [Offenbach and Tobolsky,
JACS, 79, 278 (1957).]

2,6-Di-tert.-butylphenol, m.p. 37-38°.
   Cryst. from aq. ethanol or n-hexane.

Dibutyl phthalate, b.p. 340°, d_4^{25} 1.0426, n_D^{20} 1.4929, n_D^{25} 1.4901.
   Washed with dil. NaOH (to remove any butyl hydrogen phthalate), aq.
$NaHSO_3$ (charcoal), then distd. water. Dried with $CaCl_2$, distd. under
vac. and stored in a desiccator over P_2O_5.

Di-n-butyl sulphide, α form b.p. 182°, β form 190-230° (decomp.).
   Washed with aq. 5% NaOH, then water. Dried with $CaCl_2$ and distd.
from sodium.

Di-n-butyl sulphone, m.p. 43.5°.
   Purified by zone melting.

3,5-Dicarbethoxy-1,4-dihydrocollidine.
   Cryst. from hot ethanol-water.

Dichloramine-T, m.p. 83°.
   Cryst. from pet.ether (b.p. 60-80°) or CHCl$_3$-pet.ether. Dried in
air.

Dichloroacetic acid, b.p. 95.0-95.5°/17-18 mm.
   Cryst. from benzene. Dried with MgSO$_4$ and fractionally distd.

sym-Dichloroacetone, m.p. 45°, b.p. 173°.
   Cryst. from benzene. Distd. under vac.

2,4-Dichloroaniline, m.p. 63°.
   Cryst. from ethanol-water.

3,4-Dichloroaniline, m.p. 71.5°.
   Cryst. from methanol.

2,4-Dichlorobenzaldehyde, m.p. 72°.
   Cryst. from ethanol or ligroin.

2,6-Dichlorobenzaldehyde, m.p. 70.5-71.5°.
   Cryst. from ethanol-water or pet.ether (b.p. 30-60°).

o-Dichlorobenzene, b.p. 81-82°/31-32 mm, 180.5°/760 mm, n_D^{20} 1.55145,
n_D^{25} 1.54911.
   Contaminants may include the p-isomer and trichlorobenzene. Shaken
with conc. or fuming H$_2$SO$_4$, washed with water, dried with CaCl$_2$, and
distd. from CaH$_2$ or sodium in a glass helices-packed column. Low
conductivity material ($\sim$10^{-10} mhos) has been obtained by refluxing
with P$_2$O$_5$, fractionally distilling and passing through a column
packed with silica gel or activated alumina: it was stored in a
dry-box under nitrogen or with activated alumina.

m-Dichlorobenzene, b.p. 173.0°, n_D^{20} 1.54586, n_D^{25} 1.54337.
   Washed with aq. 10% NaOH, then with water until neutral, dried and
distd. Conductivity material ($\leq$ 10^{-10} mhos) has been prepared by

m-Dichlorobenzene (continued)

refluxing over P_2O_5 for 8 hr, then fractionally distilling, and storing with activated alumina. m-Dichlorobenzene attacks rubber stoppers.

p-Dichlorobenzene, m.p. 53.0°, b.p. 174.1°, n_D^{60} 1.52849.

o-Chlorobenzene is a common impurity. Has been purified by steam distn., crystn. from ethanol or boiling methanol, air-dried and dried in the dark under vac. Also purified by zone melting.

2,2'-Dichlorobenzidine, m.p. 165°.

Cryst. from ethanol.

3,3'-Dichlorobenzidine, m.p. 132-133°.

Cryst. from ethanol or benzene. Carcinogenic.

2,4-Dichlorobenzoic acid, m.p. 163-164°.

Cryst. from aq. ethanol (charcoal), then benzene (charcoal). Can also be recryst. from water.

2,5-Dichlorobenzoic acid, m.p. 154°, b.p. 301°/760 mm.

Cryst. from water.

2,6-Dichlorobenzoic acid, m.p. 141-142°.

Cryst. from ethanol and sublimes in vacuo.

3,4-Dichlorobenzoic acid, m.p. 206-207°.

Cryst. from aq. ethanol (charcoal) or acetic acid.

3,5-Dichlorobenzoic acid, m.p. 188°.

Cryst. from ethanol and sublimes in vacuo.

2,6-Dichlorobenzonitrile, m.p. 145°.

Cryst. from acetone.

3,4-Dichlorobenzyl alcohol, m.p. 38-39°.

Cryst. from water.

2,3-Dichloro-1:3-butadiene, b.p. 41-43°/85 mm, 98°/760 mm.

Cryst. from pentane to constant m.p. about -40°. A mixture of meso and d,l forms was separated by gas chromatography on an 8 m. stainless steel column (8 mm i.d.) with 20% DEGS on Chromosorb W (60-80 mesh) at 60° and 80 ml He/min. [Su and Ache, JPC, 80, 659 (1976).]

4,6-Dichloro-o-cresol, see 2,4-Dichloro-6-methylphenol.

2,3-Dichloro-5,6-dicyano-p-benzoquinone, m.p. 203° (decomp.)

Cryst. from $CHCl_3$, or $CHCl_3$-benzene (4:1).

β,β'-Dichlorodiethyl ether, b.p. 79-80°/20 mm, 176-177.0°/743 mm.

Peroxide formation occurs rapidly, especially if distn. is attempted at atmospheric pressure. After drying with NaOH pellets for 2 days, the ether was distd. under nitrogen at reduced pressure. The distillate was made 10^{-6}M in catechol to diminish peroxide formation, and was redistd. immediately before use.

1,2-Dichloro-1,2-difluoroethane.

For purification of diastereoisomeric mixture, with resolution into meso and rac forms, see Machulla and Stocklin [JPC, 78, 658 (1974)].

Dichlorodifluoromethane, b.p. -25°.

Passage through satd. aq. KOH, then conc. H_2SO_4, and a tower packed with activated copper on kieselguhr at 200° removed CO_2 and O_2. A trap cooled to -29° removed a trace of high-boiling material.

2,5-Dichloro-3,6-dihydroxy-p-benzoquinone, see Chloranilic acid.

1,3-Dichloro-5,5-dimethylhydantoin, m.p. 132°.

Cryst. from $CHCl_3$.

Dichlorodimethylsilane, m.p. -76°.

Purified by zone melting.

1,1-Dichloroethane, b.p. 57.3°, d^{15} 1.18350, d^{30} 1.16010,
n_D^{15} 1.41975, n_D^{25} 1.4146.

Shaken with conc. H_2SO_4 or aq. $KMnO_4$, then washed with water,
satd. aq. $NaHCO_3$, again with water, dried with K_2CO_3 and distd.
from CaH_2 or $CaSO_4$. Stored above silica gel.

1,2-Dichloroethane, b.p. 83.4°, n_D^{15} 1.44759.

Usually prepared by chlorinating ethylene, so that likely impurities include higher chloro derivatives and other chloro compounds
depending on the impurities originally present in the ethylene. It
forms azeotropes with water, methanol, ethanol, trichloroethylene,
CCl_4 and isopropanol. Its azeotrope with water (containing 8.9%
water, and b.p. 77°) can be used to remove gross amounts of water
prior to final drying. As a preliminary purification step, it can
be steam distd.

Shaken with conc. H_2SO_4 (to remove alcohol added as an oxidation
inhibitor), washed with water, then dil. KOH or aq. Na_2CO_3 and again
with water. After an initial drying with $CaCl_2$, $MgSO_4$ or by distn.,
it is refluxed with P_2O_5, $CaSO_4$ or CaH_2 and fractionally distd.
Carbonyl-containing impurities can be removed as described for
Chloroform.

1,2-Dichloroethylene, b.p. 60° (cis), 48° (trans).

Shaken successively with conc. H_2SO_4, water, aq. $NaHCO_3$ and water.
Dried with $MgSO_4$ and distd. Repeated fractional distn. separates the
cis- and trans-isomers.

cis-1,2-Dichloroethylene, b.p. 60.4°, d_4^{20} 1.2830, n_D^{15} 1.44903,
n_D^{20} 1.4495.

Purified by careful fractional distn., followed by passage through
neutral activated alumina. Also by shaking with mercury, drying with
K_2CO_3 and distn. from $CaSO_4$.

trans-1,2-Dichloroethylene, b.p. 47.7°, d_4^{20} 1.2551, n_D^{15} 1.45189,
n_4^{20} 1.4462.

Dried with $MgSO_4$, and fractionally distd. under CO_2. Fractional
crystn. at low temperatures has also been used.

5,7-Dichloro-8-hydroxyquinoline, m.p. 180-181°.
    Cryst. from acetone-ethanol.

Dichloromethane, b.p. 40.0°, n_D^{20} 1.42456, n_D^{25} 1.4201.

    Shaken with portions of conc. H_2SO_4 until the acid layer remained
colourless, then washed with water, aq. 5% Na_2CO_3, $NaHCO_3$ or NaOH,
then water again. Pre-dried with $CaCl_2$, and distd. from $CaSO_4$, CaH_2
or P_2O_5. Stored away from bright light in a brown bottle with Linde
type 4A molecular sieve. Other purification steps include washing
with aq. $Na_2S_2O_3$, passage through a column of silica gel, and removal
of carbonyl-containing impurities as described under Chloroform.

3,9-Dichloro-7-methoxyacridine, m.p. 160-161°.
    Cryst. from benzene.

5,7-Dichloro-2-methyl-8-hydroxyquinoline, m.p. 114-115°.
    Cryst. from ethanol.

2,4-Dichloro-6-methylphenol, m.p. 55°, b.p. 129-132°/40 mm.
    Cryst. from water.

2,4-Dichloro-1-naphthol, m.p. 106-107°.
    Cryst. from methanol.

2,3-Dichloro-1,4-naphthoquinone, m.p. 193°.
    Cryst. from ethanol.

2,5-Dichloro-4-nitroaniline, m.p. 157-158°.
    Cryst. from ethanol, then sublimed.

2,6-Dichloro-4-nitroaniline, m.p. 193°.
    Cryst. from aq. ethanol or benzene-ethanol.

2,5-Dichloro-1-nitrobenzene, m.p. 56°.
3,4-Dichloro-1-nitrobenzene, m.p. 43°.
    Cryst. from absolute ethanol.

2,4-Dichloro-6-nitrophenol, m.p. 122-123°.
    Cryst. from acetic acid.

2,6-Dichloro-4-nitrophenol, m.p. 125°.
    Cryst. from ethanol and dried in vacuo over anhydrous MgSO$_4$.

Dichlorophen, b.p. 177-178°.
    Cryst. from toluene.

2,3-Dichlorophenol, m.p. 57°.
    Cryst. from ether.

2,4-Dichlorophenol, m.p. 42-43°.
    Cryst. from pet. ether (b.p. 30-40°). Purified by repeated zone
melting, using a P$_2$O$_5$ guard tube to exclude moisture. Very hygroscopic when quite dry.

2,5-Dichlorophenol, m.p. 58°, b.p. 211°/744 mm.
    Cryst. from ligroin and sublimed.

3,4-Dichlorophenol, m.p. 68°, b.p. 253.5°/767 mm.
3,5-Dichlorophenol, m.p. 68°, b.p. 233-234°/760 mm, 122-124°/8 mm.
    Cryst. from pet. ether-benzene mixture.

2,6-Dichlorophenol-indophenol, ε = 2.1 x 10^4 at 600 mm and pH 8.
    Dissolved in 0.001 M phosphate buffer, pH 7.5 (alternatively,
about 2 g of the dye was dissolved in 80 ml of M HCl), and extracted
into ethyl ether. The extract was washed with water, extracted with
aq. 2% NaHCO$_3$, and the sodium salt of the dye was pptd. by adding
NaCl (30 g/100 ml of NaHCO$_3$ soln.), then filtered off, washed with
dil. NaCl soln. and dried.

2,4-Dichlorophenoxyacetic acid, m.p. 146°.
α-(2,4-Dichlorophenoxy)propionic acid, m.p. 117°.
    Cryst. from methanol.

2,4-Dichlorophenylacetic acid, m.p. 131°.

2,6-Dichlorophenylacetic acid, m.p. 157-158°.

   Cryst. from aq. ethanol.

4,5-Dichlorophthalic acid, m.p. 200° (decomp. to anhydride).

   Cryst. from water.

1,2-Dichloropropane, b.p. 95.9-96.2°.

   Distd. from CaH_2.

2,2-Dichloropropane, b.p. 69.3°.

   Washed with aq. Na_2CO_3 soln., then distd. water, dried over $CaCl_2$
and fractionally distd.

1,3-Dichloro-2-propanone, see sym-Dichloroacetone.

2,6-Dichloropyridine, m.p. 87-88°.

3,5-Dichloropyridine, m.p. 64-65°.

   Cryst. from aq. ethanol.

4,7-Dichloroquinoline, m.p. 86.4-87.4°, b.p. 148°/10 mm.

   Cryst. from methanol.

4,7-Dichloroquinoline, m.p. 86-87°.

   Cryst. from 95% ethanol.

5,7-Dichloro-8-quinolinol, see 5,7-Dichloro-8-hydroxyquinoline.

2,5-Dichlorostyrene, b.p. 72-73°/2 mm, d_4^{20} 1.4045, n_D^{20} 1.5798.

   Purified by fractional crystn. from the melt and by distn.

p,α-Dichlorotoluene, see p-Chlorobenzyl chloride.

α,α'-Dichloro-p-xylene, m.p. 100°.

   Cryst. from benzene, and dried under vac.

Dicinnamalacetone, m.p. 146°.

   Cryst. from benzene-isooctane (1:1).

Dicumyl peroxide, m.p. 39-40°.
   Cryst. from 95% ethanol (charcoal).

p-Dicyanobenzene, m.p. 222°.
   Cryst. from ethanol.

Dicyanodiamide, see Cyanoguanidine.

Dicyclohexyl-18-crown-6.
   Purified by chromatography on neutral alumina and eluting with an ether-hexane mixture (see Inorg. Chem., 14, 3132 (1975)

Di-n-decylamine, m.p. 34°, b.p. 153°/1 mm, 359°/760 mm.
   Dissolved in benzene. Pptd. as its bisulphate by shaking with 4 M H_2SO_4. Filtered. Washed with benzene, separating by centrifuging, then the free base was liberated by treating with aq. NaOH. [McDowell and Allen, JPC, 65, 1358 (1961).]

Didodecylamine, m.p. 51.8°.
   Cryst. from ethanol-benzene under nitrogen.

Dienoesterol, m.p. 227-228°.
   Cryst. from aq. ethanol.

Diethanolamine, m.p. 28°, b.p. 154-155°/10 mm, 270°/760 mm.
   Fractionally distd. twice, then fractionally cryst. from its melt.

1,2-Diethoxyethane, see Ethylene glycol diethyl ether.

Diethyl acetal, b.p. 102.5°, d_4^{25} 0.8198.
   Distd. from sodium.

Diethyl acetamidomalonate, m.p. 96°.
   Cryst. from benzene-pet. ether.

Diethylamine, b.p. 55.5°, n_D^{20} 1.38637.
   Dried with $LiAlH_4$ or KOH pellets. Refluxed with, and distd. from BaO or KOH. Converted to the p-toluenesulphonamide and cryst. to constant m.p. from dry pet.ether (b.p. 90-120°), then hydrolyzed with

Diethylamine (continued)

HCl, excess NaOH added, and the amine distd. through a tower of activated alumina, redistd. and dried with activated alumina before use. [Swift, JACS, 64, 115 (1942).]

Diethylamine hydrochloride, m.p. 223.5°.

Cryst. from absolute ethanol.

Diethylammonium chloride.

Cryst. from dichloroethane-methanol.

N,N-Diethylaniline, b.p. 216.5°.

Refluxed for 4 hr with half its weight of acetic anhydride, then fractionally distd. at reduced pressure.

5,5-Diethylbarbituric acid, m.p. 188-192°.

Cryst. from water or ethanol. Dried in vac. over P_2O_5.

N,N'-Diethylcarbanilide, m.p. 79°.

Cryst. from ethanol.

Diethyl carbitol, see Diethylene glycol diethyl ether.

Diethyl carbonate, b.p. 126.8°, n_D^{15} 1.38654, n_D^{25} 1.38287.

Washed (100 ml) with an aq. 10% Na_2CO_3 (20 ml), satd. $CaCl_2$ (20 ml), then water (30 ml). Dried by standing with solid $CaCl_2$ (5 g) for 2 hr, filtering, then standing once more with 5 g $CaCl_2$ for 1 hr. (Prolonged contact should be avoided because slow combination with $CaCl_2$ occurs.) Fractionally distd.

1,1'-Diethyl-2,2'-cyanine chloride.

Cryst. from ethabol and dried in vac. oven at 80° for 4 hr.

N,N-Diethylcyclohexylamine, b.p. 193°/760 mm.

Dried with BaO and fractionally distd.

O,O-Diethyl-S-2-diethylaminoethyl phosphorothiolate, m.p. 98-99°.

Cryst. from isopropyl alcohol-ethyl ether.

<u>sym-Diethyldiphenylurea</u>, m.p. 79°.

Cryst. from ethanol.

<u>Diethyl disulphide</u>, b.p. 154.0-155.0°.

Dried with silica gel or $MgSO_4$ and distd. under reduced pressure (optionally from $CaCl_2$).

<u>Diethylene glycol</u>, f.p. -10.5°, b.p. 244.3°, n_D^{15} 1.4490, n_D^{20} 1.4475.

Fractionally distd. under reduced pressure then fractionally cryst. by partial freezing.

<u>Diethylene glycol diethyl ether</u>, b.p. 85-86°/10 mm, 188.2-188.3°/751 mm.

Dried with $MgSO_4$, then CaH_2 or $LiAlH_4$, and distd. at reduced pressure from sodium or $LiAlH_4$, under nitrogen. If sodium is used, the ether should be redistd. alone to remove any products which may be formed by the action of sodium on the ether.

As a preliminary purification, the crude ether (2 l.) can be refluxed for 12 hr with 25 ml of conc. HCl in 200 ml of water, under reduced pressure, with slow passage of nitrogen to remove aldehydes and other volatile substances. After cooling, addition of sufficient solid KOH pellets, slowly and with shaking until no more dissolve, gives two liquid phases. The upper of these is decanted, dried with fresh KOH pellets, decanted, then refluxed over, and distd. from, sodium.

<u>Diethylene glycol dimethyl ether</u>, see Diglyme.

<u>Diethylene glycol mono-n-butyl ether</u>, b.p. 69-70°/0.3 mm, 230.5°/760 mm, n_D^{20} 1.4286.

Dried with anhydrous K_2CO_3 or $CaSO_4$, filtered and fractionally distd. Peroxides can be removed by refluxing with stannous chloride or a mixture of $FeSO_4$ and $KHSO_4$ (or, less completely, by filtration under slight pressure through a column of activated alumina).

<u>Diethylene glycol monoethyl ether</u>, b.p. 201.9°, n_D^{20} 1.4273, n_D^{25} 1.4254.

Ethylene glycol can be removed by extracting 250 g in 750 ml of

Diethylene glycol monoethyl ether (continued)
benzene with 5 ml portions of water, allowing 10 min for phase separation, until successive aqueous portions show the same volume increase. Dried, and freed from peroxides, as described for Diethylene glycol mono-n-butyl ether.

Diethylene glycol monomethyl ether, b.p. 194°.
      Purified as for Diethylene glycol mono-n-butyl ether.

Diethylenetriamine, see 2,2'-Diaminodiethylamine.

Diethylenetriaminepenta-acetic acid.
      Cryst. from water. Dried under vac. or at 110°.

Diethyl ether, see Ethyl ether.

Diethyl fumarate, b.p. 218°.
      Washed with aq. 5% Na_2CO_3, then with satd. $CaCl_2$ soln., dried with $CaCl_2$ and distd.

Di-(2-ethylhexyl)phosphoric acid.
      Dissolved in n-heptane to give an 0.8 M soln. Washed with an equal vol. of M HNO_3, then with satd. $(NH_4)_2CO_3$ soln., with 3 M HNO_3, and twice with water. [Petrow and Allen, AC, 33, 1303 (1961).] Similarly, the impure sodium salt, after scrubbing with pet. ether, has been acidified with HCl and the free organic acid has been extracted into pet. ether and purified as above.

Di-(2-ethylhexyl)phthalate, d_4^{25} 0.9803, n_D^{20} 1.4863.
      Washed with Na_2CO_3 soln., then shaken with water. After the resulting emulsion had been broken by adding ether, the ethereal soln. was washed twice with water, dried ($CaCl_2$), and evapd. The residual liquid was distd. several times under reduced pressure, then stored in a vac. desiccator over P_2O_5. [French and Singer, JCS, 1424 (1956).]

Diethyl ketone, b.p. 102.1°, $d^{24.6}$ 0.8099.
      Dried with anhydrous $CaSO_4$ or $CuSO_4$, and distd. from P_2O_5 under nitrogen or under reduced pressure. Further purification by

Diethyl ketone (continued)
conversion to the semicarbazone (recryst. to constant m.p. 139°,
from ethanol) which, after drying under vac. over $CaCl_2$ and paraffin
wax, was refluxed for 30 min with excess oxalic acid, then steam
distd. and salted out with K_2CO_3. Dried with Na_2SO_4 and distd.
[Cowan, Jeffrey and Vogel, JCS, 171 (1940).]

Diethyl phthalate, b.p. 295°/760 mm, 172°/12 mm, d_4^{25} 1.1160,
n_D^{20} 1.5022.
   Washed with aq. Na_2CO_3, then distd. water, dried ($CaCl_2$), and
distd. under reduced pressure. Stored in a vac. desiccator over P_2O_5.

2,2-Diethyl-1,3-propanediol, m.p. 61.4-61.8°.
   Cryst. from pet. ether (b.p. 65-70°).

Diethylstilboestrol, m.p. 169-172°.
   Cryst. from benzene.

Diethyl succinate, b.p. 105°/15 mm, n_D^{20} 1.4199.
   Dried with $MgSO_4$, and distd. at 15 mm pressure.

Diethyl sulphate, b.p. 118°/40 mm, 96°/15 mm.
   Washed with aq. 3% Na_2CO_3 (to remove acidic material), then
distd. water, dried ($CaCl_2$), filtered and distd.

Diethyl sulphide, b.p. 0°/15 mm, 90.1°/760 mm.
   Washed with aq. 5% NaOH, then water, dried with $CaCl_2$ and distd.
from sodium. Can also be dried with $MgSO_4$ or silica gel. Alternative
purification is via the Hg(II) chloride complex (See Dimethyl
sulphide.) $(C_2H_5)_2S.2$ $HgCl_2$.

Diethyl terephthalate
   Cryst. from toluene.

sym-Diethylthiourea, m.p. 76-77°.
   Cryst. from benzene.

Digitonin, m.p. 240°.
 Cryst. from aq. 85% ethanol or methanol-ethyl ether.

Digitoxigenin, m.p. 253°.
 Cryst. from aq. 40% ethanol.

D(+)-Digitoxose, m.p. 112°.
 Cryst. from methanol-ethyl ether, or ethyl acetate.

Diglycolic acid, m.p. 148° (monohydrate).
 Cryst. from water.

Diglycyl glycine, m.p. 246° (decomp.).
 Cryst. from water or water-ethanol.

Diglyme, b.p. 62°/17 mm, 75°/35 mm, 160°/760 mm, n_D^{20} 1.4087.
 Dried with NaOH pellets or CaH_2, then refluxed with, and distd.
under reduced pressure from, sodium, CaH_2, $LiAlH_4$, $NaBH_4$ or sodium
hydride. These operations were carried out under nitrogen. The amine-
like odour of diglyme has been removed by shaking with a weakly
acidic ion-exchange resin (Amberlite IR-120) before drying and distn.
Addition of 0.01% $NaBH_4$ to the distillate inhibits peroxidation.
Purification as for dioxane.

Digoxin, m.p. 265° (decomp.).
 Cryst. from aq. ethanol or aq. pyridine.

4,4'-Di-n-heptyloxyazoxybenzene.
 Cryst. from hexane and dried by heating under vac.

9,10-Dihydroanthracene, m.p. 110-110.5°.
Dihydrochloranil.
 Cryst. from ethanol.

Dihydrocinnamic acid, m.p. 48-49°.
 Cryst. from pet. ether (b.p. 60-80°).

Dihydropyran, b.p. 84.4°/742 mm.

Partially dried with Na_2CO_3, then fractionally distd. The fraction, b.p. 84-85°, was refluxed with sodium until hydrogen was no longer evolved when fresh sodium was added. It was then dried.

Dihydrotachysterol, m.p. 125-127°.

Cryst. from 90% methanol.

1,2-Dihydroxyanthraquinone, see Alizarin.

1,4-Dihydroxyanthraquinone, see Quinizarin.

1,5-Dihydroxyanthraquinone, see Anthrarufin.

1,8-Dihydroxyanthraquinone, m.p. 193-197°.

Cryst. from ethanol. Sublimes in vac.

2,4-Dihydroxyazobenzene, m.p. 228°.

Cryst. from hot ethanol (charcoal).

2,3-Dihydroxybenzaldehyde, m.p. 135-136°.
2,4-Dihydroxybenzoic acid, m.p. 213°.

Cryst. from water.

2,5-Dihydroxybenzoic acid, m.p. 200°.

Cryst. from benzene-acetone. Dried in vac. desiccator over silica gel.

2,6-Dihydroxybenzoic acid, m.p. 167°.

Dissolved in aq. $NaHCO_3$ and the soln was washed with ether to remove non-acidic material. The acid was pptd. by adding H_2SO_4, and recryst. from water. Dried under reduced pressure and stored in the dark. [Lowe and Smith, JCS Faraday Trans I, 69, 1934 (1973).]

2,5-Dihydroxybenzyl alcohol, m.p. 100°.

Cryst. from $CHCl_3$. Sublimes.

2,2'-Dihydroxybiphenyl, m.p. 108.5-109.5°

Repeatedly cryst. from toluene, then sublimed at $60°/10^{-4}$ mm.

3α,7α-Dihydroxycholanic acid, m.p. 143°.
   Cryst. from ethyl acetate.

6,7-Dihydroxycoumarin, m.p. 268-270°.
   Cryst. from glacial acetic acid.

7,8-Dihydroxycoumarin, m.p. 256° (decomp.).
   Cryst. from aq. ethanol. Sublimed.

(N,N-Dihydroxyethyl)glycine, m.p. 193° (decomp.).
   Dissolved in a small amount of hot water and pptd. with ethanol,
twice. Repeated once more but with charcoal treatment of the aq.
soln., and filtering before adding the ethanol.

4',7-Dihydroxyisoflavone, m.p. 234-236°.
   Cryst. from aq. 50% ethanol.

Dihydroxymaleic acid, decomp. at 155°.
   Cryst. from water.

5,7-Dihydroxy-4'-methoxyflavone, m.p. 261°.
   Cryst. from 95% ethanol.

1,8-Dihydroxy-3-methylanthraquinone, m.p. 196°.
   Cryst. from ethanol or benzene. Sublimes.

1,5-Dihydroxynaphthalene, m.p. 265°.
   Cryst. from nitromethane.

1,6-Dihydroxynaphthalene, m.p. 138-139° (with previous softening).
   Cryst. from benzene-absolute ethanol after treatment with charcoal.

2,5-Dihydroxyphenylacetic acid, m.p. 152°.
   Cryst. from ethanol-CHCl$_3$.

S-β-(3,4-Dihydroxyphenyl)alanine, m.p. 285.5° (decomp.),
$[\alpha]_D^{25}$ = -12.0° (1 M HCl).
   Likely impurities: vanillin, hippuric acid, 3-methoxytyrosine,

S-β-(3,4-Dihydroxyphenyl)alanine   (continued)
tyrosine, 3-aminotyrosine. Cryst. by dissolving in dil. HCl and
adding dilute ammonia to give pH 5, under nitrogen. Alternatively,
cryst. from aq. ethanol. Unstable in aq. alkali.

3,4-Dihydroxytoluene.
   Cryst. from benzene. Purity checked by paper chromatography.

p-Diiodobenzene, m.p. 128-129°.
   Cryst. from ethanol or boiling methanol, then air dried.

5,7-Diiodo-8-hydroxyquinoline, m.p. 195° (decomp.).
   Cryst. from xylene.

Diiodomethane, m.p. 6.1°, b.p. 66-70°/11-12 mm.
   Fractionally distd. at 5 mm, then fractionally cryst. by partial
freezing.

5,7-Diiodo-8-quinolinol, see 5,7-Diiodo-8-hydroxyquinoline.

S-3,5-Diiodotyrosine, m.p. 204° (decomp.), $[\alpha]_D^{25}$ +1.5° (in 1 M HCl).
   Likely impurities: tyrosine, 3-iodotyrosine, iodide. Cryst. from
cold dil. ammonia by adding acetic acid to give pH 6. Can also be
cryst. from aq. 70% ethanol.

Diisooctyl phenylphosphonate, n_D^{25} 1.1.4780.
   Vac. distd., percolated through a column of alumina, then passed
through a packed column maintained at 150° to remove residual traces
of volatile materials in a countercurrent stream of nitrogen at
reduced pressure. [Dobry and Keller, JPC, 61, 1448 (1957).]

Diisopropanolamine, m.p. 41-44°.
   Repeatedly cryst. from dry ethyl ether.

Diisopropylamine, b.p. 83.5°/760 mm, n_D^{20} 1.39236.
   Distd. from NaOH, or refluxed over Na wire or NaH for three
minutes and distd. into a dry receiver under nitrogen.

Diisopropyl ether, see Isopropyl ether.

Diisopropyl ketone, b.p. 123-125°.
    Dried with $CaSO_4$, shaken with chromatographic alumina and fractionally distd.

Diketene, m.p. -7°, b.p. 66-68°/90 mm, n_D^{20} 1.4376, n_D^{25} 1.4348.
    Diketene polymerizes violently in the presence of alkali. Distd. at reduced pressure, then fractionally cryst. by partial freezing (using as a cooling bath a 1:1 soln. of $Na_2S_2O_3$ in water, cooled with Dry Ice until slushy, and stored in a Dewar flask). Freezing proceeds slowly, and takes about a day for half completion. The crystals are separated and stored in a refrigerator under nitrogen.

2,2'-Diketospirilloxanthin, m.p. 225-227°, $E_{1cm}^{1\%}$ = 550 (349 nm), 820 (422 nm), 2125 (488 nm), 2725 (516 nm), 2130 (551 nm), in hexane.
    Purified by chromatography on column of partially deactivated neutral alumina. Cryst. from acetone-pet.ether. Stored in the dark, in inert atmosphere, at 0°.

Dilauroyl peroxide.
    Cryst. from n-hexane.

Dilituric acid, see 5-Nitrobarbituric acid.

Dimedone, m.p. 148-149°.
    Cryst. from acetone (about 8 ml/g), water or aq. ethanol. Dried in air.

1,2-Dimercapto-3-propanol, b.p. 82-84°/0.8 mm.
1,3-Dimercapto-2-propanol, b.p. 82°/1.5 mm.
    Pptd. as the mercury mercaptide [see Sjöberg, Ber., 75, 13 (1942)], regenerated with H_2S, and distd. at 2.7 mm. [Rosenblatt and Jean, AC, 27, 951 (1955).]

4,4'-Dimethoxyazobenzene, m.p. 162.7-164.7°.
    Chromatographed on basic alumina, eluted with benzene. Cryst. from 2:2:1 (v/v) methanol-ethanol-benzene.

4,4'-Dimethoxyazoxybenzene, m.p. 165°.
    Cryst. from hot 95% ethanol, dried, then sublimed in vacuo onto a
cold finger.

3,4-Dimethoxybenzaldehyde, see Veratraldehyde.

o-Dimethoxybenzene, m.p. 23°, b.p. 208.5-208.7°, n_D^{25} 1.53232.
    Steam distd. Fractionally distd. from BaO, CaH_2, or Na. Cryst.
from benzene or low-boiling pet. ether at 0°. Fractionally cryst.
from its melt. Stored over anhydrous Na_2SO_4.

m-Dimethoxybenzene, b.p. 212-213°, n_D^{20} 1.5215.
    Extracted with aq. NaOH, and water, then dried. Fractionally
distd. from BaO or sodium.

p-Dimethoxybenzene, m.p. 57.2-57.8°.
    Steam distd. Cryst. from benzene, methanol or ethanol. Dried under
vac.

2,4-Dimethoxybenzoic acid, m.p. 109°.
2,6-Dimethoxybenzoic acid, m.p. 186-187°.
    Cryst. from water.

3,4-Dimethoxybenzoic acid, m.p. 181-182°.
    Cryst. from water or aq. acetic acid.

3,5-Dimethoxybenzoic acid, m.p. 185-186°.
    Cryst. from water or ethanol.

p,p'-Dimethoxybenzophenone, m.p. 144.5°.
    Cryst. from absolute ethanol.

2,6-Dimethoxybenzoquinone, m.p. 256°.
    Cryst. from acetic acid. Sublimed.

1,2-Dimethoxyethane, b.p. 84°.
    Traces of water and acidic materials have been removed by refluxing
with sodium, potassium or CaH_2, decanting and distilling from sodium,
potassium, CaH_2 or $LiAlH_4$. Reaction has been speeded up by using

1,2-Dimethoxyethane (continued)
vigorous high-speed stirring and molten potassium. For virtually
complete elimination of water, 1,2-dimethoxyethane has been dried
with sodium-potassium alloy until a characteristic blue colour was
formed in the solvent at Dry Ice-Cellosolve temperatures: the solvent was kept with the alloy until distd. for use. [Ward, JACS, 83,
1296 (1961).] Alternatively, solvent refluxed with benzophenone and
NaK was dry enough if, on distn., it gave a blue colour of the
ketyl immediately on addition to benzophenone and sodium. [Ayscough
and Wilson, JCS, 5412 (1963).]

5,6-Dimethoxy-1-indanone.
    Cryst. from methanol, then sublimed.

Dimethoxymethane, b.p. 42.3°, n_D^{15} 1.35626, n_D^{20} 1.35298.
    The chief impurity is methanol, which can be removed by treatment
with sodium wire, followed by fractional distn. from sodium. The
solvent is kept dry by storing in contact with molecular sieves.
Alternatively, technical dimethoxymethane was stood with paraformaldehyde and a few drops of H_2SO_4 for 24 hrs, then distd. It could
also be purified by shaking with an equal volume of 20% NaOH,
leaving for 30 mins, and distilling. Methods of purification used
for Acetal are probably applicable.

1,4-Dimethoxynaphthalene, m.p. 87-88°.
1,5-Dimethoxynaphthalene, m.p. 183-184°.
    Cryst. from ethanol.

2,6-Dimethoxyphenol, m.p. ca 50°.
    Purified by zone refining.

3,4-Dimethoxyphenylacetic acid, m.p. 82°.
    Cryst. from water or benzene-ligroin.

Dimethyl acetal, see 1,1-Dimethoxyethane.

N,N-Dimethylacetamide, b.p. 58.0-58.5°/11.4 mm.
    Shaken with BaO for several days, refluxed with BaO for 1 hr,
then fractionally distd. under reduced pressure.

β,β-Dimethylacrylic acid, m.p. 68°.
    Cryst. from hot water or pet. ether (b.p. 60-80°).

Dimethylamine, b.p. 0°/563 mm.
    Dried by passage through a KOH-filled tower, or using sodium at 0°
during 18 hr.

Dimethylamine hydrochloride, m.p. 171°.
    Cryst. from hot CHCl₃ or absolute ethanol.

p-Dimethylaminoazobenzene, m.p. 118-119°.
    Cryst. from acetic acid or isooctane, or from 95% ethanol by adding
hot water and cooling. Dried over KOH under vac. at 50°. Carcinogenic.

p-Dimethylaminobenzaldehyde, m.p. 74-75°.
    Cryst. from water, or hexane, or from ethanol (2 ml/g), after
charcoal treatment, by adding water in excess.

p-Dimethylaminobenzoic acid, m.p. 242.5-243.5°.
    Cryst. from ethanol-water.

p-Dimethylaminobenzophenone, m.p. 92-93°.
    Cryst. from ethanol.

DL-4-Dimethylamino-2,2-diphenylvaleramide, m.p. 183-184°.
    Cryst. from aq. ethanol.

L-4-Dimethylamino-2,2-diphenylvaleramide, m.p. 136.5-137.5°.
    Cryst. from pet. ether, or ethanol.

2-Dimethylaminoethanol, b.p. 134.5-135.5°, n_D^{20} 1.4362.
    Dried with anhydrous K_2CO_3 or KOH, and fractionally distd.

N,N-Dimethylaniline, f.p. 2°, b.p. 84°/15 mm, 193°/760 mm, n_D^{25} 1.5556.
    Primary and secondary amines (including aniline and monomethylanil-
ine) can be removed by refluxing for some hours with excess acetic
anhydride, and then fractionally distilling. Crocker and Jones [JCS,
1808 (1959)] used four volumes of acetic anhydride, then distd. off
the greater part of it, and took up the residue in ice-cold dil. HCl.

N,N-Dimethylaniline (continued)

Non-basic materials were removed by ether extraction, then the dimethylaniline was liberated with ammonia, extracted with ether, dried, and distd.under reduced pressure. Metzler and Tobolsky [JACS, 76, 5178 (1954)] refluxed with only 10% (w/w) of acetic anhydride, then cooled and poured into excess 20% HCl, which, after cooling, was extracted with ethyl ether. (The amine hydrochloride remains in the aq. phase.) The HCl soln. was cautiously made alkaline to phenolphthalein, and the amine layer was drawn off, dried over KOH and fractionally distd. under reduced pressure, under nitrogen. Suitable drying agents for dimethylaniline include NaOH, BaO, $CaSO_4$ and CaH_2.

Other purification procedures include the formation of the picrate, prepared in benzene soln. and cryst. to constant m.p., then decomposed with warm aq. 10% NaOH and extracted into ether: the extract was washed with water, dried, and distd. under reduced pressure. The oxalate has also been used. The base has been fractionally cryst. by partial freezing and also first from aq. 80% ethanol, then from absolute ethanol. It has also been distd. from zinc dust, under nitrogen.

2,6-Dimethylaniline, f.p. 11.0°, b.p. 210-211°/736 mm, d_{20}^{20} 0.9819, n_D^{20} 1.5604.

Converted to its hydrochloride which, after crystn., was decomposed with alkali to give the free amine. Dried over KOH and fractionally distd.

3,4-Dimethylaniline, m.p. 51°, b.p. 226°/760 mm, 116-118°/25 mm.
Cryst. from ligroin.

9,10-Dimethylanthracene, m.p. 180-181°.
Cryst. from ethanol.

Dimethyl 2,2'-azobis(isobutyrate).
Cryst. from methanol.

1,3-Dimethylbarbituric acid, m.p. 123°.
Cryst. from water and sublimes in vacuo.

7,12-Dimethylbenz[α]anthracene, m.p. 122-123°.
    Purified by chromatography on alumina. Cryst. from acetone-ethanol.

5,6-Dimethylbenzimidazole, m.p. 205-206°.
    Cryst. from ethyl ether. Sublimed at 140°/3 mm.

2,3-Dimethylbenzoic acid, m.p. 146°.
    Cryst. from ethanol and is volatile in steam.

2,4-Dimethylbenzoic acid, m.p. 126-127°, b.p. 267°/727 mm.
    Cryst. from ethanol, and sublimes in vacuo.

2,5-Dimethylbenzoic acid, m.p. 134°, b.p. 268°/760 mm.
2,6-Dimethylbenzoic acid, m.p. 117°.
    Steam distd. and cryst. from ethanol.

3,4-Dimethylbenzoic acid, m.p. 166°.
    Cryst. from water and sublimed in vacuo.

3,5-Dimethylbenzoic acid, m.p. 170°.
    Distd. in steam, cryst. from water or ethanol and sublimed in vacuo.

2,5-Dimethyl-p-benzoquinone, see Phorone.

2,6-Dimethyl-p-benzoquinone, m.p. 72° (sealed tube).
    Cryst. from water-ethanol (8:1).

2,3-Dimethylbenzothiophene, b.p. 123-124°/10 mm, n_D^{19} 1.6171.
    Fractionated through a 90 cm Monel spiral column.

N,N-Dimethylbenzylamine, b.p. 66-67°/15 mm, 181°/760 mm.
    Refluxed with acetic anhydride for 24 hr, then fractionally distd.
The middle fraction was dried with KOH, distd. under reduced pressure,
and stored under vac. Distn. of the amine with zinc dust, at reduced
pressure, under nitrogen, has also been used.

2,3-Dimethyl-1,3-butadiene, m.p. 69-70°, n_D^{20} 1.4385.
    Distd. from NaBH$_4$, and purified by zone melting.

1,3-Dimethylbutadiene sulphone, m.p. 40.4-41.0°.
   Cryst. from ethyl ether.

2,2-Dimethylbutane, b.p. 49.7°, n_D^{20} 1.36876, n_D^{25} 1.36595.
   Distd. azeotropically with methanol, then washed with water, dried
and redistd.

2,3-Dimethylbutane, b.p. 58.0°, n_D^{20} 1.37495, n_D^{25} 1.37231.
   Distd. from sodium, passed through a column of silica gel (activated by heating in nitrogen to 350° before use) to remove unsaturated
impurities, and again distd. from sodium. Also distilled azeotropically with methanol, then washed with water, dried and redistd.

Dimethylcarbamyl chloride, b.p. 34°/1 mm, n_D^{20} 1.4511.
   Must distil under vac. to avoid decomposition.

3,3'-Dimethylcarbanilide, m.p. 225°.
   Cryst. from ethyl acetate.

Dimethylcarbonate, b.p. 90-91°, m.p. 4.65°.
   Contains small amounts of water and alcohol which form azeotropes.
Stood for several days in contact with Linde type 4A molecular sieve,
then fractionally distd. The middle fraction was frozen slowly at 2°,
several times, retaining 80% of the solvent at each cycle.

1,4-Dimethylcyclohexane, b.p. 120°.
   Freed from olefins by shaking with conc. H_2SO_4, washing with
water, drying and fractionally distilling.

5,5-Dimethyl-1,3-cyclohexanedione, see Dimedone.

Dimethyldihydroresorcinol, see Dimedone.

Dimethyl disulphide, f.p. -98°, b.p. 110°/760 mm, 40°/12 mm,
d_4^{21} 1.0647, n_D^{20} 1.5260.
   Passed over neutral alumina before use.

Dimethyl ether, see Methyl ether.

P.L.C.—HH

2,2-Dimethylethyleneimine, b.p. 70.5-71.0°.
   Freshly distd. from sodium before use.

N,N-Dimethylformamide, b.p. 76°/39 mm, 153.0°/760 mm, n_D^{20} 1.4297, n_D^{25} 1.4269.

   Decomposes slightly at its normal b.p. to give small amounts of dimethylamine and carbon monoxide. The decomposition is catalyzed by acidic or basic materials, so that even at room temperature DMF is appreciably decomposed if allowed to stand for several hours with solid KOH, NaOH or CaH_2. If these reagents are used as dehydrating agents, therefore, they should not be refluxed with the DMF. Use of $CaSO_4$, $MgSO_4$, silica gel or Linde type 4A molecular sieves is preferable, followed by distn. under reduced pressure. This procedure is adequate for most laboratory purposes. Larger amounts of water can be removed by azeotropic distn. with benzene (10% (v/v) previously dried over CaH_2), at atmospheric pressure: water and benzene distil below 80°. The liquid remaining in the distn. flask is further dried by adding $MgSO_4$ (previously ignited overnight at 300-400°) to give 25 g/l. After shaking for a day, a further quantity of $MgSO_4$ is added, and the DMF distd. at 15-20 mm pressure through a 3 ft vac.-jacketed column packed with steel helices. However, $MgSO_4$ is an inefficient drying agent, leaving about 0.01 M water in the final DMF. More efficient drying (to around 0.001-0.007 M water) is achieved by standing with powdered BaO, followed by decanting before distn., with alumina powder (50 g/l.; previously heated overnight to 500-600°), and distilling from more of the alumina; or by refluxing at 120-140° for 24 hr with triphenylchlorosilane (5-10 g/l.), then distilling at about 5 mm pressure. [Thomas and Rochow, JACS, 79, 1843 (1957).] Free amine in DMF can be detected by colour reaction with 1-fluoro-2,4-dinitrobenzene.

   [For review of purification, tests for purity and physical properties, see Juillard, PAC, 49, 885 (1977).]

d,l-2,4-Dimethylglutaric acid, m.p. 144-145°.
   Distd. in steam and cryst. from ether-pet. ether.

3,3-Dimethylglutaric acid, m.p. 103-104°, b.p. 126-127°/4.5 mm, 89-90°/2 mm.
  Cryst. from water, benzene or ether-ligroin. Dried in vacuo.

N,N-Dimethylglycinehydrazide hydrochloride, m.p. 181°.
  Cryst. by adding ethanol to a conc. aq. soln.

Dimethylglyoxime, m.p. 240°.
  Cryst. from ethanol (10 ml/g) or aq. ethanol.

2,5-Dimethyl-2,4-hexadiene, f.p. 14.5°, n_D^{20} 1.4796.
  Dist., then repeatedly fractionally cryst. by partial freezing.

5,5-Dimethylhydantoin, m.p. 175°.
  Cryst. from ethanol and sublimed in vacuo.

1,1-Dimethylhydrazine, b.p. 60.1°/702 mm.
  Fractionally distd. through a 4 ft column packed with glass helices. Pptd. as its oxalate from ethyl ether soln. After crystn. from 95% ethanol, the salt was decomposed with aq. satd. NaOH, and the free base was distd., dried over BaO and redistd. [McBride and Kruse, JACS, 79, 572 (1957).] Distn. and storage should be under nitrogen.

1,2-Dimethylimidazole, b.p. 206°/760 mm.
  Cryst. from benzene.

Dimethyl itaconate, m.p. 38°.
  Cryst. from methanol by cooling to -78°.

Dimethylmaleic anhydride, m.p. 96°, b.p. 225°/760 mm.
  Distd. in steam, cryst. from benzene-ligroin and sublimed in vacuo.

Dimethylmalonic acid, m.p. 192-193°.
  Cryst. from benzene-ligroin and sublimes in vacuo with slight decomp.

1,5-Dimethylnaphthalene, m.p. 81-82°.
  Cryst. from 85% aq. ethanol.

2,3-Dimethylnaphthalene, m.p. 104-104.5°
2,6-Dimethylnaphthalene, m.p. 110-111°, b.p. 261-262°/760 mm,
122.5-123.5°/10 mm.
   Dist. in steam and cryst. from ethanol.

N,N-Dimethyl-m-nitroaniline, m.p. 60°.
   Cryst. from ethanol.

N,N-Dimethyl-p-nitroaniline, m.p. 164.5-165.2°.
   Cryst. from ethanol or aq. ethanol. Dried under vac.

N,N-Dimethyl-p-nitrosoaniline, m.p. 85.8-86.0°.
   Cryst. from pet. ether or $CHCl_3$-CCl_4. Dried in air.

N,N-Dimethyl-p-nitrosoaniline hydrochloride, m.p. 177°.
   Cryst. from hot water in the presence of a little dil. HCl.

2,6-Dimethyl-2,4,6-octatriene, b.p. 80-82°/15 mm, $\varepsilon_{278 \ nm}$ = 42,870.
   Repeated distn. at 15 mm through a long column of glass helices,
the final distn. being from sodium under nitrogen.

Dimethylolurea, m.p. 123°.
   Cryst. from aq. 75% ethanol.

Dimethyl oxalate, m.p. 54°, b.p. 163-165°.
   Cryst. repeatedly from ethanol. Degassed under nitrogen high vac.,
and distd.

2,3-Dimethylpentane, b.p. 89.8°, n_D^{20} 1.39197, n_D^{25} 1.38946.
   Purified by azeotropic distn. with ethanol, followed by washing
out the ethanol with water, drying and distn. [Streiff et al.,
J. Res. Nat. Bur. Stand., 37, 331 (1946).]

2,4-Dimethylpentane, b.p. 80.5°, n_D^{20} 1.38145, n_D^{25} 1.37882.

   Extracted repeatedly with conc. H_2SO_4, washed with water, dried and
distd. Percolated through silica gel (previously heated in nitrogen
to 350°). Purified by azeotropic distn. with ethanol, followed by
washing out the ethanol with water, drying and distn.

2,3-Dimethylphenol, m.p. 75°, b.p. 218°/760 mm, 120°/20 mm.
   Cryst. from aq. ethanol.

2,5-Dimethylphenol, m.p. 75°, b.p. 211.5°/762 mm.
   Cryst. from ethanol-ether.

2,6-Dimethylphenol, m.p. 49°, b.p. 203°/760 mm.
   Fractionally distilled under nitrogen, cryst. from benzene or
hexane and sublimed at 38°/10 mm.

3,4-Dimethylphenol, m.p. 65°, b.p. 225°/757 mm.
3,5-Dimethylphenol, m.p. 68°, b.p. 219°.
   Heated with equal weight of conc. H_2SO_4 at 103-105° for 2-3 hr,
then diluted with four volumes of water, refluxed for 1 hr, and
either steam dist. or extracted repeatedly with ethyl ether after
cooling to room temp. The steam distillate was also extracted and
evapd. to dryness. (The purification process depends on the much
slower sulphonation of 3,5-dimethylphenol than most of its likely
contaminants.) [Kester, IEC, 24, 770 (1932).] Can also be cryst.
from water or pet. ether, and vac. sublimed.

N,N-Dimethyl-2-(α-phenyl-o-tolyloxy)ethylamine hydrochloride, see
Phenyltoloxamine hydrochloride.

N,N-Dimethyl-p-phenylazoaniline, see p-Dimethylaminoazobenzene.

Dimethyl phthalate, b.p. 282°, d_4^{25} 1.1865, n_D^{20} 1.5149.
   Washed with aq. Na_2CO_3, then distd. water, dried ($CaCl_2$) and
distd. under reduced pressure.

2,2-Dimethyl-1,3-propanediol, m.p. 128.4-129.4°, b.p. 208°/760 mm.
   Cryst. from benzene or acetone-water (1:1).

2,2-Dimethyl-1-propanol, m.p. 52°, b.p. 113.1°/760 mm.
   Difficult to distil because it is a solid at ambient temp. Purified by fractional crystallization and sublimation.

N,N-Dimethylpropionamide, b.p. 175-178°.
   Shaken over BaO for 1-2 days, then distd. at reduced pressure.

2,5-Dimethylpyrazine, b.p. 156°.
  Purified via its picrate (m.p. 150°). [Wiggins and Wise, JCS,
4780 (1956).]

3,5-Dimethylpyrazole, m.p. 107-108°.
  Cryst. from cyclohexane.

Dimethylpyridine, see Lutidine.

2,5-Dimethyl-p-quinone, m.p. 124-125°.
  Cryst. from ethanol.

2,3-Dimethylquinoxaline, m.p. 106°.
  Cryst. from distd. water.

2,4-Dimethylresorcinol, m.p. 149-150°.
  Cryst. from pet. ether (b.p. 60-80°).

meso-α,β-Dimethylsuccinic acid, m.p. 211°.
  Cryst. from distd. water.

2,2-Dimethylsuccinic acid, m.p. 141°.
  Cryst. from ethanol-ether or ethanol-chloroform.

d,l-2,3-Dimethylsuccinic acid, m.p. 129°.
  Cryst. from water.

Dimethyl sulphide, f.p. -98.27°, b.p. 0°/172 mm, 37.5-38°/760 mm,
d_4^{21} 0.8458, n_D^{25} 1.4319.
  Purified via the Hg(II) chloride complex by dissolving 1 mole of
Hg(II)Cl$_2$ in 1250 ml of ethanol and slowly adding the boiling alcoholic soln. of dimethyl sulphide to give the right ratio for
2(CH$_3$)$_2$S.3HgCl$_2$. After recrystn. of the complex to constant m.p.,
500 g of complex is heated with 250 ml conc. HCl in 750 ml water.
The sulphide is separated, washed with water, and dried with CaCl$_2$
and CaSO$_4$. Finally, it is distd. at reduced pressure from sodium.

2,4-Dimethylsulpholane, b.p. 128°/77 mm, d_4^{25} 1.1314, d^{50} 1.1106.
   Vac. distd.

Dimethyl sulphone, m.p. 109°.
   Cryst. from water. Dried over P_2O_5.

Dimethyl sulphoxide, m.p. 18.0-18.5°, b.p. 75.6-75.8°/12 mm,
190°/760 mm.
   Dried with Linde types 4A or 13X molecular sieve, by prolonged
contact and by passage through a column of the material, then distd.
under reduced pressure. Other drying agents include CaH_2, CaO, BaO
and $CaSO_4$. Can also be fractionally cryst. by partial freezing.
More extensive purification is achieved by standing overnight with
freshly heated and cooled chromatographic grade alumina. It is then
refluxed for 4 hours over CaO, dried over CaH_2, and then fractionally distilled at low pressure.

Dimethyl terephthalate, m.p. 150°.
   Purified by zone melting.

N,N-Dimethylthiocarbamayl chloride.
   Cryst. twice from pentane.

N,N-Dimethyl-o-toluidine, b.p. 68°/10 mm, 211-211.5°/760 mm,
n_D^{20} 1.53664.
   Isomers and other bases have been removed by heating in a water
bath for 100 hr with two equivalents of 20% HCl and two and a half
volumes of 40% aq. formaldehyde, then making the soln. alkaline and
separating the free base. After washing well with water it was distd.
at 10 mm pressure and redistd. at ordinary pressure [von Braun and
Aust. Ber, 47, 260 (1914)]. Other procedures include drying with
NaOH, distilling from zinc dust in an atmosphere of nitrogen under
reduced pressure, and refluxing with excess acetic anhydride in the
presence of conc. H_2SO_4 as catalyst, followed by fractional distn.
under vac.

N,N-Dimethyl-m-toluidine, b.p. 211.5-212.5°.

N,N-Dimethyl-p-toluidine, b.p. 93-94°/11 mm, n_D^{20} 1.5469.
   Methods described for N,N-Dimethylaniline are applicable.

1,3-Dimethyluracil, m.p. 121-122°.
   Cryst. from ethanol-ether.

sym-Dimethylurea, m.p. 106°.
   Cryst. from acetone-ethyl ether by cooling in an ice-bath.

Di-β-naphthol, m.p. 218°.
   Cryst. from toluene or benzene (10 ml/g).

β,β'-Dinaphthylamine, m.p. 170.5°.
   Cryst. from benzene.

2,4-Dinitroaniline, m.p. 180°, $\varepsilon_{348 \text{ nm}}$ = 12,300 in dil. aq. HClO$_4$.
   Cryst. from boiling ethanol by adding one-third volume of water
and cooling slowly. Dried in a steam oven.

2,6-Dinitroaniline, m.p. 139-140°.
   Purified by chromatography on alumina, then cryst. from benzene or
ethanol.

2,4-Dinitroanisole, m.p. 94-95°.
   Cryst. from aq. ethanol.

o-Dinitrobenzene, m.p. 116.5°.
   Cryst. from ethanol.

m-Dinitrobenzene, m.p. 90.5-91°.
   Cryst. from alkaline ethanol soln. (20 g in 750 ml 95% ethanol at
40°, plus 100 ml 2 M NaOH) by cooling and adding 2.5 l. of water.
The ppted., after filtering off, washing with water and sucking dry,
was cryst. from 120, then 80, ml of absolute ethanol. [Callow,
Callow and Emmens, BJ, 32, 1312 (1938).] Has also been cryst. from
methanol or CCl$_4$.

p-Dinitrobenzene, m.p. 173°.
   Cryst. from ethanol or benzene. Dried under vac. over P_2O_5.

2,4-Dinitrobenzenesulphenyl chloride, m.p. 96°.
   Cryst. from CCl_4.

2,4-Dinitrobenzenesulphonyl chloride, m.p. 102°.
   Cryst. from benzene or benzene-pet.ether.

3,3'-Dinitrobenzidine, see 4,4'-Diamino-3,3'-dinitrobiphenyl.

2,4-Dinitrobenzoic acid, m.p. 183°.
   Cryst. from aq. 20% ethanol (10 ml/g), dried at 100°.

2,5-Dinitrobenzoic acid, m.p. 179.5-180°.
   Cryst. from distd. water. Dried in a vac. desiccator.

2,6-Dinitrobenzoic acid, m.p. 202-203°.
   Cryst. from water.

3,4-Dinitrobenzoic acid, m.p. 166°.
   Cryst. from ethanol by addition of water.

3,5-Dinitrobenzoic acid, m.p. 205°.
   Cryst. from distd. water or 50% ethanol (4 ml/g). Dried in a vac.
desiccator or at 70° over BaO under vac. for 6 hr.

4,4'-Dinitrobenzoic anhydride, m.p. 189-190°.
   Cryst. from acetone.

3,5-Dinitrobenzoyl chloride, m.p. 69.5°.
   Cryst. from CCl_4 or pet.ether (b.p. 40-60°). It reacts readily
with water, and should be kept in sealed tubes or under pet.ether.

2,2'-Dinitrobiphenyl, m.p. 123-124°.
2,4'-Dinitrobiphenyl, m.p. 92.7-93.7°.
   Cryst. from ethanol.

4,4'-Dinitrobiphenyl, m.p. 240.9-241.8°.
  Cryst. from benzene, ethanol (charcoal) or acetone. Dried under
vac. over P_2O_5.

2,4-Dinitrochlorobenzene, m.p. 51° (stable form), 43° (unstable form),
b.p. 315°/760 mm.
  Cryst. from ethanol (stable form), or from ether (unstable form).

4,6-Dinitro-o-cresol, m.p. 85-86°.
2,4-Dinitrodiphenylamine, m.p. 157°.
  Cryst. from aq. ethanol.

4,4'-Dinitrodiphenylurea, m.p. 312° (decomp.).
  Cryst. from ethanol. Sublimed.

2,4-Dinitrofluorobenzene, m.p. 25-27°, b.p. 133°/2 mm.
  Cryst. from ether. If it is to be purified by distillation in
vacuo, the distillation unit must be allowed to cool before air is
allowed into it otherwise the residue carbonizes spontaneously and
an explosion may occur. The material is a skin irritant and may
cause serious dermatitis.

3,4-Dinitro-2-methylbenzoic acid, see 3,5-Dinitro-o-toluic acid.

1,8-Dinitronaphthalene, m.p. 170-172°.
  Cryst. from benzene.

2,4-Dinitro-1-naphthol, m.p. 81-82°.
  Cryst. from benzene or aq. ethanol.

2,4-Dinitrophenetole, m.p. 85-86°.
  Cryst. from aq. ethanol.

2,4-Dinitrophenol, m.p. 114°.
  Cryst. from benzene, ethanol, ethanol-water or water acidified with
dil. HCl, then recryst. from CCl_4. Dried in oven and stored in vac.
desiccator over $CaSO_4$.

2,5-Dinitrophenol, m.p. 108°.

    Cryst. from water containing a little ethanol.

2,6-Dinitrophenol, m.p. 63.0-63.7°.

    Cryst. from benzene-cyclohexane, aq. ethanol, water or
benzene-pet.ether (b.p. 60-80°, 1:1).

3,4-Dinitrophenol, m.p. 138°.

    Steam distd. and cryst. from water and air-dried.
Caution - explosive when dry, store with 10% water.

3,5-Dinitrophenol, m.p. 126°.

    Cryst. from benzene or chloroform-pet.ether. Should be stored with
10% water because it is explosive when dry.

2,4-Dinitrophenylacetic acid, m.p. 179° (decomp.).

    Cryst. from water.

2,4-Dinitrophenylamine, m.p. 156.6-157°.

    Cryst. from ethanol.

2,4-Dinitrophenylhydrazine, m.p. 200° (decomp.).

    Cryst. from n-butyl alcohol, dioxane, ethanol or ethyl acetate.

2,2-Dinitropropane, m.p. 53.5°.

    Cryst. from ethanol or methanol. Dried over $CaCl_2$ or under vac.
for 1 hr just above the m.p.

2,4-Dinitroresorcinol, m.p. 160°.
2,6-Dinitrothymol, m.p. 53-54°.

    Cryst. from aq. ethanol.

2,3-Dinitrotoluene, m.p. 63°.

    Distil in steam and cryst. from water or benzene-pet.ether. Store
with 10% water.

2,4-Dinitrotoluene, m.p. 70.5-71.0°.
  Cryst. from acetone, isopropyl alcohol or methanol. Dried under
vac. over H_2SO_4. Purified by zone melting.

2,5-Dinitrotoluene, m.p. 51.2°.
  Cryst. from benzene.

2,6-Dinitrotoluene, m.p. 64.3°.
  Cryst. from acetone.

3,4-Dinitrotoluene, m.p. 61°.
  Distil in steam and cryst. from benzene-pet.ether. Store with 10%
water.

3,5-Dinitro-o-toluic acid, m.p. 206°.
  Cryst. from water.

2,4-Dinitro-m-xylene, m.p. 83-84°.
  Cryst. from ethanol.

Dinonyl phthalate, d_4^{25} 0.9640, n_D^{20} 1.4825 (mainly 3,5,5-trimethyl-
hexyl isomer).
  Washed with aq. Na_2CO_3, then shaken with water. Ether was added to
break the emulsion, and the soln. was washed twice with water, and
dried ($CaCl_2$). After evaporating the ether, the residual liquid was
distd. three times under reduced pressure. It was stored in a vac.
desiccator over P_2O_5. [French and Singer, JCS, 1424 (1956).]

Dioctyl phenylphosphonate, n_D^{25} 1.4780.
  Purified as described under Diisooctyl phenylphosphonate.

Diosgenin, m.p. 204-207°.
  Cryst. from acetone.

1,3-Dioxane, b.p. 104.5°/751 mm.
  Dried with sodium and fractionally distd.

1,4-Dioxane, f.p. 11.8°, b.p. 101.3°, n_D^{15} 1.4236, n_D^{25} 1.42025, d_4^{25} 1.0292.

Usual impurities are acetaldehyde, ethylene acetal, acetic acid, water and peroxides. Peroxides can be removed (and the aldehyde content decreased) by percolation through a column of activated alumina (80 g per 100-200 ml of solvent), by refluxing with NaBH$_4$ or anhydrous stannous chloride and distilling, or by acidification with conc. HCl, shaking with ferrous sulphate and leaving in contact with it for 24 hr before filtering and purifying further.

Hess and Frahm [Ber., 71, 2627 (1938)] refluxed 2 l. of dioxane with 27 ml conc. HCl and 200 ml of water for 12 hr, with slow passage of nitrogen to remove acetaldehyde. After cooling the soln., KOH pellets were added slowly and with shaking until no more would dissolve and a second layer had separated. The dioxane was decanted, treated with fresh KOH pellets to remove any aq. phase, then transferred to a clean flask where it was refluxed for 6-12 hr with sodium, then distd. from it.

Alternatively, Kraus and Vingee [JACS, 56, 511 (1934)] heated on a steam bath with solid KOH until fresh addition of KOH gave no more resin formation (due to the acetaldehyde). After filtering through paper, the dioxane was refluxed over sodium until the surface of the metal was not further discoloured during several hours. It was then distd. from sodium.

The acetal is removed during fractional distn. (its b.p. is 82.5°). Traces of benzene, if present, can be removed as the benzene-methanol azeotrope by distn. in the presence of methanol. Distn. from LiAlH$_4$ removes aldehydes, peroxides and water. Dioxane can be dried using Linde type 4X molecular sieve. Other purification procedures include distn. from excess C$_2$H$_5$MgBr, refluxing with PbO$_2$ to remove peroxides, fractional crystn. by partial freezing and the addition of KI to dioxane-acidified with aq. HCl. Dioxane should be stored out of contact with air, preferably under nitrogen.

A detailed purification procedure is as follows:

Dioxane was stood over iron(II) sulphate for at least 2 days, under nitrogen. Then water (100 ml) and conc. HCl (14 ml) per litre of dioxane were added (giving a pale yellow colour). After refluxing for 8-12 hours with vigorous nitrogen bubbling, pellets of KOH were added to the warm soln. to form two layers and to discharge the

<u>1,4-Dioxane</u> (continued)

colour. The soln. was cooled rapidly with more KOH pellets being
added (magnetic stirring) until no more dissolved in the cooled soln.
After 4-12 hours, if the lower phase was not black, the upper phase
was decanted rapidly into a clean flask containing sodium, and
refluxed over sodium (until freshly added sodium remained bright) for
one hour. The first fraction to distil was discarded (head temperature
below 101°). The middle fraction was collected (and checked for minimum absorbency below 250 nm). The distillate was fractionally frozen
three times by cooling in a refrigerator, with occasional shaking or
stirring. This material was stored in a refrigerator. For use it was
thawed, refluxed over sodium for 48 hours, and distilled into a
container for use.

All joints were clad with Teflon tape.

One of the best tests of purity of dioxane is the formation of the
purple disodium benzophenone complex during reflux and its persistence
on cooling. (Benzophenone is better than fluorenone for this purpose,
and for the storing of the solvent.) [Carter, McClelland and Warhurst,
TFS, <u>56</u>, 343 (1960).]

<u>1,3-Dioxolane</u>, b.p. 75.0-75.2°, d_4^{20} 1.0600, n_D^{21} 1.3997.

Dried with solid NaOH, KOH or CaSO$_4$, and distd. from sodium or
sodium amalgam. Barker <u>et al</u>. [JCS, 802 (1959)] heated 34 ml of
dioxolane under reflux with 3 g of PbO$_2$ for 2 hr, then cooled and
filtered. After adding xylene (40 ml) and PbO$_2$ (2 g) to the filtrate,
the mixture was fractionally distd. Addition of xylene (20 ml) and
sodium wire to the main fraction (b.p. 70-71°) led to vigorous reaction, following which the mixture was again fractionally distd.
Xylene and sodium additions were made to the main fraction
(b.p. 73-74°) before it was finally distd.

<u>S-1,2-Dipalmitin</u>, m.p. 68-69°, $[\alpha]_D$ -2.9° (c = 8 in CHCl$_3$),
$[\alpha]_{546}^{20}$ +1.0° (c = 10 in CHCl$_3$-MeOH, 9:1).

Cryst. from chloroform-pet.ether.

<u>2,5-Di-tert.-pentylhydroquinone</u>, see 2,5-Di-tert.-amylhydroquinone.

4,4'-Di-n-pentyloxyazoxybenzene.
   Cryst. from acetone, and dried by heating under vac.

Dipentyl phthalate, see Di-n-amylhydroquinone.

Diphenic acid, m.p. 228-229°.
   Cryst. from water.

Diphenic anhydride, m.p. 217°.
   After removing free acid by extraction with cold aq. Na_2CO_3, the
residue has been cryst. from acetic anhydride.

N,N'-Diphenylacetamidine, m.p. 131°.
   Cryst. from ethanol, then sublimed under vac. at about 96° onto a
"finger" cooled in solid CO_2-methanol, with continuous pumping to
free it from occluded solvent.

Diphenylacetic acid, m.p. 147.4-148.4°.
   Cryst. from benzene or aq. 50% ethanol.

Diphenylacetonitrile, m.p. 73-75°.
   Cryst. from ethanol or pet.ether (b.p. 90-100°).

Diphenylacetylene, m.p. 62.5°, b.p. 90-97°/0.3 mm.
   Cryst. from ethanol.

Diphenylamine, m.p. 62.0-62.5°.
   Cryst. from pet.ether, methanol, or ethanol-water. Dried under vac.

Diphenylamine-2-carboxylic acid, m.p. 184°.
Diphenylamine-2,2'-dicarboxylic acid, m.p. 296-297°.
   Cryst. from ethanol.

9,10-Diphenylanthracene, n.p. 248-249°.
   Cryst. from acetic acid.

N-Diphenylanthranilic acid, see Diphenylamine-2-carboxylic acid.

trans-trans-1,4-Diphenylbuta-1,3-diene, m.p. 153-153.5°.

Its soln. in pet.ether (b.p. 60-70°) was chromatographed on an alumina-Celite column (4:1) and the column was washed with the same solvent. The main zone was cut out, eluted with ethanol and transferred to pet.ether which was then dried and evaporated. [Pinckard, Wille and Zechmesiter, JACS, 70, 1938 (1948).] Recryst. from hexane.

N,N'-Diphenylbenzidine, m.p. 245-247°.

Cryst. from toluene or ethyl acetate. Stored in the dark.

sym-Diphenylcarbazide, m.p. 171°.

Cryst. from ethanol or glacial acetic acid.

sym-Diphenylcarbazone, m.p. 118-120°.

Cryst. from ethanol (about 5 ml/g), and dried at 50°. A commercial sample, nominally sym-diphenylcarbazone but of m.p. 154-156°, was a mixture of diphenylcarbazide and diphenylcarbazone. The former was removed by dissolving 5 g of the crude material in 75 ml of warm ethanol, then adding 25 g Na_2CO_3 dissolved in 400 ml of distd. water. The alkaline soln. was cooled and extracted six times with 50 ml portions of the ethyl ether (discarded). Diphenylcarbazone was then pptd. by acidifying the alkaline soln. with 3 M HNO_3 or glacial acetic acid. It was filtered on a Büchner funnel, air dried, and stored in the dark. [Gerlach and Frazier, AC, 30, 1142 (1958).]

Diphenylcarbinol, see Benzhydrol.

1,5-Diphenylcarbohydrazide, see sym-Diphenylcarbazide.

Diphenyl carbonate, m.p. 80°.

Purified by sublimation by preparative gas chromatography with 20% Apiezon on Embacel, and by cryst. from ethanol.

1,2-Diphenyl-3,5-dioxo-4-n-butylpyrazolidine, see Phenylbutazone.

Diphenyl disulphide, m.p. 60.5°.

Cryst. from methanol. Cryst. repeatedly from hot ethyl ether, then vac.dried at 30° over P_2O_5, fused under nitrogen and redried, the whole procedure being repeated, with a final drying under vac. for 24 hr.

sym-Diphenylethane, see Bibenzyl.

Diphenyl ether, see Phenyl ether.

1,1-Diphenylethylene, b.p. 268-270°, n_D^{20} 1.6088.
   Distd. under reduced pressure from KOH. Dried with CaH_2 and redistd.

N,N'-Diphenylethylenediamine, m. p. 67.5°.
   Cryst. from aq. ethanol.

N,N'-Diphenylformamidine, m.p. 142° (137°).
   Cryst. from ethanol, gives the hydrate with aq. ethanol.

Diphenylglycollic acid, see Benzilic acid.

1,3-Diphenylguanidine, m. p. 148°.
   Cryst. from toluene, aq. acetone or ethanol, and vac. dried.

5,5-Diphenylhydantoin, m.p. 293-295°.
   Cryst. from ethanol.

1,1-Diphenylhydrazine, m.p. 126°.
   Cryst. from hot ethanol containing a little ammonium sulphide or
H_2SO_3 (to prevent atmospheric oxidation), preferably under nitrogen.
Dried in a vac. desiccator.

Diphenyl hydrogen phosphate, m.p. 99.5°.
   Cryst. from $CHCl_3$-pet.ether.

Diphenylmethane, m.p. 25.4°.
   Sublimed under vac., or distd. at 72-75°/4 mm. Cryst. from
ethanol. Purified by fractional crystn. of the melt.

Diphenylmethanol, see Benzhydrol.

1,1-Diphenylmethylamine, m.p. 34°.
   Cryst. from water.

Diphenylmethyl chloride, m.p. 17.0°, b.p. $167^{\circ}/17$ mm, n_D^{20} 1.5960.
   Dried with Na_2SO_4 and fractionally distd. at reduced pressure.

Diphenylnitrosamine, see N-Nitrosodiphenylamine.

1,9-Diphenyl-1,3,6,8-nonatetraen-5-one, see Dicinnamalacetone.

2,5-Diphenyl-1,3,4-oxadiazole, (PPD), m.p. 70° (hydrate),
$139-140^{\circ}$ (anhydr.), b.p. $248^{\circ}/16$ mm, $231^{\circ}/13$ mm.
   Cryst. from ethanol and sublimed in vacuo.

2,5-Diphenyloxazole, m.p. 74°.
   Distd. in steam and cryst. from ligroin.

N,N'-Diphenyl-p-phenylenediamine, m.p. $148-149^{\circ}$.
   Cryst. from chlorobenzene-pet.ether or benzene. Has also been cryst.
from aniline, then extracted three times with absolute ethanol.

1,1-Diphenyl-2-picrylhydrazyl, m.p. $138-139.5^{\circ}$.
   Cryst. from $CHCl_3$, or benzene-pet.ether (1:1), then degassed at
100° and $<10^{-5}$ mm Hg for about 50 hr to decompose the 1:1 molecular
complex formed with benzene.

1,3-Diphenyl-1,3-propanedione, see Dibenzoylmethane.

1,3-Diphenyl-2-propanone, see Dibenzyl ketone.

2,2-Diphenylpropionic acid, m.p. $173-174^{\circ}$.
3,3-Diphenylpropionic acid, m.p. 155°.
   Cryst. from aq. ethanol.

Diphenyl sulphide, b.p. $145^{\circ}/8$ mm, d_4^{20} 1.114, n_D^{20} 1.633.
   Washed with aq. 5% NaOH, then water. Dried with $CaCl_2$, then with
sodium. The sodium was filtered off and the diphenyl sulphide was
distd. under reduced pressure.

Diphenyl sulphone, m.p. 125°, b.p. 378° (decomp.).
   Cryst. from ethyl ether. Purified by zone melting.

Diphenylthiocarbazone, see Dithizone.

sym-Diphenylthiourea, m.p. 154°.
   Cryst. from boiling ethanol by adding hot water and allowing to cool.

1,3-Diphenyltriazene, see Diazoaminobenzene.

1,1-Diphenylurea, m.p. 238-239°.
   Cryst. from methanol.

sym-Diphenylurea, see Carbanilide.

Diphosphopyridine nucleotide, $\varepsilon_{340\ nm}$ = 6.22 x 10^6.
   Purified by chromatography on Dowex-1 ion-exchange resin. The column was prepared by washing with 3 M HCl until free of material absorbing at 260 nm, then with water, 2 M sodium formate until free of chloride ion and, finally, with water. DPN, as an 0.2% soln. in water, adjusted with NaOH to pH 8, was adsorbed on the column, washed with water, and eluted with 0.1 M formic acid. Fractions with strong absorption at 260 nm were combined, acidified to pH 2.0 with 2 M HCl, and cold acetone (approx. 5 l./g of DPN) was added slowly and with constant agitation. It was left overnight in the cold, then the ppte. was collected on the centrifuge, washed with pure acetone and dried under vac. over $CaCl_2$ and paraffin wax shavings. [Kornberg, in Methods in Enzymology, Eds. Colowick and Kaplan, Academic Press, New York, vol.3, p.876 (1957).]

Dipicolinic acid, m.p. 252° (decomp.).
   Cryst. from water.

N,N-Di-n-propylaniline, b.p. 238-241°.
   Refluxed for 3 hr with acetic anhydride, then fractionally distd. at reduced pressure.

Dipropylene glycol, b.p. 109-110°/8 mm.
   Fractionally distd. at 10-15 mm pressure, using a packed column and taking precautions to avoid absorption of water.

Di-n-propyl ether, see n-Propyl ether.

Di-n-propyl ketone, b.p. 143.5°, $d^{24.6}$ 0.8143, n_D^{20} 1.40732.
   Dried with $CaSO_4$, then distd. from P_2O_5 under nitrogen.

Di-n-propyl sulphide, b.p. 141-142°.
   Washed with aq. 5% NaOH, then water. Dried with $CaCl_2$ and distd.
from sodium. [Dunstan and Griffiths, JCS, 1344 (1962).]

Di-(4-pyridoyl)hydrazine, m.p. 254-255°.
   Cryst. from water.

α,α'-Dipyridyl, see α,α'-Bipyridyl.

2,2'-Dipyridylamine, m.p. 84° and remelts at 95° after solidifying,
b.p. 307-308°, 176-178°/13 mm.
   Cryst. from benzene.

1,2-Di-(4-pyridyl)-ethane.
   Cryst. from cyclohexane-benzene (3:1).

trans-1,2-Di-(4-pyridyl)-ethylene, m.p. 153-154°.
   Cryst. from water, solubility (1.6 g per 100 ml at 100°).

1,3-Di-(4-pyridyl)-propane, m.p. 60.5-61.5°.
   Cryst. from n-hexane-benzene (5:1).

α,α'-Diquinolyl, see α,α'-Biquinolyl.

S-1,2-Distearin, m.p. 76-77°, $[\alpha]_D$ -2.8° (c = 6.3 in $CHCl_3$),
$[\alpha]_{546}^{20}$ +1.4° (c = 10 in $CHCl_3$/MeOH, 9:1).
   Cryst. from chloroform-pet.ether.

1,3-Dithiane, m.p. 54°.
   Cryst. from 1.5 times its weight of methanol at 0°, and sublimes
at 40-50°/0.1 mm.

2,2'-Dithiobis(benzothiazole), m.p. 180°.
   Cryst. from benzene.

4,4'-Dithiodimorpholine, m.p. 124-125°.
    Cryst. from hot aq. dimethylformamide.

1,4-Dithioerythritol (DTE), (erythro-2,3-dihydroxy-1,4-dithiobutane),
m.p. 82-84°.
    Cryst. from ether-hexane.

Dithiooxamide, m.p. 41°.
    Cryst. from ethanol or sublimed.

1,4-Dithio-R,S-threitol (Cleland's reagent), m.p. 42-43°.
    Cryst. from ether and sublimes at 37°/0.005 mm.

Dithizone. Ratio of $D_{620 \text{ nm}}/D_{450 \text{ nm}}$ should be $\geq$ 1.65.
    Purified by shaking a CCl_4 soln. of the reagent with 1:100 ammonia
(which extracts it), separating the layers, acidifying the aq. layer
with HCl over a fresh portion of CCl_4 and extracting the dithizone
into the organic phase. [Cooper and Hibbits, JACS, 75, 5084 (1933).]
Instead of CCl_4, $CHCl_3$ can be used, and the final extract, after
washing with water, can be evapd. in air at 40-50° and dried in a
desiccator.

Di-p-tolyl carbonate, m.p. 115°.
    Purified by g.l.c. with 20% Apiezon on Embacel followed by
sublimation in vacuo.

N,N'-Di-o-tolylguanidine, m.p. 179° (175-176°).
    Cryst. from aq. ethanol.

Di-p-tolylphenylamine, m.p. 108.5°.
    Cryst. from ethanol.

Ditolyl phenylphosphonate, n_D^{25} 1.5758.
    Purified as described under Diisooctyl phenylphosphonate.

Di-p-tolyl sulphone, m.p. 158-159°, b.p. 405°.
    Cryst. repeatedly from ethyl ether. Purified by zone melting.

Djenkolic acid, m.p. 300-350° (decomp.).
    Cryst. from a large volume of water.

Docosane, m.p. 47°, b.p. 224°/15 mm.
    Cryst. from ethanol or ether.

Docosanoic acid, m.p. 81-82°.
    Cryst. from ligroin.

1-Docosanol, m.p. 70.8°.
    Cryst. from ether or chloroform-ether.

n-Dodecane, b.p. 97.5-99.5°/5 mm, 216°/760 mm, n_D^{20} 1.42156.
    Passed through a column of Linde type 13X molecular sieve. Stored
in contact with, and distd. from, sodium. Passed through a column of
activated silica gel. Has been cryst. from ethyl ether at -60°. Unsaturated dry material which remained after passage through silica
gel has been removed by catalytic hydrogenation (PtO_2) at
45 lb./in.^{2}, followed by fractional distn. at reduced pressure.
[Zook and Goldey, JACS, 75, 3975 (1953).] Purified by partial
crystn. from the melt.

Dodecane-1,10-dioic acid, m.p. 129°, b.p. 245°/10 mm.
    Cryst. from water or acetic acid.

Dodecane-1,12-dioic acid, m.p. 125-126°.
    Cryst. from 75% ethanol-water.

1-Dodecanethiol, b.p. 111-112°/3 mm, 153-155°/24 mm.
    Dried with CaO for several days, then distd. from CaO.

Dodecanoic acid, see Lauric acid.

1-Dodecanol, m.p. 24°, d_4^{24} 0.8309 (liquid).
    Cryst. from aq. ethanol, and vac. distd. in a spinning-band
column.

Dodecyl alcohol, see 1-Dodecanol.

Dodecyldimethylamine oxide.
    Cryst. from acetone.

Dodecyl ether, m.p. 33°.
    Vac. distd., then cryst. from methanol-benzene.

Dodecyltrimethylammonium chloride.
    Cryst. repeatedly from ethanol-ethyl ether or from methanol.

Dulcin, see p-Phenetylurea.

Dulcitol, m.p. 188-189°.
    Cryst. from water by addition of ethanol.

Durene, m.p. 79.5-80.5°.
    Chromatographed on alumina, and recryst. from aq. ethanol or benzene. Dried under vac.

Duroquinone, m.p. 110-111°.
    Cryst. from 95% ethanol.

Ecdyson, m.p. 242°, $[\alpha]_D$ +64.7° (c = 1 in EtOH).
    Cryst. from water or tetrahydrofuran-pet.ether.

Echinenone, m.p. 178-179°, $E_{1cm}^{1\%}$ = 2160 (458 nm) in pet.ether.
    Purified by chromatography on partially deactivated alumina or magnesia, or by using a thin layer of silica gel G with 4:1 cyclo-hexane in ethyl ether as the developing solvent. Stored in the dark at -20°.

Eicosane, m.p. 36-37°, b.p. 205°/15 mm, $d^{36,7}$ 0.7779, n_D^{40} 1.43453.
    Cryst. from ethanol.

Elaidic acid, m.p. 44.5°.
    Cryst. from acetic acid, then ethanol.

Ellagic acid, m.p. >360°.
    Cryst. from pyridine.

Elymoclavine.
    Cryst. from methanol.

Enniatin A, m.p. 122-122.5°.
    Cryst. from ethanol + water.

Eosin.
    Freed from inorganic halides by repeated crystn. from n-butyl
alcohol.

(−)Ephedrine, m.p. 40°, b.p. 225°.
    Cryst. from aq. 70% ethanol. Dehydrated by vac. distn., the dis-
tillate being allowed to crystallize in a vac. to prevent the uptake
of CO_2 and water vapour. The anhydrous base was then recryst. from
dry ethyl ether. [Fleming and Saunders, JCS, 4150 (1955).]

(−)Ephedrine hydrochloride, m.p. 218°, $[\alpha]_{546}^{20}$ −48±3° (c = 5 in 2 N HCl).
    Cryst. from water.

Epichlorohydrin, b.p. 115.5°.
    Distd. at atmospheric pressure, heated on a steam bath with one-
quarter its weight of CaO, then decanted and fractionally distd.

Epinephrine, m.p. 215° (decomp.).
    Dissolved in dil. aq. acid, then pptd. by addition of dil. aq.
ammonia or alkali carbonates. (Epinephrine readily oxidizes in
neutral or alkaline soln. This can be diminished if sulphite is
added.)

3,4-Epoxy-1-butane, b.p. 66.4-66.6°, n_D^{25} 1.4152.
    Dried with $CaSO_4$, and fractionally distd. through a long (126 cm)
glass helices-packed column. The first fraction contains a water
azeotrope.

Equilenin, m.p. 258-259°, $[\alpha]_D^{16}$ +87° (c = 7.1 in water).
    Cryst. from aq. ethanol.

Ergocornine, m.p. 182-184°.
    Cryst. with solvent of crystn., from methanol.

Ergocristine, m.p. 165-170°.
    Cryst. with 2 moles of solvent of crystn., from benzene.

Ergocryptine, m.p. 212-214°.
    Cryst., with solvent of crystn., from acetone, benzene or methanol.

Ergosterol, m.p. 165-166°, $[\alpha]_{546}^{20}$ -171° (in chloroform).
    Cryst. from ethyl acetate, then from ethylene dichloride.

Ergotamine, m.p. 212-214° (decomp.).
    Cryst. from benzene, then dried by prolonged heating in high vac.
Very hygroscopic.

Ergotamine tartrate, m.p. 203° (decomp.).
    Cryst. from methanol.

Eriochrome blue black R.
    Freed from metallic impurities by three pptns. from aq. soln. by
addition of HCl. The pptd. dye was dried at 60° under vac.

Eriochrome black T, $E_{1cm}^{1\%}$ = 656 at 620 nm, pH 10, using the dimethylammonium salt.
    The sodium salt (200 g) was converted to the free acid by stirr-
ing with 500 ml of 1:5 HCl, and, after several minutes, the slurry
was filtered on a sintered-glass funnel. The process was repeated and
the material was air dried after washing with acid. It was extracted
with benzene for 12 hr in a Soxhlet extractor, then the benzene was
evaptd. and the residue was air dried. A further desalting with 1:5
HCl (1 l.) was followed by crystn. from dimethylformamide (in which
it is very soluble) by forming a satd. soln. at the b.p. and allowing
to cool slowly. The crystalline dimethylammonium salt so obtained was
washed with benzene and treated repeatedly with dil. HCl to give the
insoluble free acid which, after air drying, was dissolved in
alcohol, filtered and evapd. The final material was air dried, then
dried in a vac. desiccator over $Mg(ClO_4)_2$. [Diehl and Lindstrom,
AC, 31, 414 (1959).]

Erucic acid, m.p. 33.8°.
    Cryst. from methanol.

Erythritol, m.p. 122°.
    Cryst. from distd. water and dried at 60° in a vac. oven.

Erythrityl tetranitrate, m.p. 61°.
β-Erythroidine, m.p. 99.5-100°.
    Cryst. from ethanol.

Erythrosin.
    Cryst. from water.

1,3,5-Estratrien-3-ol-17-one, m.p. 260-261°, polymorphic also
m.p. 254° and 256°, $[\alpha]_{546}^{20}$ +198° (c = 1 in dioxane).
    Cryst. from ethanol.

1,3,5-Estratrien-3β,16α,17β-triol, m.p. 283° $[\alpha]_{46}^{20}$ +66° (c = 1 in
dioxane).
    Cryst. from ethanol-ethyl acetate.

Estriol, see 1,3,5-Estratrien-3β,16α,17β-triol.

Estrone, see 1,3,5-Estratrien-3-ol-17-one.

Ethane, f.p. -172°, b.p. -88°, d_4^0 1.0493 (air = 1).
    Ethylene can be removed by passing the gas through a sintered-
glass disc into fuming H_2SO_4 and then slowly through a column of
charcoal satd. with bromine. Bromine and HBr were removed by passage
through firebrick coated with N,N-dimethyl-p-toluidine. The ethane
was also passed over KOH pellets (to remove CO_2) and dried with
$Mg(ClO_4)_2$. Further purification was by several distns. of liquified
ethane, using a condensing temperature of -195°. Yang and Gant [JPC,
65, 1861 (1961)] treated ethane by standing it for 24 hr at room
temperature in a steel bomb containing activated charcoal treated
with bromine. They then immersed the bomb in a Dry Ice-acetone bath
and transferred the ethane to an activated charcoal trap cooled in
liquid nitrogen. (The charcoal had previously been degassed by

Ethane (continued)
pumping for 24 hr at 450°.) By allowing the trap to warm slowly,
ethane was distd., retaining only the middle third. Removal of methane was achieved using Linde type 13X molecular sieve (previously
degassed by pumping for 24 hr at 450°) in a trap which, after
cooling in Dry Ice-acetone, was satd. with the ethane. After pumping
for 10 min, the ethane was recovered by warming the trap to room
temperature. (The final gas contained less than 10^{-4} mole % of
either ethylene or methane.)

Ethanethiol, b.p. $32.9^{\circ}/704$ mm, d_4^{25} 0.83147.

    Dissolved in aq. 20% NaOH, extracted with a small amount of
benzene and then steam distd. until clear. After cooling, the
alkaline soln. was acidified slightly with 15% H_2SO_4 and the thiol
was distd. off, dried with $CaSO_4$ or $CaCl_2$, and fractionally distd.
under nitrogen. [Ellis and Reid, JACS, 54, 1674 (1932).]

Ethanol, b.p. 78.3°, d_4^{15} 0.79360, d_4^5 0.78506, n_D^{20} 1.36139,
n_D^{25} 1.35941.

    Usual impurities of fermentation alcohol are fusel oils (mainly
higher alcohols, especially pentanols), aldehydes, esters, ketones
and water. With synthetic ethanol, likely impurities are water,
aldehydes, aliphatic esters, acetone and diethyl ether. Traces of
benzene are likely in ethanol dehydrated with azeotropic distn.
using benzene. Anhydrous ethanol is very hygroscopic. Water (down to
0.05%) can be detected by formation of a voluminous ppte. when
aluminium ethoxide in benzene is added to a test portion. Rectified
spirit (95% ethanol) is converted to "absolute" (99.5%) ethanol by
refluxing with freshly ignited CaO (250 g/l.) for 6 hr, standing
overnight and distilling with precautions to exclude moisture.

    Numerous methods are available for further dehydration of
"absolute" ethanol. Lund and Bjerrum [Ber., 64, 210 (1931)] used
reaction with magnesium ethoxide, prepared by placing 5 g of clean
dry magnesium turnings and 0.5 g iodine (or a few drops of CCl_4) in
a 2 l. flask, followed by 50-75 ml of absolute ethanol, and warming
the mixture until a vigorous reaction occurs. When this abates,
heating is continued until all of the magnesium is converted to the
ethoxide. Up to 1 l. of ethanol is added and, after an hour's reflux,

Ethanol (continued)

it is distilled off. The water content should then be below 0.05%.
Walden, Ulich and Laun [Z. physik. Chem., 114, 275 (1925)] used
amalgamated aluminium chips, prepared by degreasing aluminium chips,
treating with alkali until hydrogen was vigorously evolved, washing
with water until the washings were weakly alkaline and then stirring
with 1% $HgCl_2$ soln. After 2 min, the chips were washed quickly with
water, then alcohol, then ether, and dried with filter paper. (The
amalgam becomes warm.) These chips are added to the ethanol, which is
then gently warmed for several hours until evolution of hydrogen
ceases. The alcohol is distd. and aspirated for some time with pure
dry air. Smith [JCS, 1288 (1927)] reacted 1 l. of "absolute" ethanol
in a 2 l. flask with 7 g of clean dry sodium, and added 25 g of pure
ethyl succinate (27.5 g of pure ethyl phthalate is an alternative),
refluxing the mixture for 2 hr in a system protected from moisture
and then distilling. A modification uses 40 g of ethyl formate,
instead, so that sodium formate separates out and, during reflux,
the excess ethyl formate decomposes to CO and ethanol.

Dehydrating agents suitable for use with ethanol include Linde
type 4X molecular sieve, calcium metal, and CaH_2. The calcium hydride
(2 g) is crushed to a powder and dissolved in 100 ml absolute ethanol
by gently boiling. About 70 ml of the ethanol is distilled off to
remove any ammonia before the remainder is poured into 1 l. of approx.
99.9% ethanol in a still, where it is boiled under reflux for 20 hr,
while a slow stream of pure, dry hydrogen is passed through. It is
then distd. [Rüber, Z. Elektrochem., 29, 334 (1923).] If calcium is
used for drying, about ten times the theoretical amount should be
taken, and traces of ammonia are removed by passing dry air into the
vapour during reflux.

Ethanol can be freed from traces of basic materials by distn.
from a little 2,4,6-trinitrobenzoic acid or sulphanilic acid. Benzene
can be removed by fractional distn. after adding a little water (the
benzene-water-ethanol azeotrope distils at 64.9°); the alcohol is
then redried using one of the methods given above. Alternatively,
careful fractional distn. can separate benzene as the benzene-
ethanol azeotrope (b.p. 68.2°). Aldehydes can be removed from ethanol
by digesting with 8-10 g dissolved KOH and 5-10 g aluminium or zinc
per l., followed by distn. Another method is to heat under reflux

Ethanol (again continued)

with KOH (20 g/l.) and AgNO$_3$ (10 g/l.) or to add 2.5-3 g of lead
acetate in 5 ml of water to 1 l. of ethanol, followed (slowly and
without stirring) by 5 g of KOH in 25 ml of ethanol: after 1 hr the
flask is shaken thoroughly, then set aside overnight before filtering
and distilling. The residual water can be removed by standing the
distillate over activated aluminium amalgam for 1 week, then filtering and distilling. Distn. of ethanol from Raney nickel eliminates
catalyst poisons.

Other purification procedures include pre-treatment with conc.
H$_2$SO$_4$ (3 ml/l.) to eliminate amines, and with KMnO$_4$ to oxidize aldehydes, refluxing with KOH to resinify aldehydes, passage through
silica gel, and drying over CaSO$_4$. Water can be removed by azeotropic
distn. with dichloromethane (azeotrope boils at 38.1° and contains
1.8% water) or 2,2,4-trimethylpentane.

Ethanolamine, see 2-Aminoethanol.

m-Ethoxy-N,N-diethylaniline, b.p. 141-142°/15 mm.
    Refluxed for 3 hr with acetic anhydride, then fractionally distd.
at reduced pressure.

S-Ethionine, m.p. 272-274° (decomp.), [α]$_D^{25}$ +23.7° (in 5 M HCl).
    Likely impurities: N-acetyl-(R and S)-ethionine, S-methionine,
R-ethionine. Cryst. from water by adding 4 volumes of ethanol.

2-Ethoxyethanol, b.p. 134.8°, n$_D^{20}$ 1.40751.
    Dried with CaSO$_4$ or K$_2$CO$_3$, filtered and fractionally distd. Perox-
ides can be removed by refluxing with anhydrous SnCl$_2$ or by filtration under slight pressure through a column of activated alumina.

2-(2-Ethoxyethoxy)ethanol, see Diethylene glycol monoethyl ether.

Bis-2-ethoxyethyl ether, b.p. 76°/32 mm.
    Refluxed with LiAlH$_4$ for several hours, distd. at reduced pressure
and stored with CaH$_2$ under nitrogen. Passed through (alkaline)alumina.

β-Ethoxyethyl methacrylate.
    Purified as described under Methyl methacrylate.

α-Ethoxynaphthalene, b.p. 282°.

Fractionally distd. (twice) under vac., then dried with, and distd. under vac. from, sodium.

β-Ethoxynaphthalene, m.p. 35.6-36.0°, b.p. 142-143°/12 mm, 274-275°/730 mm.

Cryst. from pet.ether. Dried under vac.

Ethyl acetate, b.p. 77.1°, n_D^{20} 1.37239, n_D^{25} 1.36979.

The commonest impurities - water, ethanol and acetic acid - can be removed by washing with aq. 5% Na_2CO_3, then with satd. aq. $CaCl_2$ or NaCl, and drying with K_2CO_3, $CaSO_4$ or $MgSO_4$. More efficient drying is achieved if the solvent is further dried with P_2O_5, CaH_2 or molecular sieve before distn. Calcium oxide can also be used. Alternatively, ethanol can be converted to ethyl acetate by refluxing with acetic anhydride (about 1 ml per 10 ml of ester); the liquid is then fractionally distd., dried with K_2CO_3 and redistd.

Ethyl acetoacetate, b.p. 71°/12 mm, 100°/80 mm.

Shaken with small amounts of satd. aq. $NaHCO_3$ (until no further effervescence), then with water. Dried with $MgSO_4$ or $CaCl_2$. Distd. under reduced pressure.

Ethyl acrylate, b.p. 99.5°.

Washed repeatedly with aq. NaOH until free from inhibitors such as hydroquinone, then washed with satd. aq. $CaCl_2$, dried with $CaCl_2$ and distd. under reduced pressure.

Ethyl alcohol, see Ethanol.

Ethylamine, b.p. 16°.

Condensed in an all-glass apparatus cooled by circulating ice-water, and stored with KOH pellets.

Ethylamine hydrochloride, m.p. 109-110°.

Cryst. from absolute ethanol or methanol-$CHCl_3$.

Ethyl p-aminobenzoate, m.p. 92°.

Cryst. from ethanol-water. Air dried.

p-Ethylaniline, b.p. 88°/8 mm.

Dissolved in benzene, then acetylated. The acetyl derivative was cryst. repeatedly from benzene-pet.ether, and hydrolyzed by refluxing 50 g with 500 ml of water and 115 ml of conc. H_2SO_4 until the soln. became clear. The amine sulphate was isolated, suspended in water and solid NaOH was added to regenerate the free base, which was separated, dried and distd. from zinc dust under vac. [Berliner and Berliner, JACS, 76, 6179 (1954).]

Ethylbenzene, b.p. 136.2°, n_D^{20} 1.49594, n_D^{25} 1.49330.

Shaken with cold conc. H_2SO_4 until a fresh portion of acid remained colourless, then washed with aq. 10% NaOH or $NaHCO_3$, followed by distd. water until neutral. Dried with $MgSO_4$ or $CaSO_4$, then dried further with, and distd. from, sodium, sodium hydride or CaH_2. Can also dry by passing through silica gel. Sulphur-containing compounds have been removed by prolonged shaking with mercury. Also purified by fractional freezing.

Ethyl benzoate, b.p. 98°/19 mm, 212.4°/760 mm, n_D^{15} 1.50748, n_D^{25} 1.5043.

Washed with aq. 5% Na_2CO_3, then satd. $CaCl_2$, dried with $CaSO_4$ and distd. under reduced pressure.

Ethyl bixin, m.p. 138°.

Cryst. from ethanol.

Ethyl bromide, b.p. 0°/165 mm, 38°/745 mm, n_D^{20} 1.4241.

The main impurities are usually ethanol and water, with both of which it forms azeotropes. Ethanol and unsaturated compounds can be removed by washing with conc. H_2SO_4 until no further coloration is produced. The ethyl bromide is then washed with water, aq. Na_2CO_3, and water again, then dried with $CaCl_2$, $MgSO_4$ or CaH_2, and distd. from P_2O_5. Olefinic impurities can also be removed by storing the ethyl bromide in daylight with elementary bromine, later removing the free bromine by extraction with dil. aq. Na_2SO_3, drying the ethyl bromide with $CaCl_2$ and fractionally distilling. Alternatively, unsaturated compounds can be removed by bubbling oxygen containing about 5% ozone through the liquid for an hour, then washing with

Ethyl bromide (continued)

aq. Na_2SO_3 to hydrolyze ozonides and remove hydrolysis products, followed by drying and distn.

Ethyl bromoacetate, b.p. 158-158.5°/758 mm.

Ethyl α-bromopropionate, b.p. 69-70°/25 mm.

   Washed with satd. aq. Na_2CO_3 (three times), 50% aq. $CaCl_2$ (three times) and satd. aq. NaCl (twice). Dried with $MgSO_4$, $CaCl_2$ or $CaCO_3$, and distd.

2-Ethyl-1-butanol, b.p. 146.3°, n_D^{15} 1.4243, n_D^{25} 1.4205.

   Dried with $CaSO_4$ for several weeks, filtered and fractionally distd.

2-Ethylbut-1-ene, b.p. 66.6°.

   Washed with satd. aq. NaOH, then water. Dried with $CaCl_2$, filtered and fractionally distd.

Ethyl n-butyrate, b.p. 49°/50 mm, 119-120°/760 mm.

   Dried with anhydrous $CuSO_4$ and distd. under dry nitrogen.

Ethyl carbamate, m.p. 48.0-48.6°.

   Cryst. from benzene.

Ethyl carbonate, b.p. 124-125°.

   Washed with aq. 10% Na_2CO_3, then aq. satd. $CaCl_2$. Dried with $MgSO_4$ and distd.

N-Ethylcarbazole, m.p. 67-68°.

   Cryst. from ethanol-water, or from isopropanol. Dried below 55°.

Ethyl chloride, b.p. 12.4°.

   Passed through absorption towers containing, successively, conc. H_2SO_4, NaOH pellets, P_2O_5 on glass wool, or soda-lime, $CaCl_2$, P_2O_5. Condensed into a flask containing CaH_2 and fractionally distd. Has also been purified by illumination in the presence of bromine at 0° using a 1000 W lamp, followed by washing, drying and distn.

Ethyl chloroacetate, b.p. 143-143.2°, n_D^{25} 1.4192.

   Shaken with satd. aq. Na_2CO_3 (three times), aq. 50% $CaCl_2$ (three times) and sat. aq. NaCl (twice). Dried with Na_2SO_4 or $MgSO_4$ and distd.

Ethyl cinnamate, f.p. 6.7°, b.p. 127°/6 mm, 272.7°/768 mm, n_D^{20} 1.55983.

   Washed with aq. 10% Na_2CO_3, then with water, dried ($MgSO_4$), and distd. The purified ester was saponified with aq.KOH, and, after acidifying the soln., cinnamic acid was isolated, washed and dried. The ester was reformed by refluxing for 15 hr the cinnamic acid (25 g) with absolute ethanol (23 g), conc. H_2SO_4 (4 g) and dry benzene (100 ml), after which it was isolated, washed, dried and distd. at reduced pressure. [Jeffery and Vogel, JCS, 658 (1948).]

Ethyl trans-crotonate, b.p. 137°.

   Washed with aq. 5% Na_2CO_3, then with satd. aq. $CaCl_2$, dried with $CaCl_2$ and distd.

Ethyl cyanoacetate, b.p. 206.0°, n_D^{20} 1.41751.

   Shaken several times with aq. 10% Na_2CO_3, washed well with water, dried with Na_2SO_4 and fractionally distd.

Ethylcyclohexane, b.p. 131.8°, n_D^{20} 1.43304, n_D^{25} 1.43073.

   Purified by azeotropic distn. with 2-ethoxyethanol, then the alcohol was washed out with water and, after drying, the ethylcyclo-hexane was redistd.

Ethylcyclohexanecarboxylate, b.p. 92-93°/34 mm, 76-77°/10 mm.

   Washed with N sodium hydroxide soln., then water dried with Na_2SO_4 and distd.

Ethyl dibromoacetate, b.p. 81-82.5°/14.5 mm, n_D^{22} 1.4973.

   Washed briefly with conc. aq. $NaHCO_3$, then with aq. $CaCl_2$. Dried with $CaSO_4$ and distd. at reduced pressure.

Ethyl α,β-dibromo-β-phenylpropionate, m.p. 75°.

   Cryst. from pet.ether (b.p. 60-80°).

Ethyl dichloroacetate, b.p. 131.0-131.5°/401 mm.

Shaken with aq. 3% $NaHCO_3$ to remove free acid, washed with distd. water, dried for 3 days with $CaSO_4$ and distd. under reduced pressure.

Ethylene, m.p. -169.4°, b.p. -102.4°/700 mm.

Purified by passage through a series of towers containing either asbestos coated with NaOH, then molecular sieve (or anhydrous $CaSO_4$), or a cuprous ammonia soln., then conc. H_2SO_4, followed by KOH pellets. Alternatively, ethylene has been condensed in liquid nitrogen, with melting, freezing and pumping to remove air before passage through an activated charcoal trap, followed by a further condensation in liquid air. A sputtered sodium trap has also been used, to remove oxygen.

[Ethylene bis(oxyethylenenitrilo)]tetraacetic acid, see Ethylene glycol bis(β-aminoethylether)tetraacetic acid.

Ethylene carbonate, m.p. 37°, n_D^{40} 1.4199.

Dried over P_2O_5 then fractionally distd. at 10 mm pressure. Cryst. from dry ethyl ether.

Ethylene chlorohydrin, see 2-Chloroethanol.

Ethylenediamine, f.p. 11.0°, b.p. 117.0°, n_D^{20} 1.45677, n_D^{30} 1.4513.

Forms a constant boiling (118.5°) mixture with water (15% water). Material containing 30% water was dried with solid NaOH (600 g/l.), heating on a waterbath for 10 hr. Above 60°, separation took place into two phases. The hot ethylenediamine layer was decanted, digested with 150 g of NaOH for several hours, then decanted, refluxed with 40 g of sodium for 2 hr and distd. [Putnam and Kobe, Trans. Electro-chem. Soc., 74, 609 (1938).] Ethylenediamine is usually distd. under nitrogen. Type 5A Linde molecular sieve (70 g/l.), then a mixture of 50 g CaO and 15 g KOH/l., with further dehydration of the supernatant with molecular sieve has also been used for drying this diamine, followed by distn. from molecular sieve and, finally, from sodium metal. A spectroscopically improved material was obtained by shaking with freshly baked alumina (20 g/l.) before distn. The base should be protected from CO_2.

N,N'-Ethylenediaminediacetic acid.
Ethylenediamine dihydrochloride.
   Cryst. from water.

Ethylenediaminetetraacetic acid.
   Dissolved in aq. KOH or ammonium hydroxide, and pptd. with dil.HCl
or HNO$_3$, twice. Boiled with distd. water twice to remove mineral
acid, then recryst. from water or N,N-dimethylformamide. Dried at
110°.

Ethylene dibromide, see 1,2-Dibromoethane.

Ethylene dichloride, see 1,2-Dichloroethane.

Ethylene dimethacrylate.
   Distd. through a short Vigreux column at about 1 mm pressure, in
the presence of 3% (w/w) of phenyl-β-naphthylamine.

Ethylene dimyristate, m.p. 61.7^o.
   Cryst. from benzene-methanol or ethyl ether-methanol, and dried
in a vac. desiccator.

(Ethylenedinitrilo)tetraacetic acid, see Ethylenediaminetetraacetic
acid.

Ethylene dipalmitate, m.p. 69.1^o.
Ethylene distearate, m.p. 75.3^o.
   Cryst. from benzene-methanol or ethyl ether-methanol and dried
in a vac. desiccator.

Ethylene glycol, b.p. 68°/4 mm, 197.9^o/760 mm, d$_4^{25}$ 1.10986,
n$_D^{15}$ 1.43312, n$_D^{25}$ 1.43056.
   Very hygroscopic, and also likely to contain higher diols. Dried
with CaO, CaSO$_4$, MgSO$_4$ or NaOH and distd. under vac. Further dried
by reaction with sodium under nitrogen, refluxing for several hours
and distilling. The distillate was then passed through a column of
Linde type 4A molecular sieve and finally distd. under nitrogen,
from more molecular sieve. Fractionally crystd.

Ethylene glycol bis(β-aminoethyl ether)-N,N'-tetraacetic acid.
   Dissolved in aq. NaOH, pptd. by addition of aq. HCl, washed with water, and distd.

Ethylene glycol diacetate, b.p. 190.2°, n_D^{20} 1.4150, n_D^{26} 1.4188.
   Dried with $CaCl_2$, filtered (taking precautions to exclude moisture), and fractionally distd. under reduced pressure.

Ethylene glycol dibutyl ether.
   Shaken with aq. 5% Na_2CO_3, dried with $MgSO_4$ and stored with chromatographic alumina to prevent peroxide formation.

Ethylene glycol diethyl ether, b.p. 121.5°.
   After refluxing for 12 hr a mixture of the ether (2 l.), conc. HCl (27 ml) and water (200 ml), with slow passage of nitrogen, the soln. was cooled, and KOH pellets were added slowly and with shaking until no more dissolved. The organic layer was decanted, treated with some KOH pellets and again decanted. It was refluxed with, and distd. from, sodium immediately before use. Alternatively, after removal of peroxides by treatment with activated alumina, the ether has been refluxed in the presence of the blue ketyl formed by sodium-potassium alloy with benzophenone, then distd.

Ethylene glycol dimethyl ether, b.p. 85°.
   Purified by distn. from $LiAlH_4$ or sodium.

Ethylene glycol monobutyl ether, see 2-Butoxyethanol.

Ethylene glycol monoethyl ether, see 2-Ethoxyethanol.

Ethylene glycol monomethyl ether, see 2-Methoxyethanol.

Ethylene oxide, b.p. 10.9°.
   Dried with $CaSO_4$, then distd. from crushed NaOH. Has also been purified by its passage, as a gas, through towers containing solid NaOH.

Ethylene thiourea, m.p. 203-204°.
   Cryst. from ethanol or amyl alcohol.

Ethylene urea, m.p. 131°.
  Cryst. from methanol (charcoal).

Ethylenimine, b.p. 55.5°.
  Dried with BaO, and distd. from sodium under nitrogen. (Toxic.)

Ethyl ether, b.p. 34.6°, d_4^{20} 0.714, n_D^{15} 1.35555, n_D^{20} 1.35272.
  Usual impurities are water, ethanol, diethyl peroxide (which is
explosive when concentrated), and aldehydes. Peroxides (detected by
liberation of iodine from weakly acid (HCl) solutions of KI, or by
the blue colour in the ether layer when 1 mg of $Na_2Cr_2O_7$ and 1 drop
of dil. H_2SO_4 in 1 ml of water is shaken with 10 ml of ether) can be
removed in several different ways. The simplest method is to pass dry
ether through a column of activated alumina (80 g Al_2O_3/700 ml of
ether). More commonly, 1 l. of ether is shaken repeatedly with 5-10
ml of a soln. comprising 6.0 g ferrous sulphate and 6 ml conc. H_2SO_4
in 110 ml of water. Aqueous 10% Na_2SO_3 or stannous chloride can also
be used. The ether is then washed with water, dried for 24 hr with
$CaCl_2$, filtered and dried further by adding sodium wire until it
remains bright. The ether is stored in the dark in a cool place,
until distd. from sodium just before use. Peroxides can also be
removed by wetting ether with a little water, then adding excess
$LiAlH_4$ or CaH_2 and leaving to stand for several hours. (This also
dries the ether.)
  Werner [Analyst, 58, 335 (1933)] removed peroxides and aldehydes
by adding 8 g $AgNO_3$ in 60 ml of water to 1 l. of ether, then 100 ml
of 4% NaOH and shaking for 6 min. Fierz-David [Chimia, 1, 246 (1947)]
shook 1 l. of ether with 10 g. of a copper-zinc couple. (This reagent
was prepared by suspending 10 g of zinc dust in 50 ml of hot water,
adding 5 ml of 2 M HCl, and decanting after 20 sec, washing twice
with water, covering with 50 ml of water and 5 ml of 5% cuprous
sulphate with swirling. The liquid was decanted and discarded, and
the residue was washed three times with 20 ml of ethanol and twice
with 20 ml of ethyl ether.)
  Aldehydes can be removed from ethyl ether by distn. from hydrazine
hydrogen sulphate, phenylhydrazine or thiosemicarbazide. Peroxides
and oxidizable impurities have also been removed by shaking with
strongly alkaline satd. $KMnO_4$ (with which the ether was left to

Ethyl ether (continued)

stand in contact, for 24 hr), followed by washing with water, conc.
H_2SO_4, and water again, then drying ($CaCl_2$) and distn. from sodium.
Other purification procedures include distn. from sodium triphenyl-
methide or butyl magnesium bromide, and drying with solid NaOH or
P_2O_5.

2-Ethylethylenimine, b.p. 88.5-89°. (Toxic.)
  Freshly distd. from sodium before use.

Ethyl formate, b.p. 54.2°, d_4^{30} 0.9096, n_D^{20} 1.35994, n_D^{25} 1.3565.

  Free acid or alcohol is removed by standing with anhydrous K_2CO_3,
with occasional shaking, then decanting and distilling from P_2O_5.
Alternatively, the ester can be stood with CaH_2 for several days,
then distd. from fresh CaH_2. Cannot be dried with $CaCl_2$ because it
reacts rapidly with the ester to form a crystalline compound.

2-Ethyl-1-hexanol, b.p. 184.3°.
  Dried with sodium, then fractionally distd.

2-Ethylhexyl vinyl ether.
  Usually contains amines as polymerization inhibitors. These are
removed by fractional distn.

Ethylidene chloride, see 1,1-Dichloroethane.

Ethyl iodide, b.p. 72.4°, n_D^{15} 1.51682, n_D^{25} 1.5104.
  Drying with P_2O_5 is unsatisfactory, and with $CaCl_2$ is incomplete.
It is probably best to dry with sodium wire and distil. [Hammond, et
al., JACS, 82, 704 (1960).] Exposure of ethyl iodide to light leads
to rapid decomposition, with the liberation of iodine. Free iodine
can be removed by shaking with several portions of dil. aq. $Na_2S_2O_3$
or Na_2SO_3 (until the colour is discharged), followed by washing with
water, drying (with $CaCl_2$, then sodium), and distn. The distilled
ethyl iodide is stored, over mercury, in a dark bottle away from
direct sunlight. Other purification procedures include passage through
a 60 cm column of silica gel, followed by distn.; and treatment with
elemental bromine, extraction of free halogen with $Na_2S_2O_3$ solution,

Ethyl iodide (continued)
followed by washing with water, drying and distn. Free iodine and HI
have also been removed by direct distn. through a LeBel-Henninger
column containing copper turnings. Purification by shaking with
alkaline solns., and storage over silver, are reported to be
unsatisfactory.

Ethyl isobutyrate, b.p. 110°.
   Washed with aq. 5% Na_2CO_3, then with satd. aq. $CaCl_2$. Dried with
$CaSO_4$ and distd.

3-Ethylisothionicotinamide, m.p. $164-166^\circ$ (decomp.).
   Cryst. from ethanol.

Ethyl isovalerate, b.p. 134.7°, n_D^{20} 1.39621, n_D^{25} 1.3975.
   Washed with aq. 5% Na_2CO_3, then satd. aq. $CaCl_2$. Dried with $CaSO_4$
and distd.

Ethyl malonate, b.p. $198-199^\circ$, d_{25}^{25} 1.0507.
   The ester (250 g) has been heated on a steam bath for 36 hr with
absolute ethanol (125 ml) and conc. H_2SO_4 (75 ml), then fractionally
distd. under reduced pressure.

Ethyl mercaptan, see Ethanethiol.

Ethyl methacrylate, b.p. 59°/100 mm.
   Washed successively with aq. 5% $NaNO_2$, 5% $NaHSO_3$, 5% NaOH, then
water. Dried with $MgSO_4$, added 0.2% (w/w) of phenyl-β-naphthylamine,
and distd. through a short Vigreux column. [Schultz, JACS, 80, 1854
(1958).]

3-Ethyl-4-methylpyridine, b.p. 76°/12 mm, 194.5°/750 mm.
   Dried with solid NaOH, and fractionally distd.

5-Ethyl-2-methylpyridine, b.p. 178.5°/765 mm.
   Purified by conversion to the picrate, crystn. and regeneration of
the free base.

N-Ethylmorpholine, b.p. 138-139°/763 mm.

Distd. twice, then converted by HCl gas into the hydrochloride (extremely deliquescent) which was cryst. from anhydrous ethanol-acetone (1:2). [Herries, Mathias and Rabin, BJ, 85, 127 (1962).]

Ethyl p-nitrobenzoate, m.p. 56°.

Dissolved in ethyl ether and washed with aq. alkali, then the ether was evaporated and the solid recryst. from ethanol.

Ethyl orthoformate, b.p. 144°.

Shaken with aq. 5% NaOH, dried with solid KOH and distd. from sodium through a 20 cm Vigreux column.

o-Ethylphenol, f.p. -34°.
p-Ethylphenol, m.p. 45.1°, b.p. 218.0°/762 mm.

Non-acidic impurities were removed by passing steam through a boiling soln. containing 1 mole of the phenol and $1^3/4$ moles of NaOH (as aq. 10% soln.). The residue was cooled and acidified with 30% (v/v) H_2SO_4, and the free phenol was extracted into ethyl ether. The extract was washed with water, dried with $CaSO_4$ and the ether was evaporated. The phenol was distd. at 100 mm pressure through a Stedman gauze-packed column. It was further purified by fractional crystn. by partial freezing, and by zone refining, under nitrogen. [Biddiscombe et al., JCS, 5764 (1963).]

Alternative purification is via the benzoate, as for Phenol.

Ethyl phenylacetate, b.p. 99-99.3°/14 mm.

Shaken with sat. aq. Na_2CO_3 (three times), aq. 50% $CaCl_2$ (twice) and satd. aq. NaCl (twice). Dried with $CaCl_2$ and distd. under reduced pressure.

2-Ethyl-2-phenylglutarimide, m.p. 84°.

Cryst. from ether or ethyl acetate + pet.ether.

3-Ethyl-5-phenylhydantoin, m.p. 94°.

Cryst. from water.

3-Ethyl-3-phenyl-2,6-piperidinedione, m.p. 84°.

Cryst. from ethyl ether or ethyl acetate-pet.ether.

Ethyl propionate, b.p. 99.1^0, n$_D^{15}$ 1.38643, n$_D^{20}$ 1.38394.
   Treated with anhydrous CuSO$_4$ and distd. under nitrogen.

2-Ethylpyridine, b.p. 148.6^0.
4-Ethylpyridine, b.p. 168.2-168.3^0.
   Dried with BaO, and fractionally distd. Purified by conversion to
the picrate, recrystn. and regeneration of the free base.

4-Ethylpyridine 1-oxide, m.p. 109-110^0.
   Cryst. from acetone-ether.

1-Ethylquinolinium iodide, see Quinoline ethiodide.

Ethyl red, m.p.150-152^0.
   Cryst. from ethanol-ethyl ether.

Ethyl stearate, m.p. 33^0, b.p. 213-215^0/15 mm.
   The solid portion was separated from the partially solid starting
material, then cryst. twice from ethanol, dried by azeotropic distn.
with benzene, and fractionally distd. in a spinning-band column at
1 mm pressure. [Welsh, TFS, 55, 52 (1959).]

Ethyl sulphide, b.p. 92.1^0, n$_D^{15}$ 1.44550.
   Fractionally distd. from sodium.

N-Ethylthiourea, m.p.110^0.
   Cryst. from ethanol, methanol or ether.

Ethyl trichloroacetate, b.p. 100-100.5^0/30 mm.
   Shaken with satd. aq. Na$_2$CO$_3$ (three times), aq. 50% CaCl$_2$ (three
times), satd. aq. NaCl (twice), then dried with CaCl$_2$ and distd.
under reduced pressure.

Ethyl vinyl ether, b.p. 35.5^0.
   Contains polymerization inhibitors (usually amines) which can be
removed by fractional distn. Redistd. from sodium.

Ethynylbenzene, see Phenylacetylene.

Etiocholane, m.p. 78-80°.
  Cryst. from acetone.

Etiocholanic acid, m.p. 228-229°.
  Cryst. from glacial acetic acid.

Etioporphyrin, m.p. 360-363°.
  Cryst. from pyridine or $CHCl_3$-pet.ether.

Farnesyl pyrophosphate.
  Purified by chromatography on Whatman No.3 MM paper in a system of
isopropyl alcohol-isobutyl alcohol-ammonia-water (40:20:1:30) (v/v).
Stored as Li or NH_4 salt at 0°.

Ferulic acid, see 4-Hydroxy-3-methoxycinnamic acid.

Flavin adenine dinucleotide.
  Small quantities, purified by paper chromatography using
tert.-butyl alcohol-water, cutting out the main spot and eluting with
water. Larger amounts can be pptd. from water as the uranyl complex
by adding a slight excess of uranyl acetate to a soln. at pH 6.0,
dropwise and with stirring. The soln. is stood overnight in the cold,
and the ppte. is centrifuged off, washed with small portions of cold
ethanol, then with cold, peroxide-free ethyl ether. It is dried in
the dark under vac. over P_2O_5 at 50-60°. The uranyl complex is
suspended in water and, after adding sufficient 0.01 M NaOH to adjust
to pH 7.0, the ppte. of uranyl hydroxide is removed by centrifuging.
[Huennekens and Felton, in Methods in Enzymology, Eds. Colowick and
Kaplan, Academic Press, New York, vol.3, p.954 (1957).] Can also
cryst. from water.

Flavin mononucleotide.
  Purified by paper chromatography using tert.-butyl alcohol-water,
cutting out the main spot and eluting with water.

Flavone, m.p. 99-100°.
  Cryst. from pet.ether.

Fluoranthrene, m.p. 110-111°.
  Purified by chromatography of CCl_4 solns. on alumina, with benzene as eluant. Cryst. from ethanol or benzene. Purified by zone melting.

2-Fluorenamine, m.p. 131-132°.
  Cryst. from ethanol.

9-Fluorenamine, m.p. 64-65°.
  Cryst. from hexane.

Fluorene, m.p. 114.7-115.1°, b.p. 160°/15 mm.
  Purified by chromatography of CCl_4 or pet.ether (b.p. 40-60°) soln. on alumina, with benzene as eluant. Cryst. from 95% ethanol, 90% acetic acid and again from ethanol. Crystn. using glacial acetic acid retained an impurity which was removed by partial mercuration and pptn. with LiBr. [Brown, Dubeck and Goldman, JACS, 84, 1229 (1962).] Has also been cryst. from hexane, or benzene-ethanol, distd. under vac., and zone melted.

9-Fluorenone, m.p. 82.5-83.0°, b.p. 341°/760 mm.
  Distd. under high vac. Cryst. from absolute ethanol or benzene-pentane.

N-2-Fluorenylacetamide, m.p. 194-195°.
  Cryst. from ethanol-water. Carcinogenic.

Fluorescein, ε_{495nm} = 7.84 x 10^4 (in 10^{-3} M NaOH).
  Dissolved in dil. aq. NaOH, filtered and pptd. by adding dil. (1:1) HCl. The process was repeated twice more and the fluorescein was dried at 100°. Alternatively, it has been cryst. from acetone by allowing the soln. to evaporate at 37° in an open beaker.

Fluoroacetamide, m.p. 108°.
  Cryst. from chloroform.

Fluorobenzene, b.p. 84.8°, n_D^{20} 1.46573, n_D^{30} 1.4610.
  Dried for several days with P_2O_5, then fractionally distd.

o-Fluorobenzoic acid, m.p. 127°.
  Cryst. from 50% aq. ethanol, then zone melted or vac. sublimed at 130-140°.

m-Fluorobenzoic acid, m.p. 124°.
  Cryst. from 50% aq. ethanol, then vac. sublimed at 130-140°.

p-Fluorobenzoic acid, m.p. 182°.
  Cryst. from 50% aq. ethanol, then zone melted or vac. sublimed at 130-140°.

1-Fluoro-2,4-dinitrobenzene, m.p. 25-26°, b.p. 140-141°/5 mm.
  Vac. distd. through a Todd column. Cryst. from absolute ethanol.

3-Fluoro-4-hydroxyphenylacetic acid, m.p. 33°.
  Cryst. from water.

1-Fluoro-4-nitrobenzene, m.p. 27° (stable form), 21.5° (unstable form), b.p. 205.3°/735 mm, 95-97.5°/22 mm, 86.6°/14 mm.
  Cryst. from ethanol.

o-Fluorophenol, m.p. 16°, b.p. 53°/14 mm.
  Passed at least twice through a gas chromatographic column.

p-Fluorophenoxyacetic acid, m.p. 106°.
  Cryst. from ethanol.

p-Fluorophenylacetic acid, m.p. 86°.
  Cryst. from heptane.

p-Fluorophenyl-o-nitrobenzyl ether, m.p. 62°.
  Cryst. from ethanol.

o-Fluorotoluene, b.p. 114.4°.
m-Fluorotoluene, b.p. 116.5°, n_D^{27} 1.46524.
p-Fluorotoluene, b.p. 116.6°, n_D^{20} 1.46884.
  Dried with P_2O_5 or $CaSO_4$ and fractionally distd. through a silvered vacuum-jacketed glass column with $^1/8$-in. glass helices. A high reflux ratio is necessary because of the closeness of the b.p.s of

o-, m-, and p-Fluorotoluene (continued)
the three isomers. [Potter and Saylor, JACS, 37, 90 (1951).]

Folic acid. Decomp. >250°. $[\alpha]_D^{25}$ +23° (c = 0.5 in 0.1 N NaOH).
   Cryst. from hot water. See Blakley [BJ, 65, 331 (1957)] and also
Kalifa, Furrer, Bieri and Viscontini [Helv. Chim. Acta, 61, 2739
(1978)].

Formaldehyde.
   Commonly contains added methanol. Addition of KOH soln. (1 mole
KOH:100 moles HCHO) to 40% formaldehyde soln., or evapn. to dryness,
gives paraformaldehyde polymer which, after washing with water, is
dried in a vac. desiccator over P_2O_5 or H_2SO_4. Formaldehyde is regenerated by heating the paraformaldehyde to 120° under vac. or by
decomposing it with barium peroxide. The monomer, a gas, is passed
through a glass-wool filter cooled to -48° in $CaCl_2$-ice mixture to
remove particles of polymer, then dried by passage over P_2O_5 and
either condensed in a bulb immersed in liquid nitrogen or absorbed
in ice-cold conductivity water.

Formamide, f.p. 2.6°, b.p. 103°/9 mm, 210.5°/760 mm (decomp.),
n_D^{20} 1.44754, n_D^{25} 1.44682.
   Formamide is easily hydrolyzed by acids and bases. It also reacts
with peroxides, acid halides, acid anhydrides, esters and (on heating) alcohols, while strong dehydrating agents convert it to a
nitrile. It is very hygroscopic. Commercial material often contains
acids and ammonium formate. Verhoek [JACS, 58, 2577 (1956)] added
some bromothymol blue to formamide and then neutralized with NaOH
before heating to 80-90° under reduced pressure to distil off ammonia
and water. The amide was again neutralized and the process was
repeated until the liquid remained neutral on heating. Sodium formate
was added, and the formamide was distd. under reduced pressure at
80-90°. The distillate was again neutralized and redistd. It was then
fractionally cryst. in the absence of CO_2 and water by partial freezing.
   Formamide (specific conductance 2 x 10^{-7} $ohm^{-1}cm^{-1}$) of low water
content was dried by passage through a column of 3A molecular sieve,
then deionized by treatment with a mixed-bed ion-exchange resin
loaded with H^+ and $HCONH^-$ ions (using sodium formamide in formamide).

Formamide (continued)
[Notley and Spiro, JCS, (B), 362 (1966).]

Formamidine sulphinic acid.
   Dissolved in five parts of aq. 1:1% $NaHSO_3$ at 60-63° (charcoal),
then cryst. slowly, with agitation, at 10°. Filtered. Dried immediately at 60°. [Koniecki and Linch, AC, 30, 1134 (1958).]

Formanilide, m.p. 50°, b.p. 216°/120 mm, 166°/14 mm.
   Cryst. from ligroin-xylene.

Formic acid (anhydr.), f.p. 8.3°, b.p. 25°/40 mm, 100.7°/760 mm,
n_D^{20} 1.37140, n_D^{25} 1.36938.
   Anhydrous formic acid can be obtained by direct fractional distn.
under reduced pressure, the receiver being cooled in ice-water. The
use of P_2O_5 or $CaCl_2$ as dehydrating agent is unsatisfactory.
Reagent grade 88% formic acid can be satisfactorily dried by refluxing with phthalic anhydride for 6 hr and then distilling. Alternatively, if it is left in contact with freshly prepared anhydrous
$CuSO_4$ for several days about half of the water is removed from 88%
formic acid: distn. removes the remainder. Boric anhydride (prepared
by melting boric acid in an air oven at a high temperature, cooling
in a desiccator, and powdering) is a suitable dehydrating agent for
98% formic acid; after prolonged stirring with the anhydride the
formic acid is distd. under vac. Formic acid can be further purified by fractional crystn. using partial freezing.

D-Fructose, $[\alpha]_D^{20}$ -94.4°.
   Dissolved in an equal weight of water (charcoal, previously washed
with water to remove any soluble matter), filtered and evapd. under
reduced pressure at 45-50° to give a syrup containing 90% fructose.
After cooling to 40°, the syrup was seeded and kept at this temperature for 20-30 hr with occasional stirring. The crystals were
separated by centrifuging, washed with a small quantity of water and
dried to constant weight under vac. over conc. H_2SO_4. For higher
purity, this material was recryst. from 50% aq. ethanol. [Tsuzuki,
Yamazaki and Kagami, JACS, 72, 1071 (1950).]

Fructose-1,6-diphosphate.

For purification via the acid strychnine salt, see Neuberg, Lustig and Rothenberg, Arch. Biochem., 3, 33 (1943). The calcium salt can be partially purified by soln. in ice-cold M HCl (1 g per 10 ml) and repptn. by dropwise addition of 2 M NaOH: the ppte. and supernatant are heated on a boiling waterbath for a short time, then filtered and the ppte. is washed with hot water. The magnesium salt can be pptd. from cold aq. soln. by adding four volumes of ethanol.

Fructose-6-phosphate.

Cryst. as the barium salt from water by adding four volumes of ethanol. The barium can be removed by passage through the H^+ form of a cation-exchange resin.

α-Fucose, m.p. 144°.

Cryst. from ethanol.

Fumagillin, m.p. 194-195°, $[α]_D^{25}$ -26.2° (in 95% ethanol).

Forty grams of a commercial sample containing 42% fumagillin, 45% sucrose, 10% antifoam agent and 3% other impurities was digested with 150 ml of $CHCl_3$. The insoluble sucrose was filtered off and washed with $CHCl_3$. The combined $CHCl_3$ extracts were evapd. almost to dryness at room temperature under reduced pressure. The residue was triturated with 20 ml of methanol and the fumagillin was filtered off under suction. It was cryst. twice from 500 ml of hot methanol by standing overnight in a refrigerator. (The long-chain fatty ester used as antifoam agent was still present, but was then removed by repeated digestion, on the steam-bath, with 100 ml portions of ethyl ether.) For further purification, the fumagillin (10 g) was dissolved in 150 ml of 0.2 M cold ammonia, and the insoluble residue was filtered off. The ammonia soln. (cooled with running cold water) was brought to pH 4 by careful addition of 1 M HCl with constant shaking in the presence of 150 ml of $CHCl_3$. (Fumagillin is acid-labile and must be removed rapidly from acid aq. soln.) The $CHCl_3$ extract was washed several times with distd. water, dried (Na_2SO_4) and evapd. under reduced pressure. The solid residue was washed with 20 ml of methanol. The fumagillin was filtered by suction, then cryst. from 200 ml. of hot methanol. [Tarbell et al., JACS, 77, 5610 (1955).]

Fumagillin (continued)

Alternatively, 10 g fumagillin in 100 ml of $CHCl_3$ was passed through a silica gel (5 g) column to remove tarry material, and the $CHCl_3$ was evapd. to leave an oil which gave fumagillin on crystn. from amyl acetate. Recryst. from methanol (charcoal). The fumagillin was stored in dark bottles in the absence of oxygen and at low temperature. [Schenck, Hargie and Isarasena, JACS, 77, 5606 (1955).]

Fumaric acid, m.p. 289.5-291.5° (sealed tube).
Cryst. from hot M HCl or water. Dried at 100°.

2-Furaldehyde, see Furfural.

Furan, b.p. 31.3°, n_D^{20} 1.4214.
Shaking with aq. 5% KOH, dried with $CaSO_4$ or Na_2SO_4, then distd. under N, from KOH or sodium immediately before use. A trace of hydroquinone might be added as an inhibitor.

2-Furanacrylic acid, m.p. 141°.
Cryst. from water.

Furan-2-carboxylic acid, m.p. 133-134°, b.p. 230-232°/760 mm, 141-144°/20 mm.
Cryst. from water and sublimes at 130-140°/50-60 mm.

Furan-3-carboxylic acid, m.p. 122-123°.
Furan-3,4-dicarboxylic acid, m.p. 217-218°.
Cryst. from water.

Furfural, b.p. 59-60°/15 mm, 161.8°/760 mm, n_D^{20} 1.52608.
Unstable to air, light and acids. Impurities include formic acid, β-formylacrylic acid and furan-α-carboxylic acid. Distd. over an oil-bath from 7% (w/w) Na_2CO_3 (added to neutralize acids, especially pyromucic acid). Redistd. from 2% (w/w) Na_2CO_3, and then, finally fractionally distd. under vac. (6 mm). Stored in the dark. [Evans and Aylesworth, IEC, 18, 24 (1926).]

Impurities resulting from storage can be removed by passage through chromatographic grade alumina. Furfural can be separated from impurities other than carbonyl compounds by the bisulphite addition compound.

Furfuryl alcohol, b.p. 170.0°, n_D^{20} 1.4873, n_D^{30} 1.4801.

Distd. at reduced pressure to remove tarry material, shaken with aq. $NaHSO_3$, dried with Na_2SO_4 and fractionally distd. under reduced pressure from Na_2CO_3. Further dried by shaking with Linde type 5A molecular sieve.

Furfurylamine, b.p. 142.5-143°/735 mm.

Distd. under nitrogen from KOH through a column packed with glass helices.

Furil, m.p. 165-166°.

Cryst. from methanol or benzene (charcoal).

2-Furoic acid, m.p. 131-132°.

Cryst. from hot water (charcoal), dried at 120° for 2 hr, then recryst. from $CHCl_3$ and again dried at 120° for 2 hr. For use as a standard in volumetric analysis, good quality commercial 2-furoic acid should be cryst. from $CHCl_3$ and dried as above.

Furoin, m.p. 135-136°.

Cryst. from methanol (charcoal).

Furylacrylic acid, m.p. 140°.

Cryst. from pet.ether (b.p. 80-100°)(charcoal).

Galactaric acid, see Mucic acid.

Galactono-1,4-lactone, m.p. 134-137°, $[\alpha]_D^{20}$ -78° (in water).

Cryst. from ethanol.

D-Galactosamine hydrochloride, $[\alpha]_D^{25}$ +96.4° (c = 3.2 in water).

Dissolved in a small volume of water. Then added three volumes of ethanol, followed by acetone until faintly turbid and stood overnight in a refrigerator. [Roseman and Ludoweig, JACS, 76, 301 (1954).]

α-D-Galactose, m.p. 167-168°, $[\alpha]_D^{20}$ +80.4° (c = 4 in water).

Cryst.twice from aq. 80% ethanol at -10°, then dried under vac. over P_2O_5.

Galanthamine hydrobromide.

Gallic acid, m.p. 253° (decomp.).
   Cryst. from water.

Genistein, m.p. 297-298°.
   Cryst. from 60% ethanol-water or ether.

Genistin, m.p. 256°.
   Cryst. from 80% ethanol-water.

α-Gentiobiose, m.p. 86°.
   Cryst. from methanol. (Retains solvent of crystn.)

β-Gentiobiose, m.p. 190-195.
   Cryst. from ethanol.

Geraniol, b.p. 230°, n_D^{20} 1.4766.
   Purified by ascending paper chromatography or by thin layer chromatography on plates of kieselguhr G with acetone-water-liquid paraffin
(130:70:1) as solvent system. Hexane-ethyl acetate (1:4) is also
suitable. Also purified by gas-liquid chromatography on a silicone-
treated column of Carbowax 20M (10%) on Chromosorb W (60-80 mesh).
[Porter, PAC, 20, 499 (1969).] Stored in full, tightly sealed
containers in the cool, protected from light.

Geranylgeranyl pyrophosphate.
   Purified by counter-current distribution between the two phases of
a butyl alcohol-isopropyl ether-ammonia and water mixture (15:5:1:19)
(v/v), or by chromatography on DEAE-cellulose (linear gradient of
0.02 M KCl in 1 mM tris buffer, pH 8.9.) Stored as powder at 0°.

Geranyl pyrophosphate.
   Purified by paper chromatography on Whatman no. 3 MM paper in a
system of isopropyl alcohol-isobutyl alcohol-ammonia-water
(40:20:1:39), R_f = 0.77-0.82. Stored in the dark as NH_4 salt at 0°.

Gibberellic acid, m.p. 233-235° (decomp.).
   Cryst. from ethyl acetate.

Girard reagent T, m.p. 192°.
  Cryst. from absolute ethanol.

Glucamine, m.p. 127°.
  Cryst. from methanol.

D-Gluconamide, m.p. 144° $[\alpha]_D^{23}$ +31° (c = 2.0 in water).
  Cryst. from ethanol.

D-Glucono-δ-lactone, m.p. 152-153°.
  Cryst. from ethylene glycol monomethyl ether and dried for 1 hr at
110°.

Glucosamine, m.p. 110° (decomp.).
  Cryst. from methanol.

D-Glucosamine hydrochloride, $[\alpha]_D^{25}$ +71.8° (c = 2.0 in water).
  Cryst. from 3 M HCl, water, and finally water-ethanol-acetone as
for Galactosamine hydrochloride.

α-D-Glucose, m.p. 146°, $[\alpha]_D^{20}$ +52.5° (c = 4 in water).
  Recryst. slowly from aq. 80% ethanol, then vac. dried over P_2O_5.
Alternatively, cryst. from water at 55°, then dried for 6 hr in a
vac. oven between 60-70° at 2 mm.

β-D-Glucose, m.p. 148-150°.
  Cryst. from hot glacial acetic acid.

α-D-Glucose penta-acetate, m.p. 110-111°.
β-D-Glucose penta-acetate, m.p. 131-132°.
  Cryst. from methanol or ethanol.

D-Glucose phenylosazone, m.p. 208°.
  Cryst. from aq. ethanol.

Glucose-1-phosphate.
  Two litres of 5% aq. soln. was brought to pH 3.5 with glacial
acetic acid (+ 3 g charcoal, and filtered). An equal volume of ethanol was added, the pH was adjusted to 8.0 (glass electrode) and the

Glucose-1-phosphate (continued)

soln. was stored at 3° overnight. The ppte. was filtered off, dissolved in 1.2 l. of distd. water, filtered and an equal volume of ethanol
was added. After standing at 0° overnight, the crystals were collected
at the centrifuge, and washed with 95% ethanol, then absolute ethanol,
ethanol-ethyl ether (1:1), and ethyl ether. [Sutherland and Wosilait,
JBC, 218, 459 (1956).] Its barium salt can be recryst. from water and
ethanol. Heavy metal impurities can be removed by passage of an aq.
soln. (about 1%) through an Amberlite IR-120 column (in the appropriate H^+, Na^+ or K^+ form).

Glucose-6-phosphate.

Can be freed from metal impurities as described for Glucose-1-
phosphate. Its barium salt can be purified by soln. in dilute HCl and
pptn. by neutralizing the soln. The ppte. is washed with small volumes
of cold water and dried in air.

D-Glucuronic acid, m.p. 165°.

Cryst. from ethanol or ethyl acetate.

D-Glucuronolactone, m.p. 175-177°.

Cryst. from water.

S-Glutamic acid, $[\alpha]_D^{25}$ -31.2°.

Cryst. from water acidified to pH 3.2 by adding 4 vols. of ethanol,
and dried at 110°. Likely impurities: aspartic acid, cysteine.

S-Glutamine, m.p. 184-185°, $[\alpha]_D^{25}$ +31.8° (1 M HCl).

Likely impurities: glutamic acid, ammonium pyroglutamate, tyrosine,
asparagine, isoglutamine, arginine. Cryst. from water.

Glutaraldehyde, b.p. 71° (10 mm Hg), as 50% aq. soln.

Likely impurities are oxidation products - acids, semialdehydes and
polymers. Can be purified by repeated washing with activated charcoal
(Norit XX) followed by vacuum filtration, using 15-20 g charcoal-100
ml of glutaraldehyde solution.

Vac. distn. at 60-65° (15 mm Hg), discarding the first 5-10%, was
followed by dilution with an equal volume of freshly distd. water at

Glutaraldehyde (continued)

70-75°, using magnetic stirring under nitrogen. Soln stored at low temperatures (3-4°) in a light-tight container. Standardized by titration with hydroxylamine. [Anderson, J. Histochem. Cytochem., 15, 652 (1967).]

Glutaric acid, m.p. 97.5-98°.

Cryst. from benzene, $CHCl_3$, distd. water or benzene containing 10% (w/w) of ethyl ether. Dried under vac.

Glutathione, m.p. 195°.

Cryst. from aq. 50% ethanol.

RS-Glyceraldehyde, m.p. 145°.

Cryst. from ethanol-ethyl ether.

Glycerol, m.p. 18.2°, b.p. 182°/20 mm, 290.0°/760 mm, n_D^{25} 1.47352.

Dissolved in an equal volume of n-butyl alcohol (or n-propyl alcohol, amyl alcohol or liquid ammonia) in a watertight container, then cooled and seeded while slowly revolving in an ice-water slurry. The crystals were collected by centrifuging, then washed with cold acetone or isopropyl ether. [Hass and Patterson, IEC, 33, 615 (1941).] Coloured impurities can be removed from substantially dry glycerol by extraction with 2,2,4-trimethylpentane. Alternatively, glycerol can be decolorized and dried by treatment with activated charcoal and alumina, followed by filtering. Glycerol can be distd. at 15 mm in a stream of dry nitrogen, and stored in a desiccator over P_2O_5. Crude glycerol can be purified by digestion with hot conc. H_2SO_4 and saponification with a lime paste, then re-acidified with H_2SO_4, filtered, treated with an anion adsorbent resin and fractionally distd. under vac.

Glycinamide hydrochloride, m.p. 186-189° (207-208°).

Cryst. from ethanol.

Glycine, see Aminoacetic acid.

Glycine ethyl ester hydrochloride.
Glycine hydrochloride.
Glycine methyl ester hydrochloride, m.p. 144°.
   Cryst. from absolute ethanol.

Glycocholic acid, m.p. 154-155°.
   Cryst. from hot water. Dried at 100°.

Glycogen.
   A 5% aq. soln. (charcoal) was filtered and an equal volume of
ethanol was added. After standing overnight at 3° the ppte. was
collected at the centrifuge and washed with absolute ethanol, then
ethanol-ethyl ether (1:1), and ethyl ether. [Sutherland and
Wosilait, JBC, 218, 459 (1956).]

Glycol dimethyl ether, see 1,2-Dimethoxyethane.

Glycollic acid, m.p. 81°.
   Cryst. from ethyl ether.

N-Glycylaniline.
   Cryst. from water.

Glycylglycine, m.p. 260-262° (decomp.).
   Cryst. from aq. 50% ethanol or water at 50-60° by addition of
ethanol. Dried at 110°.

Glycylglycine hydrochloride.
   Cryst. from 95% ethanol.

Glycylglycylglycine, m.p. 246° (decomp.).
   Cryst. from water, and dried at 110°.

Glycyl-S-proline, m.p. 185°.
   Cryst. from water at 50-60° by addition of ethanol.

RS-Glycylserine, m.p. 207° (decomp.).
   Cryst. from water (charcoal) by addition of ethanol.

Glycyrrhizic acid. Decomp. at 220°.
  Cryst. from glacial acetic acid.

Glyoxaline, see Imidazole.

Glyoxylic acid, m.p. 98° (anhydr.).
  Cryst. from water.

Gramicidin S, m.p. 268-270°.
  Cryst. from ethanol.

Gramine, m.p. 134°.
  Cryst. from ethyl ether, ethanol or acetone.

Griseofulvin, m.p. 220°.
  Cryst. from benzene.

Guaiacic acid, m.p. 99-100.5°.
  Cryst. from ethanol.

Guaiacol, m.p. 32°, b.p. 106°/24 mm, 205°/746 mm.
  Cryst. from benzene-pet.ether or distd.

Guaiacol carbonate, m.p. 88.1°.
  Cryst. from ethanol.

Guanidine.
  Cryst. from water-ethanol. Deliquescent.

Guanidine carbonate, m.p. 197°.
  Cryst. from methanol.

Guanidine hydrochloride.
  Cryst. from hot methanol by chilling to about -10°, with vigorous
stirring. The fine crystals were filtered throught fritted glass,
washed with cold (-10°) methanol, and dried at 50° under vac. for 5
hr. (The product is more pure than that obtained by crystn. at room
temp. from methanol by adding large amounts of ethyl ether.)
[Kolthoff et al., JACS, 79, 5102 (1957).]

Guanosine.
Guanylic acid, m.p. 208° (decomp.).
   Cryst. from water. Dried at 110°.

Haematin. Decomp. at 200°.
   Cryst. from pyridine. Dried at 40° under vac.

Haematoporphyrin dimethyl ester, m.p. 212°.
   Cryst. from $CHCl_3$-methanol.

Haematoxylin, m.p. 100-120°.
   Cryst. from dil. aq. $NaHSO_3$ until colourless.

Haemin.
   Cryst. from glacial acetic acid or $CHCl_3$-pyridine-acetic acid.

Harmine, m.p. 261° (decomp.).
   Cryst. from methanol.

Harmine hydrochloride, m.p. 280° (decomp.).
   Cryst. from water.

Hecogenin acetate, m.p. 265-268°.
   Cryst. from methanol.

Heparin.
   Dissolved in 0.1 M NaCl (1 g in 100 ml) and pptd. by addition of
ethanol (150 ml).

Heptadecanoic acid, m.p. 60-61°, b.p. 227°/100 mm.
   Cryst. from methanol or pet.ether.

1-Heptadecanol, m.p. 54°.
   Cryst. from acetone.

n-Heptaldehyde, b.p. 40.5°/12 mm, 152.8°/760 mm, n_D^{25} 1.4130.
   Dried with $CaSO_4$ or Na_2SO_4 and fractionally distd. under reduced
pressure. More extensive purification by pptn. as the bisulphite
compound (formed by adding the aldehyde to sat. aq. $NaHSO_3$) which

n-Heptaldehyde (continued)
was filtered off and recryst. from hot water. The crystals, after
being filtered and washed well with water, were hydrolyzed by adding
700 ml of aq. Na_2CO_3 (12½% w/w of anhydrous Na_2CO_3) per 100 g of
aldehyde. The aldehyde was then steam distd., separated, dried with
$CuSO_4$ and distd. under reduced pressure in a slow stream of nitrogen.
[McNesby and Davis, JACS, 76, 2148 (1954).]

n-Heptaldoxime, m.p. 53-55°.
  Cryst. from aq. 60% ethanol.

n-Heptane, b.p. 98.4°, n_D^{20} 1.38765, n_D^{25} 1.38512.
  Passage through a silica gel column greatly reduces the ultraviolet absorption of n-heptane. (The silica gel is previously heated
to 350°.) For more extensive purification, heptane is shaken with
successive small portions of conc. H_2SO_4 until the lower (acid) layer
remains colourless. The heptane is then washed successively with
water, aq. 10% Na_2CO_3, water (twice), and dried with $CaSO_4$, $MgSO_4$ or
$CaCl_2$. It is distd. from sodium. n-Heptane can be distd. azeotropically with methanol, then the methanol can be washed out with water
and, after drying, the heptane is redistd. Other purification procedures include passage through activated basic alumina, drying with
CaH_2, storage with sodium, and stirring with 0.5 N $KMnO_4$ in 6 N-H_2SO_4
for 12 hr after treatment with conc. H_2SO_4. Carbonyl-containing
impurities have been removed by percolation through a column of
impregnated Celite made by dissolving 0.5 g of 2,4-dinitrophenyl-
hydrazine in 6 ml of 85% H_3PO_4 by grinding together, then adding
4 ml of distd. water and 10 g of Celite. [Schwartz and Parks, AC,
33, 1396 (1961).]

4-Heptanone, see Dipropyl ketone.

Hept-1-ene, b.p. 93°/771 mm.
  Distd. from sodium, then carefully fractionally distd. using an
18-in. gauze-packed column. Can be purified by azeotropic distn.
with ethanol.

n-Heptyl alcohol, b.p. 175.6°.

Shaken with successive lots of alkaline $KMnO_4$ until the colour persisted for 15 min, then dried with K_2CO_3 or CaO, and fractionally distd.

n-Heptylamine, b.p. 155°.

Dried in contact with KOH pellets for 24 hr, then decanted and fractionally distd.

n-Heptyl bromide, b.p. 70.6°/19 mm, 180°/760 mm.

Shaken with conc. H_2SO_4, washed with water, dried with K_2CO_3, and fractionally distd.

Heptyloxoazoxybenzene.

Recryst. from toluene.

Hesperetin, m. p. 227-228°.

Cryst. from ethyl acetate.

Hesperidin, m.p. 258-262°.

Dissolved in dil. aq. alkali and pptd. by adjusting the pH to 6-7.

Hexachlorobenzene, m.p. 230.2-231.0°.

Cryst. repeatedly from benzene. Dried under vac. over P_2O_5.

Hexachloro-1,3-butadiene, m.p. 39°, b.p. 283-284° (decomp.)/733mm.

Vac. distd. at less than 15 mm pressure.

1,2,3,4,5,6-Hexachlorocyclohexane, m.p. 158°(α-), 312° (β-), 112.5° (γ-isomer).

Cryst. from ethanol. Purified by zone melting.

Hexachlorocyclopentadiene, b.p. 80°/1 mm, n_D^{25} 1.5628.

Dried with $MgSO_4$. Distd. under vac. in nitrogen.

Hexachloroethane, m.p. 187°.

Steam distd., then cryst. from 95% ethanol. Dried in the dark under vac.

Hexacosane, m.p. 56.4°, b.p. 262°/15 mm, 205°/1 mm, 169°/0.05 mm.
  Cryst. from ether.

Hexacosanoic acid, m.p. 88-89°.
  Cryst. from ethanol.

n-Hexadecane, m.p. 18.2°, b.p. 105°/0.1 mm, n_D^{20} 1.4345, n_D^{25} 1.4325.
  Passed through a column of silica gel and distd., under vac. in a
column packed with Pyrex helices. Stored over silica gel. Cryst.
from acetone, or fractionally cryst. by partial freezing.

1,14-Hexadecanedioic acid, m.p. 126°.
  Cryst. from ethanol or ethyl acetate.

1-Hexadecyl-, see Cetyl-.

1,5-Hexadiene, b.p. 59.6°, n_D^{20} 1.4039.
  Distd. from $NaBH_4$.

Hexaethylbenzene, m.p. 128.7-129.5°.
  Cryst. from benzene or benzene-ethanol.

Hexafluoroacetone, m.p. -129°, b.p. -28°.
  Dehydrated by passage of the vapour over P_2O_5. Ethylene was
removed by passing the dried vapour through a tube containing Pyrex
glass wool moistened with conc. H_2SO_4. Further purification was by
low temperature distn. using Ward-Le Roy stills. Stored in the dark
at -78°. [Holmes and Kutschke, TFS, 58, 333 (1962).]

Hexafluorobenzene, m.p. 5.1°.
  Main impurities are incompletely fluorinated benzenes. Purified by
standing in contact with oleum for 4 hr at room temperature, repeating until the oleum does not become discoloured. Washed several
times with water, then dried with P_2O_5. Final purification was by
repeated fractional crystn.

Hexafluoroethane, b.p. -79°.

Purified for pyrolysis studies by passage through a copper vessel containing CoF_3 at about 270°, and held for 3 hr in a bottle with a heated (1300°) platinum wire. It was then fractionally distd. [Steunenberg and Cady, JACS, 74, 4165 (1962).]

1,1,1,3,3,3-Hexafluoropropan-2-ol, b.p. 57-58°/760 mm, $D^{20.5}$ 1.4563, n_D^{22} 1.2750.

Distd. from molecular sieve 3A, retaining the middle fraction.

Hexamethylbenzene, m.p. 165-165.5°.

Sublimed, then cryst. from absolute ethanol, benzene or ethanol-benzene. Also purified by zone melting. Dried under vac. over P_2O_5.

Hexamethylenediamine, m.p. 42°, b.p. 204-205°, 100°/20 mm, 84.9°/9 mm, 46-47°/1 mm.

Distd. in stream of nitrogen or CO_2, sublimes in vac.

Hexamethylenediamine dihydrochloride, m.p. 248°.

Cryst. from water or ethanol.

Hexamethylene glycol, m.p. 41.6°.

Fractionally cryst. from its melt.

Hexamethylenetetramine.

Cryst. from ethanol.

Hexamethylphosphoric triamide, f.p. 7.2°, b.p. 68-70°/1 mm, 235°/760 mm.

Impurities are water, dimethylamine and its hydrochloride. Refluxed over BaO or CaO at about 4 mm Hg in an atmosphere of nitrogen, for several hrs, then distd. from sodium at same pressure. The middle fraction (b.p. about 90°) was collected, refluxed over sodium under reduced pressure under nitrogen, and distd. Kept in the dark, in a nitrogen atmosphere, and stored in solid CO_2. Can also be stored over 4A molecular sieve.

Alternatively, distd. under vac. from CaH_2 at 60° and cryst. twice in cold room at 0°, seeding the liquid with crystal obtained by

Hexamethylphosphoric triamide (continued)
cooling in liq. nitrogen. After about two-thirds frozen, the remaining liquid was drained away. [Fujinaga, Izutsu and Sakura, PAC, 44,
117 (1975).]

Hexanamide, see n-Caproamide.

n-Hexane, b.p. 68.7°, n_D^{20} 1.37486, n_D^{25} 1.37226.
   Purification as for n-Heptane. Modifications include the use of
chlorosulphonic acid or 35% fuming H_2SO_4 instead of conc. H_2SO_4 in
washing the alkane, and final drying and distn. from sodium hydride.
Unsatd. compounds can be removed by shaking the hexane with nitrating
acid (58% H_2SO_4, 25% conc. HNO_3, 17% water, or 50% HNO_3, 50% H_2SO_4),
then washing the hydrocarbon layer with conc. H_2SO_4, followed by
water, drying, and distg. over sodium or n-butyl lithium.

Hexanenitrile, see Capronitrile.

Hexanoic acid, see Caproic acid.

1-Hexane, b.p. 63°.
   Fractionally distd. from sodium under nitrogen.

Hexoestrol, m.p. 185-188°.
   Cryst. from benzene or aq. ethanol.

n-Hexyl alcohol, b.p. 157.5°, n_D^{15} 1.4198, n_D^{25} 1.4158.
   Commercial material usually contains other alcohols which are
difficult to remove. A suitable method is to esterify with hydroxy-
benzoic acid, recrystallize the ester and saponify. [Olivier,
Rec. Trav. chim., 55, 1027 (1936).] Drying agents include K_2CO_3 and
$CaSO_4$, followed by filtration and distn. (Some decomposition to the
olefin occurred when Al amalgam was used as the drying agent at room
temperature, even though the amalgam was removed prior to distn.) If
the alcohol is required anhydrous, the redistd. material can be
refluxed with the appropriate alkyl phthalate or succinate, as
described under Ethanol.

n-Hexylamine, b.p. 131°.
  Dried with, and fractionally distd. from, KOH or CaH_2.

n-Hexyl bromide, b.p. 87-88°/90 mm, 155°/743 mm.
  Shaken with sulphuric acid, washed with water, dried with K_2CO_3
and fractionally distd.

n-Hexyl methacrylate.
  Purified as described for Methyl methacrylate.

1-Hexyne, b.p. 12.5°/75 mm, 71°/760 mm.
2-Hexyne, b.p. 83.8°/760 mm.
3-Hexyne, b.p. 81°/760 mm.
  Distd. from $NaBH_4$ to remove peroxides. Stood with sodium for 24 hr,
then fractionally distd. under reduced pressure.

Hippuric acid, m.p. 187.2°.
  Cryst. from water. Dried over P_2O_5.

Histamine, m.p. 86° (sealed tube).
  Cryst. from benzene or $CHCl_3$.

Histamine dihydrochloride, m.p. 244-246°.
  Cryst. from aq. ethanol.

S-Histidine, $[\alpha]_D^{25}$ -39.7° (water), 13.0° (6 M HCl).
  Likely impurity: arginine. Adsorbed from aq. soln. on to Dowex 50-H^+
ion-exchange resin, washed with 1.5 M HCl (to remove other amino
acids), then eluted with 4 M HCl as the dihydrochloride. Histidine is
also purified as the dihydrochloride which is finally dissolved in
water, the pH is adjusted to 7.0 and the free base is cryst. by
addition of ethanol.

Histidine dihydrochloride, m.p. 252°.
  Cryst. from water or aq. ethanol, and washed with acetone, then
ethyl ether. Converted to histidine di-(3,4-dichlorobenzenesulphon-
ate) by dissolving 3,4-dichlorobenzenesulphonic acid (1.5 g/10 ml) in
the histidine soln. by heating, and then the soln. is cooled in ice.

<u>Histidine dihydrochloride</u> (continued)
The resulting crystals can be recryst. from 5% 3,4-dichlorobenzene-sulphonic acid, then dried over $CaCl_2$ under vac., and washed with ethyl ether to remove excess reagent. The dihydrochloride can be regenerated by passing the soln. through a Dowex-1 (Cl$^-$ form) ion-exchange column. The solid is obtained by evapn. of the soln. on a steam bath.

<u>S-Histidine monohydrochloride</u> (monohydrate), $[\alpha]_D^{25}$ +13.0° (in 6 M HCl).
<u>Homocysteine</u>, m.p. 232-233°.
  Cryst. from aq. ethanol.

<u>Homocystine</u>. Decomp. at 260-265°.
  Cryst. from water.

<u>Homophthalic acid</u>, m.p. 182-183° (but varies with rate of heating).
  Cryst. from boiling water (25 ml/g). Dried at 100°.

<u>S-Homoserine</u>, m.p. 203° $[\alpha]_D^{26}$ +18.3° (in 2 M HCl).
  Likely impurities:N-(chloroacetyl)-S-homoserine, N-(chloroacetyl)-R-Homoserine ,S-homoserine, homoserine lactone, homoserine anhydride (formed in strong solns. of homoserine if slightly acidic). Cyclizes to lactone in strongly acidic soln. Cryst. from water by adding 9 volumes of ethanol.

<u>Hordenine</u>, m.p. 117-118°.
  Cryst. from ethanol or water.

<u>Humulon</u>, m.p. 65-66.5°.
  Cryst. from ethyl ether.

<u>Hyamine 1622</u>.
  Cryst. from boiling acetone after filtering. The ppte. was filtered off, washed with ethyl ether and dried for 24 hr in a vac. desiccator.

<u>Hydantoin</u>, m.p. 220°.
  Cryst. from methanol.

Hydrazine N,N'-dicarboxylic acid diamide, m.p. 248°.

4-Hydrazinobenzoic acid, m.p. 215° (decomp.).
  Cryst. from water.

1-Hydrazinophthalazine, m.p. 172-173°.
  Cryst. from methanol.

Hydrazobenzene, m.p.128°.
  Cryst. from pet.ether (b.p. 60-110°) to constant absorption
spectrum.

Hydrobenzamide, m.p. 101-102°.
  Cryst. from absolute ethanol or cyclohexane-benzene. Dried under
vac. over P_2O_5.

RS-Hydrobenzoin, m.p. 120°.
  Cryst. from ethyl ether-pet.ether.

meso-Hydrobenzoin, m.p. 139°.
  Cryst. from ethanol or water.

Hydrocinnamic acid, m.p. 48.0-48.5°.
  Cryst. from benzene, $CHCl_3$, or pet.ether (b.p. 40-60°). Dried in
vac.

Hydroquinone, m.p. 175.4-176.6°.
  Cryst. from acetone, benzene, ethanol, ethanol-benzene, water or
acetonitrile (25 g in 30 ml), preferably under nitrogen. Dried under
vac.

Hydroquinone dimethyl ether, m.p. 56°.
  Cryst. from ethanol.

Hydroquinone monobenzyl ether, see p-(Benzyloxy)phenol.

Hydroquinone monomethyl ether, m.p. 53°, b.p. 243°.
  Cryst. from water. Dried over P_2O_5 under vac.

p-Hydroxyacetophenone, m.p. 109°.
  Cryst. from ethyl ether, aq. ethanol or benzene-pet.ether
(b.p. 60-80°).

4'-Hydroxyacetanilide, m.p. 169-170.5°.
  Cryst. from water.

4-Hydroxyacridine, m.p. 116.5°.
  Cryst. from ethanol.

erythro-3-Hydroxy-RS-aspartic acid.
  Likely impurities: 3-chloromalic acid, ammonium chloride, threo-
3-hydroxyaspartic acid. Cryst. from water.

p-Hydroxyazobenzene, see p-Phenylazophenol.

o-Hydroxybenzaldehyde, see Salicylaldehyde.

m-Hydroxybenzaldehyde, m.p. 108°.
  Cryst. from water.

p-Hydroxybenzaldehyde, m.p. 115-116°.
  Cryst. from water (containing some H_2SO_3). Dried over P_2O_5 under
vac.

m-Hydroxybenzoic acid, m.p. 200.8°.
  Cryst. from absolute methanol.

p-Hydroxybenzoic acid, m.p. 213-214°.
p-Hydroxybenzonitrile, m.p. 113-114°.
  Cryst. from water.

4-Hydroxybenzophenone, m.p. 135°.
1-Hydroxybenzotriazole, m.p. 159-160°.
  Cryst. from aq. ethanol.

o-Hydroxybenzyl alcohol, m.p. 87°.
  Cryst. from water or benzene.

P.L.C.—KK

2-Hydroxybiphenyl, m.p. 56°, b.p. 275°/760 mm, 145°/14 mm.
   Cryst. from pet.ether.

4-Hydroxybiphenyl, m.p. 164-165°, b.p. 305-308°/760 mm.
   Cryst. from aq. ethanol.

3-Hydroxy-2-butanone, b.p. 144-145°.
   Washed with ethanol until colourless, then with ethyl ether or
acetone to remove biacetyl. Air dried by suction and further dried
in a vac. desiccator.

2-Hydroxycaprylic acid, see 2-Hydroxyoctanoic acid.

p-Hydroxycinnamic acid, m.p. 214-215°.
   Cryst. from hot water (charcoal).

4-Hydroxycoumarin, m.p. 206°.
R-2-Hydroxy-3,3-dimethyl-γ-butyrolactone, m.p. 89-91°, $[\alpha]_{546}^{20}$ -62°
(c = 3 in H_2O).
   Cryst. from water.

2-Hydroxy-4,6-dimethylpyrimidine, m.p . 198-199°.
   Cryst. from absolute ethanol (charcoal).

p-Hydroxydiphenylamine, m.p. 72-73°.
   Cryst. from chlorobenzene-pet.ether.

N-[2-Hydroxyethyl]ethylenediamine, b.p. 91.2°/5 mm, 238-240°/752 mm.
   Distd. twice through a Vigreux column. Redistd. from solid NaOH,
then from CaH_2. Alternatively, converted to the dihydrochloride and
recryst. from water. Dried. Mixed with excess solid NaOH and the free
amine distd. from the mixture. Redistd. from CaH_2. [Drinkard, Bauer
and Bailar, JACS, 82, 2992 (1960).]

N-(2-Hydroxyethyl)ethylenediaminetriacetic acid.
   Cryst. from warm water, after filtering, by adding 95% ethanol and
allowing to cool. The crystals, filtered on a sintered-glass funnel,
were washed three times with cold absolute ethanol, then again cryst.

N-(2-Hydroxyethyl)ethylenediaminetriacetic acid (continued)
from water. After leaching with cold water, the crystals were dried
at 100° under vac. [Spedding, Powell and Wheelwright, JACS, 78, 34
(1956).]

N-Hydroxyethyliminodiacetic acid.
  Cryst. from water.

2-Hydroxyethylimino-tris(hydroxymethyl)methane (Mono-tris), m.p.91°.
  Cryst. twice from ethanol. Dried under vac. at 25°.

2-Hydroxyethyl methacrylate.
  Dissolved in water and extracted with n-heptane to remove ethylene
glycol dimethacrylate (checked by gas-liquid chromatography) and
distd. twice under reduced pressure. [Strop, Mikes and Kalal, JPC,
80, 694 (1976).]

N-2-Hydroxyethylpiperazine-N'-2-ethanesulphonic acid, (HEPES).
  Cryst. from hot ethanol and water.

β-Hydroxyglutamic acid. Decomp. at 100°.
  Cryst. from water.

4-Hydroxyindane, m.p. 49-50°, b.p. 120°/12 mm.
5-Hydroxyindane, m.p. 55°, b.p. 255°/760 mm.
  Cryst. from pet.ether.

2-Hydroxy-5-iodobenzoic acid, see 5-Iodosalicylic acid.

α-Hydroxyisobutyric acid, see 2-Hydroxy-2-methylpropionic acid.

5-Hydroxy-S-lysine monohydrochloride, $[\alpha]_D^{25}$ +17.8° (in 6 M HCl).
  Likely impurities: 5-allo-hydroxy-(S and R)-lysine, histidine,
lysine, ornithine. Cryst. from water by adding 2-9 volumes of
ethanol stepwise.

4-Hydroxy-3-methoxyacetophenone, m.p. 115°.
  Cryst. from water, or ethanol-pet.ether.

4-Hydroxy-3-methoxycinnamic acid, m.p. 174°.
  Cryst. from water.

17β-Hydroxy-17α-methyl-3-androstanone, m.p. 192-193°.
  Cryst. from ethyl acetate.

3-Hydroxy-4-methylbenzaldehyde, m.p. 116-117°, b.p. 179°/15 mm.
  Cryst. from water.

RS-2-Hydroxy-2-methylbutyric acid, m.p. 72-73°.
  Cryst. from benzene, and sublimes at 90°.

2-Hydroxy-3-methylbutyric acid, m.p. 86°.
  Cryst. from ether-pentane.

7-Hydroxy-4-methylcoumarin, m.p. 185-186°.
  Cryst. from absolute ethanol.

5-(Hydroxymethyl)furfural, m.p. 33.5°.
  Cryst. from ethyl ether-pet.ether.

2-Hydroxymethyl-2-methyl-1,3-propanediol, m.p. 198°.
  Cryst. from acetone-water (1:1).

dl-3-Hydroxy-N-methylmorphinan, m.p. 251-253°.
  Cryst. from anisole + aq. ethanol.

5-Hydroxy-2-methyl-1,4-naphthoquinone, see Plumbagin.

6-Hydroxy-2-methyl-1,4-naphthoquinone.
  Cryst. from aq. ethanol. Sublimes on heating.

2-(Hydroxymethyl)-2-nitro-1,3-propanediol, m.p. 214°.
  Cryst. from $CHCl_3$-ethyl acetate or ethyl acetate-benzene.

4-Hydroxy-4-methyl-2-pentanone, b.p. 166°, n_D^{20} 1.4235, n_D^{25} 1.4213.
  Loses water when heated. Can be dried with $CaSO_4$, then fractionally
distd. at reduced pressure.

17α-Hydroxy-6α-methylprogesterone, m.p. 220-223.5°, $[\alpha]_D^{25}$ +75°.
   Cryst. from CHCl$_3$.

2-Hydroxy-2-methylpropionic acid, m.p. 79°, b.p. 212°/760 mm,
114°/12 mm, 84°/15 mm.
   Distd. in steam, cryst. from ether and sublimed at 50°.

8-Hydroxy-2-methylquinoline, m.p. 74-75°, b.p. 266-267°.
   Cryst. from ethanol.

2-Hydroxymyristic acid, see 2-Hydroxytetradecanoic acid.

2-Hydroxy-1-naphthaldehyde, m.p. 82°, b.p. 192°/27 mm.
   Cryst. from ethanol or ethyl acetate.

β-Hydroxy-α-naphthaldehyde, m.p. 80-81°.
   Cryst. from water or ethanol (1.5 ml/g).

2-Hydroxy-1-naphthaleneacetic acid.
   Treated with activated charcoal and cryst. from ethanol-water
(1:9, v/v). Dried under vac., over silica gel, in the dark. Stored
in the dark at -20°. [Gafni, Modlin and Brand, JPC, 80, 898 (1976).]
Forms a lactone (m.p. 107°) readily.

6-Hydroxy-2-naphthalenepropionic acid, m.p. 180-181°.
   Cryst. from aq. ethanol or aq. methanol.

3-Hydroxy-2-naphthanilide, m.p. 248.0-248.5°.
3-Hydroxy-2-naphtho-4'-chloro-o-toluidide, m.p. 243.5-244.5°.
3-Hydroxy-2-naphthoic-α-naphthalide, m.p. 217.5-218.0°.
3-Hydroxy-2-naphthoic-β-naphthalide, m.p. 243.5-244.5°, and other
naphthol AS derivatives.
   Cryst. from xylene. [Schnopper, Broussard and La Forgia, AC, 31,
1542 (1959).]

1-Hydroxy-2-naphthoic acid, see 1-Naphthol-2-carboxylic acid.

3-Hydroxy-2-naphthoic acid, see 3-Naphthol-2-carboxylic acid.

2-Hydroxy-1,4-naphthoquinone, m.p. 192° (decomp.).
  Cryst. from benzene.

5-Hydroxy-1,4-naphthoquinone, m.p. 155°.
  Cryst. from benzene-pet. ether.

6-Hydroxynicotinic acid, see 2-Hydroxypyridine-5-carboxylic acid.

2-Hydroxy-5-nitrobenzylbromide, m.p. 147°.
  Cryst. from benzene or benzene-ligroin.

2-Hydroxyoctanoic acid, m.p. 69.5^O.
  Cryst. from ethanol-pet.ether or ether-ligroin.

1-Hydroxyphenazine, m.p. 157-158°.
  Chromatographed on acidic alumina with benzene-ether. Cryst. from
benzene-heptane, and sublimed.

2-Hydroxyphenylacetic acid, m.p. 148-149°, b.p. 240-243°/760 mm.
  Cryst. from ether or chloroform.

3-Hydroxyphenylacetic acid, m.p. 137°.
  Cryst. from benzene-ligroin.

4-Hydroxyphenylacetic acid, m.p.150-151°.
  Cryst. from water.

3-Hydroxy-2-phenylcinchoninic acid, m.p. 206-207° (decomp.).
  Cryst. from ethanol.

N-(p-Hydroxyphenyl)glycine, m.p. 200° (decomp.).
  Cryst. from water.

N-(4-Hydroxyphenyl)-3-phenylsalicylamide, m.p. 183-184°.
  Cryst. from aq. methanol.

L-2-Hydroxy-3-phenylpropionic acid, m.p. 125-126°, $[\alpha]_D^{12}$ -18.7^O
(in EtOH).
  Cryst. from water, methanol, ethanol or benzene.

R,S-2-Hydroxy-3-phenylpropionic acid, m.p. 97-98°,
b.p. 148-150°/15 mm.
  Cryst. from benzene or chloroform.

3-Hydroxy-2-phenylpropionic acid, see Tropic acid.

3-p-Hydroxyphenylpropionic acid, m.p. 129-130°.
  Cryst. from ether.

p-Hydroxyphenylpyruvic acid, m.p. 220°.
  Cryst. three times from 0.1 M HCl-ethanol (4:1, v/v) immediately
before use. [Rose and Powell, BJ, 87, 541 (1963).]

3-β-Hydroxy-5-pregnen-2-one, m.p. 189-190°, $[\alpha]_D$ + 30° (in EtOH),
$[\alpha]_{546}^{20}$ +34° (c = 1 in EtOH).
  Cryst. from methanol.

17α-Hydroxyprogesterone, m.p. 222-223°.
  Cryst. from acetone or ethanol.

21-Hydroxyprogesterone, see 11-Deoxycorticosterone.

S-4-Hydroxyproline, m.p. 274°, $[\alpha]_D$ -76.0°.
  Cryst. from methanol-ethanol (1:1).

4'-Hydroxypropiophenone, m.p. 149°.
  Cryst. from water.

2-(α-Hydroxypropyl)piperidine, m.p. 121°, b.p. 226°.
  Cryst. from ether.

7-(2-Hydroxypropyl)theophylline, m.p. 135-136°.
  Cryst. from ethanol.

6-Hydroxypurine. Decomp. at 150°.
  Cryst. from hot water. Dried at 105°.

2-Hydroxypyridine, m.p. 105-107°, b.p. 181-185°/24 mm,
ε_{293nm} = 5900, in water.
  Distd. under vac. to remove coloured impurity, then cryst. from
benzene, CCl_4 or ethanol.

3-Hydroxypyridine, m.p. 129°.
  Cryst. from water or ethanol.

4-Hydroxypyridine, m.p. 65° (hydrate), 148.5° (anhydr.),
b.p. >350°/760 mm.
  Cryst. from water. Loses H_2O of cryst. on drying in vacuo over
H_2SO_4, and stored over KOH because it is hygroscopic.

2-Hydroxypyridine-5-carboxylic acid, m.p. 304° (decomp.).
4-Hydroxypyridine-2,6-dicarboxylic acid, m.p. 254° (decomp.).
  Cryst. from water.

2-Hydroxypyrimidine, m.p. 179-180°.
  Cryst. from ethanol or ethyl acetate.

4-Hydroxypyrimidine, m.p. 164-165°.
  Cryst. from benzene or ethyl acetate.

2-Hydroxypyrimidine hydrochloride, m.p. 205° (decomp.).
  Cryst. from ethanol.

2-Hydroxyquinoline, m.p. 199-200°.
  Cryst. from methanol.

8-Hydroxyquinoline, m.p. 75-76°.
  Cryst. from hot ethanol, acetone, pet.ether (b.p. 60-80°) or water.

8-Hydroxyquinoline-5-sulphonic acid.
  Cryst. from water or hot dil. HCl (about 2% by weight).

5-Hydroxysalicylic acid, m.p. 204.5-205°.
  Cryst. from hot water.

4-Hydroxystilbene, m.p. 189°.
   Cryst. from benzene or acetic acid.

d,1-2-Hydroxytetradecanoic acid, m.p. 81-82°.
1-2-Hydroxytetradecanoic acid, m.p. 88.2-88.5°, $[\alpha]_D$ -31° (in $CHCl_3$).
   Cryst. from chloroform.

4-Hydroxy-2,2,6,6-tetramethylpiperidine, m.p. 130-131°.
   Cryst. from water as hydrate, and cryst. from ether as anhydrous.

5-Hydroxy-S-tryptophan, m.p. 273° (decomp.), $[\alpha]_D^{22}$ -32.5°,
$[\alpha]_{546}^{20}$ -73.5°(c = 1 in H_2O).
   Likely impurities: 5-hydroxy-R-tryptophan, 5-benzyloxytryptophan.
Cryst. under nitrogen from water by adding ethanol. Stored under
nitrogen.

Hydroxyurea, m.p. 70-72° (unstable form), 141°.
   Cryst. from water by addition of ethanol.

α-Hyodeoxycholic acid, m.p. 196-197°, $[\alpha]_{546}^{20}$ +8° (c = 2 in EtOH).
   Cryst. from ethyl acetate.

Hyoscine, $[\alpha]_D^{20}$ -18° (in Et_2OH), -28° (in H_2O), $[\alpha]_{546}^{20}$-30°
(c = 5 in $CHCl_3$).
   Cryst. from benzene-pet.ether. Racemate has m. p. 56-57° (1 H_2O),
37-38° (2 H_2O), syrup anhydr., (+) and (-) are syrups.

Hypericin, m.p. 320° (decomp.).
   Cryst. from pyridine by addition of methanolic HCl.

Hypoxanthine, see 6-Hydroxypurine.

Ibogaine, m.p. 152-153°.
   Cryst. from absolute ethanol.

Imidazole, m.p. 89.5-90°.
   Cryst. from benzene, CCl_4, ethanol or pet.ether. Dried at 40°
under vac. Distd. at low pressure. Purified by zone melting.

2-Imidazolidinethione, see Ethylene thiourea.

2-Imidazolidone, see Ethylene urea.

Iminodiacetic acid, m.p. 225° (decomp.).
   Cryst. from water.

2,2'-Iminodiethanol, see Diethanolamine.

Indane, b.p. 177°.
   Shaken with conc. H_2SO_4, then water, dried and fractionally distd.

Indanthrone. Decomp. at 470-500°.
   Cryst. repeatedly from 1,2,4-trichlorobenzene.

Indazole, m.p. 147°.
   Cryst. from water, then from pet.ether (b.p. 60-80°).

Indene, f.p. -1.5°, b.p. 114.5°/100 mm, n_D^{20} 1.5763.
   Shaken with 6 M HCl for 24 hr (to remove basic nitrogenous material), then refluxed with 40% NaOH for 2 hr (to remove benzonitrile).
Fractionally distd., and then fractionally cryst. by partial freezing. The higher-melting portion was converted to its sodium salt by
adding a quarter of its weight of sodamide under nitrogen and
stirring for 3 hr at 120°. Unreacted organic material was distd. off
at 120°/1 mm. The sodium salts were hydrolyzed with water, and the
organic fraction was separated by steam distn., followed by fractional distn. Before use, the distillate was passed, under nitrogen,
through a column of activated silica gel. [Russell, JACS, 78, 1041
(1956).]

Indigo and halogen-substituted indigo dyes.
   Reduced in alkaline soln. with sodium hydrosulphite, and filtered.
The filtrate was then oxidized by air, and the resulting ppte. was
filtered off, dried at 65-70°, ground to a fine powder, and extracted with $CHCl_3$ in a Soxhlet extractor. Evaporation of the $CHCl_3$ gave
the purified dye. [Brode, Pearson and Wyman, JACS, 76, 1034 (1954).
Spectral characteristics are listed.]

Indole, m.p. 52°, b.p. 253-254°/760 mm, 124°/5 mm.
   Cryst. from benzene, hexane, water or ethanol-water (1:10).
Purified by sublimation and zone melting.

Inosine, m.p. 215°.
   Cryst. from aq. 80% ethanol.

i-Inositol, m.p. 228°.
   Cryst. from water or aq. 50% ethanol. Dried under vac.

meso-Inositol, m.p. 223-225°.
   Cryst. from aq. ethanol.

Inositol monophosphate, m.p. 195-197° (decomp.).
   Cryst. from water and ethanol.

Inulin.
   Cryst. from water.

Iodoacetamide, m.p. $\sim$143° (decomp.).
   Cryst. from water or CCl$_4$.

Iodoacetic acid, m.p. 78°.
   Cryst. from pet.ether (b.p. 60-80°) or CHCl$_3$-CCl$_4$.

2-Iodoaniline, m.p. 60-61°.
   Distd. with steam and cryst. from benzene-pet.ether.

4-Iodoaniline, m.p. 62-63°.
   Cryst. from pet.ether (b.p. 60-80°) by refluxing, then cooling in
an ice-salt freezing mixture. Dried in air.

4-Iodoanisole, m.p. 51-52°, b.p. 237°/726 mm, 139°/35 mm.
   Cryst. from aq. ethanol.

Iodobenzene, b.p. 188°, n$_D^{25}$ 1.6169.
   Washed with dil. aq. Na$_2$S$_2$O$_3$, then water. Dried with CaCl$_2$ or
CaSO$_4$. Decolorized with charcoal. Distd. under reduced pressure and
stored with mercury or silver powder.

o-Iodobenzoic acid, m.p. 162°.

m-Iodobenzoic acid, m.p. 186.6-186.8°.

p-Iodobenzoic acid, m.p. 271-272°.

Cryst. repeatedly from water and ethanol. Sublimed under vac. at 100°.

4-Iodobiphenyl, m.p. 113.7-114.3°.

Cryst. from ethanol-benzene and dried under vac. over P_2O_5.

1-Iodobutane, see n-Butyl iodide.

2-Iodobutane, b.p. 120.0°, n_D^{25} 1.4973.

Purified by shaking with conc. H_2SO_4, then washing with water, aq. Na_2SO_3 and again with water. Dried with $MgSO_4$ and distd. Alternatively, passed through a column of activated alumina before distn., or treated with elemental bromine, followed by extraction of the free halogen with aq. $Na_2S_2O_3$, thorough washing with water, drying and distn. Stored over silver powder and distd. before use.

4-Iodo-1,3-dinitrobenzene, m.p. 88°.

Cryst. from ethyl ether.

Iodoethane, see Ethyl iodide.

Iodoform, m.p. 119°.

Cryst. from methanol, ethanol or ethanol-ethyl acetate.

1-Iodo-2-methylpropane, see Isobutyl iodide.

1-Iodo-4-nitrobenzene, m.p. 171-172°.

Pptd. from acetone by addition of water, then recryst. from ethanol.

o-Iodophenol, m.p. 42°.

Cryst. from $CHCl_3$ or ethyl ether.

p-Iodophenol, m.p. 94°.

Cryst. from pet.ether (b.p. 80-100°).

1-Iodopropane, see n-Propyl iodide.

2-Iodopropane, see Isopropyl iodide.

3-Iodopropene, see Allyl iodide.

5-Iodosalicylic acid, m.p. 197°.
  Cryst. from water.

o-Iodosobenzoic acid. Decomp. >200°.
  Cryst. from ethanol.

N-Iodosuccinimide, m.p. 200-201°.
  Cryst. from dioxane-CCl$_4$.

α-Iodotoluene, see Benzyl iodide.

p-Iodotoluene, m.p. 35°, b.p. 211-212°.
  Cryst. from ethanol.

3-Iodo-S-tyrosine, $[\alpha]_D^{25}$ -4.4° (in 1 M HCl).
  Likely impurities: tyrosine, diiodotyrosine, iodide. Cryst. by
soln. in dil. ammonia, at room temperature, followed by addition of
dilute acetic acid to pH 6. Stored in the cold.

β-Ionone, b.p. 150-151°/24 mm, n_D^{20} 1.5211, ε_{296nm} = 10,700.

  Converted to the semicarbazone (m.p. 149°) by adding 50 g of semicarbazide hydrochloride and 44 g of potassium acetate in 150 ml of
water to a soln. of 85 g of β-ionone in ethanol. (More ethanol was
added to redissolve any β-ionone that pptd.) The semicarbazone
cryst. on cooling in an ice-bath and was recryst. from ethanol or
75% methanol to constant m.p. (148-149°). The semicarbazone (5 g)
was shaken at room temp. for several days with 20 ml of pet.ether
and 48 ml of M H$_2$SO$_4$, then the ether layer was washed with water and
dil. aq. NaHCO$_3$, dried and the solvent was evapd. The β-ionone was
distd. (The customary steam distn. of β-ionone semicarbazone did not
increase the purity.) [Young et al., JACS, 66, 855 (1944).]

Iproniazid phosphate, m.p. 178-179°.
  Cryst. from water and acetone.

Isatin, m.p. 200°.
  Cryst. from amyl alcohol.

Isoamyl acetate, b.p. 142.0°, n_D^{20} 1.40535.
  Dried with finely divided K_2CO_3 and fractionally distd.

Isoamyl alcohol, b.p. 132°/760 mm.
  Dried with K_2CO_3 or $CaSO_4$, then fractionally distd. If more nearly
anhydrous alcohol is required, the distillate can be refluxed with
the appropriate alkyl phthalate or succinate as described for
Ethanol.

Isoamyl bromide, f.p. -112°, b.p. 119.2°/737 mm.
  Shaken with conc. H_2SO_4, washed with water, dried with K_2CO_3 and
fractionally distd.

Isoamyl chloride, b.p. 99°/734 mm.
  Shaken vigorously with 95% H_2SO_4 until the acid layer no longer
became coloured during 12 hr, then washed with water, saturated aq.
Na_2CO_3, and more water. Dried with $MgSO_4$, filtered and fractionally
distd. Alternatively, a stream of oxygen containing about 5% ozone
was passed through the chloride for a time, three times longer than
was necessary to cause the first coloration of starch iodide paper
by the exit gas. Subsequent washing of the liquid with aq. $NaHCO_3$
hydrolyzed the ozonides and removed organic acids. After drying and
filtering, the isoamyl alcohol was distd. [Chien and Willard, JACS,
75, 6160 (1953).]

Isoamyl ether, b.p. 173.4°, n_D^{20} 1.40850.
  Refluxing with sodium for 5 hr, then distilling under reduced
pressure, removes alcohols. Isoamyl ether can also be dried with
$CaCl_2$ and fractionally distd. from P_2O_5.

Isoascorbic acid, m.p. 174° (decomp.).
  Cryst. from water or dioxane.

RS-Isoborneol, m.p. 212° (sealed tube).

Cryst. from ethanol or pet. ether (b.p. 60-80°). Sublimed.

Isobutane, b.p. -10.2°.

Olefines and moisture can be removed by passage at 65° through a bed of silica-alumina catalyst which had previously been evacuated at about 400°. Alternatively, water and CO_2 can be taken out by passage through P_2O_5 then asbestos impregnated with NaOH. Treatment with anhydrous $AlBr_3$ at 0° then removes traces of olefines. Inert gases can be separated by freezing the isobutane at -195° and pumping out the system.

Isobutene, b.p. -6.6°/760 mm.

Dried by passage through anhydrous $CaSO_4$ at 0°. Purified by freeze-pump-thaw cycles and trap-to-trap distn.

Isobutyl alcohol, b.p. 108°/760 mm.

Dried with K_2CO_3, $CaSO_4$ or $CaCl_2$, filtered and fractionally distd. For further drying, the redist. alcohol can be refluxed with the appropriate alkyl phthalate or succinate as described under Ethanol.

Isobutyl bromide, b.p. 91.2°.

Partially hydrolyzed to remove any tertiary alkyl halide, then fractionally distd., washed with conc. H_2SO_4, water and aq. K_2CO_3, then redist. from dry K_2CO_3. [Dunbar and Hammett, JACS, 72, 109 (1950).]

Isobutyl chloride, b.p. 68.8°/760 mm.

Same methods as described under Isoamyl chloride.

Isobutyl formate, b.p. 98.4°, n_D^{20} 1.38546.

Washed with satd. aq. $NaHCO_3$ in the presence of satd. NaCl, until no further reaction occurred, then with satd. aq. NaCl, dried ($MgSO_4$) and fractionally distd.

Isobutyl iodide, b.p. 83°/250 mm, 120°/760 mm.

Shaken with conc. H_2SO_4, and washed with water, aq. Na_2SO_3, and water, dried with $MgSO_4$ and distd. Alternatively, passed through a

Isobutyl iodide (continued)

column of activated alumina before distn. Stored under nitrogen with mercury in a brown bottle or in the dark.

Isobutyl mercaptan, see 2-Methylpropane-1-thiol.

Isobutyl vinyl ether, b.p. 108-110°.

Washed three times with equal volumes of aq. 1% NaOH, dried with CaH_2, refluxed with sodium for several hours, then fractionally distd. from sodium.

Isobutyraldehyde, b.p. 62.0°.

Dried with $CaSO_4$ and used immediately after distn. under nitrogen because of the great difficulty in preventing oxidation. Can be purified through its acid bisulphite derivative.

Isobutyramide, m.p. 128-129°.

Cryst. from acetone, benzene, $CHCl_3$ or water, then dried under vac. over P_2O_5 or 99% H_2SO_4. Sublimed under vac.

Isobutyric acid, b.p. 154-154.5°.

Distd. from $KMnO_4$, then redistd. from P_2O_5.

Isobutyronitrile, b.p. 103.6°, d^{25} 0.7650.

Shaken with conc. HCl (to remove isonitriles), then with water and aq. $NaHCO_3$. After a preliminary drying with silica gel or Linde type 4A molecular sieve, it is shaken or stirred with CaH_2 until hydrogen evolution ceases, then decanted and distd. from P_2O_5 (not more than 5 g/l., to minimize gel formation). Finally it is refluxed with, and slowly distd. from, CaH_2 (5 g/l.), taking precautions to exclude moisture.

S-Isoleucine, m.p. 285-286° (decomp.), $[\alpha]_D^{20}$ +40.6° (in 6 M HCl).

Cryst. from water by addition of 4 volumes of ethanol.

Isolysergic acid, m.p. 218° (decomp.), $[\alpha]_D^{20}$ +281° (c = 1 in pyridine).

Cryst. from water.

Isonicotinamide, m.p. 155.5-156°.
  Recryst. from hot water.

Isonicotinic acid, m.p. 320°.
  Cryst. repeatedly from water. Dried under vac. at 110°.

Isonicotinic acid hydrazide, m.p. 172°.
  Cryst. from 95% ethanol.

Isonicotinic acid 2-isopropylhydrazide, m.p. 112.5-113.5°.
  Cryst. from benzene-pet.ether.

1-Isonicotinyl-2-salicylidenehydrazine, m.p. 232-233°.
  Cryst. from ethanol.

Isonitrosoacetone, m.p. 69°.
  Cryst. from ether + pet.ether, or CCl_4.

Isonitrosoacetophenone, m.p. 126-128°.
  Cryst. from water.

5-Isonitrosobarbituric acid.
  Cryst. from water or ethanol.

Isononane, b.p. 142°/760 mm.
  Passed through columns of activated silica gel and basic alumina
(activity 1). Distd. under high vac. from sodium-potassium alloy.

Isooctane, see 2,2,4-Trimethylpentane.

Isopentane, see 2-Methylbutane.

Isopentyl-, see Isoamyl-.

Isopentenyl pyrophosphate.
  Purified by chromatography on Whatman No.1 paper using tert.-butyl
alcohol-formic acid-water (20:5:8, R_f = 0.60) or 1-propanol-ammonia-
water (6:3:1, R_f = 0.48).   [continued]

Isopentenyl pyrophosphate (continued)

Also by chromatography on DEAE-cellulose column or by Dowex-1 (formate form) ion-exchanger using formic acid and ammonium formate as eluants. A further purification step is to convert to monocyclohexylammonium salt by passage through a column of Dowex-50 (cyclohexylammonium form) ion-exchange resin. Can also be converted into its lithium salt.

Isophorone, b.p. $94^{\circ}/16$ mm, n_D^{18} 1.4778.

Washed with aq. 5% Na_2CO_3 and then distd. under reduced pressure, immediately before use. Alternatively, can be purified via the semicarbazone. [Erskine and Waight, JCS, 3425 (1960).]

Isophthalic acid, m.p. $345-348^{\circ}$.

Cryst. from aq. ethanol.

Isoprene, b.p. $34.5-35^{\circ}/762$ mm, n_D^{25} 1.4225.

Refluxed with sodium. Distd. from sodium or $NaBH_4$ under dry nitrogen, then passed through a column containing KOH, $CaSO_4$ and silica gel. tert.-Butylcatechol (0.02% w/w) was added, and the isoprene was stored in this way until redistd. just before use.

The inhibitor (tert.-butylcatechol) in isoprene can be removed by several washings with dil. NaOH and water. The isoprene is then dried over CaH_2, distd. under nitrogen at atmospheric pressure, and the fraction distilling at 32° is collected. Stored under nitrogen at -15°.

Isopropanol, b.p. 82.5°, $n_D^{25.8}$ 1.3739.

Forms a constant-boiling mixture, b.p. 80.3°, with water. Most of the water can be removed from this 91% isopropanol by refluxing with CaO (200 g/l.) for several hours, then distilling. The distillate can be dried further with CaH_2, magnesium ribbon, BaO, $CaSO_4$, calcium, anhydrous $CuSO_4$ or Linde type 5A molecular sieve. Distn. from sulphanilic acid removes ammonia and other basic impurities. Peroxides (indicated by liberation of iodine from weakly acid (HCl) solns. of 2% KI) can be removed by refluxing with solid stannous chloride. To obtain isopropanol only 0.002 M in water, sodium (8 g/l.) has been dissolved in material dried by distn. from $CaSO_4$, 35 ml of isopropyl

Isopropanol (continued)

benzoate has been added and, after refluxing for 3 hr, the alcohol
has been distd. through a 50-cm Vigreux column. [Hine and Tanabe,
JACS, 80, 3002 (1958).] Other purification steps for isopropanol
include refluxing with solid aluminium isopropoxide, and the removal
of acetone by treatment with, and distn. from 2,4-dinitrophenyl-
hydrazine. Peroxides re-form in isopropanol if it is stood for
several days.

Isopropyl acetate, b.p. 88.4°, n_D^{20} 1.37730.

   Washed with 50% aq. K_2CO_3 (to remove acid), then with satd. aq.
$CaCl_2$ (to remove any alcohol). Dried with $CaCl_2$ and fractionally distd.

Isopropyl alcohol, see Isopropanol.

Isopropyl benzene, see Cumene.

Isopropyl bromide, b.p. 0°/69.2 mm, 59.4°/760 mm, n_D^{15} 1.42847,
n_D^{20} 1.4251.

   Washed with 95% H_2SO_4 (conc. acid partially oxidized it) until a
fresh portion of acid did not become coloured after several hours,
then with water, aq. 10% $NaHSO_3$, aq. 10% Na_2CO_3 and again with water.
(The H_2SO_4 can be replaced by conc. HCl.) Prior to this treatment,
isopropyl bromide has been purified by bubbling a stream of oxygen
containing 5% ozone through it for 1 hr, followed by shaking with 3%
hydrogen peroxide soln., neutralizing with aq. Na_2CO_3, washing with
distilled water and drying: alternatively, it has been treated with
elementary bromine and stored for 4 weeks, then extracted with aq.
$NaHSO_3$ and dried with $MgSO_4$. After the acid treatment, isopropyl
bromide can be dried with Na_2SO_4, $MgSO_4$ or CaH_2, and fractionally
distd.

N-Isopropylcarbazole.

   Cryst. from isopropanol. Sublimed under vac. Zone refined.

Isopropyl chloride, b.p. 34.8°, n_D^{20} 1.3779, n_D^{25} 1.3754.

   Purified with 95% H_2SO_4 as described for isopropyl bromide, then
dried with $MgSO_4$, P_2O_5 or CaH_2, and fractionally distd. from Na_2CO_3

<u>Isopropyl chloride</u> (continued)

or CaH_2. Alternatively, a stream of oxygen containing about 5% ozone
has been passed through the chloride for about three times as long as
was necessary to obtain the first coloration of starch iodide paper
by the exit gas, and the liquid was then washed with $NaHCO_3$ soln. to
hydrolyze ozonides and remove organic acids before drying and distilling.

<u>Isopropyl ether</u>, b.p. 68.3^o, n_D^{20} 1.36888, n_D^{25} 1.36618.

Common impurities are water and peroxides (detected by the liberation of iodine from weakly acid (HCl) solutions of 2% KI). Peroxides
can be removed by shaking with aq. Na_2SO_3 or with acidified ferrous
sulphate (6.0 g ferrous sulphate + 6 ml conc. H_2SO_4 in 110 ml of
water, using 5-10 g of soln. per l. of ether). The ether is then
washed with water, dried with $CaCl_2$ and distd. Alternatively, refluxing with $LiAlH_4$ or CaH_2, or drying with $CaSO_4$, then passage through
an activated alumina column, can be used to remove water and peroxides. Other dehydrating agents used with isopropyl ether include
P_2O_5, sodium amalgam and sodium wire. (The ether is often stored in
brown bottles, or in the dark, with sodium wire.) Bonner and Goishi
[<u>JACS</u>, <u>83</u>, 85 (1961)] treated isopropyl ether with dilute $Na_2Cr_2O_7$-
H_2SO_4 soln., followed by repeated shaking with a 1:1 mixture of
6 M NaOH and satd. $KMnO_4$. The ether was washed several times with
water, dilute aq. HCl, and water, with a final washing with, and
storage over, ferrous ammonium sulphate acidified with H_2SO_4.
Blaustein and Gryder [<u>JACS</u>, <u>79</u>, 540 (1957)], after washing with
alkaline $KMnO_4$, then water, treated the ether with ceric nitrate in
nitric acid, and again washed with water. Hydroquinone was added
before drying with $CaCl_2$ and $MgSO_4$, and refluxing with sodium amalgam (108 g Hg/100 g Na) for 2 hr under nitrogen. The distillate
(nitrogen atmosphere) was made 2×10^{-5}M in hydroquinone to inhibit
peroxide formation (which was negligible if the ether was stored in
the dark). Pyrocatechol and resorcinol are alternative inhibitors.

<u>4,4'-Isopropylidenediphenol</u>, m.p. 158°.

Cryst. from acetic acid-water (1:1).

Isopropyl iodide, b.p. 88.9°, n_D^{20} 1.4987.

Treated with elemental bromine, followed by extraction of free
halogen with aq. $Na_2S_2O_3$ or $NaHSO_3$, washing with water, drying
($MgSO_4$ or $CaCl_2$) and distn. (The treatment with bromine is optional.)
Other purification methods include passage through activated alumina,
or shaking with copper powder or mercury to remove iodine, drying
with P_2O_5 and distn. Washing with conc. H_2SO_4, water, aq. Na_2SO_3,
water and aq. Na_2CO_3 has also been used. Treatment with silica gel
causes some liberation of iodine. Distillations should be carried
out at reduced pressure. Purified isopropyl iodide is stored in the
dark with mercury.

Isopropyl mercaptan, see Propane-2-thiol.

Isopropyl p-nitrobenzoate, m.p. 105-106°.

Dissolved in ethyl ether, washed with aq. alkali, then water and
dried. Evaporated the ether and recryst. from ethanol.

Isopropyl peroxydicarbonate.

Cryst. from pentane.

p-Isopropyltoluene, b.p. 176.9°/744 mm, n_D^{20} 1.4902.

Dried with CaH_2 and fractionally distd. Stored with CaH_2.

Isoquinoline, m.p. 24°, b.p. 120°/18 mm, $n_D^{25.1}$ 1.62239.

Dried with Linde type 5A molecular sieve or Na_2SO_4 and fractionally
distd. at reduced pressure. Alternatively, refluxed with, and distd.
from, BaO. Also purified by fractional crystn. from its melt and
distd. from zinc dust. Converted to its phosphate (m.p. 135°) or
picrate (m.p. 223°), which were purified by recrystn. and then
reconverted to the free amine. [Packer, Vaughan and Wong, JACS, 80,
905 (1958).]

The procedure for purifying via the picrate comprises the addition
of isoquinoline to picric acid dissolved in the minimum volume of
95% ethanol to yield yellow crystals which are washed with ethanol
and air dried before being recryst. from acetonitrile. The crystals
are dissolved in dimethyl sulphoxide (previously dried over 4A
molecular sieve) and passed through a basic alumina column, on which

Isoquinoline (continued)
picric acid is adsorbed. The free base in the effluent is extracted
with n-pentane and distd. under vac. Traces of solvent are removed
by vapour phase chromatography. [Mooman and Anton, JPC, 80, 2243
(1976).]

Isovaleric acid, b.p. $176.5^\circ/762$ mm, n_D^{15} 1.4064, n_D^{20} 1.40331.
Dried with Na_2SO_4, then fractionally distd.

S-Isovaline, m.p. $\sim300^\circ$.
Cryst. from aq. acetone.

Isovanillin, see 3-Hydroxy-4-methoxybenzaldehyde.

Isoviolanthrone, m.p. $510-511^\circ$ (uncorrected).
Dissolved in 98% H_2SO_4 and pptd. by adding water to reduce the acid
concentration to about 90%. Sublimed. [Parkyns and Ubbelohde, JCS,
4188 (1960).]

Itaconic acid, m.p. $165-166^\circ$.
Cryst. from ethanol, ethanol-ether or ethanol-benzene.

Itaconic anhydride, m.p. $66-68^\circ$, b.p. $139-140^\circ/301$ mm.
Cryst. from chloroform-pet.ether.

Janus red B.
Cryst. from ethanol-water (1:1 v/v).

Jervine, m.p. $241-243^\circ$.
Cryst. from methanol by addition of water.

Juglone, see 2-Hydroxy-1,4-naphthaquinone.

Kerosene.
Stirred with conc. H_2SO_4 until a fresh portion of acid remains
colourless, then washed with water, dried with solid KOH and distd.
in a Claisen flask. For more complete drying, the kerosene can be
refluxed with, and distilled from, sodium.

α-Ketoglutaric acid, m.p. 111-113°.
   Cryst. from acetone-benzene.

2-Keto-L-gulonic acid, m.p. 171°.
   Cryst. from water and washed with acetone.

Khellin, m.p. 154-155°, b.p. 180-200°/0.05 mm.
   Cryst. from methanol or ether.

Kinetin, m.p. 266-267° (sealed tube).
   Cryst. from absolute ethanol.

Kojic acid, m.p. 154.0-155.0°.
   Cryst. from methanol (charcoal) by adding ethyl ether. Sublimed at
0.1 mm pressure.

Kynurenic acid, m.p. 282-283°.
   Cryst. from absolute ethanol.

S-Kynurenine, m.p. 190° (decomp.).
   Cryst. from water.

S-Kynurenine sulphate, m.p. 194°. Monohydrate, m.p. 178° $[\alpha]_D^{25}$ +9.6°
(in H_2O).
   Cryst. from water by addition of ethanol.

S-Lactic acid, m.p. 52.8°, b.p. 105°/0.1 mm, $[\alpha]_D^{15}$ +3.82 (in H_2O).
   Purified by fractional distn. at 0.1 mm pressure, followed by
fractional crystn. from ethyl ether-isopropyl ether (1:1, dried with
sodium). [Borsook, Huffman and Liu, JBC, 102, 449 (1933).] The
solvent mixture, benzene-ethyl ether (1:1) containing 5% pet.ether
(b.p. 60-80°) has also been used.

Lactobionic acid, m.p. 128-130°, $[\alpha]_{546}^{20}$ +28° (c = 3 in H_2O after 24 hr.)
   Cryst. from water by addition of ethanol.

α-Lactose (monohydrate).
   Cryst. from water below 93.5°.

Lactulose, m.p. 167-169° (decomp.), $[\alpha]_{546}^{20}$ -57° (c = 1 in H$_2$O).

Lanatoside A.
  Cryst. from methanol.

Lanatoside B.
Lanatoside C.
  Cryst. from ethanol.

Lanosterol, m.p. 138-140°, $[\alpha]_D^{20}$ +62.0°(c = 1 in CHCl$_3$).
  Recryst. from anhydrous methanol. Dried in vacuo over P$_2$O$_5$ for
3 hrs at 90°. Purity checked by proton magnetic spectral measurements.

Lapachol, m.p. 140°.
  Cryst. from ethanol or ether.

d,1-Laudanosine, m.p. 114-115°.
  Cryst. from ethanol.

Lauraldehyde, b.p. 99.5-100°/3.5 mm, n$_D^{24.7}$ 1.4328.
  Converted to the addition compound by shaking with satd. aq.
NaHSO$_3$ for 1 hr. The ppte. was filtered off, washed with cold water,
ethanol and ether, then decomposed with aq. Na$_2$CO$_3$. The aldehyde was
extracted into ethyl ether which, after drying and evaporation, gave
an oil which was then fractionally distd. under vac.

Lauric acid, m.p. 44.1°, b.p. 225°/100 mm, 141-142°/0.6-0.7 mm.
  Vac. distd. Cryst. from absolute ethanol, or from acetone at -25°.
Alternatively, purified via its methyl ester (b.p. 140.0°/15 mm), as
described under Capric acid. Purified by zone melting.

Lauroyl peroxide.
  Cryst. from benzene.

S-Leucine, m.p. 293-295° (decomp.), $[\alpha]_D^{25}$ +15.6° (in 5 M HCl).
  Likely impurities: isoleucine, valine, methionine. Cryst. from
water by adding 4 volumes of ethanol.

Leucomalachite green, m. p. 92-93°.
 Cryst. from 95% ethanol (10 ml/g), then from benzene-ethanol, and
finally from pet.ether.

Lissamine green BN.
 Cryst. from ethanol-water (1:1 v/v).

Lithocholic acid, m.p. 184-186°.
 Cryst. from ethanol or acetic acid.

Lumazine, see 2,4(1H,3H)-Pteridinedione.

Luminol, m.p. 329-332°.
 Dissolved in KOH soln., treated with Norit (charcoal), filtered
and pptd. with conc. HCl. [Hardy, Seitz and Hercules, Talanta, 24,
297 (1977).] It is chemiluminescent.

RS-Lupanine, m.p. 98-99°.
 Cryst. from acetone.

Lupulon, m.p. 92-94°.
 Cryst. from 90% methanol.

Lutein, m.p. 151°, $E_{1cm}^{1\%}$ 1750 (423 nm), 2560 (446.5 nm), 2340
(477.5 nm) in ethanol.
 Purified by chromatography on columns of magnesia or calcium
hydroxide, and cryst. from CS_2-ethanol. May also be purified via
the dipalmitate ester. Stored in the dark, in an inert atmosphere,
at -20°.

Lutidine.
 For the preparation of pure 2,3-, 2,4- and 2,5-lutidine from
commercial "2,4-/2,5-lutidine", see Coulson et al., JCS, 1934 (1959)
and Kyte, Jeffery and Vogel, JCS, 4454 (1960).

2,3-Lutidine, f.p. -14.8°, b.p. 160.6°, d_4^{20} 0.9464, n_D^{20} 1.50857.
 Steam distd. from a soln. containing about 1.2 equivalents of 20%
H_2SO_4, until about 10% of the base had been carried over with the
non-basic impurities. The acid soln. was then made alkaline, and the

P.L.C.—L

2,3-Lutidine (continued)

base was separated, dried with NaOH or BaO, and fractionally distd. The distd. lutidine was converted to its urea complex by stirring 100 g with 40 ml of urea in 75 ml of water, cooling to 5°, filtering at the pump., and washing with 75 ml of water. The complex, dissolved in 300 ml of water, was steam distd. until the distillate gave no turbidity with a little solid NaOH. The distillate was then treated with excess solid NaOH, and the upper layer was removed: the aqueous layer was also extracted with ethyl ether (80 ml). The upper layer and the ether extract were combined, dried (K_2CO_3), and distd. through a short column. Final purification was by fractional crystn. using partial freezing. [Kyte, Jeffery and Vogel, JCS, 4454 (1960).]

2,4-Lutidine, b.p. 157.8°, d_4^{20} 0.9305, n_D^{20} 1.50087, n_D^{25} 1.4985.

Dried with Linde type 5A molecular sieve, BaO or sodium, and fractionally distd. The distillate (200 g) was heated with benzene (500 ml) and conc. HCl (150 ml) in a Dean and Stark apparatus on a water-bath until water no longer separated and the temperature just above the liquid reached 80°. When cold, the supernatant benzene was decanted and the 2,4-lutidine hydrochloride, after washing with a little benzene, was dissolved in water (350 ml). After removing any benzene by steam distn., an aq. soln. of NaOH (80 g) was added, and the free 2,4-lutidine was steam distd. It was isolated by saturating the distillate with solid NaOH, and distd. through a short column. The pptn. cycle was repeated, then the final distillate was partly frozen in an apparatus at -67.8 to -68.5° (cooled by acetone-CO_2). The crystals were then melted and distd. [Kyte, Jeffery and Vogel, JCS, 4454 (1960).] Alternative purifications are via the picrate [Clarke and Rothwell, JCS, 1885 (1960)] or the hydrobromide [Warnhoff, JOC, 27, 4587 (1962)]. The latter is pptd. from a soln. of lutidine in benzene by passing HBr gas: the salt is recryst. from $CHCl_3$-methyl ethyl ketone, then decomp. with NaOH, and the free base is extracted into ethyl ether.

2,5-Lutidine, m.p. -15.3°, b.p. 156.7°/759 mm, n_D^{25} 1.4982.

Steam distd. from a soln. containing 1-2 equivalents of 20% H_2SO_4 until about 10% of the base had been carried over with the non-basic impurities, then the acid soln. was made alkaline, and the base was

2,5-Lutidine (continued)
separated, dried with NaOH and fractionally distd. twice. Dried with
sodium and fractionally distd. through a Todd column packed with
glass helices.

2,6-Lutidine, m.p. -5.9°, b.p. 144.0°, d_4^{20} 0.92257, n_D^{20} 1.49779.
   Likely contaminants include 3- and 4-picoline (similar b.p.s.).
However, they can be removed using BF_3, with which they react preferentially, by adding 4 ml of BF_3 to 100 ml of dry, fractionally distd.
2,6-lutidine and redistilling. Distn. of commercial material from
$AlCl_3$ (14 g per 100 ml) can also be used to remove picolines (and
water). Alternatively, lutidine (100 ml) can be refluxed with ethyl-
benzenesulphonate (20 g) or ethyl p-toluenesulphonate (20 g) for 1 hr,
then the upper layer is cooled, separated and distd. The distillate
is refluxed with BaO or CaH_2, then fractionally distd. through a
glass helices-packed column.
   2,6-Lutidine can be dried with KOH or sodium, or by refluxing with
(and distilling from) BaO, prior to distn. For purification via its
picrate, 2,6-lutidine, dissolved in absolute ethanol, is treated
with an excess of warm ethanolic picric acid. The ppte. is filtered
off, recryst. from acetone (to give m.p. 163-164.5°), and partitioned
between ammonia and $CHCl_3$-ethyl ether. The organic soln., after
washing with dil. aq. KOH, is dried with Na_2SO_4 and fractionally
distd. [Warnhoff, JOC, 27, 4587 (1962).] Alternatively, 2,6-lutidine
can be purified via its urea complex, as described under 2,3-Lutidine.
Other purification procedures include azeotropic distn. with phenol
[Coulson et al., J. Appl. Chem. (London), 2, 71 (1952)], fractional
crystn. by partial freezing, and vapour-phase chromatography using a
180-cm column of polyethylene glycol-400 (Shell, 5%) on Embacel (May
and Baker) at 100°, with argon as carrier gas. [Bamford and Block,
JCS, 4989 (1961).]

3,5-Lutidine, f.p. -6.3°, b.p. 172.0°/767 mm, d_4^{20} 0.9419,
n_D^{20} 1.50613, n_D^{25} 1.5035.
   Dried with sodium and fractionally distd. through a Todd column
packed with glass helices. Dissolved (100 ml) in dil. HCl (1:4) and
steam distd. until 1 l. of distillate had collected. Excess conc.NaOH
was added to the residue which was again steam distd. The base was

3,5-Lutidine (continued)

extracted from the distillate, using ethyl ether. The extract was dried with K_2CO_3, and distd. It was then fractionally cryst. by partial freezing.

Lycopene, m.p. 172-173°, $E_{1cm}^{1\%}$=2250 (446 nm), 3450 (472 nm), 3150 (505 nm), in pet.ether.

 Cryst. from CS_2-methanol, ethyl ether-pet.ether, or acetone-pet. ether, and purified by column chromatography on deactivated alumina, $CaCO_3$, calcium hydroxide or magnesia. Stored in the dark, in an inert atmosphere, at -20°.

Lycorine, m.p. 275-280° (decomp.).

 Cryst. from ethanol.

Lycoxanthin, m.p. 173-174°, $E_{1cm}^{1\%}$ = 3360 (472.5 nm), also λ_{max} 444, 503 nm, in pet.ether.

 Cryst. from ethyl ether-light petroleum, benzene-pet.ether or CS_2. Purified by chromatography on columns of $CaCO_3$, $Ca(OH)_2$ or deactivated alumina, developing with benzene and eluting with 3:1 benzene-methanol. Stored in the dark, in inert atmosphere, at -20°.

Lysergic acid, m.p. 240° (decomp.).

 Cryst. from water.

S-Lysine. Decomp. >210°.

 Cryst. from aq. ethanol.

S-Lysine dihydrochloride, m.p. 193°, $[\alpha]_D^{25}$ +25.9° (in 5 M HCl).

 Cryst. from methanol, in the presence of excess HCl, by adding ethyl ether.

S-Lysine monohydrochloride, $[\alpha]_D^{25}$ +25.9° (in 5 M HCl).

 Likely impurities: arginine, R-lysine, 2,6-diaminoheptane-dioic acid, glutamic acid. Cryst. from water at pH 4-6 by adding 4 volumes of ethanol. Above 60% relative humidity, forms the dihydrate.

β-D-Lyxose, m.p. 118-119°, $[\alpha]_D^{20}$ -14° (c = 4, in water).

   Cryst. from ethanol or aq. 80% ethanol. Dried under vac. at 60° and stored in a vac. desiccator over P_2O_5 or $CaSO_4$.

Malachite green, m.p. 121-121.5° (carbinol).

   The oxalate was recryst. from hot water and dried in air. The carbinol was pptd. from the oxalate (1 g) in distd. water (100 ml) by adding M NaOH (10 ml). The ppte. was filtered off, recryst. from 95% ethanol containing a little dissolved KOH, then washed with ether, and cryst. from pet.ether. An acid soln. (2 x 10^{-5} M in 6 x 10^{-5} M H_2SO_4) rapidly reverted to the dye. [Swain and Hedberg, JACS, 72, 3373 (1950).]

Maleamic acid, m.p. 172-173°.

   Cryst. from ethanol.

Maleic acid, m.p. 143.5°.

   Cryst. from acetone-pet.ether (b.p. 60-80°) or hot water. Dried at 100°.

Maleic acid hydrazide, m.p. 144° (decomp.).

   Cryst. from water.

Maleic anhydride, m.p. 54°, b.p. 94-96°/20 mm, 199°/760 mm.

   Cryst. from benzene, $CHCl_3$ or CCl_4. Sublimed under reduced pressure.

Maleuric acid, m.p. 167-168° (decomp.).

   Cryst. from hot water.

RS-Malic acid, m.p. 128-129°.

   Cryst. from acetone, then from acetone-CCl_4, or from ethyl acetate by adding pet.ether (b.p. 60-70°). Dried at 35° under 1 mm pressure to avoid formation of the anhydride.

S-Malic acid, m.p. 104.5-106°, $[\alpha]_D$ -2.3° (c = 8.5 in water).

   Cryst. (charcoal) from ethyl acetate-pet.ether (b.p. 55-65°), keeping the temperature below 65°. Or, dissolved by refluxing in fifteen parts of anhydrous ethyl ether, decanted, concentrated to one-third volume and cryst. at 0°, repeatedly, to constant m.p.

Malonamide, m.p. 170°.
  Cryst. from water.

Malonic acid, m.p. 136°.
  Cryst. from benzene-ethyl ether (1:1) containing 5% of pet.ether
(b.p. 60-80°), washed with ethyl ether, then recryst. from water or
acetone. Dried under vac. over conc. H_2SO_4.

Malononitrile, m.p. 32.1°.
  Cryst. from water. Distd. from P_2O_5.

Maltol, m.p. 161-162°.
  Cryst. from $CHCl_3$ or aq. 50% ethanol. Volatile in steam. Can be
readily sublimed.

Maltose, m.p. 118°.
  Purified by chromatography from aq. soln. on to a charcoal-Celite
(1:1) column, washed with water to remove glucose and other monosaccharides, then eluted with aq. 75% ethanol. Cryst. from water, aq.
ethanol or ethanol containing 1% nitric acid. Dried as the monohydrate at room temperature under vac. over H_2SO_4 or P_2O_5.

S-Mandelic acid, m.p. 133°, $[\alpha]_{546}^{20}$ +188° (c = 5, in water).
RS-Mandelic acid, m.p. 118°.
  Purified by Soxhlet extraction with benzene (about 6 ml/g), allow-
ing the extract to crystallize. Dried at room temperature under vac.

Mannitol, m.p. 166.1°
  Cryst. from ethanol or distd. water, drying at 100°.

Mannitol hexanitrate, m.p. 106-108°.
  Cryst. from ethanol.

α-D-Mannose, m.p. 132°, $[\alpha]_D^{20}$ +14.1° (c = 4, in water).
  Cryst. repeatedly from ethanol or aq. 80% ethanol, then dried under
vac. over P_2O_5 at 60°.

Margaric acid, see Heptadecanoic acid.

Meconic acid.
   Cryst. from water and dried at 100° for 20 min.

Melamine, m.p. 353°.
   Cryst. from water or dil. aq. NaOH.

Melezitose, m.p. 153-154°.
   Cryst. as dihydrate from water, then dried at 110° (anhydr.).

Melibiose, m.p. 84-85°.
   Cryst. as dihydrate from water or aq. ethanol.

p-Mentha-1,5-diene, see α-Phellandrene.

1-Menthol, m.p. 44-46.5°, $[\alpha]_D^{30}$ -48.8°.
   Cryst. from $CHCl_3$, pet.ether or ethanol-water.

Meprobamate, m.p. 104-106°.
   Cryst. from hot water.

2-Mercaptobenzimidazole, m. p. 302-304°.
   Cryst. from aq. ethanol or aq. ammonia.

2-Mercaptobenzothiazole, m.p. 182°.
   Cryst. repeatedly from 95% ethanol, or purified by incomplete pptn.
by dil. H_2SO_4 from a basic soln., followed by several crystns. from
acetone-water or benzene.

2-Mercaptoethylamine, m.p. 97-98.5°.
   Sublimed under vac. and stored under nitrogen.

2-Mercaptoimidazole, m.p. 221-222°.
   Cryst. from water.

6-Mercaptopurine.
   Cryst. from pyridine (30 ml/g), washed with pyridine, then triturated with water (25 ml/g), adjusting to pH 5 by adding M HCl. Recryst.
by heating, then cooling, the soln. Filtered, washed with water and

6-Mercaptopurine (continued)
dried at 110°. Has also been cryst. from water (charcoal).

Mercaptosuccinic acid, see Thiomalic acid.

Mesaconic acid, m.p. 204-205°.
Mescaline sulphate, m.p. 183-186°.
   Cryst. from water.

Mesitylene, b.p. 99.0-99.8°/100 mm, 166.5-167°/760 mm, n_D^{20} 1.4962, n_D^{25} 1.4967.
   Dried with $CaCl_2$ and distd. from sodium in a glass helices-packed column. Treated with silica gel and redistd. Alternative purifications include vapour-phase chromatography, or fractional distn. followed by azeotropic distn. with 2-methoxyethanol (which is subsequently washed out with water), drying and fractional distn. More exhaustive purification uses sulphonation by dissolving in two volumes of conc. H_2SO_4, precipitating with four volumes of conc. HCl at 0°, washing with cold conc. HCl, and recrystallizing from $CHCl_3$. The mesitylene sulphonic acid is hydrolyzed with boiling 20% HCl and steam distd. The separated mesitylene is dried ($MgSO_4$ or $CaSO_4$) and distd. Can also be fractionally cryst.

Mesityl oxide, b.p. 112°, n_D^{24} 1.4412.
   Purified via the semicarbazone (m.p. 165°). [Erskine and Waight, JCS, 3425 (1960).]

Metalphthalein, (monohydrate), m.p. 186°.
   Dissolved in sodium acetate soln. and fractionally pptd. with HCl. This removed unsubstituted and monosubstituted cresol phthaleins (which separated at lower acidities). Washed with cold water, dried to monohydrate at 30° in vacuo.

Metanilic acid (decomposes on heating).
   Cryst. (as the hydrate) from water, under CO_2 in a semi-darkened room. (The soln. is photosensitive.) Dried over 90% H_2SO_4 in a vac. desiccator.

α-Methacraldehyde, b.p. 68.4°.

Fractionally distd. under nitrogen through a short Vigreux column. Stored in sealed ampoules. (Slight polymerization may occur.)

Methacrylamide, m.p. 111-112°.

Cryst. from benzene or ethyl acetate and dried under vac. at room temperature.

Methacrylic acid, b.p. 72°/14 mm, 160°/760 mm.

Aq. methacrylic acid (90%) was satd. with NaCl (to remove the bulk of the water), then the organic phase was dried with $CaCl_2$ and distd. under vac. Polymerization inhibitors include 0.025% p-methoxyphenol, 0.1% hydroquinone, and 0.05% N,N'-diphenyl-p-phenylenediamine.

Methacrylic anhydride, b.p. 65°/2 mm.

Distd. at 2 mm pressure, immediately before use, in the presence of hydroquinone.

Methacrylonitrile, b.p. 90.3°, n_D^{20} 1.4007, n_D^{30} 1.3954.

Washed (to remove inhibitors such as p-tert-butylcatechol) with satd. aq. $NaHSO_3$, 1% NaOH in satd. NaCl and then with satd. aq. NaCl. Dried with $CaCl_2$ and fractionally distd. under nitrogen to separate from impurities such as methacrolein and acetone.

Methadone hydrochloride, m.p. 241-242°.

Cryst. from ethanol.

Methane, m.p. -184°, b.p. -164°/760 mm, -130.9°/6.7 atm, d^{-164} 0.466 (air = 1).

Dried by passage over $CaCl_2$ and P_2O_5, then passed through a Dry-Ice trap and fractionally distd. from a liquid-nitrogen trap. Oxygen can be removed by prior passage in a stream of hydrogen over reduced copper oxide at 500°, and higher hydrocarbons can be removed by pre-chlorinating about 10% of the sample: the hydrocarbons, chlorides and HCl are readily separated from the methane by condensing the sample in the liquid-nitrogen trap and fractionally distilling it.

Methanesulphonic acid, b.p. 134.5-135°/3 mm.

Dried, either by azeotropic removal of water with benzene or toluene, or by stirring 20 g of P_2O_5 with 500 ml of the acid at 100° for ½ hr. Then distd. under vac. and fractionally cryst. by partial freezing. Sulphuric acid, if present, can be removed by prior addition of $Ba(OH)_2$ to a dilute soln., filtering off the $BaSO_4$ and concentrating under reduced pressure.

Methanesulphonyl chloride, b.p. 161°/730 mm.

Distd. from P_2O_5.

Methanol, b.p. 64.5°, d_4^{15} 0.79609, d_4^{25} 0.78675, n_D^{15} 1.33057, n_D^{25} 1.32663.

Almost all methanol is now obtained synthetically. Likely impurities are water, acetone, formaldehyde, ethanol, methyl formate and traces of dimethyl ether, methylal, methyl acetate, acetaldehyde, carbon dioxide and ammonia. Most of the water (down to about 0.01%) can be removed by fractional distn. Drying with CaO is unnecessary and wasteful. Anhydrous methanol can be obtained from "absolute" material by passage through Linde type 4A molecular sieve, or by drying with CaH_2, $CaSO_4$, or with just a little more sodium than required to react with the water present; in all cases the methanol is then distd. Two treatments with sodium reduces the water content to about 5 x 10^{-5}%. [Friedman, Gill and Doty, JACS, 83, 4050 (1961).] Lund and Bjerrum [Ber. 64, 210 (1931)] warmed clean dry magnesium turnings (5 g) and iodine (0.5 g) with 50-75 ml of "absolute" methanol in a flask until the iodine disappeared and all of the magnesium was converted to the methoxide. Up to 1 l. of methanol was added and, after refluxing for 2-3 hr, it was distd. off, excluding moisture from the system. Redistn. from tribromobenzoic acid removed basic impurities and traces of magnesium methoxides, and left conductivity-quality material. The method of Hartley and Raikes [JCS, 127, 524 (1925)] gives a slightly better product. This consists of an initial fractional distn., followed by distn. from aluminium methoxide, and then ammonia and other volatile impurities are removed by refluxing for 6 hr with freshly dehydrated $CuSO_4$ (2 g/l.) while dry air is passed through: the methanol is finally distd. (The aluminium methoxide is prepared by warming with aluminium amalgam (3 g/l.) until all the aluminium has

Methanol (continued)

reacted. The amalgam is obtained by warming pieces of sheet aluminium with a soln. of $HgCl_2$ in dry methanol.) This treatment also removes aldehydes.

If acetone is present in the methanol, it is usually removed prior to drying. Bates, Mullaly and Hartley [JCS, 401 (1923)] dissolved 25 g of iodine in 1 l. of methanol and then poured the soln., with constant stirring, into 500 ml of M NaOH. Addition of 150 ml of water pptd. iodoform. The soln. was stood overnight, filtered, then boiled under reflux until the smell of iodoform disappeared, and fraction-ally distd. (This treatment also removes formaldehyde.) Morton and Mark [IECAE, 6, 151 (1934)] refluxed methanol (1 l.) with furfural (50 ml) and 10% NaOH soln. (120 ml) for 6-12 hr, the resulting resin carrying down with it the acetone and other carbonyl-containing impurities. The alcohol was then fractionally distd. Evers and Knox [JACS, 73, 1739 (1951)], after refluxing 4.5 l. of methanol for 24 hr with 50 g of magnesium, distd. off 4 l. of it, which they then refluxed with $AgNO_3$ for 24 hr in the absence of moisture or CO_2. The methanol was again distd., shaken for 24 hr with activated alumina before being filtered through a glass sinter and distd. under nitrogen in an all-glass still. Material suitable for conductivity work was obtained.

Variations of the above methods have also been used. For example, a sodium hydroxide soln. containing iodine has been added to methanol and, after standing for 1 day, the soln. has been poured slowly into about a quarter of its volume of 10% $AgNO_3$, shaken for several hours, then distd. Sulphanilic acid has been used instead of tribromobenzoic acid in Lund and Bjerrum's method. A soln. of 15 g of magnesium in 500 ml of methanol has been heated under reflux, under nitrogen, with hydroquinone (30 g), before degassing and distilling the methanol, which was subsequently stored with magnesium (2 g) and hydroquinone (4 g per 100 ml). Refluxing for about 12 hr removes the bulk of the formaldehyde from methanol: further purification has been obtained by subsequent distn., refluxing for 12 hr with dinitrophenylhydrazine (5 g) and H_2SO_4 (2 g/l.), and again fractionally distilling.

A simple purification procedure consists of adding 2 g $NaBH_4$ to 1.5 l. methanol, gently bubbling with argon and refluxing for 1 day at 30°, then adding 2 g freshly cut sodium (washed with methanol) and refluxing for 1 day before distilling. The middle fraction is taken. [Jou and Freeman, JPC, 81, 909 (1977).]

RS-Methionine, m.p. 281° (decomp.).
  Cryst. from hot water.

S-Methionine, m.p. 283° (decomp.), $[\alpha]_D^{25}$ +21.2° (in 0.2 M HCl).
  Cryst. from aq. ethanol.

RS-Methionine sulphoxide.
  Likely impurities: RS-methionine sulphone, RS-methionine. Cryst.
from water by adding ethanol in excess.

Methoxyacetic acid, b.p. 97°/13-14 mm.
  Fractionally cryst. by repeated partial freezing, then fractionally
distd. under vac. through a vacuum-jacketed Vigreux column 20 cm long.

p-Methoxyacetophenone, m.p. 39°, b.p. 139°/15 mm, 264°/736 mm.
  Cryst. from ethyl ether-pet.ether.

Methoxyamine hydrochloride, m.p. 149-151°.
  Cryst. from ethanol by addition of ethyl ether.

p-Methoxyazobenzene, m.p. 54-56°.
  Cryst. from ethanol.

3-Methoxybenzanthrone.
  Cryst. from benzene.

m-Methoxybenzoic acid, m.p. 110°.
  Cryst. from ethanol-water.

p-Methoxybenzoic acid, m.p. 184.0-184.5°.
  Cryst. from ethanol, water, ethanol-water or toluene.

"Methoxychlor", 1,1-Bis(p-methoxyphenyl)-2,2,2-trichloroethane
(Dimorphic), m.p. 78-78.2°, or 86-88°.
  Freed from 1,1-bis(p-chlorophenyl)-2,2,2-trichloroethane by
crystn. from ethanol.

trans-p-Methoxycinnamic acid, m.p. 173.4-174.8°.
  Cryst. from methanol to constant m.p. and u.v. spectrum.

2-Methoxyethanol, b.p. 124.4°, n_D^{20} 1.4017.

Peroxides can be removed by refluxing with stannous chloride or by filtration under slight pressure through a column of activated alumina. 2-Methoxyethanol can be dried with K_2CO_3, $CaSO_4$, $MgSO_4$ or silica gel, with a final distn. from sodium. Aliphatic ketones (and water) can be removed by making the solvent 0.1% in 2,4-dinitrophenylhydrazine and allowing to stand overnight with silica gel before fractinally distilling.

β-Methoxyethylamine, b.p. 94°.

An aq. 70% soln. was dehydrated by azeotropic distn. with benzene and the amine was distd. twice.

6-Methoxy-1-indanone, m.p. 151-153°.

Cryst. from methanol, then sublimed.

5-Methoxyindole, m.p. 55°, b.p. 176-178°/17 mm.

Cryst. from cyclohexane or pet.ether.

1-Methoxynaphthalene, b.p. 268.4-268.5°.

Fractionally distd. from CaH_2.

2-Methoxynaphthalene, m.p. 73.0-73.6°, b.p. 273°/760 mm.

Fractionally distd. under vac. Cryst. from absolute ethanol, benzene or n-heptane, and dried under vac. in an Abderhalden pistol.

4-Methoxynitrobenzene, m.p. 55°.

Cryst. from hexane and dried under vac.

p-Methoxyphenol, m.p. 54-55°.

Cryst. from benzene, pet.ether or water, and dried under vac. over P_2O_5 at room temperature. Sublimes in vacuo.

(m-Methoxyphenyl)acetic acid, m.p. 71.0-71.2°.

Cryst. from water (charcoal), or aq. ethanol.

(p-Methoxyphenyl)acetic acid, m.p. 85-87°.

Cryst. from ethanol-water.

5-(p-Methoxyphenyl)-1,2-dithiole-3-thione, m.p. 111°.
  Cryst. from butyl acetate.

N-(p-Methoxyphenyl)-p-phenylenediamine, m.p. 102°, b.p. 238°/12 mm.
  Cryst. from ligroin.

8-Methoxypsoralen, m.p. 148°.
  Cryst. from ethanol + ether or benzene + pet.ether.

N-Methylacetamide, m.p. 30°, b.p. 70-71°/2.5-3 mm.
  Fractionally distd. under vac., then fractionally cryst. twice from
its melt. Impurities include acetic acid, methylamine and water. For
detailed purification procedure, see Knecht and Kolthoff, Inorg. Chem.
1, 195 (1962).

N-Methylacetanilide, m.p. 102-104°.
  Cryst. from water.

Methyl acetate, b.p. 56.7-57.2°, n_D^{20} 1.36193, n_D^{25} 1.3538.
  Methanol in methyl acetate can be detected by measuring solubility
in water. At 20°, the solubility of methyl acetate in water is about
35 g per 100 ml, but 1% methanol confers miscibility. Methanol can be
removed by conversion to methyl acetate, using refluxing for 6 hr with
acetic anhydride (85 ml/l.), followed by fractional distn. Acidic
impurities can be removed by shaking with anhydrous K_2CO_3 and distilling. An alternative treatment is with acetyl chloride, followed by
washing with conc. NaCl and drying with CaO or $MgSO_4$. (Solid $CaCl_2$
cannot be used because it forms a crystalline addition compound.)
Distn. from copper stearate destroys peroxides. Free alcohol or acid
can be eliminated from methyl acetate by shaking with strong aq.
Na_2CO_3 or K_2CO_3 (three times), then with aq. 50% $CaCl_2$ (three times),
satd. aq. NaCl (twice), drying with K_2CO_3 and distn. from P_2O_5.

p-Methylacetophenone, b.p. 93.5°/7 mm, 110°/14 mm, n_D^{20} 1.5335.
  Impurities, including the o- and m-isomers, were removed by forming
the semicarbazone which, after repeated crystn., was hydrolyzed to the
ketone. [Brown and Marino, JACS, 84, 1236 (1962).] Also purified by
distn. under reduced pressure, followed by low temperature crystn.
from isopentane.

Methyl acetylsalicylate, m.p. 51-52°.
  Cryst. from pet.ether.

Methyl acrylate, b.p. 80°, d_4^{20} 0.9535, n_D^{20} 1.4040.
  Washed repeatedly with aq. NaOH until free from inhibitors (such
as hydroquinone), then washed with distd. water, dried ($CaCl_2$) and
fractionally distd. under reduced pressure in an all-glass apparatus.
Sealed under nitrogen and stored at 0° in the dark.

Methylal, see Dimethoxymethane.

2-Methylalanine, see α-Aminoisobutyric acid.

Methylamine (gas), b.p. -7.55°/719 mm.
  Dried with sodium or BaO.

Methylamine hydrochloride, m.p. 231.8-233.4°.
  Cryst. from n-butanol, absolute ethanol or methanol-$CHCl_3$. Washed
with $CHCl_3$ to remove traces of dimethylamine hydrochloride. Dried
under vac. Deliquescent.

1-Methylaminoanthraquinone, m.p. 166.5°.
  Cryst. to constant m.p. from butanol, then cryst. from ethanol.
Can be sublimed under vac.

N-Methyl-o-aminobenzoic acid, m.p. 178.5°.
  Cryst. from ethanol or water.

p-Methylaminophenol sulphate, m.p. 260° (decomp.).
  Cryst. from water. Stored in the dark under nitrogen.

p-(2-Methylaminopropyl)phenol, m.p. 162-163°.
  Cryst. from methanol.

2-Methyl-2-amino-1,3-propanediol.
  Cryst. from acetone.

Methylammonium chloride, m.p. 225-226°, b.p. 225-230°/15 mm.
  Cryst. from ethanol.

N-Methylaniline, b.p. $57^{\circ}/4$ mm, $81-82^{\circ}/14$ mm.

  Dried with KOH pellets and fractionally distd. under vac. Acetylated and the acetyl derivative was recryst. to constant m.p.
($101-102^{\circ}$), then hydrolyzed with aq. HCl and distd. under reduced
pressure. [Hammond and Parks, JACS, 77, 340 (1955).]

o-, m- and p-Methylaniline, see o-, m- and p-Toluidine.

N-Methylaniline hydrochloride, m.p. $123.0-123.1^{\circ}$.
  Cryst. from dry benzene-CHCl$_3$ and dried under vac.

Methyl anisate, m.p. 48°.
  Cryst. from ethanol.

p-Methyl anisole, b.p. $175-176^{\circ}$, d_{15}^{15} 0.9757.
  Purified by zone refining.

N-Methylanthranilic acid, see N-Methyl-o-aminobenzoic acid.

2-Methylanthraquinone, m.p. 176°.
  Cryst. from ethanol, then sublimed.

Methyl benzoate, b.p. $104-105^{\circ}/39$ mm, $199.5^{\circ}/760$ mm, n_D^{15} 1.52049,
n_D^{20} 1.51701.
  Washed with dil. aq. NaHCO$_3$, then with water, dried with Na$_2$SO$_4$
and fractionally distd. under reduced pressure.

p-Methylbenzophenone, m.p. 57°.
  Cryst. from methanol and pet.ether.

Methyl benzoquinone, m.p. $68-69^{\circ}$.
  Cryst. from heptane.

2-Methyl-3,4-benzphenanthrene, m.p. 70°.
  Cryst. from ethanol.

RS-α-Methylbenzyl alcohol, b.p. $60.5-61.0^{\circ}/3$ mm.
  Dried with MgSO$_4$ and distd. under vac.

p-Methylbenzyl alcohol, see p-Tolyl carbinol.

p-Methylbenzyl bromide, m.p. 35°, b.p. 218-220°/740 mm.
   Cryst. from pentane.

p-Methylbenzyl chloride, b.p. 80°/2 mm.
   Dried with $CaSO_4$ and fractionally distd. under vac.

Methyl benzylpenicillinate, m.p. 97°, $[\alpha]_D^{20}$ +328° (c = 1 in methanol).
   Cryst. from CCl_4.

Methylbixin, m.p. 163°.
   Cryst. from ethanol-$CHCl_3$.

Methyl bromide, b.p. 3.6°.
   Purified by bubbling through conc. H_2SO_4, followed by passage
through a tube containing glass beads coated with P_2O_5. Also purified by distn. from $AlBr_3$ at -80°, by passage through a tower of KOH
pellets and by partial condensation.

Methyl o-bromobenzoate, b.p. 243-244°, 122°/17 mm.
   Soln. in ether is washed with 10% aq. Na_2CO_3, H_2O, dried and distd.

Methyl p-bromobenzoate, m.p. 79.5-80.5°.
   Cryst. from methanol.

2-Methyl-1,3-butadiene, see Isoprene.

2-Methylbutane, b.p. 27.9°, n_D^{20} 1.35373, n_D^{25} 1.35088.
   Stirred for several hours in the cold with conc. H_2SO_4 (to remove
olefinic impurities), then washed with water, aq. Na_2CO_3 and water
again. Dried with $MgSO_4$ and fractionally distd. using a Todd column
packed with glass helices. Material transparent down to 180 nm was
obtained by distilling from sodium wire, and passing through a
column of silica gel which had previously been dried in place at
350° for 12 hr before use. [Potts, JCP, 20, 809 (1952).]

2-Methyl-1-butanol, b.p. 128.6°, n_D^{25} 1.4082.

Refluxed with CaO, distd., refluxed with magnesium and again fractionally distd. A small sample of highly purified material was obtained by fractional crystn. after conversion to a suitable ester such as the trinitrophthalate or the 3-nitrophthalate. The latter was converted to the cinchonidine salt in acetone and recryst. from $CHCl_3$ by adding pentane. The salt was saponified, extracted with ether, and fractionally distd. [Terry et al., J. Chem. Eng. Data, 5, 403 (1960).]

2-Methyl-2-butanol, see tert.-Amyl alcohol.

3-Methyl-1-butanol, b.p. 132.0°, n_D^{15} 1.40853.

Dried by heating with CaO and fractionally distilling, then heating with BaO and redistilling. Alternatively, boiled with conc. KOH, washed with dil. H_3PO_4, and dried with K_2CO_3, then anhydrous $CuSO_4$, before fractionally distilling.

3-Methyl-1-butanol, b.p. 132°/760 mm, 128.5°/750 mm, d_4^{15} 0.8129, n_D^{20} 1.4075, n_D^{15} 1.4085.

Separated from 2-methyl-1-butanol by fractional distn., fractional crystn. and preparative gas chromatography.

3-Methyl-2-butanol, b.p. 111.5°, n_D^{20} 1.4095, n_D^{25} 1.4076.

Refluxed with magnesium, then fractionally distd.

3-Methyl-2-butanone, b.p. 93-94°/752 mm.

Refluxed with a little $KMnO_4$. Fractionated on a spinning-band column. Dried with $CaSO_4$ and distd.

2-Methyl-2-butene, f.p. -133.78°, b.p. 38.4°/760 mm, d^{20} 0.6783, d_4^{15} 0.66708, d_4^{25} 0.65694, n_D^{15} 1.3908.

Distd. from sodium.

1-Methylbutyl-, see sec.-Amyl-.

Methyl-tert.-butyl ketone, b.p. 106°/760 mm, 105°/746 mm.

Refluxed with a little $KMnO_4$. Dried with $CaSO_4$ and distd.

Methyl n-butyrate, b.p. $102.3°/760$ mm.
Treated with anhydrous $CuSO_4$, then distd. under dry nitrogen.

Methyl carbamate, m.p. $54.4-54.8°$.
Cryst. from benzene.

9-Methylcarbazole, m.p. $89°$.
Purified by zone melting.

Methyl carbitol, see Diethylene glycol monomethyl ether.

3-Methylcatechol, m.p. $68°$, b.p. $112°/3$ mm, $241°/760$ mm.
Cryst. from high-boiling pet.ether and distd.

Methylcellosolve, see 2-Methoxyethanol.

Methyl chloride, b.p. $-24.1°$.
Bubbled through a sintered-glass disc dipping into conc. H_2SO_4, then washed with water, condensed at low temperature and fractionally distd. Has been distd. from $AlCl_3$ at $-80°$. Alternatively, passed through towers containing $AlCl_3$, soda-lime and P_2O_5, then condensed and fractionally distd. Stored as a gas.

Methyl chloroacetate, b.p. $129-130°$.
Shaken with sat. aq. Na_2CO_3 (three times), aq. 50% $CaCl_2$ (three times), sat. aq. NaCl (twice), dried (Na_2SO_4) and fractionally distd.

20-Methylcholanthrene, m.p. $179-180°$.
Cryst. from benzene and ethyl ether.

Methyl cyanide, see Acetonitrile.

Methyl cyanoacetate, f.p. $-13°$, b.p. $205°$.
Purified by shaking with 10% Na_2CO_3 soln., washing well with water, drying with anhydrous Na_2SO_4, and distg.

Methylcyclohexane, b.p. 100.9°, d^{25} 0.7650, n_D^{20} 1.4231, n_D^{25} 1.42058.
   Passage through a column of activated silica gel gives material
transparent down to 220 nm. Can also be purified by passage through
a column of activated basic alumina, or by azeotropic distn. with
methanol followed by washing out the methanol with water, drying
and distn. Methylcyclohexane can be dried with $CaSO_4$, CaH_2 or sodium.
Has also been purified by shaking with a mixture of conc. HNO_3 and
H_2SO_4 in the cold, washing with water, drying with $CaSO_4$ and fractionally distilling from potassium. Percolation through a Celite
column impregnated with 2,4-dinitrophenylhydrazine, phosphoric acid
and water (prepared by grinding 0.5 g DNPH with 6 ml 85% H_3PO_4, then
mixing with 4 ml of distd. water and 10 g of Celite) removes
carbonyl-containing impurities.

2-Methylcyclohexanol, b.p. 65°/20 mm, 167.6°/760 mm, $n_D^{13.4}$ 1.46585,
n_D^{20} 1.46085.

3-Methylcyclohexanol, b.p. 69°/16 mm, 172°/760 mm, n_D^{20} 1.45757,
$n_D^{25.5}$ 1.45444.
   Dried with Na_2SO_4 and distd. under vac.

4-Methylcyclohexanone, b.p. 165.5°/743 mm, n_D^{20} 1.44506.
   Dried with $CaSO_4$, then fractionally distd.

1-Methylcyclohexene, b.p. 107.4-108°.
   Freed from hydroperoxides by refluxing with cupric stearate,
filtered and fractionally distd. from sodium.

Methylcyclopentane, b.p. 71.8°, n_D^{20} 1.40970, n_D^{25} 1.40700.
   Purification procedures include passage through columns of silica
gel (prepared by heating in nitrogen to 350° prior to use) and activated basic alumina, distn. from sodium-potassium alloy, and azeotropic distn. with methanol, followed by washing out the methanol
with water, drying and distn. Has also been purified by repeated
extraction with small portions of conc. H_2SO_4, washing with water,
drying and distn. Can be stored with CaH_2 or sodium.

3'-Methyl-1,2-cyclopentenophenanthrene, m.p. 126-127°.
   Cryst. from acetic acid.

Methyl cyclopropyl ketone, b.p. 111.6-111.8°/752 mm.
  Stored with anhydrous CaSO$_4$, distd. under nitrogen. Redist. under
vac.

S-Methyl-L-cysteine, m.p. 207-211°, $[\alpha]_D^{26}$ -32.0° (in water).
  Likely impurities: cysteine, cystine, S-methyl-DL-cysteine.
Cryst. from water by adding 4 volumes of ethanol.

5-Methylcytosine, m.p. 270° (decomp.).
  Cryst. from water.

Methyl decanoate, b.p. 224°/760 mm, 114°/15 mm.
  Passed through alumina before use.

3-Methyl-4,6-di-tert.-butylphenol, m.p. 62.1°.
  Cryst. from pet.ether (b.p. 25-65°) and dried on a porous plate.

4-Methyl-2,6-di-tert.-butylphenol, m.p. 71.0°.
  Cryst. from methanol and pet.ether (b.p. 60-80°).

Methyl 2,4-dichlorophenoxyacetate, m.p. 43°.
  Cryst. from methanol.

m-Methyl-N,N-dimethylaniline, b.p. 72-74°/5 mm, 215°/760 mm.
p-Methyl-N,N-dimethylaniline, b.p. 76.5-77.5°/4 mm, 211°/760 mm.
  Refluxed for 3 hr with 2 gram-equivalents of acetic anhydride,
then fractionally distd. at reduced pressure.

Methyl dodecanoate, m.p. 5°, b.p. 141°/15 mm, n_D^{50} 1.4199.
  Passed through alumina before use.

N-Methyleneaminoacetonitrile, m.p. 129°.
  Cryst. from ethanol or acetone.

Methylene-bis-(acrylamide).
  Cryst. from methanol.

p,p'-Methylene-bis-(N,N-dimethylaniline), m.p. 89.5°.
   Cryst. from 95% ethanol (charcoal) (approx. 12 ml/g).

Methylene blue.
   Cryst. from 0.1 M HCl (16 ml/g), the crystals separated by centrifuging, washed with chilled ethanol and ethyl ether and dried under
vac. Cryst. from 50% aq. ethanol, washed with absolute ethanol, and
dried at 50-55° for 24 hr. Also cryst. from benzene-methanol (3:1).
Salted out with NaCl from a commercial conc. aq. soln., then cryst.
from water.

Methylene chloride, see Dichloromethane.

4,4'-Methylenedianiline, see p,p'-Diaminodiphenylmethane.

3,4-Methylenedioxyaniline, m.p. 45-46°, b.p. 144°/14 mm.
   Cryst. from pet.ether.

3,4-Methylenedioxycinnamic acid, m.p. 238°.
   Cryst. from glacial acetic acid.

5,5'-Methylenedisalicylic acid, m.p. 238° (decomp.).
   Cryst. from acetone and benzene.

Methylene green.
   Cryst. three times from water (18 ml/g).

Methylene iodide, see Diodomethane.

Methyl ether, b.p. -63.5°/96.5 mm.
   Dried by passing over alumina and then BaO, or over CaH_2, followed
by fractional distn. at low temperatures.

N-Methylethylamine hydrochloride, m.p. 126-130°.
   Cryst. from absolute ethanol and ethyl ether.

Methyl ethyl ether, b.p. 10.8°.
   Dried with $CaSO_4$, passed through an alumina column (to remove
peroxides), then fractionally distd.

Methyl ethyl ketone, see 2-Butanone.

N-Methylformamide, m.p. -3.5°, b.p. 100.5°/25 mm, n_D^{25} 1.4306.
  Dried with molecular sieve for 2 days, then distd. at reduced
pressure through a column packed with glass helices. Fractionally
cryst. by partial freezing and the solid portion was vac. distd.

Methyl formate, b.p. 31.5°, n_D^{15} 1.34648, n_D^{20} 1.34332.
  Washed with strong aq. Na_2CO_3, dried with solid Na_2CO_3 and distd.
from P_2O_5. (Procedure removes any free alcohol or acid.)

2-Methylfuran, b.p. 62.7-62.8°/731 mm.
  Washed with acidified sat. ferrous sulphate soln. (to remove peroxides), separated, dried with $CaSO_4$ or $CaCl_2$, and fractionally
distd. from KOH immediately before use. To reduce the possibility
of spontaneous polymerization, addition of about one-third its
volume of heavy mineral oil to 2-methylfuran prior to distn. has
been recommended.

Methyl gallate, m.p. 202°.
N-Methylglucamine, m.p. 128-129°.
Methyl α-D-glucoside, m.p. 165°, $[\alpha]_D^{25}$ +157.8° (c = 3.0, in water).
  Cryst. from methanol.

α-Methylglutaric acid, m.p. 79°.
β-Methylglutaric acid, m.p. 87°.
  Cryst. from distd. water, then dried under vac. over conc. H_2SO_4.

N-Methylglycamine.
  Cryst. from absolute ethanol and dried under vac.

Methylglyoxal, b.p. ca 72°/760 mm.
  Commercial 30% (w/v) aq. soln. was diluted to about 10% and distd.
twice, taking the fraction boiling below 50°/20 mm Hg. (This
treatment does not remove lactic acid.)

Methyl green.
  Cryst. from hot water.

1-Methylguanine.
   Cryst. from 50% aq. acetic acid.

7-Methylguanine.
   Cryst. from water.

2-Methylhexane, b.p. 90.1°, n_D^{20} 1.38485, n_D^{25} 1.38227.
3-Methylhexane, b.p. 91.9°, n_D^{20} 1.38864, n_D^{25} 1.38609.
   Purified by azeotropic distn. with methanol, then washed with water
(to remove the methanol), dried and distd.

Methyl hexanoate, b.p. 150°/760 mm, 52°/15 mm.
   Passed through alumina before use.

Methylhydrazine, b.p. 87°/745 mm.
   Dried with BaO, then vac. distd. Stored under nitrogen.

2-Methyl-4-hydroxyazobenzene, m.p. 100-101°.
3-Methyl-4-hydroxyazobenzene, m.p. 125-126°.
   Cryst. from hexane.

Methyl p-hydroxybenzoate, m.p. 127.5°.
   Fractionally cryst. from its melt, recryst. from benzene, then from
benzene-methanol and dried over $CaCl_2$ in a vac. desiccator.

Methyl 3-hydroxy-2-naphthoate, m.p. 73-74°.
   Cryst. from methanol (charcoal) containing a little water.

2-Methyl-8-hydroxyquinoline, m.p. 72-73°.
   Cryst. from ethanol-water.

2-Methylimidazole, m.p. 140-141°, b.p. 267°/760 mm.
   Recryst. from benzene.

2-Methylindole, m.p. 61°.
3-Methylindole, m.p. 95°.
   Cryst. from benzene. Purified by zone melting.

Methyl iodide, b.p. 42.8°, n_D^{20} 1.5315.

Deteriorates rapidly with liberation of iodine if exposed to light. Usually purified by shaking with dil. aq. $Na_2S_2O_3$ or $NaHSO_3$ until colourless, then washed with water, dil. aq. Na_2CO_3, and more water, dried with $CaCl_2$ and distd. It is stored in a brown bottle away from sunlight in contact with a small amount of mercury, powdered silver or copper. (Prolonged exposure of mercury to methyl iodide forms methylmercuric iodide.) Methyl iodide can be dried further using $CaSO_4$ or P_2O_5. An alternative purification is by percolation through a column of silica gel or of activated alumina, then distn.

Methyl isobutyl ketone, see 4-Methyl-2-pentanone.

Methyl isopropyl ether, b.p. 32.5°/777 mm.

Purified by drying with $CaSO_4$, passage through a column of alumina (to remove peroxides) and fractional distn.

Methyl isopropyl ketone, see 3-Methyl-2-butanone.

1-Methyl-2-mercaptoimidazole, m.p. 142°.

Cryst. from ethanol.

Methyl methacrylate, f.p. -50°, b.p. 46°/100 mm.

Washed twice with aq. 5% NaOH (to remove inhibitors such as hydroquinone) and twice with water. Dried with $CaCl_2$, Na_2CO_3, Na_2SO_4 or $MgSO_4$, then with CaH_2 and distd. from CaH_2 under nitrogen at reduced pressure. The distillate is stored at low temperatures and redist. before use. Prior to distn., inhibitors such as phenyl-β-naphthylamine (0.2%) or di-β-naphthol are sometimes added. Also purified by zone melting. Boiling for 12-14 hr with powdered KOH, then filtering and distilling under reduced pressure in oxygen-free nitrogen removed carbonyl compounds.

α-Methylmethionine.

Cryst. from aq. ethanol.

S-Methyl-L-methionine chloride, $[\alpha]_D^{23}$ +33° (in 0.2 M HCl).

Likely impurities: methionine, methionine sulphoxide, methionine sulphone. Cryst. from water by adding a large excess of ethanol. Stored in a cool, dry place, protected from light.

N-Methylmorpholine, b.p. 116-117°/764 mm.

Dried by refluxing with BaO or sodium, then fractionally distd. through a helices-packed column.

1-Methylnaphthalene, f.p. -30°, b.p. 244.6°, n_D^{20} 1.6108.

Dried for several days with CaCl$_2$ or by prolonged refluxing with BaO. Fractionally distd. through a glass helices-packed column from sodium. Purified further by soln. in methanol and pptn. of its picrate complex by adding to a satd. soln. of picric acid in methanol. The picrate, after crystn. to constant m.p. (140-141°) from methanol, was dissolved in benzene and extracted with aq. 10% LiOH until the extract was colourless. Evaporation of the benzene under vac. gave 1-methylnaphthalene. [Kloetzel and Herzog, JACS, 72, 1991 (1950).] However, neither the picrate nor the styphnate complexes satisfactorily separates 1- and 2-methylnaphthalenes. To achieve this, 2-methylnaphthalene (10.7 g) in 95% ethanol (50 ml) has been pptd. with 1,3,5-trinitrobenzene (7.8 g) and the complex has been cryst. from methanol to m.p. 153-153.5° (m.p. of the 2-methyl isomer is 124°). [Alternatively, 2,4,7-trinitrofluorene in hot glacial acetic acid could be used, and the derivative (m.p. 163-164°) recryst. from glacial acetic acid.] The 1-methylnaphthalene was regenerated by passing a soln. of the complex in dry benzene through a 15-in. column of activated alumina and washing with benzene-pet.ether (b.p. 35-60°) until the coloured band of the nitro compound had moved down near the end of the column. The complex could also be decomposed using tin and acetic-hydrochloric acids, followed by extraction with ethyl ether and benzene; the extracts were washed successively with dil. HCl, strongly alkaline sodium hypophosphite, water, dil. HCl and water. [Soffer and Stewart, JACS, 74, 567 (1952).] Can be purified from anthracene by zone melting.

2-Methylnaphthalene, m.p. 34.7-34.9°, b.p. 129-130°/25 mm.

Fractionally cryst. repeatedly from its melt, then fractionally distd. under reduced pressure. Cryst. from benzene and dried under

2-Methylnaphthalene (continued)
vac. in an Abderhalden pistol. Purified via its picrate (m.p. 114-
115°) as described for 1-Methylnaphthalene.

6-Methyl-2-naphthol, m.p. 128-129°.
7-Methyl-2-naphthol.
   Cryst. from ethanol or ligroin.

2-Methyl-1,4-naphthoquinone (vitamin K$_3$), m.p. 106°.
   Cryst. from ethanol or pet.ether.

Methyl 1-naphthyl ether, b.p. 90-91°/2 mm, n$_D^{26}$ 1.6210.
   Steam distd. from alkali. The distillate was extracted with ethyl
ether. After drying the extract and evaporating the ethyl ether,
the methyl naphthyl ether was distd. under reduced pressure.

Methyl nitrate, b.p. 65°/760 mm, d^5 1.2322, d^{15} 1.2167, d^{25} 1.2032.
   Distd. at -80°. Middle fraction subjected to several freeze-pump-
thaw cycles. Vapour explodes on heating.

N-Methyl-p-nitroaniline, m.p. 152.2°.
   Cryst. from aq. ethanol.

2-Methyl-5-nitroaniline, m.p. 109°.
   Acetylated, and the acetyl derivative cryst. to constant m.p.,
then hydrolyzed with 70% H$_2$SO$_4$ and the free base regenerated by
treatment with ammonia. [Bevan, Fayiga and Hirst, JCS, 4284 (1956).]

4-Methyl-3-nitroaniline, m.p. 81.5°.
   Cryst. from hot water (charcoal), then ethanol and dried in a
vac. desiccator.

Methyl m-nitrobenzoate, m.p. 78°.
   Cryst. from methanol (1 g/g).

Methyl p-nitrobenzoate, m.p. 95-95.5°.
   Dissolved in ethyl ether, then washed with aq. alkali, the ether
was evaporated and the ester was recryst., from ethanol.

2-Methyl-2-nitro-1,3-propanediol, m.p. 145°.
  Cryst. from n-butanol.

2-Methyl-2-nitro-1-propanol, m.p. 87-88°.
  Cryst. from pet. ether.

N-Methyl-4-nitrosoaniline, m.p. 118°.
  Cryst. from benzene.

N-Methyl-N-nitroso-p-toluenesulphonamide, m.p. 62°.
  Cryst. from benzene by addition of pet.ether.

Methyl octanoate, b.p. 193-194°/760 mm, 83°/15 mm, d_0^0 0.8942
  Passed through alumina before use.

Methyl oleate, f.p. -19.9°, b.p. 217° (decomp.)/16 mm.
  Purified by fractional distn. under reduced pressure, and by
crystn. from acetone.

Methyl orange.
  Cryst. from hot water, then washed with a little ethanol followed
by ethyl ether.

3-Methyl-2-oxazolidone.
  Purified by successive fractional freezing, then dried in dry-box
over Linde type 4A molecular sieves for 2 days.

Methylpentane (mixture of isomers).
  Passage through a long column of activated silica gel gave
material transparent down to 200 nm.

2-Methylpentane, b.p. 60.3°, n_D^{20} 1.37145, n_D^{25} 1.36873.
  Purified by azeotropic distn. with methanol, followed by washing
out the methanol with water, drying ($CaCl_2$, then sodium), and distn.
[Forziati et al., J. Res. Nat. Bur. Stand., 36, 129 (1946).]

3-Methylpentane, b.p. 63.3°, n_D^{20} 1.37652, n_D^{25} 1.37384.
  Purified by azeotropic distn. with methanol, as for 2-methylpen-
tane. Purified for ultraviolet spectroscopy by passage through

3-Methylpentane (continued)

columns of activated silica gel or alumina activated by heating for 8 hr at 210° under a stream of nitrogen. Has also been treated with conc. (or fuming) H_2SO_4, then washed with water, aq. 5% NaOH, water again, then dried ($CaCl_2$, then sodium), and distd. through a long, glass helices-packed column.

2-Methyl-2,4-pentanediol, b.p. 107.5-108.5°/25 mm, n_D^{25} 1.4265.

Dried with Na_2SO_4, then CaH_2 and fractionally distd. under reduced pressure through a packed column, taking precautions to avoid absorption of water.

2-Methyl-1-pentanol, b.p. 146-147°, 65-66°/60 mm.

Dried with Na_2SO_4 and distd.

4-Methyl-2-pentanol, b.p. 131-132°.

Washed with aq. $NaHCO_3$, dried and distd. Purified by conversion to the phthalate ester by adding 120 ml of dry pyridine and 67 g of phthalic anhydride per mole of alcohol, purifying the ester and steam distilling it in the presence of NaOH. The distillate was extracted with ether, and the extract was dried and fractionally distd. [Levine and Walti, JBC, 94, 367 (1931).]

3-Methyl-3-pentanol carbamate, m.p. 56-58.5°.

Cryst. from 30% ethanol.

4-Methyl-2-pentanone, b.p. 115.7°, n_D^{20} 1.3958, n_D^{25} 1.3933.

Refluxed with a little $KMnO_4$, washed with aq. $NaHCO_3$, dried with $CaSO_4$ and distd. Acidic impurities were removed by passage through a small column of activated alumina.

2-Methyl-1-pentene, b.p. 61.5-62°.

Water was removed, and peroxide formation prevented, by several vac. distns. from sodium, followed by storage with sodium-potassium alloy.

cis-4-Methyl-2-pentene, b.p. 57.7-58.5°.

trans-4-Methyl-2-pentene.

Dried with CaH_2.

3-Methyl-1-pentyn-3-ol carbamate, m.p. 55.8-57°.
   Cryst. from ether-pet.ether or cyclohexane.

5-Methyl-1,10-phenanthroline, m.p. 113° (anhydr.).
   Cryst. from benzene-pet.ether.

N-Methylphenazonium methosulphate, m.p. 155-157°.
   Cryst. from ethanol.

p-Methylphenylacetic acid, m.p. 94°.
   Cryst. from heptane.

1-Methyl-1-phenylhydrazine sulphate.
   Cryst. from hot aq. soln. by addition of hot ethanol.

N-Methyl-2-phenylsuccinimide, see Phensuximide.

2-Methylpiperazine, m.p. 66°.
   Purified by zone melting.

3-Methylpiperidine, b.p. 125°/763 mm, n_D^{25} 1.4448.
   Purified via the hydrochloride, m.p. 172°. [Chapman, Isaacs and
Parker, JCS, 1925 (1959).]

4-Methylpiperidine, b.p. 124.4°/755 mm, n_D^{25} 1.4430.
   Purified via the hydrochloride, m.p. 189°. Freed from 3-methyl-
pyridine by zone melting.

2-Methylpropane-1,2-diamine, b.p. 47-48°/17 mm.
   Dried with sodium for 2 days, then distd. under reduced pressure
from sodium.

2-Methylpropane-1-thiol, b.p. 41.2°/142 mm, d_4^{25} 0.82880, n_D^{25} 1.43582.
   Dissolved in ethanol, and added to 0.25 M Pb(OAc)$_2$ in 50% aq.
ethanol. The pptd. lead mercaptide was filtered off, washed with a
little ethanol, and impurities were removed from the molten salt by
steam distn. After cooling, dil. HCl was added dropwise to the
residue, and the mercaptan was distd. directly from the flask. Water
was separated from the distillate, and the mercaptan was dried

2-Methylpropane-1-thiol (continued)

(Na_2CO_3) and distd. under nitrogen. [Mathias, JACS, 72, 1897 (1950).]

2-Methylpropane-2-thiol, b.p. 61.6°/701 mm, d_4^{25} 0.79426, n_D^{25} 1.41984.
    Dried for several days with CaO, then distd. from CaO. Purified
as for 2-Methylpropane-1-thiol.

2-Methyl-1-propanol, b.p. 107.9°, n_D^{15} 1.39768, n_D^{25} 1.3939.
    Dried by refluxing with CaO and BaO for several hours, followed
by treatment with calcium or aluminium amalgam, then fractional
distn. from sulphanilic or tartaric acids. More exhaustive purifications involve formation of phthalate or borate esters. Heating with
phthalic anhydride gives the acid phthalate which, after crystn. to
constant m.p. (65°) from pet.ether, is hydrolyzed with aq. 15% KOH.
The alcohol is distd. as the water azeotrope and dried with K_2CO_3,
then anhydrous $CuSO_4$, and finally magnesium turnings, followed by
fractional distn. [Hückel and Ackermann, J. prakt. Chem., 136, 15
(1933).] The borate ester is formed by heating the dried alcohol for
6 hr in an autoclave at 160-175° with a quarter of its weight of
boric acid. After several fractional distns. under vac. the ester is
hydrolyzed by heating for a short time with aq. alkali and the
alcohol is dried with CaO and distd. [Michael, Scharf and Voigt,
JACS, 38, 653 (1916).]

2-Methyl-2-propanol, see tert.-Butyl alcohol.

N-Methylpropionamide, f.p. -30.9°, b.p. 103°/12-13 torr, n_D^{25} 1.4350.
    Dried over CaO. Water and unreacted propionic acid removed as
their xylene azeotropes. Vac. dried.
    Material for use as an electrolyte solvent (specific conductance
less than 10^{-6} ohm^{-1}cm^{-1} was obtained by fractional distn. at
reduced pressure, and was stored over BaO or molecular sieves. Karl
Fischer titration of water could be performed directly. [Hoover,
PAC, 37, 581 (1974).]

Methyl propionate, b.p. 79.7°.
    Washed with satd. aq. NaCl, then dried with Na_2CO_3 and distd.
from P_2O_5. (This procedure removes any free acid or alcohol.) Can
also be dried with anhydrous $CuSO_4$.

Methyl n-propyl ether, b.p. 39°.

Dried with $CaSO_4$, then passed through a column of alumina (to remove peroxides) and fractionally distd.

Methyl n-propyl ketone, b.p. 102.4°, n_D^{20} 1.3903.

Refluxed with a little $KMnO_4$, dried with $CaSO_4$ and distd. Converted to its bisulphite addition compound by shaking with excess satd. aq. $NaHSO_3$ at room temperature, cooling to 0°, filtering, washing with ethyl ether and drying. Steam distd. The ketone was recovered from the distillate, washed with aq. $NaHCO_3$ and distd. water, dried (K_2CO_3) and fractionally distd. [Waring and Garik, JACS, 78, 5198 (1956).]

2-Methylpyrazine, b.p. 136-137°.

Purified via the picrate. [Wiggins and Wise, JCS, 4780 (1956).]

2-Methylpyridine, b.p. 129.4°, d_4^{20} 0.9444, n_D^{20} 1.50102.

Biddiscombe and Handley [JCS, 1957 (1954)] steam distd. a boiling soln. of the base in 1.2 equivalents of 20% H_2SO_4 until about 10% of the base had been carried over, along with non-basic impurities. Excess aq. NaOH was then added to the residue, the free base was separated, dried with solid NaOH and fractionally distd.

2-Methylpyridine can also be dried with BaO, CaO, CaH_2, $LiAlH_4$, sodium or Linde type 5A molecular sieve. An alternative purification is via the $ZnCl_2$ adduct, which is formed by adding 2-methylpyridine (90 ml) to a soln. of anhydrous $ZnCl_2$ (168 g) and 42 ml conc. HCl in absolute ethanol (200 ml). Crystals of the complex are filtered off, recryst. twice from absolute ethanol (to give m.p. 118.5-119.5°), and the free base is liberated by addition of excess aq. NaOH. It is steam distd., and solid NaOH is added to the distillate to form two layers, the upper one of which is then dried with KOH pellets, stored for several days with BaO and fractionally distd. Instead of $ZnCl_2$, $HgCl_2$ (430 g in 2.4 l. of hot water) can be used. The complex, which separates on cooling, can be dried at 110° and recryst. from 1% HCl (to m.p. 156-157°).

3-Methylpyridine, m.p. -18.5°, b.p. 144.0°/767 mm, d_4^{20} 0.9566,
n_D^{20} 1.50685.

In general, the same methods of purification that are described
for 2-Methylpyridine can be used. However, 3-methylpyridine often
contains 4-methylpyridine and 2,6-lutidine, neither of which can be
removed satisfactorily by drying and fractionation, or by using the
ZnCl$_2$ complex. Biddescombe et al. [JCS, 1957 (1954)], after steam
distn. as for 2-Methylpyridine, treated the residue with urea to
remove 2,6-lutidine, then azeotropically distd. with acetic acid
(the azeotrope had b.p. 114.5°/212 mm), and recovered the base by
adding excess aq. 30% NaOH, drying with solid NaOH and carefully
fractionally distilling. The distillate was then fractionally cryst.
by slow partial freezing. An alternative treatment [Reithof et al.,
IECAE, 18, 458 (1946)] is to reflux the crude base (500 ml) for 20-
24 hr with a mixture of acetic anhydride (125 g) and phthalic
anhydride (125 g) followed by distn. until phthalic anhydride begins
to pass over. The distillate was treated with NaOH (250 g in 1.5 l.
of water) and then steam distd. Addition of solid NaOH (250 g) to
this distillate (about 2 l.) led to the separation of 3-methylpyrid-
ine which was removed, dried (K$_2$CO$_3$, then BaO) and fractionally
distd. (Subsequent fractional freezing would probably be advantageous.)

4-Methylpyridine, m.p. 4.25°, b.p. 145.0°/765 mm, d_4^{20} 0.95478,
n_D^{20} 1.50584.

Can be purified as for 2-Methylpyridine. Biddescombe et al's
method for 3-methylpyridine is also applicable. Lidstone [JCS, 242
(1940)] purified via the oxalate (m.p. 137-138°) by heating 100 ml
of 4-methylpyridine to 80° and adding slowly 110 g of anhydrous oxalic acid, followed by 150 ml of boiling ethanol. After cooling and
filtering, the ppte. was washed with a little ethanol, then recryst.
from ethanol, dissolved in the minimum quantity of water and distd.
with excess 50% KOH. The distillate was dried with solid KOH and
again distd. Hydrocarbons can be removed from 4-methylpyridine by
converting the latter to its hydrochloride, crystallizing from
ethanol-ethyl ether, regenerating the free base by adding alkali and
distilling. As a final purification step, 4-methylpyridine can be
fractionally cryst. by partial freezing to effect a separation from

<u>4-Methylpyridine</u> (continued)
3-methylpyridine. Contamination by 2,6-lutidine is detected by its strong absorption at 270 nm.

<u>4-Methylpyridine-1-oxide</u>, m.p. 184°.
   Cryst. from acetone-ether.

<u>N-Methylpyrrole</u>, b.p. $115-116^{O}/756$ mm.
   Dried with $CaSO_4$, then fractionally distd. from KOH immediately before use.

<u>1-Methyl-2-pyrrolidinone</u>, f.p. -24.4^{O}, b.p. $202^{O}/760$ mm.
   Dried by removing water as benzene azeotrope. Fractionally distd. at 10 torr through a 100 cm column packed with glass helices.

<u>2-Methylquinoline</u>, b.p. $86-87^{O}/1$ mm, $115^{O}/14$ mm, $246-247^{O}/760$ mm, n_D^{20} 1.6126.
   Dried with Na_2SO_4 or by refluxing with BaO, then fractionally distd. at reduced pressure. Redistd. from zinc dust. Purified by conversion to its phosphate (m.p. 220°) or picrate (m.p. 192°) from which after recrystn., the free base was regenerated. [Packer, Vaughan and Wong, <u>JACS</u>, <u>80</u>, 905 (1958).] Its $ZnCl_2$ complex can be used for the same purpose.

<u>4-Methylquinoline</u>, b.p. 265.5^{O}, n_D^{20} 1.61995.
   Refluxed with BaO, then fractionally distd. Purified <u>via</u> its recryst. dichromate complex (m.p. 138°). [Cumper, Redford and Vogel, <u>JCS</u>, 1176 (1962).]

<u>6-Methylquinoline</u>, b.p. 258.6^{O}, n_D^{20} 1.61605.
   Refluxed with BaO, then fractionally distd. Purified <u>via</u> its recryst. $ZnCl_2$ complex (m.p. 190°). [Cumper, Redford and Vogel, <u>JCS</u>, 1176 (1962).]

<u>7-Methylquinoline</u>, m.p. 38°, b.p. $255-260^{O}$, n_D^{20} 1.61481.
   Purified <u>via</u> its dichromate complex (m.p. 149° after five recrystns. from water). [Cumper, Redford and Vogel, <u>JCS</u>, 1176 (1962).]

8-Methylquinoline, b.p. 122.5°/16 mm, 247.8°/760 mm, n_D^{20} 1.61631.

Purified as for 2-Methylquinoline. The phosphate and picrate have m.p.s. of 158° and 201°, respectively.

2-Methyl-8-quinolinol, see 2-Methyl-8-hydroxyquinoline.

Methyl red, m.p. 181-182°.

Extracted with boiling toluene in a Soxhlet flask. The crystals which separated on slow cooling to room temperature are filtered off, washed with a little toluene and recryst. from glacial acetic acid, benzene or toluene followed by pyridine-water. Alternatively, diss- olved in aq. 5% $NaHCO_3$ soln., and pptd. from hot soln. by dropwise addition of aq. HCl. Repeated until extinction coefficients did not increase.

α-Methylstyrene, b.p. 57°/15 mm, n_D^{20} 1.5368.

Washed three times with aq. 10% NaOH (to remove inhibitors such as quinol), then six times with distd. water, dried with $CaCl_2$ and distd. under vac. The distillate is kept under nitrogen, in the cold, and redistd. if kept more than 48 hr before use. It can also be dried with CaH_2.

Methylsuccinic acid, m.p. 115.0°.

Cryst. from water.

17α-Methyltestosterone, m.p. 164-165°, $[\alpha]_{546}^{20}$+87° (c = 1 in dioxane).

Cryst. from hexane-benzene.

Methyl tetradecanoate, m.p. 18.5°, b.p. 155-157°/7 mm.

Passed through alumina before use.

2-Methyltetrahydrofuran, b.p. 80.0°, n_D^{20} 1.405.

Likely impurities: 2-methylfuran, methyldihydrofurans and hydro- quinone (stabilizer). Washed with 10% aq. NaOH, dried, vac. distd. from CaH_2, passed through freshly activated alumina under nitrogen, and refluxed over sodium metal under vac. Stored over sodium. [Ling and Kevan, JPC, 80, 592 (1976).]

Vac. distd. from sodium, and stored with sodium-potassium alloy.

2-Methyltetrahydrofuran (continued)

(Treatment removes water and prevents formation of peroxides.) Alternatively, it can be freed from peroxides by treatment with ferrous sulphate and sodium bisulphate, then solid KOH, followed by drying with, and distn. from, sodium. It may be difficult to remove benzene if it is present as an impurity (can be readily detected by its ultraviolet absorption in the 249-268 nm region).

3-(Methylthio)alanine, see S-Methyl-L-cysteine.

3-Methylthiophen, b.p. 111-113°.
   Dried with Na_2SO_4, then distd. from sodium.

6-Methyl-2-thiouracil, m.p. 330° (decomp.).
   Cryst. from a large volume of water.

Methyltricaprylylammonium chloride, see Aliquat 336.

N-Methyltryptophan, m.p. 295° (decomp.), $[\alpha]_D^{21}$ +44.4° (c = 2.8 in 0.5 M HCl).
   Cryst. from water.

5-Methyltryptophan, m.p. 275° (decomp.).
   Cryst. from aq. ethanol.

4-Methyluracil, m.p. 270-280° (decomp.), λ_{max} 260 nm, log ε 3.97.
   Cryst. from ethanol or acetic acid.

3-Methyluric acid, m.p. >350°.
7-Methyluric acid, m.p. >380°.
9-Methyluric acid, m.p. >400°.
   Cryst. from water.

Methyl vinyl ketone, b.p. 79-80°, 62-68°/400 mm.
   Forms an 85% azeotrope with water.After drying with K_2CO_3 and $CaCl_2$ (with cooling), the ketone is distd. at 2-3 mm pressure.

Methyl vinyl sulphone.

    Passed through a column of alumina, then degassed and distd. on a
vac. line and stored at $-190°$ until required.

Methyl violet.

    Cryst. from absolute ethanol by pptn. with ethyl ether during
cooling in an ice-bath. Filtered off and dried at $105°$.

1-Methylxanthine, m.p. $>360°$.
3-Methylxanthine, m.p. $>360°$.
7-Methylxanthine, m.p. $>380°$ (decomp.).
8-Methylxanthine, m.p. $292-293°$ (decomp.).
9-Methylxanthine, m.p. $384°$ (decomp.).
    Cryst. from water.

Metrazole, m.p. $61°$.

    Cryst. from ethyl ether. Dried under vac. over P_2O_5.

Mevalonic acid.

    Purified via dibenzyl-ethylenediammonium salt (m.p. $124-125°$)
[Hoffman et al., JACS, 79, 2316 (1957)], or by chromatography on
paper or on a Dowex-1 (formate) column.[Bloch et al., JBC, 234,
2595 (1959)]. Stored as DBED salt, or as lactone in sealed container,
in refrigerator.

Mevalonic acid 5-phosphate.

    Purified by conversion to tricyclohexylammonium salt by treatment
with cyclohexylamine (m.p. $154-156°$). Cryst. from water-acetone at
$-15°$. Alternatively, the phosphate was chromatographed by ion-
exchange or paper (Whatman No.1) in a system isobutyric acid-ammonia-
water (66:3:30, R_f = 0.42). Stored as cyclohexylammonium salt.

Mevalonic acid 5-pyrophosphate.

    Purified by ion-exchange chromatography on Dowex-1 formate [Bloch
et al., JBC, 234, 2595 (1959)], DEAE-cellulose [Skilletar and Kekwick,
Anal. Biochem., 20, 171 (1967)], or by paper chromatography [Rogers
et al., BJ, 99, 381 (1966)]. Likely impurities are ATP and mevalonic
acid phosphate. Stored as dry powder or as slightly alkaline
(pH 7-9) soln. at $-20°$.

Michler's ketone, m.p. 179°.

Dissolved in dil. HCl, filtered and pptd. by adding ammonia (to remove water-insoluble impurities such as benzophenone). Then cryst. from ethanol or pet.ether·[Suppan, JCS Faraday Trans I, 71, 539 (1975).]

Milling red SWB.
Milling yellow G.

Salted out three times with sodium acetate, then repeatedly extracted with ethanol. [McGrew and Schneider, JACS, 72, 2547 (1950).] See entry under Chlorazol sky blue FF.

Morin, m.p. 289-292°.

Stirred at room temperature with ten times its weight of absolute ethanol, then left overnight to settle. Filtered, and evapd. under a heat lamp to one-tenth its volume. An equal volume of water was added, and the pptd. morin was filtered off, dissolved in the minimum amount of ethanol and again pptd. with an equal volume of water. The ppte. was filtered, washed with water and dried at 110° for 1 hr. (Yield about 2½%) [Perkins and Kalkwarf, AC, 28, 1989 (1956).]

Morphine (monohydrate), m.p. 230° (decomp.), $[\alpha]_D^{23}$ -130.9° (in MeOH).
Cryst. from methanol.

Morpholine, f.p. -4.9°, b.p. 128.9°, d^{20} 1.0007, n_D^{20} 1.4540, n_D^{25} 1.4533.

Dried with KOH, fractionally distd., then refluxed with sodium, and again fractionally distd. Dermer and Dermer [JACS, 59, 1148 (1937)] pptd. as the oxalate by adding slowly to slightly more than 1 molar equivalent of oxalic acid in ethanol. The ppte. was filtered and recryst. twice from 60% ethanol. Addition to conc. aq. KOH regenerated the base, which was separated and dried with solid KOH, then sodium, before being fractionally distd.

2-(N-Morpholino)ethanesulphonic acid (MES).
Cryst. from hot ethanol containing a little water.

Mucic acid, m.p. 212-213° (decomp.).

Dissolved in the minimum volume of dil. aq. NaOH, and pptd. by adding dil. HCl. The temperature should be kept below 25°.

Mucochloric acid, m.p. 124-126°.

Cryst. twice from water (charcoal).

trans, trans-Muconic acid, m.p. 298°.

Murexide.

Cryst. from hot water.

Myristic acid, m.p. 58°.

Purified via the methyl ester (b.p. 153-154°/10 mm, n_D^{25} 1.4350), as for Capric acid. [Trachtman and Miller, JACS, 84, 4828 (1962).] Also by zone melting. Cryst. from pet.ether.

Naphthacene, m.p. 340-341°.

Cryst. from ethanol or benzene. Dissolved in sodium-dried benzene and passed through a column of alumina. The benzene was evaporated under vac., and the chromatography was repeated using fresh benzene. Finally, the naphthacene was sublimed under vac. [Martin and Ubbelohde, JCS, 4948 (1961).]

2-Naphthaldehyde, m.p. 59°, b.p. 160°/19 mm.

Distilled with steam and cryst. from water.

Naphthalene, m.p. 80.3°, b.p. 218.0°, n_D^{85} 1.5898.

Cryst. from one or more of the following solvents: ethanol, methanol, CCl_4, benzene, glacial acetic acid, acetone or ethyl ether-followed by drying at 60° in an Abderhalden drying apparatus. Also purified by vac. sublimation and by fractional crystn. of its melt. Other purification procedures include refluxing in ethanol over Raney nickel, and chromatography of a CCl_4 sol. on alumina with benzene as elutriant. Baly and Tuck [JCS, 1902 (1908)] purified naphthalene for spectroscopy by heating with conc. H_2SO_4 and MnO_2, followed by steam distn. (repeating the process), and formation of the picrate which, after recrystn., was decomposed and the naphthalene was steam distd. It was then recryst. from dil. ethanol. It

Naphthalene (continued)
can be dried over P_2O_5 under vac.

Washed at 85° with 10% NaOH to remove phenols, with 50% NaOH to remove nitriles, with 10% H_2SO_4 to remove organic bases, and with 0.8 g $AlCl_3$ to remove thianaphthalenes and various alkyl derivatives. Then treated with 20% H_2SO_4, 15% Na_2CO_3 and finally distilled.

Zone melting purified naphthalene from anthracene, 2,4-dinitro-phenylhydrazine, methyl violet, benzoic acid, methyl red, chrysene, and induline.

Naphthalene-1,8-diamine, m.p. 66.5^O.
Cryst. from aq. ethanol.

Naphthalenediol, see Dihydroxynaphthalene.

Naphthalene-2,7-disulphonic acid.
Cryst. from conc. HCl.

Naphthalene scarlet 4R.
Dissolved in the minimum quantity of boiling water, filtered and enough alcohol was added to ppte. about 80% of the dye. This process was repeated until a soln. of the dye in aq. 20% pyridine had a constant extinction coefficient.

Naphthalene-2-sulphonic acid, m.p. 91°.
Cryst. from conc. HCl.

Naphthalene-2-sulphonyl chloride, m.p. 79°.
Cryst. (twice) from benzene-pet.ether (1:1 v/v).

Naphthalene-2-thiol, m.p. $81-82^O$.
Cryst. from ethanol.

Naphthalic anhydride, m.p. 274°.
Extracted with cold aq. Na_2CO_3 to remove free acid, then cryst. from acetic anhydride.

2-Naphthamide, m.p. 195°.
Cryst. from ethanol.

Naphthionic acid, see 4-Aminonaphthalene-1-sulphonic acid.

α-Naphthoic acid, m.p. 162.5-163.0°.
    Cryst. from toluene (3 ml/g) (charcoal), pet.ether (b.p. 80-100°), or aq. 50% ethanol.

β-Naphthoic acid, m.p. 184-185°.
    Cryst. from ethanol (4 ml/g), or aq. 50% ethanol. Dried at 100°.

α-Naphthol, m.p. 95.5-96°.
    Sublimed, then cryst. from aq. methanol (charcoal), aq. 25% ethanol, benzene, cyclohexane, heptane, CCl_4 or boiling water. Dried over P_2O_5 under vac.

β-Naphthol, m.p. 122.5-123.5°.
    Cryst. from aq. 25% ethanol, water, benzene or CCl_4, e.g. by repeated extraction with small amounts of ethanol, followed by dissolution in a minimum amount of ethanol and pptn. with distd. water, then drying over P_2O_5 under vac. Has also been dissolved in aq. NaOH, and pptd. by adding acid (repeating several times), then pptd. from benzene by addition of heptane. Final purification can be by zone melting.

1-Naphtholbenzein.
    Cryst. from ethanol.

p-Naphtholbenzein, m.p. 236-241°.
    Cryst. from glacial acetic acid, or alcohol and water.

1-Naphthol-2-carboxylic acid, m.p. 203-204°.
    Successively cryst. from ethanol-water, ethyl ether and acetonitrile, with filtration through a column of charcoal and Celite. [Tong and Glesmann, JACS, 79, 583 (1957).]

2-Naphthol-3-carboxylic acid, m.p. 222-223°.
    Cryst. from water or acetic acid.

1,2-Naphthoquinone, m.p. 126-132° (decomp.).
    Cryst. from ether (red needles) or benzene (orange leaflets).

1,4-Naphthoquinone, m.p. 125-125.5°.

Cryst. from ethyl ether (charcoal). Steam distd. Cryst. from benzene or ethanol-water. Sublimed.

β-Naphthoxyacetic acid, m.p. 156°.

Cryst. from hot water or benzene.

β-Naphthoyltrifluoroacetone.

Cryst. from ethanol.

β-Naphthyl acetate, m.p. 71°.

Cryst. from pet.ether (b.p. 60-80°) or dil. aq. ethanol.

1-Naphthylacetic acid, m.p. 132°.

Cryst. from water or ethanol.

2-Naphthylacetic acid, m.p. 143.1-143.4°.

Cryst. from water or benzene.

α-Naphthylamine, m.p. 50.8-51.2°.

Sublimed at 120° in a stream of nitrogen, then cryst. from pet. ether (b.p. 60-80°), or absolute ethanol then ethyl ether. Dried under vac. in an Abderhalden pistol. Has also been purified by crystn. of its hydrochloride from water, followed by liberation and distn. of the free base which was finally purified by zone melting. Carcinogenic.

β-Naphthylamine, m.p. 110.2-110.7°.

Sublimed at 180° in a stream of nitrogen. Cryst. from hot water or benzene. Dried under vac. in an Abderhalden pistol. Carcinogenic.

α-Naphthylamine hydrochloride.

Cryst. from water (charcoal).

1-Naphthylamine-4-sulphonic acid.
1-Naphthylamine-5-sulphonic acid.
2-Naphthylamine-1-sulphonic acid.

Cryst. under nitrogen from boiling water and dried in a steam oven.

2-Naphthylamine-6-sulphonic acid.
    Cryst. from a large volume of hot water.

2-Naphthylethylene, m.p. 66°, b.p. 135-137°/18 mm, 95-96°/2.1 mm.
    Cryst. from aq. ethanol.

N-(α-Naphthyl)ethylenediamine dihydrochloride, m.p. 188-190°.
    Cryst. from water.

β-Naphthyl lactate.
    Cryst. from ethanol.

Naphthyl methyl ether, see Methoxynaphthalene.

2-(β-Naphthyloxy)ethanol, m.p. 76.7°.
    Cryst. from benzene-pet.ether.

N-1-Naphthylphthalamic acid, m.p. 203°.
    Cryst. from ethanol.

β-Naphthyl salicylate, m.p. 95°.
α-Naphthyl thiourea, m.p. 198°.
    Cryst. from ethanol.

1-Naphthylurea, m.p. 215-220°.
2-Naphthylurea, m.p. 219-220°.
    Cryst. from ethanol.

1,5-Naphthyridine, m.p. 75°, b.p. 112°/15 mm.
    Purified by repeated sublimation.

Narceine, m.p. 176-177° (145° anhydr.).
    Cryst. (as trihydrate) from water.

Naringenin, m.p. 251°.
    Cryst. from aq. ethanol.

Naringin, m.p. 171° (dihydrate), $[\alpha]_D^{19}$ -82.1° (in EtOH), $[\alpha]_{546}^{20}$ -107° (c = 1 in EtOH).
    Cryst. from water. Dried at 110° (to give dihydrate).

Neopentyl glycol, see 2,2-Dimethyl-1,3-propanediol.

Neostigmine bromide, m.p. 167° (decomp.).
    Cryst. from ethanol-ethyl ether.

Neostigmine methyl sulphate, m.p. 142-145°.
    Cryst. from ethanol.

Nerolidol, m.p. of semicarbazide 134-135°.
    Purified by thin layer chromatography on plates of kieselguhr G [McSweeney, J. Chromatog., 17, 183 (1965)] or silica gel plates impregnated with $AgNO_3$, using 1,2-dichloromethane-$CHCl_3$-ethyl acetate-propanol (10:10:1:1) as solvent system. Also by gas-liquid chromatography on butanediol succinate (20%) on Chromsorb W. Stored in cool place, in an inert atmosphere, in the dark.

Neutral red (CI 825).
    Cryst. from benzene-methanol (1:1).

"New methylene blue N" (CI 927).
    Cryst. from benzene-methanol (3:1).

Nicotinaldehyde thiosemicarbazone, m.p. 222-223°.
    Cryst. from water.

Nicotinamide, m.p. 128-131°.
    Cryst. from benzene.

Nicotinic acid, m.p. 232-234°.
    Cryst. (repeatedly) from water.

Nicotinic acid hydrazide, m.p. 158-159°.
    Cryst. from aq. ethanol or benzene.

Nile blue A, m.p. 183°.
    Cryst. from pet.ether.

Ninhydrin, m.p. 241-243° (decomp.).
    Cryst. from hot water (charcoal). Dried under vac. and stored in
a sealed brown container.

Nioxime, see Cyclohexanedionedioxime.

Nisin.
    Cryst. from ethanol.

Nitrilotriacetic acid, m.p. 247° (decomp.).
    Cryst. from water. Dried at 110°.

2,2',2"-Nitrilotriethanol hydrochloride, see Triethanolamine hydrochloride.

Nitrin, m.p. 227-229°.
    Cryst. from acetone.

o-Nitroacetanilide, m.p. 93-94°.
    Cryst. from water.

p-Nitroacetanilide, m.p. 214°.
    Pptd. from 80% H_2SO_4 by adding ice, then washed with water, and
cryst. from ethanol. Dried in air.

m-Nitroacetophenone, m.p. 81°, b. p. 202°/760 mm, 167°/18 mm.
    Distilled in steam and cryst. from ethanol.

p-Nitroacetophenone, m.p. 80-81°, b.p. 145-152°/760 mm.
    Cryst. from ethanol, or aq. ethanol.

3-Nitroalizarin, m.p. 244° (decomp.).
    Cryst. from acetic acid.

o-Nitroaniline, m.p. 72.5-73.0°.

Cryst. from hot water (charcoal), then recryst. from water, aq. 50% ethanol, or ethanol, and dried in a vac. desiccator. Has also been chromatographed on alumina, then recryst. from benzene.

m-Nitroaniline, m.p. 114°.

Purified as for o-Nitroaniline. Warning: Absorbed through skin.

p-Nitroaniline, m.p. 148-148.5°.

Purified as for o-Nitroaniline. Cryst. from acetone. Freed from o- and m-isomers by zone melting and sublimation.

o-Nitroanisole, f.p. 9.4°, b.p. 265°/737 mm.

Purified by repeated vac. distn. in the absence of oxygen.

p-Nitroanisole, m.p. 54°.

Cryst. from pet.ether.

5-Nitrobarbituric acid, m.p. 176°.

Cryst. from water.

o-Nitrobenzaldehyde, m.p. 44-45°, b.p. 120-144°/3-6 mm.

Cryst. from toluene (2-2.5 ml/g) by addition of pet.ether (b.p. 40-60°)(7 ml/ml of soln.). Can also be distd. at reduced pressure.

m-Nitrobenzaldehyde, m.p. 58°.
p-Nitrobenzaldehyde, m.p. 106°.

Cryst. from water or ethanol-water, then sublimed twice at 2 mm pressure at a temperature slightly above its m.p.

Nitrobenzene, f.p. 5.8°, b.p. 84-86.5°/6.5-8 mm, 210.8°/760 mm, n_D^{15} 1.55457, n_D^{20} 1.55257.

Common impurities include nitrotoluene, dinitrothiophen, dinitro-benzene and aniline. Most impurities can be removed by steam distn. in the presence of dil. H_2SO_4, followed by drying with $CaCl_2$, and shaking with, then distn. at low pressure from, BaO, P_2O_5, $AlCl_3$ or activated alumina. It can also be purified by fractional crystn. by partial freezing, or by crystn. from absolute ethanol (by refrigeration).

Nitrobenzene (continued)

Another purification process includes extraction with aq. 2 M NaOH, then water, dil. HCl, and water, followed by drying (CaCl$_2$, MgSO$_4$, or CaSO$_4$) and fractional distn. under reduced pressure. The pure material is stored in a brown bottle, in contact with silica gel, or CaH$_2$. It is very hygroscopic.

p-Nitrobenzene-azo-resorcinol, m.p. 199-200°.
    Cryst. from methanol. Air dried.

p-Nitrobenzhydrazide, m.p. 213-214°.
    Cryst. from ethanol.

o-Nitrobenzoic acid, m.p. 146-148°.
    Cryst. from benzene (twice), n-butyl ether (twice), then water (twice). Dried and stored in a vac.desiccator. [Le Noble and Wheland, JACS, 80, 5397 (1958).] Has also been cryst. from ethanol-water.

m-Nitrobenzoic acid, m.p. 143-143.5°.
p-Nitrobenzoic acid, m.p. 241-242°.
    Cryst. from benzene, water, ethanol (charcoal), glacial acetic acid or methanol-water. Dried and stored in a vac. desiccator.

p-Nitrobenzoyl chloride, m.p. 75°, b.p. 155°/20 mm.
    Cryst. from dry pet.ether (b.p. 60-80°) or CCl$_4$. Distilled under vac.

p-Nitrobenzyl alcohol, m.p. 93°.
    Cryst. from aq. ethanol.

p-Nitrobenzyl bromide, m.p. 98.5-99.0°.
    Recryst. four times from absolute ethanol, then once from cyclo-hexane—hexane-benzene (1:1:1), followed by vac. sublimation at 0.1 mm and a final recrystn. from the same solvent mixture. [Lichtin and Rao, JACS, 83, 2417 (1961).] Has also been cryst. from pet.ether (b.p. 80-100°, 10 ml/g, charcoal). It slowly decomposes even when stored in a desiccator in the dark.

m-Nitrobenzyl chloride, m.p. 45°.
    Cryst. from pet.ether (b.p. 90-120°).

p-Nitrobenzyl chloride, m.p. 72.5-73°.
    Cryst. from CCl_4, dry ethyl ether, 95% ethanol or n-heptane, and
dried under vac.

p-Nitrobenzyl cyanide, m.p. 117°.
    Cryst. from ethanol.

γ-(4-Nitrobenzyl)pyridine, m.p. 70-71°.
    Cryst. from cyclohexane.

2-Nitrobiphenyl, m.p. 36.7°.
    Cryst. from ethanol (seeding required). Sublimed under vac.

m-Nitrocinnamic acid, m.p. 200-201°.
    Cryst. from benzene or ethanol.

p-Nitrocinnamic acid, m.p. 143° (cis), 286° (trans).
    Cryst. from ethanol-water.

p-Nitrodiphenylamine, m.p. 133-134°.
    Cryst. from ethanol.

Nitroethane, b.p. 115°, n_D^{20} 1.3920, n_D^{25} 1.39015.
    Purified as described for Nitromethane. A spectroscopic impurity
has been removed by shaking with activated alumina, decanting and
rapidly distilling.

2-Nitrofluorene, m.p. 156°.
    Cryst. from aq. acetic acid.

Nitroguanidine, m.p. 232° (decomp.).
    Cryst. from water (20 ml/g).

Nitromesitylene, m.p. 44°, b.p. 255°.
    Cryst. from ethanol.

Nitromethane, f.p. -28.5°, b.p. 101.3°, d^{20} 1.13749, d^{30} 1.12398, n_D^{20} 1.38197, n_D^{30} 1.37730.

Usual impurities include aldehydes, nitroethane, water and small amounts of alcohols. Most of these can be removed by drying with $CaCl_2$ or by distn. to remove the water-nitromethane azeotrope, followed by drying with $CaSO_4$. Phosphorus pentoxide is not suitable as a drying agent. [Wright et al., JCS, 199 (1936).] The purified material should be stored in dark bottles, away from strong light, in a cool place. Purifications using extraction are commonly used. For example, Van Looy and Hammett [JACS, 81, 3872 (1959)] mixed about 150 ml of conc. H_2SO_4 with 1 l. of nitromethane and allowed it to stand for 1 or 2 days. The solvent was washed with water, aq. Na_2CO_3, and again with water, then dried for several days with $MgSO_4$, filtered and stood with $CaSO_4$. It was fractionally distd. before use. Smith, Fainberg and Winstein [JACS, 83, 618 (1961)] washed successively with aq. $NaHCO_3$, aq. $NaHSO_3$, water, 5% H_2SO_4, water and dil. $NaHCO_3$. The solvent was dried with $CaSO_4$, then percolated through a column of Linde type 4A molecular sieve, followed by distn. from some of this material (in powdered form). Buffagni and Dunn (JCS, 5105 (1961)] refluxed for 24 hr with activated charcoal while bubbling a stream of nitrogen through the liquid. The suspension was filtered, dried (Na_2SO_4) and distd., then passed through an alumina column and redistd.

Can be purified by zone melting or by distn. under vac. at 0°, subjecting the middle fraction to several freeze-pump-thaw cycles. An impure sample containing higher nitroalkanes and traces of cyano-alkanes was purified (on basis of n.m.r. spectrum) by crystn. from ethyl ether at -60° (cooling in Dry Ice) [Parrett and Sun, J. Chem. Educ., 54, 448 (1977)].

Fractional crystn. was more effective than fractional distn. from Drierite in purifying nitromethane for conductivity measurements. [Coetzee and Cunningham, JACS, 87, 2529 (1965).] Specific conductances around 5 x 10^{-9} $ohm^{-1}cm^{-1}$ were obtained.

Nitron, m.p. 189° (decomp.).

Cryst. from ethanol or $CHCl_3$.

1-Nitronaphthalene, m.p. 57.3-58.3°.

Fractionally distd. at reduced pressure, then cryst. from ethanol, aq. ethanol or heptane. Chromatographed on alumina from benzene-pet. ether.

2-Nitronaphthalene, m.p. 79°.

Cryst. from aq. ethanol.

1-Nitro-2-naphthol, m.p. 103°.

Cryst. (repeatedly) from benzene-pet.ether (b.p. 60-80°)(1:1).

2-Nitro-1-naphthol, m.p. 127-128°.

Cryst. (repeatedly) from ethanol.

5-Nitro-1,10-phenanthroline, m.p. 197-198°.

Cryst. from benzene-pet.ether, until anhydrous.

o-Nitrophenol, m.p. 44.5-45.5°.

Cryst. from ethanol-water, water, ethanol, benzene or methanol-pet.ether (b.p. 70-90°). Can be steam distd. Petrucci and Weygandt [AC, 33, 275 (1961)] cryst. from hot water (twice), then ethanol (twice), followed by fractional crystn. from the melt (twice), drying over $CaCl_2$ in a vac. desiccator and then in an Abderhalden drying pistol.

m-Nitrophenol, m.p. 96°, b.p. 160-165°/12 mm.

Cryst. from water, $CHCl_3$, CS_2, ethanol or pet.ether (b.p. 80-100°), and dried under vac. over P_2O_5 at room temperature. Can also be distd. at low pressure.

p-Nitrophenol, m.p. 113-114°.

Cryst. from water (which may be acidified, e.g. N H_2SO_4 or 0.5 N HCl), ethanol, aq. methanol, $CHCl_3$, benzene or pet.ether, then dried under vac. over P_2O_5 at room temperature. Can also be sublimed at $60°/10^{-4}$ mm.

2-Nitrophenoxyacetic acid, m.p. 156.5°.

Cryst. from water.

m-Nitrophenylacetic acid, m.p. 120°.
  Cryst. from ethanol-water.

p-Nitrophenylacetic acid, m.p. 80.5°.
  Cryst. from ethanol-water (1:1), then from sodium-dried ethyl
ether and dried over P_2O_5 under vac.

4-Nitrophenylacetonitrile, m.p. 116-117°.
  Cryst. from ethanol.

4-(p-Nitrophenylazo)resorcinol, see p-Nitrobenzene-azo-resorcinol.

4-Nitro-o-phenylenediamine, m.p. 201°.
  Cryst. from water.

p-Nitrophenylhydrazine, m.p. 158° (decomp.).
  Cryst. from ethanol.

m-Nitrophenyl isocyanate, m.p. 52-54°.
p-Nitrophenyl isocyanate, m.p. 53°.
  Cryst. from pet.ether (b.p. 28-38°).

o-Nitrophenylpropiolic acid, m.p. 157° (decomp.).
  Cryst. from water.

p-Nitrophenylurea, m.p. 238°.
  Cryst. from ethanol or hot water.

3-Nitrophthalic acid, m.p. 216-218°.
  Cryst. from hot water (1.5 ml/g). Air dried.

4-Nitrophthalic acid, m.p. 165°.
  Cryst. from ether or ethyl acetate.

3-Nitrophthalic anhydride, m.p. 164°.
  Cryst. from benzene, benzene-pet.ether, acetic acid or acetone.
Dried at 100°.

1-Nitropropane, b.p. 131.4°, n_D^{20} 1.40161, n_D^{25} 1.39936.

2-Nitropropane, b.p. 120.3°, n_D^{20} 1.3949, n_D^{25} 1.39206.
   Purified as for Nitromethane.

5-Nitro-2-n-propoxyaniline, m.p. 47.5-48.5°.
   Cryst. from n-propyl alcohol-pet.ether.

5-Nitroquinoline, m.p. 70°.
   Cryst. from pentane, then from benzene.

8-Nitroquinoline, m.p. 88-89°.
   Cryst. from hot water, methanol, ethanol or ethanol-ethyl ether (3:1).

2-Nitroresorcinol, m.p. 81-82°.
   Cryst. from aq. ethanol.

4-Nitrosalicylic acid, m.p. 227-228°.
   Cryst. from water.

5-Nitrosalicylic acid, m.p. 233°.
   Cryst. from acetone (charcoal), then twice more from acetone alone.

Nitrosobenzene, m.p. 67.5-68°, b.p. 57-59°/18 mm.
   Steam distd., then cryst. from ethanol.

p-Nitrosodimethylaniline, m.p. 25-25.5°, b.p. 191-192°/100 mm,
ε(440 nm) = 3.2 x 10^4.
   Cryst. from ethyl ether-pet.ether.

N-Nitrosodiphenylamine, m.p. 68°.
   Cryst. from absolute ethanol, then dried in air.

p-Nitrosodiphenylamine, m.p. 144-145°.
   Cryst. from benzene.

1-Nitroso-2-naphthol, m. p. 110.4-110.8°.
   Cryst. from pet.ether (b.p. 60-80°, 7.5 ml/g) or from ethanol by
addition of water. Dried at room temperature.

4-Nitroso-1-naphthol, m.p. 198°.
  Cryst. from benzene.

2-Nitroso-1-naphthol-4-sulphonic acid.
  Cryst. from dil. HCl soln. Crystals dried over CaCl$_2$ in vac.
desiccator.

p-Nitrosophenol. Decomp. >124°.
  Cryst. from xylene.

N-Nitroso-N-phenylbenzylamine, m.p. 58°.
  Cryst. from absolute ethanol and dried in air.

3-Nitrostyrene, m.p. 60°.
  Cryst. from absolute ethanol, or (three times) from benzene-pet.
ether (b.p. 60-80°)(1:1).

4-Nitrostyrene, m.p. 20.5-21°.
  Cryst. from CHCl$_3$-heptane (1:3 v/v) or CCl$_4$-hexane. Purified by
addition of methanol to ppte. polymer, then cryst. at -40° from
methanol.

o-Nitrotoluene, b.p. 222.3°, m.p. -9.55° (α-form), -3.85° (β-form).
  Cryst. (repeatedly) from absolute ethanol by cooling in a Dry Ice-
alcohol mixture. Further purified by passage of an alcoholic soln.
through a column of alumina.

m-Nitrotoluene, m.p. 16°, b.p. 232.6°.
  Dried with P$_2$O$_5$ for 24 hr, then fractionally distd. at reduced
pressure.

p-Nitrotoluene, m.p. 52°.
  Cryst. from methanol-water, ethanol-water (1:1) or methanol. Air
dried, then dried in a vac.desiccator over H$_2$SO$_4$.

2-Nitro-4-sulphobenzoic acid.
  Cryst. from dil. HCl.

Nitrourea, m.p. 158.4-158.8° (decomp.).
  Cryst. from alcohol + pet.ether.

Nonactin, m.p. 147-148°, $[\alpha]_D^{20}$ 0±2° (c = 1.2 in CHCl$_3$).
  Cryst. from methanol.

n-Nonane, b.p. 150.8°, n_D^{20} 1.40542, n_D^{25} 1.40311.
  Fractionally distd., then stirred with successive portions of conc.
H$_2$SO$_4$ for 12 hr each until no further coloration was observed. Then
washed with water, dried with MgSO$_4$ and fractionally distd. Alternatively, purified by azeotropic distn. with 2-ethoxyethanol, followed
by washing out the alcohol with water, drying and distn. [Forziati
et al., J. Res. Nat. Bur. Stand., 36, 129 (1946).]

Nonanoic acid, see Pelargonic acid.

Norcamphor (bicyclo[2.2.1]heptan-2-one), m.p. 94-95°.
  Cryst. from water.

Norcholanic acid, m.p. 177°.
  Cryst. from acetic acid.

Norcodeine, m.p. 185°.
  Cryst. from acetone or ethyl acetate.

Nordihydroguaiaretic acid, m. p. 184-185°.
  Cryst. from dil. acetic acid.

Norleucine, m.p. 301° (R or S), 297-300° (RS).
  Cryst. from water.

Norvaline, decomp. at 305° (R or S), 303° (RS).
  Cryst. from aq. ethanol or water.

Novobiocin. Decomp. at 152-156°.
  Cryst. from ethanol and stored in the dark. The sodium salt can be
cryst. from methanol, then dried at 60°/0.5 mm. [Sensi, Gallo and
Chiesa, AC, 29, 1611 (1957).]

Nuclear fast red.
    A soln. of 5 g of the dye in 250 ml of warm 50% ethanol was cooled
to 15° for 36 hr, then filtered on a Büchner funnel, washed with
ethanol until the washings were colourless, then with 100 ml of
ethyl ether and dried over P_2O_5. [Kingsley and Robnett, AC, 33, 552
(1961).]

n-Octacosane, m.p. 62.5°.
    Purified by forming its adduct with urea, washing and crystn. from
acetone-water. [McCubbin, TFS, 58, 2307 (1962).] Cryst. from hot,
filtered isopropyl ether soln. (10 ml/g).

n-Octadecane, m.p. 18-18.5°, b.p. 280°.
    Cryst. from acetone and distd. under reduced pressure from sodium.

1-Octadecanol, see n-Octadecyl alcohol.

Octadecyl acetate, m.p. 32.6°.
    Distd. under vac., then cryst. from ethyl ether-methanol.

n-Octadecyl alcohol, m.p. 58.4-58.8°.
    Cryst. from methanol, or dry ethyl ether and benzene, then frac-
tionally distd. under reduced pressure. Purified by column chromatography. Freed from cetyl alcohol by zone melting.

n-Octadecylbenzyldimethylammonium chloride, m.p. 62°.
    Cryst. from acetone.

Octadecyl ether, m.p. 59.4°.
    Vac. distd., then cryst. from methanol-benzene.

Octafluoropropane, b.p. -38°.
    Purified for pyrolysis studies by passage through a copper vessel
containing CoF_3 at about 270°, then fractionally distd. [Steunenberg
and Cady, JACS, 74, 4165 (1952).]

1,2,3,4,6,7,8,9-Octahydroanthracene, m.p. 78°.
    Cryst. from ethanol, then zone refined.

Octamethylcyclotetrasiloxane, m.p. 17.3°.
  Purified by zone melting.

Octan-4,5-diol, see Dipropylene glycol.

n-Octane, b.p. 126.5°, n_D^{20} 1.39743, n_D^{25} 1.39505.
  Extracted repeatedly with conc. H_2SO_4 or chlorosulphonic acid,
then washed with water, dried and distd. Also purified by azeotropic
distn. with ethanol, followed by washing with water to remove the
ethanol, drying and distd. For further details, see n-Heptane. Also
purified by zone melting.

n-Octanoic acid, b.p. 144-145°/27 mm.
  Fractionally cryst. by partial freezing. Dried with Linde type 4A
molecular sieve and fractionally distd. at reduced pressure.

Octaphenylcyclotetrasiloxane, m.p. 201-202°, b.p. 330-340°/760 mm.
  Cryst. from benzene-ethanol or glacial acetic acid.

1-Octene, b.p. 121°/742 mm, n_D^{20} 1.4087.

2-Octene, b.p. 124-124.5° (trans), n_D^{20} 1.4132 (trans).
  Distd. under nitrogen from sodium. (Removes water and peroxides.)
Peroxides can also be removed by percolation through dried, acid-
washed alumina.

n-Octyl alcohol, b.p. 98°/19 mm, 195.3°/760 mm, n_D^{20} 1.43018.
  Fractionally distd. under reduced pressure. Dried with sodium and
again fractionally distd. or refluxed with boric anhydride and distd.
(b.p. 195 to 205°/5 mm), the distillate being neutralized with NaOH
and again fractionally distd. Also purified by distn. from Raney
nickel and by preparative g.l.c.

4-Octylbenzoic acid.
  Cryst. from ethanol has m.p. 139°; cryst. from aq. ethanol has
m.p. 99-100° (liquid crystal).

n-Octyl bromide, b.p. 201.5°, n_D^{25} 1.4503.

   Shaken with H_2SO_4, washed with water, dried with K_2CO_3 and fractionally distd.

p-tert.-Octylphenol, m.p. 85-86°.

   Cryst. from n-heptane.

1-Octyne, b.p. 126.2°/760 mm, n_D^{25} 1.4159.

   Distd. from $NaBH_4$ to remove peroxides.

"α"Oestradiol, m.p. 178°, $[α]_D^{18}$ +78° (EtOH).

   Cryst. from 80% ethanol.

Oestradiol benzoate, m.p. 192-193°.

Oestrone, m.p. 256°.

   Cryst. from ethanol.

Oleic acid, m.p. 16°, b.p. 360° (decomp.), n_D^{30} 1.4571.

   Purified by fractional crystn. from its melt, followed by molecular
distn. at 10^{-3} mm, or by conversion to its methyl ester and fractional
distn. After hydrolyzing the ester, the free acid can be cryst. from
acetone at -40 to -45° (12 ml/g). For purification by the use of
lead and lithium salts, see Keffler and McLean [JCS Ind. (London),
54, 176T (1935)].

   Purification based on direct crystn. from acetone is described by
Brown and Shinowara [JACS, 59, 6 (1937)].

Oleyl alcohol, b.p. 182-184°/1.5 mm, $n_D^{27.5}$ 1.4582.

   Purified by fractional crystn. at -40° from acetone, then distd.
under vac.

Opianic acid, m.p. 150°.

   Cryst. from water.

Orcinol, m.p. 107.5°.

   Cryst. from $CHCl_3$-benzene (2:3).

S-Ornithine

   Cryst. from water containing 1 mM EDTA (to remove metal ions).

S-Ornithine monohydrochloride, $[\alpha]_D^{25}$ +28.3° (in 5 M HCl).
    Likely impurities: citrulline, arginine, R-ornithine. Cryst. from
water by adding 4 volumes of ethanol.

Orotic acid, m.p. 345-346°.
    Cryst. from water.

Orthanilic acid.
    Cryst. from aq. soln, containing 20 ml of conc. HCl per l., then
cryst. from distd. water.

Ouabain, m.p. ∿180°.
    Cryst. from water or ethanol. Dried at 130°. Stored in the dark.

Oxalic acid, m.p. 101.5° (anhydr. 189.5°).
    Cryst. from distd. water. Dried under vac. over conc. H_2SO_4.

Oxaloacetic acid.
    Cryst. from boiling ethyl acetate, or from hot acetone by the
addition of hot benzene.

Oxamide. Decomp. >320°.
    Cryst. from water. Ground. Dried in an oven at 150°.

Oxamycin, see R-4-Amino-3-isoxazolidone.

Oxetane, b.p. 48°/760 mm.
    Distd. from sodium metal.

α-Oxoglutaric acid, m. p. 114°.
    Cryst. repeatedly from acetone-benzene.

2-Oxohexanemethyleneimine, see ε-Caprolactam.

2,2'-Oxydiethanol, see Diethylene glycol.

Palmitic acid, m.p. 63-64°.
    Cryst. from ethanol. Purified via the methyl ester (b.p. 193-194°/
12 mm, n_D^{35} 1.4359) as for Capric acid, or by zone melting.

R-Pantothenic acid sodium salt, m.p. 122-124°, $[\alpha]_D^{25}$ +27° (in H$_2$O).
   Cryst. from ethanol.

Papain.
   A suspension of 50 g of papain (freshly ground in a cold mortar)
in 200 ml of cold water was stirred at 4° for 4 hr, then filtered
through a Whatman No.1 filter paper. The clear yellow filtrate was
cooled in an ice-bath while a rapid stream of H$_2$S was passed through
it for 3 hr, and the suspension was centrifuged at 2000 rev/min for
20 min. Sufficient cold methanol was added slowly and with stirring
to the supernatant to give a final methanol conc. of 70 vol%. The
ppte., collected by centrifugation for 20 min at 2000 rev/min, was
dissolved in 200 ml of cold water, the soln. was satd. with H$_2$S,
centrifuged, and the enzyme again pptd. with methanol. This process
was repeated four times. [Bennett and Niemann, JACS, 72, 1798 (1950).]

Paraffin (oil).
   Treated with fuming H$_2$SO$_4$, then washed with water and dil. aq.
NaOH, then percolated through activated silica gel.

Paraffin (wax).
   Melted in the presence of conc. NaOH, washing with water until all
of the base had been removed. The paraffin was allowed to solidify
after each washing. Finally, 5 g of paraffin was melted by heating
on a water-bath, then shaken for 20-30 min with 100 ml of boiling
water and fractionally cryst.

Paraldehyde, m.p. 12.5°, b.p. 124°.
   Washed with water and fractionally distd.

Patulin, m.p. 111.0°.
   Cryst. from ethyl ether or CHCl$_3$.

Pectic acid.
   Citrus pectic acid (500 g) was extracted under reflux for 18 hr
with 1.5 l. of 70% ethanol and the suspension was filtered hot. The
residue was washed with hot 70% ethanol, then 95% ethanol and finally with ethyl ether. It was dried in a current of air, ground and

Pectic acid (continued)
dried for 18 hr at 80° under vac. [Morell and Link, JBC, 100, 385
(1933).] It can be further purified by dispersing in water and adding
just enough dil. NaOH to dissolve the pectic acid, then passing the
soln. through columns of cation- and anion-exchange resins [Williams
and Johnson, IECAE, 16, 23 (1944)], and precipitating with two
volumes of 95% ethanol containing 0.01% HCl. The ppte. is worked
with 95% ethanol, then ethyl ether, dried and ground.

Pectin.
   Dissolved in hot water to give a 1% soln., then cooled, and made
about 0.05 M in HCl by addition of conc. HCl, and pptd. by pouring
slowly and with vigorous stirring into two volumes of 95% ethanol.
After standing for several hours, the pectin is filtered on to nylon
cloth, then redispersed in 95% ethanol and stood overnight. The ppte.
is filtered off, washed with ethanol-ethyl ether, then ethyl ether
and air dried.

Pelargonic acid, m.p. 15°, b.p. 98.9°/1 mm, 225°/760 mm.
   Esterified with ethylene glycol and distd. (This removes dibasic
acids as undistillable residues.) The acid was regenerated by
hydrolyzing the ester.

Penicillic acid, m.p. 58-64° (monohydrate), 83-84° (anhydrous).
   Cryst. from water as monohydrate, or from pet. ether.

Pentaacetyl-α-D-glucopyranose, m.p. 112°.
Pentaacetyl-β-D-glucopyranose, m.p. 131°.
   Cryst. from ethanol.

Pentabromoacetone, m.p. 76°.
   Cryst. from ethanol or ethyl ether.

Pentacene, m.p. 300°.
   Cryst. from benzene.

Pentachloroethane, b.p. 162.0°, n_D^{15} 1.50542.
   Usual impurities include trichloroethylene. Partially decomposes if
distd. at atmospheric pressure. Drying with CaO, KOH or sodium is

Pentachloroethane (continued)

unsatisfactory because HCl splits off. Can purify by steam distn., or by washing with conc. H_2SO_4, water, and then aq. K_2CO_3, drying with solid K_2CO_3 or $CaSO_4$, and fractional distn. at reduced pressure.

Pentachloronitrobenzene, m.p. $146°$.
  Cryst. from ethanol.

Pentachlorophenol, m.p. $190-191°$.
  Sublimed under vac.

Pentachlorothiophenol.
  Cryst. from benzene.

Penta-1,3-diene, b.p. $42°$, n_D^{20} 1.4316.

Penta-1,4-diene, b.p. $25.8-26.2°/756$ mm, n_D^{20} 1.3890.
  Distd. from $NaBH_4$.

Penta-2,4-dione, see Acetylacetone.

Pentaerythritol, m.p. $260.5°$.
  Refluxed with an equal vol. of methanol, then cooled and the ppte. dried at $90°$. Cryst. from dil. aq. HCl. Sublimed under vac. at $200°$.

Pentaerythrityl tetraacetate, m.p. $78-79°$.
  Cryst. from hot water, then leached with cold water until the odour of acetic acid was no longer detectable.

Pentaerythrityl laurate, m.p. $50°$.
  Cryst. from pet.ether.

Pentaerythrityl tetranitrate, m.p. $140.1°$.
  Cryst. from acetone or acetone-ethanol. Explosive.

Pentaethylenehexamine.
  Fractionally distd. twice at 10-20 mm, the fraction boiling at $220-250°$ being collected. Its soln. in methanol (40 ml in 250 ml) was cooled in an ice-bath and conc. HCl was added dropwise with

Pentaethylenehexamine (continued)
stirring. About 50 ml was added, and the pptd. hydrochloride was
filtered off, washed with acetone and ethyl ether, then dried in a
vac. desiccator. [Jonassen et al., JACS, 79, 4279 (1957).]

2,2,3,3,3-Pentafluoropropan-1-ol.
   Shaken with alumina for 24 hr, dried with anhydrous K_2CO_3, and
distd., collecting the middle fraction (b.p. 80-81°).

2',3,4',5,7-Pentahydroxyflavone, see Morin.

Pentamethylbenzene, m.p. 53.5-55.1°.
   Successively cryst. from absolute ethanol, toluene, and methanol
and dried under vac. [Rader and Smith, JACS, 84, 1443 (1962).] Has
also been cryst. from benzene or aq. ethanol.

Pentamethylenetetrazole, m.p. 60-61°.
   Cryst. from ethyl ether.

n-Pentane, b.p. 36.1°, n_D^{20} 1.35748, n_D^{25} 1.35472.
   Stirred with successive portions of conc. H_2SO_4 until there was no
further coloration during 12 hr, then with 0.5 N $KMnO_4$ in 3M H_2SO_4
for 12 hr, washed with water and aq. $NaHCO_3$. Dried with $MgSO_4$ or
Na_2SO_4, then P_2O_5 and fractionally distd. through a column packed
with glass helices. Also purified by passage through a column of
silica gel, followed by distn. and storage with sodium hydride. An
alternative purification is by azeotropic distn. with methanol,
which is subsequently washed out from the distillate (using water),
followed by drying and distn. For removal of carbonyl-containing
impurities, see n-Heptane.
   Also purified by fractional freezing (about 40%) on a copper coil
through which cold air was passed, then washed with conc. H_2SO_4 and
fractionally distd.

2,4-Pentanedione, see Acetylacetone.

Pentane-1-thiol, b.p. 122.9°/697.5 mm, d_4^{25} 0.8375.
   Dissolved in aq. 20% NaOH, then extracted with a small amount of
ethyl ether. The soln. was acidified slightly with 15% H_2SO_4, and

Pentane-1-thiol (continued)
the thiol was distd. out, dried with $CaSO_4$ or $CaCl_2$, and fractionally
distd. under nitrogen. [Ellis and Reid, JACS, 54, 1674 (1932).]

Pentan-1-ol, see n-Amyl alcohol.

Pentan-2-ol, b.p. 119.9°, n_D^{20} 1.41787, n_D^{25} 1.4052.

Pentan-3-ol, b.p. 116.2°, n_D^{25} 1.4072.
   Refluxed with CaO, distd., refluxed with magnesium and again
fractionally distd.

Pentan-3-one, b.p. 101.7°, n_D^{20} 1.39240, n_D^{25} 1.39003.
   Refluxed with $CaCl_2$ for 2 hr, then left overnight with fresh $CaCl_2$,
filtered and distd.

Pentaquine monophosphate, m.p. 189-190°.
   Cryst. from 95% ethanol.

Pent-2-ene (mixed isomers), b.p. 36.4°, n_D^{20} 1.38003, n_D^{25} 1.3839.
   Refluxed with sodium wire, then fractionaly distd. twice through
a Fenske column.

cis-Pent-2-ene, b.p. 37.1°, n_D^{20} 1.3830, n_D^{25} 1.3798.
   Dried with sodium wire and fractionally distd., or purified by
azeotropic distn. with methanol, followed by washing out the methanol with water, drying and distn.
   (trans-isomer). Washed with water, dried over anhydrous Na_2CO_3,
and fractionally distd. The middle cut was purified by two passes
of fractional melting.

Pentyl-, see n-Amyl-.

tert.-Pentyl-, see tert.-Amyl-.

neo-Pentyl alcohol, see 2,2-Dimethyl-1-propanol.

Pent-2-yne, b.p. $26°/2.4$ mm, n^{25} 1.4005.

  Stood with, then distd. at low pressure from, sodium or $NaBH_4$. (Removes peroxides.)

Pepsin.

  Cryst. from ethanol.

Perbenzoic acid, m.p. $41-43°$.

  Cryst. from benzene. Readily sublimed.

Perchlorobutadiene, b.p. $144.1°/100$ mm, $210-212°/760$ mm, n_D^{20} 1.5556.

  Washed with four or five 1/10th volumes of methanol (or until the yellow colour had been extracted), then stirred for 2 hr with H_2SO_4, washed with distd. water until neutral and filtered through a column of P_2O_5. Distd. under reduced pressure through a packed column. [Rutner and Bauer, JACS, 82, 298 (1960).]

Perfluorobutyric acid, m.p. $-17.5°$, b.p. $120°/735$ mm, d_4^{20} 1.651, n_D^{16} 1.295.

  Fractionally distd. twice in an Oldershaw column with an automatic vapour-dividing head, the first distn. in the presence of conc. H_2SO_4 as a drying agent.

Perfluorocyclohexane, m.p. $51°$ (sublimes), b.p. $52°$.

  Extracted repeatedly with methanol, then passed through a column of silica gel (previously activated by heating at $350°$).

Perfluorodimethylcyclohexane, b.p. $101°$.

  Fractionally distd., then 35 ml was sealed with about 7 g KOH pellets in a borosilicate glass ampoule and heated at $135°$ for 48 hr. The ampoule was cooled and opened, and the liquid was resealed with fresh KOH in another ampoule and heated as before. This process was continued until no further decomposition was observed. The substance was then washed with distd. water, dried ($CaSO_4$) and distd. [Grafstein, AC, 26, 523 (1954).]

Perfluoroheptane, b.p. 82.4°, d^{25} 1.72006.

Purified as for Perfluorodimethylcyclohexane. Other procedures include shaking with H_2SO_4, washing with water, drying with sodium for 48 hr and fractionally distilling. Alternatively, it has been refluxed for 24 hr with satd. acid $KMnO_4$ (to oxidize and remove hydrocarbons), then neutralized, steam distd., dried with P_2O_5, and passed slowly through a column of dry silica gel. It has been purified by fractional crystn. using partial freezing.

Perfluoro-n-hexane.

Purified by fractional crystn. using partial freezing. The methods described for Perfluoroheptane should also be applicable.

Perfluoromethylcyclohexane, b.p. 76.3°, d^{25} 1.78777.

Refluxed for 24 hr with satd. acid $KMnO_4$ (to oxidize and remove hydrocarbons), then neutralized, steam distd., dried with P_2O_5 and passed slowly through a column of dry silica gel. [Glew and Reeves, JPC, 60, 615 (1956).]

Perfluorononane.
Perfluorotripropylamine.

Purified as for Perfluorodimethylcyclohexane.

Perylene, m.p. 273-274°.

Purified by silica-gel chromatography of its recryst. picrate. [Ware, JACS, 83, 4374 (1961).] Cryst. from benzene.

Petroleum ether.

Shaken several times with conc. H_2SO_4, then 10% H_2SO_4 and conc. $KMnO_4$ (to remove unsaturated, including aromatic, hydrocarbons) until the permanganate colour persists. Washed with water, aq. Na_2CO_3 and again with water. Dried with $CaCl_2$ or Na_2SO_4, and distd. Can be dried further using CaH_2 or sodium wire. Passage through a column of activated alumina, or treatment with CaH_2 or sodium, removes peroxides. For the elimination of carbonyl-containing impurities without using permanganate, see n-Heptane.

(−)-α-Phellandrene, b.p. 61°/11 mm, 175-176°/760 mm.
   Purified by gas chromatography on an Apiezon column.

Phenacetin, see p-Acetophenetidine.

Phenacyl bromide, see 2-Bromoacetophenone.

Phenanthrene, m.p. 98°.
   Likely contaminants include anthracene, carbazole, fluorene and
other polycyclic hydrocarbons. Purified by distn. from sodium, boiling with maleic anhydride in xylene, crystn. from acetic acid, sublimation and zone melting. Has also been cryst. repeatedly from
ethanol, benzene or pet.ether (b.p. 60-70°), with subsequent drying
under vac. over P_2O_5 in an Abderhalden pistol. Feldman, Pantages and
Orchin [JACS, 73, 4341 (1951)] separated from most of the anthracene
impurity by refluxing phenanthrene (671 g) with maleic anhydride
(194 g) in xylene (1.25 1.) under nitrogen for 22 hr, then filtered.
The filtrate was extracted with aq. 10% NaOH, the organic phase was
separated, and the solvent was evaporated. The residue, after stirring for 2 hr with 7 g of sodium, was vac. distd., then recryst. twice
from 30% benzene in ethanol, then dissolved in hot glacial acetic
acid (2.2 ml/g), slowly adding an aq. soln. of CrO_3 (60 g in 72 ml
H_2O added to 2.2 1. of acetic acid), followed by slow addition of
conc. H_2SO_4 (30 ml). The mixture was refluxed for 15 min, diluted
with an equal volume of water and cooled. The ppted. was filtered
off, washed with water, dried and distd., then recryst. twice from
ethanol. Further purification is possible by chromatography from
$CHCl_3$ soln. on activated alumina, with benzene as eluant, and by
zone melting.

Phenanthrene-9-aldehyde, m.p. 102.2-103°, b.p. 231-233°/12 mm.
   Cryst. from ethanol and sublimes at 95-98°/0.07 mm.

9,10-Phenanthrenequinone, m.p. 208°.
   Cryst. from dioxane or 95% ethanol and dried under vac.

Phenanthridine, m.p. 106.5°.
   Purified via the $HgCl_2$ addition compound formed when phenanthridine
(20 g) in 1:1 HCl (100 ml) was added to aq. $HgCl_2$ (60 g in 3 1.), the

Phenanthridine (continued)

mixture being heated to boiling and conc. HCl being added until all
of the solid had dissolved. The compound separated on cooling, and
was decomposed with strong aq. NaOH. Phenanthridine was extracted
with ethyl ether and cryst. from pet.ether (b.p. 80-100°). [Cumper,
Ginman and Vogel, JCS, 4518 (1962).]

1,10-Phenanthroline, m.p. 108-110° (monohydrate), 118° (anhydr.).
   Cryst. as its picrate (m.p. 191°) from ethanol, then the free base
was liberated, dried at 78°/8 mm over P_2O_5 and cryst. from pet.ether
(b.p. 80-100°). [Cumper, Ginman and Vogel, JCS, 1188 (1962).] Can be
purified by zone refining. Also cryst. from benzene-pet.ether (b.p.
40-60°) or sodium-dried benzene, dried and stored over H_2SO_4. The
monohydrate is obtained by crystn. from aq. ethanol.

4,7-Phenanthroline-5,6-dione, m.p. 295° (decomp.).
   Cryst. from methanol.

Phenazine, m.p. 171°.
   Cryst. from ethanol. Can also be sublimed.

Phenazine methosulphate, m.p. 155-157° (or 198° decomp.on rapid
heating).
   Cryst. from ethanol.

Phenethylamine, b.p. 87°/13 mm.
   Distd. from CaH_2, under reduced pressure, just before use.

Phenethyl bromide, b.p. 92°/11 mm.
   Washed with conc. H_2SO_4, water, aq. 10% Na_2CO_3 and water again,
then dried with $CaCl_2$ and fractionally distd. just before use.

p-Phenethylurea, m.p. 173-174°.
   Cryst. from water.

Phenetole, b.p. 170.0°, n_D^{20} 1.50735, n_D^{25} 1.50485.
   Small quantities of phenol can be removed by shaking with NaOH, but
this is not a very likely contaminant of commercial material. Fractional
distn.from sodium, at low pressure, probably gives adequate purification.

Phenobarbital, m.p. 172-174°.
  Cryst. from aq. 50% ethanol.

Phenocoll hydrochloride.
  Cryst. from water.

Phenol, m.p. 40.9°,b.p. 85.5-86.0°/20 mm, 181.8°/760 mm,
n_D^{41} 1.54178, n_D^{46} 1.53957.
  Steam was passed through a boiling soln. containing 1 mole of
phenol and 1.5-2.0 moles of NaOH in 5 l.of water until all non-acidic
material had distd. The residue was cooled, acidified with 20% (v/v)
H_2SO_4, and the phenol was separated, dried with $CaSO_4$ and fractionally distd. at reduced pressure. It was then fractionally cryst.
several times from its melt. [Andon et al., JCS, 5246 (1960).]
Purification via the benzoate has been used by Berliner, Berliner and
Nelidow [JACS, 76, 507 (1954)]. The benzoate was cryst. from 95%
ethanol, then hydrolyzed to the free phenol by refluxing with two
equivalents of KOH in aq. ethanol until the soln. became homogeneous.
It was acidified with HCl and extracted with ethyl ether. The ether
layer was freed from benzoic acid by thorough extraction with aq.
$NaHCO_3$, and, after drying and removing the ether, the phenol was
distd.
  Phenol has also been crystd. from a 75% w/w soln. in water by cooling to 11° and seeding with a crystal of the hemihydrate. The
crystals were centrifuged off, rinsed with cold water (0-2°) satd.
with phenol, and dried.
  Draper and Pollard [Science, 109, 448 (1949)] added 12% water, 0.1%
aluminium (can also use zinc), and 0.05% $NaHCO_3$ to phenol, and distd.
at atmospheric pressure until the azeotrope was removed. The phenol
was then distd. at 25 mm. Phenol has also been dried by distn. from
a benzene soln. to remove the water-benzene azeotrope and the excess
benzene, followed by distn. of the phenol at reduced pressure under
nitrogen. Processes such as this are probably adequate for analytical
grade phenol which has as its main impurity water. Has also been
cryst. from pet.ether-benzene or pet.ether (b.p. 40-60°). Purified
material is stored in a vac. desiccator over P_2O_5 or $CaSO_4$.

Phenol-2,4-disulphonic acid.
   Cryst. from alcohol-ether.

Phenolphthalein, m.p. 263°.
   Dissolved in ethanol (7 ml/g), then diluted with eight volumes of
cold water. Filtered. Heated on a water-bath to remove most of the
alcohol and the pptd. phenolphthalein is filtered off and dried
under vac.

Phenolphthalol, m.p. 201-202°.
   Cryst. from aq. ethanol.

Phenosafranine, λ_{max} 530 nm (in water).
   Cryst. from dil. HCl.

Phenothiazine, m.p. 184-185°.
   Cryst. from benzene or toluene (charcoal) after boiling for 10 min
under reflux. Filtered on a suction filter. Dried in an oven at 100°,
then in a vac. desiccator over paraffin chips.

Phenoxazine, m.p. 156°.
   Cryst. from ethanol.

Phenoxyacetic acid, m.p. 98-99°.
   Cryst. from water or aq. ethanol.

4-Phenoxyaniline, m.p. 95°.
   Cryst. from water.

Phenoxybenzamine, m.p. (hydrochloride) 137.5-140°.
   Cryst. from alcohol-ether.

2-Phenoxybenzoic acid, m.p. 113°, b.p. 355°/760 mm.
3-Phenoxybenzoic acid, m.p. 145°.
   Cryst. from aq. ethanol.

2-Phenoxypropionic acid, m.p. 115-116°, b.p. 265-266°/758 mm,
105-106°/5 mm.
   Cryst. from water.

Phensuximide, m.p. 71-73°.
  Cryst. from hot 95% ethanol.

Phenylacetamide, m.p. 158.5°.
  Cryst. repeatedly from absolute methanol. Dried under vac.over
P_2O_5.

Phenyl acetate, b.p. 78°/10 mm, n_D^{22} 1.5039.
  Freed from phenol and acetic acid by washing (either directly or as
a soln. in pentane) with aq. 5% Na_2CO_3, then with satd. aq. $CaCl_2$,
drying with $CaSO_4$ or Na_2SO_4, and fractional distn. at reduced pressure.

Phenylacetic acid, m.p. 76-77°, b.p. 140-150°/20 mm.
  Cryst. from pet.ether (b.p. 40-60°), isopropyl alcohol, aq. 50%
ethanol or hot water. Dried under vac. Can also be distd. under
reduced pressure.

Phenylacetone, b.p. 69-71°/3 mm.
  Converted to the semicarbazone and cryst. three times from ethanol
(to m.p. 186-187°). The semicarbazone was hydrolyzed with 10% phosphoric acid and the ketone was distd. [Kumler, Strait and Alpen,
JACS, 72, 1463 (1950).]

Phenylacetonitrile, see Benzyl cyanide.

p-Phenylacetophenone, m.p. 120.3-121.2°, b.p. 196-210°/18 mm.
  Cryst. from ethanol. Can also be distd. under reduced pressure.

Phenylacetylene, b.p. 75°/80 mm, n_D^{25} 1.5463.
  Distd. through a spinning-band column.

R,S-Phenylalanine, m.p. 162°.
  Cryst. from water and dried under vac. over P_2O_5.

S-Phenylalanine, m.p. 280° (decomp.), $[\alpha]_D^{25}$ -34.0° (in water),
m.p. 162° (RS).
  Likely impurities: leucine, valine, methionine, tyrosine. Cryst.
from water by adding 4 volumes of ethanol. Dried under vac. over P_2O_5.

3-Phenylallyl chloride, b.p. 92-93°/3 mm.
  Distd. under vac. three times from K_2CO_3.

Phenyl p-aminosalicylate, m.p. 153°.
  Cryst. from isopropanol.

p-Phenylanisole, m.p. 89.9-90.1°.
  Cryst. from benzene-pet.ether. Dried under vac. in an Abderhalden
pistol.

9-Phenylanthracene, m.p. 153-154°.
  Chromatographed on alumina in benzene and cryst. from acetic acid.

N-Phenylanthranilic acid, m.p. 182-183°.
  Cryst. from ethanol (5 ml/g) or acetic acid (2 ml/g) by adding hot
water (1 ml/g).

2-Phenyl-1-azaindolizine.
  Cryst. from ethanol.

p-Phenylazoaniline, see p-Aminoazobenzene.

p-Phenylazobenzoyl chloride, m.p. 93°.
  Cryst. from pet.ether (b.p. 60-80°).

4-Phenylazodiphenylamine, see Benzeneazodiphenylamine.

Phenylazo-β-naphthol, m.p. 131°.
  Cryst. from ethanol.

Phenylazo-α-naphthylamine.
  Cryst. from cyclohexane.

Phenylazo-β-naphthylamine, m.p. 99-100°.
  Cryst. from absolute ethanol or glacial acetic acid.

p-Phenylazophenacyl bromide, m.p. 103-104°.
  Purified on a column of silica gel, using pet.ether-ethyl ether
(9:1 v/v) as solvent.

p-Phenylazophenol, m.p. 155°.
  Cryst. from benzene or 95% ethanol.

Phenyl benzoate, m.p. 69.5°.
  Cryst. from ethanol using approx. twice the volume needed for
complete soln. at 69°.

Phenylbenzoquinone.
  Cryst. from heptane.

N-Phenylbenzylamine, see Benzylaniline.

Phenylbiguanide, m.p. 144-146°.
  Cryst. from water or toluene.

1-Phenyl-1,3-butanedione, see Benzoylacetone.

Phenylbutazone, m.p. 105°.
  Cryst. from ethanol.

trans-4-Phenyl-3-buten-2-one, see Benzalacetone.

α-Phenylbutyramide, m.p. 86°.
  Cryst. from water.

γ-Phenylbutyric acid, m.p. 50°.
  Cryst. from pet.ether (b.p. 40-60°).

o-(Phenylcarbamoyl)-1-scopolamine methobromide, m.p. 200.5-201.5°
(decomp.)
  Cryst. from 95% ethanol.

N-Phenylcarbazole, m.p. 94-95°.
  Cryst. from ethanol or isopropanol.

Phenyl cinnamate, m.p. 75-76°.
  Cryst. from ethanol (2 ml/g). Can also be distd. under reduced
pressure.

α-Phenylcinnamic acid, m.p. 174° (cis), 138-139° (trans).
  Cryst. from ether-pet.ether.

α-Phenyl-p-cresol, see p-Benzylphenol.

1-Phenyl-2,3-dimethyl-5-pyrazolone, m.p. 114°.
  Cryst. from ethanol-water, benzene, benzene-pet.ether or hot water
(after charcoal). Dried under vac.

o-Phenylenediamine, m.p. 100-101°.
  Cryst. from aq. 1% sodium hyposulphite (charcoal), washed with
ice-water and dried in a vac. desiccator.

p-Phenylenediamine, m.p. 140°.
  Cryst. from ethanol or benzene, and sublimed.

o-Phenylenediamine dihydrochloride, m.p. 180°.
  Cryst. from dil. HCl (60 ml conc. HCl, 40 ml water, with 2 g
stannous chloride), after treatment of the hot soln. with charcoal,
by adding an equal volume of conc. HCl and cooling in an ice-salt
mixture. The crystals were washed with a small amount of conc. HCl
and dried in a vac. desiccator over NaOH.

Phenyl ether, m.p. 27.0°, $n_D^{30.7}$ 1.57596.
  Cryst. from 90% ethanol. Melted, washed with 3 M NaOH and water,
dried with $CaCl_2$ and fractionally distd. under reduced pressure.
Fractionally cryst. from its melt and stored over P_2O_5.

1-Phenylethanol, b.p. 106-107°/22-23 mm, n_D^{25} 1.5254.
  Purified via its hydrogen phthalate. [See Houssa and Kenyon, JCS,
2260 (1930).]

Phenyl ethylbarbituric acid, m.p. 175-176°.
  Cryst. from hot water. Dried at 90-100°.

Phenyl ethyl ether, b.p. 60°/9 mm, 77.5°/31 mm.

Dissolved in ethyl ether and washed with aq. 10% NaOH (to remove traces of phenol), then with distd. water. The ethyl ether was evaporated and the phenyl ethyl ether was fractionally distd. under reduced pressure.

p-α-Phenylethylphenol, m.p. 56.0-56.3°.
    Cryst. from pet.ether.

5-(α-Phenylethyl)semioxamazide, m.p. 167-168° (1-), 157° (dl).
    Cryst. from ethanol.

S-α-Phenylglycine, m.p. 305-310°.
    Cryst. from aq. ethanol.

Phenylglycine-o-carboxylic acid, m.p. 208°.
    Cryst. from hot water (charcoal).

Phenylglyoxaldoxime, see Isonitrosoacetophenone.

Phenylhydrazine, m.p. 23°, b.p. 137-138°/18 mm, 241-242°/760 mm.
    Purified by chromatography, then crystn. from pet.ether (b.p. 60-80°)-benzene. [Shaw and Stratton, JCS, 5004 (1962).]

Phenylhydrazine hydrochloride, m.p. 244°.

One litre of boiling ethanol was added to 100 g of phenylhydrazine hydrochloride dissolved during 1-3 hr (without heating) in 200 ml of warm water (60-70°). The soln. was filtered off, while still hot, through Whatman No.2 filter paper and cooled in a refrigerator. The ppte. was collected on a medium sintered-glass filter and recryst. twice in this way, then washed with cold ethanol, dried thoroughly and stored in a stoppered brown bottle. [Peterson, Karrer and Guerra, AC, 29, 144 (1957).] Hough, Powell and Woods [JCS, 4799 (1956)] boiled the hydrochloride with three times its weight of water, filtered hot (charcoal), added one-third volume of conc. HCl and cooled to 0°. The crystals were washed with acetone, and dried over P_2O_5 under vac. The salt has also been cryst. from 95% ethanol.

Phenylhydroxylamine, m.p. 82°.
   Cryst. from water.

2-Phenyl-1,3-indandione, m.p. 149-151°.
2-Phenylindolizine.
   Cryst. from ethanol.

Phenylisocyanate, b.p. 45-47°/10 mm.
   Distd. under reduced pressure from P_2O_5.

3-Phenyllactic acid, see 2-Hydroxy-3-phenylpropionic acid.

Phenyl methanesulphonate, m.p. 61-62°.
   Cryst. from methanol.

1-Phenyl-3-methyl-5-pyrazolone, m.p. 127°.
   Cryst. from hot water, or ethanol-water (1:1).

2-Phenylnaphthalene, m.p. 103-104°.
   Chromatographed on alumina in benzene and cryst. from aq. ethanol.

N-Phenyl-1-naphthylamine, m.p. 63.7-64.0°.
   Cryst. from ethanol, pet.ether or benzene-ethanol. Dried under
vac. in an Abderhalden pistol.

N-Phenyl-2-naphthylamine, m.p. 107.5-108.5°.
   Cryst. from ethanol, acetic acid or benzene-hexane.

N-Phenyl-β-naphthylamine, m.p. 108°.
   Cryst. from methanol.

p-Phenylphenacyl bromide, m.p. 126°.
   Cryst. (charcoal) from ethanol (15 ml/g), or ethyl acetate-pet.
ether (b.p. 90-100°).

p-Phenylphenol, m.p. 166-167°.
   Cryst. from benzene or ethanol, and vac. dried in a desiccator
over $CaCl_2$.

Phenylphosphonous acid, m.p. 71°.
    Cryst. from hot water.

Phenyl-2-propanone, see Phenylacetone.

Phenylpropiolic acid, m.p. 137.8-138.4°.
    Cryst. from benzene, CCl$_4$ or aq. ethanol.

α-Phenylpropionic acid, m.p. 49°.
    Cryst. from pet.ether (b.p. 40-60°).

3-Phenylpropyl bromide, b.p. 110°/12 mm, 128-129°/29 m.
    Washed successively with conc. H$_2$SO$_4$, water, 10% aq. Na$_2$CO$_3$ and
again with water, then dried with CaCl$_2$ and fractionally distd. just
before use.

Phenyl 2-pyridyl ketoxime, m.p. 151-152°.
6-Phenylquinoline, m.p. 110.5-111.5°.
    Cryst. from ethanol (charcoal).

2-Phenylquinoline-4-carboxylic acid, m.p. 210°.
    Cryst. from ethanol (about 20 ml/g).

Phenyl salicylate, m.p. 41.8-42.6°.
    Fractionally cryst. from its melt, then recryst. from benzene.

3-Phenylsalicylic acid, m.p. 186-187.5°.
    Dissolved in approx. 1 equivalent of satd. aq. Na$_2$CO$_3$, filtered
and pptd. by adding 0.8 equivalents of M HCl. Cryst. from ethylene
dichloride (charcoal), and sublimed at 0.1 mm. [Brooks, Eglinton and
Morman, JCS, 661 (1961).]

1-Phenylsemicarbazide, m.p. 172°.
4-Phenylsemicarbazide, m.p. 122°.
Phenylsuccinic acid, m.p. 168°.
1-Phenyl-2-thiourea, m.p. 154°.
    Cryst. from water.

1-Phenyl-5-sulphanilamidopyrazole, m.p. 179-183°.

1-Phenylthiosemicarbazide, m.p. 200-201° (decomp.).

4-Phenylthiosemicarbazide, m.p. 140°.
    Cryst. from ethanol.

Phenyltoloxamine hydrochloride, m.p. 119-120°.
    Cryst. from methyl isobutyl ketone.

Phenyl p-toluenesulphonate, m.p. 94.5-95.5°.
    Cryst. from methanol.

Phenyl p-tolylcarbonate, m.p. 67°.
    Purified by preparative g.l.c. with 20% Apiezon on embacel and
sublimed in vacuo.

1-Phenyl-1,2,4-triazole-3,5-diol, m.p. 263-264°.
    Cryst. from water.

4-Phenylurazole, see 1-Phenyl-1,2,4-triazole-3,5-diol.

Phenylurea, m.p. 148°.
    Cryst. from boiling water (10 ml/g). Dried in a steam oven.

Phloretic acid, see 3-p-Hydroxyphenylpropionic acid.

Phloretin, m.p. 264-271° (decomp.).
    Cryst. from aq. ethanol.

Phlorizin, m.p. 110°.
    Cryst. as dihydrate from water.

Phloroacetophenone, m.p. 218-219°.
    Cryst. from hot water (35 ml/g).

Phloroglucinol, m.p. 117° (anhydr.), 217-219° (dihydrate).
    Cryst. from water, and stored in the dark under nitrogen.

Phorone, m.p. 28°, b.p. 197°/743 mm.
    Cryst. repeatedly from ethanol.

"Phosphine" (a dye, CI 793).
    Cryst. from benzene-methanol (1:1).

Phthalazine, m.p. 90-91°.
    Cryst. from ethyl ether and sublimed under vac.

Phthalazone, m.p. 183-184°, b.p. 337°/760 mm.
    Cryst. from water and sublimes in vacuo.

o-Phthalic acid, m.p. 211-211.5°.
    Cryst. from water.

Phthalic anhydride, m.p. 132°, b.p. 295°.
    Distd. under reduced pressure. Cryst. from $CHCl_3$, CCl_4 or benzene.
Fractionally cryst. from its melt. Dried under vac. at 100°.

Phthalide, m.p. 72-73°.
    Cryst. from water (75 ml/g) and dried in air on filter paper.

Phthalimide, m.p. 235°.
    Cryst. from ethanol (20 ml/g)(charcoal), or by sublimation.

Phthalimidoglycine, m.p. 192-193°.
    Cryst. from water or ethanol.

Phthalonitrile, m.p. 141°.
    Cryst. from benzene or ethanol. Can also be distd. under vac.

Phthalylsulphacetamide, m.p. 196°.
    Cryst. from water.

Phthiocol, m.p. 173-174°.
    Cryst. from ethyl ether-pet.ether.

Physalien, m.p. 98.5-99.5, $E_{1cm}^{1\%}$ 1410 (449 nm), 1255 (478 nm),
(in hexane).
    Purified by chromatography on water-deactivated alumina, using
hexane-ethyl ether (19:1) to develop the column. Cryst. from

Physalien (continued)
benzene-ethanol or benzene-methanol. Stored in the dark, in inert atmosphere, at $0°$.

Physostigmine, m.p. 105-106°.
   Cryst. from ethyl ether or benzene.

Phytoene, λ_{max} 275, 285, 297 nm, $E_{1cm}^{1\%}$ = 850 (287 nm), (in hexane).
   Purified by chromatography on columns of magnesia-Supercel or alumina [Rabourn et al., Arch. Biochem. Biophys., 48, 267 (1954)]. Stored as soln. in pet.ether under nitrogen at -20°.

Phytofluene, λ_{max} 331, 348, 367 nm, $E_{1cm}^{1\%}$ = 1350 (348 nm)(in pet.ether).
   Purified by chromatography on partially deactivated alumina [Kushwaha et al., JBC, 245, 4708 (1970).] Soln. in pet.ether stored under nitrogen at -20°.

Picein, m.p. 195-196°.
   Cryst. from absolute methanol or (as monohydrate) from water.

Picene, m.p. 364°.
   Cryst. from isopropylbenzene-xylene. Can also be sublimed.

Picoline, see Methylpyridine.

Picolinic acid, m.p. 138°.
   Cryst. from water or benzene.

Picric acid, m.p. 122°.
   Cryst. from acetone, benzene, $CHCl_3$, aq. 30% ethanol, 95% ethanol, methanol or water. Dried in a vac. oven at 80° for 2 hr. Alternatively, dried over $Mg(ClO_4)_2$ or fused and allowed to freeze under vac. three times. Because of its explosive nature, picric acid should be stored moistened with water, and only small portions should be dried at any one time.

Picrolonic acid, m.p. 120° (decomp.).
   Cryst. from water or ethanol.

Picrotoxin, m.p. 203°.
   Cryst. from water.

Picryl chloride, m.p. 83°.
   Cryst. from $CHCl_3$ or ethanol.

Picryl iodide, m.p. 164-165°.
   Cryst. from benzene.

Pimelic acid, m.p. 105-106°.
   Cryst. from water or from benzene containing 5% ethyl ether.

Pinacol (hexahydrate), m.p. 46.5°, b.p. 59°/4 mm.
   Distd. then cryst. repeatedly from distd. water.

Pinacol (anhydr.), m.p. 41.1°, b.p. 172°.
   The hydrate is rendered anhydrous by azeotropic distn. of
the water with benzene. Recryst. from benzene-pet.ether, absolute
ethanol or dry ethyl ether.

Pinacolone oxime, m.p. 78°.
   Cryst. from aq. ethanol.

Pinacyanol.
   Cryst. from ethanol-ethyl ether.

R-α-Pinene, b.p. 61°/30 mm, 156.2°/760 mm, n_D^{15} 1.4634, n_D^{20} 1.4658,
$[\alpha]_D^{25}$ +47.3°.
S-α-Pinene, b.p. 155-156°, n_D^{20} 1.4634, $[\alpha]_D^{20}$ -47.2°.

   Isomerized by heat, acids and certain solvents. Should be distd.
under reduced pressure under nitrogen and stored in the dark.
Purified via the nitrosochloride [Waterman et al., Rec. Trav. Chim.
Pays-Bas, 48, 1191 (1929)]. For purification of optically active
forms by crystn. see Lynn [JACS, 91, 361 (1919).]
   Small quantities (0.5 ml) have been purified by gas-liquid chromatography, using helium as a carrier and a column at 90° packed with

S-α-Pinene (continued)

20 wt.% of polypropylene sebacate on a Chromosorb support. Larger quantities were fractionally distd. under reduced presuure in a column packed with stainless steel gauze spirals. Material could be dried with CaH_2 or sodium, and stored in a refrigerator: $CaSO_4$ and silica gel were unsatisfactory because they induced spontaneous isomerization. [Bates, Best and Williams, JCS, 1521 (1962).]

Pipecolic acid, m.p. 264° (RS).
    Cryst. from water.

Piperazine, m.p. 110-112°, 44° (hydrate), b.p. 125-130°/760 mm.
    Cryst. from ethanol or anhydrous benzene, and dried at 0.01 mm. Can be sublimed under vac. and purified by zone melting.

Piperazine-N,N'-bis(2-ethanesulphonic acid)(PIPES).
    Cryst. from boiling water (maximum solubility is about 1 g. 1^{-1}) or as described for ADA.

Piperazine dihydrochloride, m.p. 82.5-83.5°.
    Cryst. from aq. ethanol. Dried at 110°.

Piperazine-2,5-dione, m.p. 309-310°.
    Cryst. from water.

Piperazine phosphate (monohydrate).
    Cryst. twice from water, air-dried and stored for several days over Drierite. Salt dehydrates slowly if heated at 70°.

Piperic acid, m.p. 217°.
    Cryst. from ethanol. Protect from light.

Piperidine, f.p. -9°, b.p. 35.4°/40 mm, 106°/760 mm, n_D^{20} 1.4535, n_D^{25} 1.4500.
    Dried with BaO, KOH, CaH_2, or sodium, and fractionally distd. (optionally from sodium, CaH_2, or P_2O_5). Purified from pyridine by zone melting.

RS-Piperidine-2-carboxylic acid, m.p. 264°.
    Cryst. from water.

Piperidine hydrochloride, m.p. 244-245°.
    Cryst. from ethanol-ethyl ether in the presence of a small amount
of HCl.

Piperidinium nitrate, m.p. 110°.
    Cryst. from acetone-ethyl acetate.

Piperine, m.p. 129-129.5°.
    Cryst. from ethanol or benzene-ligroin.

Piperonal, m.p. 37°, b.p. 140°/15 mm, 263°/760 mm.
    Cryst. from aq. 70% ethanol or ethanol-water.

Piperonylic acid, m.p. 229°.
    Cryst. from ethanol or water.

Pivalic acid, m.p. 35.4°.
    Fractionally distd. under reduced pressure, then fractionally
cryst. from its melt. Recryst. from benzene.

Plumbagin, m.p. 78-79°.
    Cryst. from aq. ethanol.

Polyacrylonitrile.
    Pptd. from dimethylformamide by methanol.

Polyethylene.
    Cryst. from thiophen-free benzene and dried over P_2O_5 under vac.

Polygalacturonic acid, see Pectic acid.

Polymethyl acrylate.
    Pptd. from a 2% soln. in acetone by addition of water.

Polystyrene.
    Pptd. from $CHCl_3$ by methanol.

Polystyrenesulphonic acid.

Purified by repeated pptn. of the sodium salt from aq. soln. by methanol, with subsequent conversion to the free acid by passage through an Amberlite IR-120 ion-exchange resin. [Kotin and Nagasawa, JACS, 83, 1026 (1961).]

Also purified by passage through cation and anion exchange resins in series (Rexyn 101 cation exchange resin and Rexyn 203 anion exchange resin), then titrated with NaOH to pH 7. Sodium form of polystyrenesulphonic acid pptd. by addition of 2-propanol. Dried in vac. oven at 80° for 24 hr, finally increasing to 120° prior to use. [Kowblansky and Ander, JPC, 80, 297 (1976).]

Polyvinyl acetate.

Pptd. from acetone by n-hexane.

Poly(N-vinylcarbazole).

Pptd. seven times from tetrahydrofuran with methanol, with a final freeze-drying from benzene. Dried under vac.

Polyvinyl chloride.

Pptd. from cyclohexanone by methanol.

Pontachrome azure blue B.

Cryst. from methanol.

Pontacyl carmine 2G.

Pontacyl light yellow GX.

Salted out three times with sodium acetate, then repeatedly extracted with ethanol. See Chlorazol sky blue FF. [McGrew and Schneider, JACS, 72, 2547 (1950).]

Pregnane, m.p. 83.5°.

Cryst. from methanol.

Pregnane-3α,20α-diol, m.p. 243-244°, $[\alpha]^{20}_{546}$ +31° (c = 1 in EtOH).

Cryst. from acetone.

Pregnane-3α,20β-diol, m.p. 244-246°, $[\alpha]^{20}_{546}$ +22° (c = 1 in EtOH).

Cryst. from ethanol.

Pregnenolone, see 3β-Hydroxy-5-pregnen-20-one.

Procaine, m.p. 51° (dihydrate), 61° (anhydr.).
    Cryst. as dihydrate from aq. ethanol or as anhydrous material
from pet.ether or ethyl ether. The latter is hygroscopic.

Progesterone, m.p. 128.5°.
    Cryst. from aq. ethanol.

Proline, m.p. 215-220° (decomp.)(R-form); 220-222° (decomp.)(S-form);
205° (decomp.)(RS-form), $[\alpha]_{1cm}^{25}$ -85.1° (in water)(S).

    Likely impurity: hydroxyproline. Purified via its picrate which
was cryst. twice from water, then decomp. with 40% H_2SO_4. The picric
acid was extracted with ethyl ether, the H_2SO_4 pptd. with Ba(OH)$_2$,
and the water was evapd. The residue was cryst. from hot absolute
ethanol [Mellan and Hoover, JACS, 73, 3879 (1951)] or ethanol-ether.
Hygroscopic. Stored in desiccator.

Prolycopene, m. p. 111°, λ_{max} 443.5, 470 nm in pet.ether.
    Purified by chromatography on deactivated alumina [Kushwaha et al.,
JBC, 245, 4708 (1970)]. Cryst. from pet.ether. Stored in the dark, in
inert atmosphere, at -20°.

S-Prolylglycine.
    Cryst. from water at 50-60° by addition of ethanol.

Proneurosporene, λ_{max} 408, 432, 461 nm, $E_{1cm}^{1\%}$ = 2040 (432 nm) (in hexane).
    Purified by chromatography on deactivated alumina [Kushwaha et al.,
JBC, 245, 4708 (1970)]. Stored in the dark, in inert atmosphere, at
0°.

Propane-1,2-diamine, b.p. 120.5°.
    Purified by azeotropic distn. with toluene. [Horton, Thomason,
and Kelley, AC, 27, 269 (1955).]

Propane-1,2-diol, b.p. 104°/32 mm.
    Dried with Na_2SO_4, decanted and distd. under reduced pressure.

<u>Propane-1,3-diol</u>, b.p. 110-122°/12 mm, $n_D^{18.5}$ 1.4398.

Dried with K_2CO_3 and distd. under reduced pressure. More extensive purification involved conversion with benzaldehyde to 2-phenyl-1,3-dioxane (m.p. 47-49°) which was subsequently decomposed by shaking vigorously with 0.5 M HCl (3 ml/g) for 15 min and standing overnight at room temperature. After neutralization with K_2CO_3, the benzaldehyde was removed by distn. and the diol was recovered from the remaining aq. soln. by continuous extraction with $CHCl_3$ for 1 day. The extract was dried with K_2CO_3, the $CHCl_3$ was evapd. and the diol was distd. [Foster, Haines and Stacey, <u>Tetrahedron</u>, <u>16</u>, 177 (1961).]

<u>Propane-1-thiol</u>, b.p. 65.3°/702 mm, d_4^{25} 0.83598, n_D^{25} 1.43511.

<u>Propane-2-thiol</u>, b.p. 49.8°/696 mm, d_4^{25} 0.80895, n_D^{25} 1.42154.

Purified by soln. in aq. 20% NaOH, extraction with a small amount of benzene and steam distn. until clear. After cooling, the soln. was acidified slightly with 15% H_2SO_4, and the thiol was distd. out, dried with anhydrous $CaSO_4$ or $CaCl_2$, and fractionally distd. under nitrogen. [Mathias and Filho, <u>JPC</u>, <u>62</u>, 1427 (1958).] Also purified by liberation of the mercaptan by adding dil. HCl to the residue remaining after steam distn. After direct distn. from the flask, and separation of the water, the mercaptan was dried (Na_2CO_3) and distd. under nitrogen.

<u>1,2,3-Propanetricarboxylic acid</u>, see Tricarballylic acid.

<u>Propan-1-ol</u>, see <u>n</u>-Propyl alcohol.

<u>Propan-2-ol</u>, see Isopropyl alcohol.

<u>Propargyl alcohol</u>, b.p. 54°/57 mm, 113.6°/760 mm.

Commercial material contains a stabilizer. An aq. soln. of propargyl alcohol can be concentrated by azeotropic distn. with butanol or butyl acetate. Dried with K_2CO_3 and distd. under reduced pressure, in the presence of about 1% succinic acid, through a glass helices-packed column.

<u>Propene</u>, m.p. -185.2°, b.p. -47.8°/750 mm, n_D 1.001.

Purified by freeze-pump-thaw cycles and trap-to-trap distn.

2-Propen-1-ol, b.p. 97.0°, n_D^{20} 1.41266, n_D^{30} 1.4090.

Dried by azeotropic distn. or by refluxing with magnesium, then fractionally distd. [Hands and Norman, Ind. Chemist, 21, 307 (1945).]

p-(1-Propenyl)phenol, m.p. 93-94°.

Cryst. from water.

β-Propiolactone, b.p. 83°/45 mm, n_D^{25} 1.4117.

Fractionally distd., under reduced pressure, from sodium. Carcinogenic.

Propionaldehyde, b.p. 48.5-48.7°, n_D^{20} 1.3733, n_D^{25} 1.37115.

Dried with $CaSO_4$ or $CaCl_2$, and fractionally distd. under nitrogen or in the presence of a trace of hydroquinone (to retard oxidation). Blacet and Pitts [JACS, 74, 3382 (1952)] repeatedly vac. distd. the middle fraction until it no longer gave a solid polymer when cooled to -80°. It was stored with $CaSO_4$.

Propionamide, m.p. 79.8-80.8°.

Cryst. from acetone, benzene, $CHCl_3$, water or acetone-water, then dried in a vac. desiccator over P_2O_5 or 99% H_2SO_4.

Propionic acid, b.p. 141°, n_D^{20} 1.3865, n_D^{25} 1.3843.

Dried with Na_2SO_4 or by fractional distn., then redistd. after refluxing with a few crystals of $KMnO_4$. An alternative purification uses conversion to the ethyl ester, fractional distn. and hydrolysis. [Bradbury, JACS, 74, 2709 (1952).] Propionic acid can also be heated for ½ hr with the amount of benzoic anhydride equivalent to the water present (in the presence of CrO_3 as catalyst), followed by fractional distn. [Chan and Israel, JCS, 196 (1960).]

Propionic anhydride, b. p. 67°/18 mm, 168°/780 mm.

Shaken with P_2O_5 for several minutes, then distd.

Propionitrile, b.p. 97.2°, n_D^{15} 1.36812, n_D^{30} 1.36132.

Shaken with 1:5 dil. HCl, or with conc. HCl until the smell of isonitrile had gone, then washed with water, and aq. K_2CO_3. After a preliminary drying with silica gel or Linde type 4A molecular sieve, it is stirred with CaH_2 until hydrogen evolution ceases, then decanted

Propionitrile (continued)

and distd. from P_2O_5 (not more than 5 g/l., to minimize gel formation). Finally, it is refluxed with, and slowly distd. from, CaH_2
(5 g/l.), taking precautions to exclude moisture.

n-Propyl acetate, b.p. 101.5°, n_D^{20} 1.38442.

   Washed with satd. aq. $NaHCO_3$ until neutral, then with satd. aq.
NaCl. Dried with $MgSO_4$ and fractionally distd.

n-Propyl alcohol, b.p. 97.2°, d_4^{25} 0.79995.

   The main impurities in n-propyl are usually water and 2-propen-1-
ol. Water can be removed by azeotropic distn. either directly (azeotrope contains 28% water) or by using a ternary system, e.g. by
adding benzene. Alternatively, for gross amounts of water, refluxing
over CaO for some hours is suitable, followed by distn. and further
drying. To obtain more nearly anhydrous alcohol, suitable drying
agents are firstly NaOH, $CaSO_4$ or K_2CO_3, then CaH_2, aluminium
amalgam, magnesium activated with iodine, or a small amount of
sodium. Alternatively, the alcohol can be refluxed with n-propyl
succinate or n-propyl phthalate in a method similar to one described
under Ethanol. Allyl alcohol is removed by adding bromine (15 ml/l.)
and then fractionally distilling from a small amount of K_2CO_3. Distn.
from sulphanilic or tartaric acids removes basic impurities.
Albrecht [JACS, 82, 3813 (1960)] obtained spectroscopically pure
material by heating with charcoal to 50-60°, filtering and adding
2,4-dinitrophenylhydrazine and a few drops of conc. H_2SO_4. After
standing for several hours, the mixture was cooled to 0°, filtered
and vac. distd. Gold and Satchell [JCS, 1938 (1963)] heated n-propyl
alcohol with 3-nitrophthalic anhydride at 76-110° for 15 hr, then
recryst. the resulting ester from water, benzene-pet.ether (b.p.100-
120°)(3:1), and benzene. The ester was hydrolyzed under reflux with
aq. 7.5 M NaOH for 45 min under nitrogen, followed by distn. (also
under nitrogen). The fraction (b.p. 87-92°) was dried with K_2CO_3 and
stored under reduced pressure in the dark over 2,4-dinitrophenyl-
hydrazine, then freshly distilled. Also purified by adding 2 g $NaBH_4$
to 1.5 l. alcohol, gently bubbling with argon and refluxing for 1 day
at 50°. Then added 2 g freshly cut sodium (washed with propanol) and
refluxed for 1 day. Distd., taking the middle fraction [Jou and
Freeman, JPC, 81, 909 (1977)].

n-Propylamine, b.p. 48.5°, n_D^{20} 1.38815.

Distd. from zinc dust, at reduced pressure, in an atmosphere of nitrogen.

n-Propyl bromide, b.p. 71.0°, n_D^{15} 1.43695, n_D^{25} 1.43123.

Likely contaminants include n-propyl alcohol and isopropyl bromide. The simplest purification procedure uses drying with $MgSO_4$ or $CaCl_2$ (with or without a preliminary wash of the bromide with aq. $NaHCO_3$, then water), followed by fractional distn. away from bright light. Chien and Willard [JACS, 79, 4872 (1957)] bubbled a stream of oxygen containing 5% ozone through n-propyl bromide for 1 hr, then shook with 3% hydrogen peroxide soln., neutralized with aq. Na_2CO_3, washed with distd. water and dried. Then followed vigorous stirring with 95% H_2SO_4 until fresh acid did not discolour within 12 hr. The propyl bromide was separated, neutralized, washed, dried with $MgSO_4$ and fractionally distd. The centre cut was stored in the dark. Instead of ozone, Schuler and McCauley [JACS, 79, 821 (1957)] added bromine and stored for 4 weeks, the bromine then being extracted with aq. $NaHSO_3$ before the sulphuric acid treatment was applied. Distd. Further purified by preparative gas chromatography on a column packed with 30% SE-30 (General Electric ethylsilicone rubber) on 42/60 Chromsorb P at 150° and 40 psig, using helium. [Chu, JCP, 41, 226 (1964).]

n-Propyl chloride, b.p. 46.6°, n_D^{20} 1.3880.

Dried with $MgSO_4$ and fractionally distd. More extensively purified using extraction with H_2SO_4 as for n-Propyl bromide. Alternatively, Chien and Willard [JACS, 75, 6160 (1953)] passed a stream of oxygen containing about 5% ozone through the n-propyl chloride for three times as long as was needed to cause the first coloration of starch iodide paper by the exit gas. After washing with aq. $NaHCO_3$ to hydrolyze ozonides and remove organic acids, the chloride was dried with $MgSO_4$ and fractionally distd.

1-Propyl-3-(p-chlorobenzenesulphonyl)urea, m.p. 127-129°.

Cryst. from aq. ethanol.

Propylene carbonate, b.p. 110°, 0.5-1 torr.

Likely impurities are water, CO_2, propylene glycol, allyl alcohol and propylene oxide. Can be purified by percolation through molecular sieves (Linde 5A, dried at 350° for 14 hr under a stream of argon), followed by distn. under vac. [Jasinski and Kirkland, AC, 39, 163 (1967).] It can be stored over molecular sieves under an inert gas atmosphere. Solvent purified in this way contains less than 2 p.p.m. water.

Propylenediamine, see Propane-1,2-diamine.

Propyleneglycol, see Propane-1,2-diol.

Propylene oxide, b.p. 34.5°, n_D^{20} 1.3664.

Dried with Na_2SO_4 or CaH_2, and fractionally distd. through a column packed with glass helices after refluxing with sodium. CaH_2, or KOH pellets.

n-Propyl ether, b.p. 90.1°, n_D^{15} 1.38296, n_D^{20} 1.3803.

Purified by drying with $CaSO_4$, by passage through an alumina column (to remove peroxides), and by fractional distn.

Propyl formate, b.p. 81.3°, n_D^{20} 1.37693.

Distd., then washed with satd. aq. NaCl, and with satd. aq. $NaHCO_3$ in the presence of solid NaCl, dried with $MgSO_4$ and fractionally distd.

n-Propyl gallate, m.p. 150°.

Cryst. from aq. ethanol.

n-Propyl iodide, b.p. 102.5°, n_D^{20} 1.5041.

Should be distd. at reduced pressure to avoid decomposition. Dried with $MgSO_4$ or silica gel and fractionally distd. Stored under nitrogen with mercury in a brown bottle. Prior to distn., free iodine can be removed by shaking with copper powder or by washing with aq. $Na_2S_2O_3$ and drying. Alternatively, the n-propyl iodide can be treated with bromine, then washed with aq. $Na_2S_2O_3$ and dried. See also n-Butyl iodide.

n-Propyl propionate, b.p. 122°.
    Treated with anhydrous CuSO$_4$, then distd. under dry nitrogen.

2-Propyn-1-ol, see Propargyl alcohol.

Protocatechualdehyde, m.p. 153°.
    Cryst. from water or toluene.

Protopine, m.p. 208°.
    Cryst. from alcohol-CHCl$_3$.

Protoporphyrin, m.p. >300°.
    Cryst. from ethyl ether.

S,S-Pseudoephedrine, m.p. 118-119°, $[\alpha]_D^{20}$ +40.0° (in water), +53.0°
(in EtOH).
    Cryst. from dry ethyl ether, or from water and dried in a vac.
desiccator.

Pseudoephedrine hydrochloride, m.p. 181-182°.
Pseudoisocyanine iodide.
    Cryst. from ethanol.

2,4-(1H,3H)-Pteridinedione, m.p. >350°.
    Cryst. from water.

Pterocarpin, m.p. 165°, $[\alpha]_D^{20}$ -220°.
    Cryst. from ethanol.

Pteroic acid.
    Cryst. from dil. HCl.

Pulegone, b.p. 69.5°/5 mm, n_D^{20} 1.4849.
    Purified via the semicarbazone. [Erskine and Waight, JCS, 3425
(1960).]

Purine, m.p. 216-217°.
    Cryst. from toluene or ethanol.

Purpurin.
 Cryst. from aq. ethanol. Dried at 100°.

Purpurogallin, m.p. 274° (rapid heating).
 Cryst. from acetic acid.

Pyocyanine, m.p. 133°.
 Cryst. from water.

Pyrazinamide, m.p. 189-191°.
 Cryst. from water or ethanol.

Pyrazine, m.p. 47°, b.p. 115.5-115.8°.
 Distilled in steam and cryst. from water. Purified by zone melting.

Pyrazinecarboxylic acid, m.p. 225-229° (decomp.).
 Cryst. from water.

Pyrazinecarboxamide, m.p. 189-191° (sublimes slowly at 159°).
 Cryst. from water or ethanol.

Pyrazine-2,3-dicarboxylic acid, m.p. 183-185° (decomp.).
 Cryst. from water. Dried at 100°.

Pyrazole, m.p. 70°.
 Cryst. from pet.ether or cyclohexane.

Pyrazole-3,5-dicarboxylic acid, m.p. 287-289° (decomp.).
 Cryst. from water or ethanol.

Pyrene, m.p. 149-150°.
 Cryst. from acetic acid or benzene. Purified by chromatography of
CCl_4 solns. on alumina, with benzene or n-hexane as eluant. Also zone
refined, and purified by sublimation. Marvel and Anderson [JACS, 76,
5434 (1954)] refluxed pyrene (35 g) in toluene (400 ml) with maleic
anhydride (5 g) for 4 days, then added 150 ml of aq. 5% KOH and refluxed for 5 hr with occasional shaking. The toluene layer was separated, washed thoroughly with water, concentrated to about 100 ml

Pyrene (continued)
and allowed to cool. Crystalline pyrene was filtered off and recryst.
twice from ethanol. The material was free from anthracene derivatives.
Another purification step involved passage of pyrene in cyclohexane
through a column of silica gel.

1-Pyrenebutyric acid.
    Cryst. from benzene, or ethanol-water (70:30%, v/v). Dried over
P_2O_5.

Pyrene-1-carboxaldehyde, m.p. 126-127°.
    Cryst. from benzene.

Pyrene-3-carboxaldehyde.
    Cryst. from 95% ethanol.

1-Pyrenesulphonic acid.
    Cryst. from ethanol-water.

1,3,6,8-Pyrenetetrasulphonic acid.
    Cryst. from water.

Pyridine, f.p. -41.8^O, b.p. 115.6^O, d_4^{20} 0.9831, n_D^{20} 1.51021.
    Likely impurities are water and amines such as the picolines and
the lutidines. Pyridine is hygroscopic. It can be dried by refluxing
with solid KOH, NaOH, CaO, BaO or sodium, followed by fractional
distn. Other methods of drying include standing with Linde type 4A
molecular sieve, CaH_2 or $LiAlH_4$, azeotropic distn. of the water with
toluene or benzene, or treatment with phenylmagnesium bromide in
ether, followed by evaporation of the ether and distn. of the pyridine. A recommended [Lindauer and Mukherjee, PAC, 27, 267 (1971)]
method dries pyridine over solid KOH (20 g/kg) for 2 weeks, and
fractionally distils the supernatant over Linde type 5A molecular
sieves and solid KOH. The product is stored under CO_2-free nitrogen.
Pyridine can be stored in contact with BaO, CaH_2 or molecular sieve.
Non-basic materials can be removed by steam distilling a soln. containing 1.2 equivalents of 20% H_2SO_4 or 17% HCl until about 10% of
the base has been carried over along with the non-basic impurities.

Pyridine (continued)

The residue is then made alkaline, and the base is separated, dried with NaOH and fractionally distd.

Alternatively, pyridine can be treated with oxidizing agents. Thus pyridine (800 ml) has been stirred for 24 hr with a mixture of ceric sulphate (20 g) and anhydrous K_2CO_3 (15 g), then filtered and fractionally dist. Hurd and Simon [JACS, 84, 4519 (1962)] stirred pyridine (135 ml), water (2.5 l.) and $KMnO_4$ (90 g) for 2 hr at $100°$, then stood for 15 hr before filtering off the pptd. manganese oxides. Addition of solid NaOH (about 500 g) caused pyridine to separate. It was decanted, refluxed with CaO for 3 hr and distd.

Separation of pyridine from some of its homologues can be achieved by crystn. of the oxalates. Pyridine is pptd. as its oxalate by adding it to a stirred soln. of oxalic acid in acetone. The ppte. is filtered, washed with cold acetone, and pyridine is regenerated and isolated. Other methods are based on complex formation with $ZnCl_2$ or $HgCl_2$. Heap, Jones and Speakman [JACS, 43, 1936 (1921)] added crude pyridine (1 l.) to a soln. of zinc chloride (848 g) in 730 ml of water, 346 ml. of conc. HCl and 690 ml of 95% ethanol. The crystalline ppte. of $ZnCl_2.2py$ was filtered off, recryst. twice from absolute ethanol, then treated with a conc. NaOH soln., using 26.7 g of NaOH to 100 g of the complex. The ppte. was filtered off, and the pyridine was dried with NaOH pellets and distd. Similarly, Kyte, Jeffery and Vogel [JCS, 4454 (1960)] added pyridine (60 ml) in 300 ml of 10% (w/v) HCl to a soln. of $HgCl_2$ (405 g) in hot water (2.3 l.). On cooling, crystals of pyridine-$HgCl_2$ (1:1) complex separated and were filtered off, cryst. from 1% HCl (to m.p. 178.5-179$°$), washed with a little ethanol and dried at $110°$. Free base was liberated by addition of excess aq. NaOH and separated by steam distn. The distillate was satd. with solid KOH, and the upper layer was removed, dried further with KOH, then BaO and distd. Another possible purification step is fractional crystn. by partial freezing.

Small amounts of pyridine have been purified by vapour-phase chromatography, using a 180-cm column of polyethyleneglycol-400 (Shell, 5%) on Embacel (May and Baker) at $100°$, with argon as carrier gas.

Pyridine-2-aldehyde, b.p. 81.5°/25 mm.
Pyridine-3-aldehyde, b.p. 89.5°/14 mm.
Pyridine-4-aldehyde, b.p. 79.5°/12 mm.
    Sulphur dioxide was bubbled into a soln. of 50 g in 250 ml of
boiled out water, under nitrogen, at 0°, until pptn. was complete.
The addition compound was filtered off rapidly and, after washing
with a little water, it was refluxed in 17% HCl (200 ml) under
nitrogen until a clear soln. was obtained. Neutralization with $NaHCO_3$
and extraction with ether separated the aldehyde which was recovered
by drying the extract, then distilling twice, under nitrogen. [Kyte,
Jeffery and Vogel, JCS, 4454 (1960).]

Pyridine-2-aldoxime, m.p. 113°.
Pyridine-3-aldoxime, m.p. 150°.
Pyridine-4-aldoxime, m.p. 129°.
    Cryst. from water.

Pyridine-α-carboxylic acid, see Picolinic acid.

Pyridine-β-carboxylic acid, see Nicotinic acid.

Pyridine-γ-carboxylic acid, see Isonicotinic acid.

2,6-Pyridinedialdoxime, m.p. 212°.
    Cryst. from water.

Pyridine-2,5-dicarboxylic acid, m.p. 254°.
Pyridine-3,4-dicarboxylic acid, m.p. 256°.
    Cryst. from dil. aq. HCl.

Pyridine hydrochloride, m.p. 144°, b.p. 218°.
    Cryst. from $CHCl_3$-ethyl acetate and washed with ethyl ether.

Pyridine N-oxide, m.p. 67°.
    Purified by vac. sublimation.

Pyridinium bromide perbromide.
    Cryst. from acetic acid.

Pyridoxal hydrochloride.

Dissolved in water, adjusted to pH 6 with NaOH and set aside overnight to crystallize. The crystals were washed with cold water, dried
in a vac. desiccator over P_2O_5 and stored in a brown bottle at room
temperature. [Fleck and Alberty, JPC, 66, 1678 (1962).]

Pyridoxamine dihydrochloride.
   Cryst. from hot methanol.

Pyridoxine hydrochloride, m.p. 209-210° (decomp.).
   Cryst. from ethanol-acetone.

Pyrocatechol, see Catechol.

Pyrogallol, m.p. 136°.
   Cryst. from ethanol-benzene.

l-Pyroglutamic acid, m.p. 162-164°.
   Cryst. from ethanol by addition of pet.ether.

Pyromellitic acid, m.p. 276°.
   Dissolved in 5.7 parts of hot dimethylformamide, decolorized and
filtered. The ppte. obtained on cooling was separated and air dried,
the solvent being removed by heating in an oven at 150-170° for
several hours. Cryst. from water.

Pyromellitic dianhydride, m.p. 286°.
   Cryst. from methyl ethyl ketone or dioxane. Dried, and sublimed
in vacuo.

Pyronin B.
   Cryst. from ethanol.

Pyronin Y, CI 739.
   Commercial material contained a large quantity of zinc. Purified
by dissolving 1 g in 50 ml hot water containing 5 g NaEDTA. Cooled
to 0°, filtered, evapd. to dryness and extracted residue with ethanol.
Soln. evapd. to 5-10 ml, filtered, and dye pptd. by addition of
excess dry ethyl ether. Centrifuged and crystals washed with dry

Pyronin Y, CI 739 (continued)

ether. The procedure was repeated, then the product was dissolved in CHCl$_3$, filtered, and evapd. The dye was stored in vac.

Pyrrole, b.p. 129-130°, n$_D^{20}$ 1.5097.

Dried with NaOH, CaH$_2$ or CaSO$_4$. Fractionally distd. under reduced pressure from sodium or CaH$_2$. Stored under nitrogen. Redistd. immediately before use.

Pyrrolidine, b.p. 87.5-88.5°.

Dried with BaO or sodium, then fractionally distd. under nitrogen, through a Todd column packed with glass helices.

2-Pyrrolidone-5-carboxylic acid, m.p. 162-163°.

Cryst. from ethanol-pet.ether.

Pyruvic acid, m.p. 13°, b.p. 65°/10 mm.

Distd. twice, then fractionally cryst. by partial freezing.

Quercetin.

Cryst. from aq. ethanol and dried at 100°.

Quercitrin, m.p. 168°.

Cryst. from aq. ethanol and dried at 135°.

Quinaldic acid, m.p. 156-157°.

Cryst. from benzene.

Quinaldine, see 2-Methylquinoline.

Quinalizarin.

Cryst. from acetic acid or nitrobenzene. Can be sublimed under vac.

Quinazoline, m.p. 48.0-48.5°, b.p. 120-121°/17-18 mm.

Distd. under reduced pressure, sublimed under vac. and cryst. from pet.ether (b.p. 40-60°).

Quinhydrone, m.p. 168°.

Cryst. from water heated to 65°, then dried in a vac. desiccator.

D-Quinic acid, m.p. 172°, $[\alpha]_{546}^{20}$ -51° (c = 20 in water).
   Cryst. from water.

Quinidine, m.p. 171°, $[\alpha]_{5461}^{20}$ + 301.1° in $CHCl_3$ contg. 2.5%(v/v)EtOH.
   Cryst. from benzene or dry $CHCl_3$-pet.ether (b.p. 40-60°), discarding the initial, oily crop of crystals. Dried under vac. at 100°
over P_2O_5.

Quinine, m.p. 177° (decomp.).
   Cryst. from absolute ethanol.

Quinine bisulphate, m.p. 160° (for anhydrous salt).
   Cryst. from 0.1 M H_2SO_4, forms heptahydrate when cryst. from water.

Quinine sulphate, m.p. 205° (in $2H_2O$).
   Cryst. from water. Dried at 110°.

Quinizarin, m.p. 200-202°.
   Cryst. from glacial acetic acid.

Quinol, see Hydroquinone.

Quinoline, m.p. -16°, b.p. 111.5°/16 mm, 236°/758 mm, d_4^{20} 1.0937.
   Dried with Na_2SO_4 and vac. distd. from zinc dust. Also dried by
boiling with acetic anhydride, then fractionally distilling. Calvin
and Wilmarth [JACS, 78, 1301 (1956)] cooled redistd. quinoline in
ice and added enough HCl to form its hydrochloride. Diazotization
removed aniline, the diazo compound being broken down by warming the
soln. to 60°. Non-basic impurities were removed by ether extraction.
Quinoline was liberated by neutralizing the hydrochloride with NaOH,
then dried with KOH and fractionally distd. at low pressure. Addition
of cuprous acetate (7 g/l. of quinoline) and shaking under hydrogen
for 12 hr at 100° removed impurities due to the nitrous acid treatment. Finally the hydrogen was pumped off and the quinoline was distd.
Other purification procedures depend on conversion to the phosphate
(pptd. from methanol soln., filtered, washed with methanol, then
dried at 55°) or the picrate which, after crystn. (m.p. of phosphate,
159°, and of picrate, 201°), were reconverted to the amine.

Quinoline (continued)

[Packer, Vaughan and Wong, JACS, 80, 905 (1958).] The method using
the picrate is as follows: quinoline is added to picric acid dissolved in a minimum volume of 95% ethanol, giving yellow crystals which
are washed with ethanol, air-dried and cryst. from acetonitrile.
Dissolved in dimethyl sulphoxide (previously dried over 4A molecular
sieves) and passed through a basic alumina column, on which picric
acid is adsorbed. The free base in the effluent is extracted with
n-pentane and distd. under vac. Traces of solvent can be removed by
vapour-phase chromatography. [Moomaw and Anton, JPC, 80, 2243
(1976).] The $ZnCl_2$ and dichromate complexes have also been used.
[Cumper, Redford and Vogel, JCS, 1176 (1962).]

Quinoline-2-aldehyde, m.p. 71°.
  Steam distd. Cryst. from water. Protected from light.

Quinoline-8-carboxylic acid, m.p. 186-187.5°.
  Cryst. from water.

Quinoline ethiodide, m.p. 158-159°.
  Cryst. from aq. ethanol.

Quinolinol, see Hydroxyquinoline.

Quinoxaline, m.p. 28° (anhydr.), 37° (monohydrate),
b.p. 108-110°/0.1 mm, 140°/40 mm.
  Cryst. from pet.ether. Crystallizes as the monohydrate on addition
of water to a pet.ether soln.

Quinoxaline-2,3-dithiol, m.p. 345° (decomp.).
  Purified by repeated dissolution in alkali and repptn. by acetic
acid.

p-Quinquephenyl, m.p. 388.5°.
  Cryst. from pyridine, then sublimed.

Quinuclidine, m.p. 158°.
  Cryst. from ethyl ether.

Raffinose, m.p. 80°.
  Cryst. from aq. ethanol.

Rauwolscine hydrochloride, m.p. 278-280°.
  Cryst. from water.

RDX, see Cyclotrimethylenetrinitramine.

Reductic acid, m.p. 213°.
  Cryst. from ethyl acetate.

Rescinnamine, m.p. 238-239° (vac.).
  Cryst. from benzene.

Reserpic acid, m.p. 241-243°.
  Cryst. from methanol.

Reserpine, m.p. 262-263°.
  Cryst. from aq. acetone.

Resorcinol, m.p. 111.2-111.6°.
  Cryst. from benzene, toluene or benzene-ethyl ether.

Retene, m.p. 99°.
  Cryst. from ethanol.

Retinal, $E_{1cm}^{1\%}$ (all-trans) 1530 (381 nm), (13-cis) 1250 (375 nm),
in ethanol.
  Separated from retinol by column chromatography on water-deactivated alumina. Eluted with 1-2% acetone in hexane, or on thin-layer
plates of silica gel G developed with 1:1 ether-hexane. Cryst. from
pet.ether, or as the semicarbazone from ethanol. Six isomers reported.
Stored in the dark, in inert atmosphere, at 0°.

Retinoic acid, $E_{1cm}^{1\%}$ (all-trans) 1500 (350 nm), (13-cis) 1320 (354 nm),
(9-cis) 1230 (345 nm), (9,13-di-cis) 1150 (346 nm), in ethanol.
  Cryst. from methanol, ethanol, isopropyl alcohol, or as methyl
ester from methanol. Purified by column chromatography on silicic

Retinoic acid (continued)
acid columns, eluting with a small amount of ethanol in hexane.
Stored in the dark, in inert atmosphere, at 0°.

Retinol, $E_{1cm}^{1\%}$ (all-trans) 1832 (325 nm), (13-cis) 1686 (328 nm),
(11-cis) 1230 (319 nm), (9-cis) 1480 (323 nm), (9,13-di-cis) 1379
(324 nm), (11,13-di-cis) 908 (311 nm), in ethanol.
   Purified by chromatography on column of water-deactivated alumina,
eluting with 3-5% acetone in hexane. Separation of isomers is by
thin-layer plates of silica gel G, developed with pet.ether (low
boiling)-methyl heptanone (11:2). Stored in the dark, under nitrogen,
at 0°, as soln. in ethyl ether, acetone or ethyl acetate.

Retinyl acetate.
   Separated from retinol by column chromatography, then cryst. from
methanol. See Kofler and Rubin [Vitamins Hormones, 18, 315 (1960)]
for review of purification methods. Stored in the dark, under inert
atmosphere, at 0°.

Retinyl palmitate, $E_{1cm}^{1\%}$ (all-trans) 1000 (325 nm), in ethanol.
   Separated from retinol by column chromatography on water-deactivated alumina with hexane containing a very small percentage of acetone. Also chromatographed on thin-layer silica gel G, using pet.ether
-isopropyl ether-acetic acid-water (180:20:2:5) or pet.ether-aceto-
nitrile-acetic acid-water (190:10:1:15) to develop the chromatogram.
Then cryst. from propylene oxide.

Rhamnetin, m.p. >300° (decomp.).
   Cryst. from ethanol.

L-α-Rhamnose, m.p. 105°, $[\alpha]_D^{15}$ +9.1° (in water).
   Cryst. from water or ethanol.

Rhodamine B, CI 749.
   Major impurities are partially dealkyated compounds not removed by
crystn. Purified by chromatography, using ethyl acetate-isopropanol-
ammonia (.880 sg)(9:7:4, R_f = 0.75 on Kieselgel G). Cryst. from conc.
soln.in methanol by slow addition of dry ethyl ether. Stored in the dark.

Rhodamine B chloride.

Cryst. from absolute ethanol containing a drop of conc. HCl by slow addition of ten volumes of dry ethyl ether. Washed with ether and air dried. The dried material has also been extracted with benzene to remove oil-soluble material prior to recrystn.

Rhodamine 6G.

Cryst. from methanol.

Rhodanine, m.p. 168.5° (capillary).

Cryst. from glacial acetic acid or water.

Riboflavin, m.p. 271° (decomp.), $[\alpha]_D$ -9.8° (in H_2O), -125° (in 0.05 N NaOH).

Cryst. from 2 M acetic acid, then extracted with $CHCl_3$ to remove the lumichrome impurity. [Smith and Metzler, JACS, 85, 3285 (1963).] Has also been cryst. from water.

Riboflavin-5'-phosphate.

Cryst. from acidic aq. soln.

Ribonucleic acid.

Martin et al. [BJ, 89, 327 (1963)] dissolved RNA (5 g) in 90 ml of 0.1 mM EDTA, then homogenized with 90 ml of 90% (w/v) phenol in water using a Teflon pestle. The suspension was stirred vigorously for 1 hr at room temperature, then centrifuged for 1 hr at 0° at 2500 rev/min. The lower (phenol) layer was extracted four times with 0.1 mM EDTA and the aq. layers were combined, then made 2% (w/v) with respect to potassium acetate and 70% (v/v) with respect to ethanol. After standing overnight at -20°, the ppte. was centrifuged down, dissolved in 50 ml of 0.1 mM EDTA, made 0.3 M in NaCl and left 3 days at 0°. The purified RNA was then centrifuged down at 10,000 g for 30 min, dissolved in 100 ml of 0.1 mM EDTA, dialyzed at 4° against water, and freeze-dried. It was stored at -20° in a desiccator. Michelson [JCS, 1371 (1959)] dissolved 10 g of RNA in water, adding 2 M ammonia to adjust to pH 7.8, then dialyzed in Visking tubing against five volumes of water for 24 hr. The process was repeated three times, then the material after dialysis was treated with 2 M HCl and ethanol to

Ribonucleic acid (continued)
ppte. the RNA which was collected, washed with ethanol and ether and
dried.

α-D-Ribose, m.p. 90°.
   Cryst. slowly from aq. 80% ethanol, dried under vac. at 60° over
P_2O_5 and stored in a vac. desiccator.

Ricinoleic acid, m.p. 7-8° (α-form), 5.0° (γ-form), n_D^{20} 1.4717.
   Purified as methyl acetylricinoleate [Rider, JACS, 53, 4130
(1931)], fractionally distilling at 180-185°/0.3 mm, then 87 g of
this ester was refluxed with KOH (56g),water (25 ml), and methanol
(250 ml) for 10 min. The free acid was separated, cryst. from
acetone at -50°, and distd. in small batches, b.p. 180°/0.005 mm.
[Bailey et al., JCS, 3027 (1957).]

Rosaniline.
   Cryst. from water.

Rubijervine, m.p. 240-246°.
   Cryst. from ethanol. Has solvent of crystn.

Rutin, m.p. 188-190°.
   Cryst. from methanol or water-ethanol, air dried, then dried for
several hours at 110°.

Saccharic acid, m.p. 125-126°.
   Cryst. from 95% ethanol.

Saccharin, m.p. 229°.
   Cryst. from hot water.

Safranine O.
   Cryst. from benzene-methanol (1:1) or water. Dried under vac. over
H_2SO_4.

Salicin, m.p. 204-208°, $[α]_D^{25}$ -63.5° (c = about 3 in water).
   Cryst. from ethanol.

Salicylaldehyde, b.p. 195-197°.

Pptd. as the bisulphite addition compound by pouring the aldehyde slowly and with stirring into a 25% soln. of $NaHSO_3$ in 30% ethanol, then standing for 30 min. The ppte. after filtering at the pump, and washing with ethanol, was decomposed with aq. 10% $NaHCO_3$, and the aldehyde was extracted into ethyl ether, dried with Na_2SO_4 or $MgSO_4$, and distd. under reduced pressure. Alternatively, salicylaldehyde can be pptd. as its copper complex by adding it to a warm, satd. soln. of copper acetate, shaking and then standing in ice. The ppte. is filtered off, washed thoroughly with ethanol, then with ethyl ether, and decomposed with 10% H_2SO_4, the aldehyde being extracted into ethyl ether, dried and distd.

Salicylaldoxime, m.p. 57°.
  Cryst. from $CHCl_3$-pet.ether (b.p. 40-60°).

Salicylamide, m.p. 138-139°.
Salicylanilide, m.p. 135°.
  Cryst. from water.

Salicylhydroxamic acid, m.p. 179-180° (decomp.).
  Cryst. from acetic acid.

Salicylic acid, m.p. 159-160°.
  Cryst. from hot water, absolute methanol, or cyclohexane. Air dried. Can also be sublimed.

Sarcosine, m.p. 212-213° (decomp.).
  Cryst. from absolute ethanol.

Sarcosine anhydride, m.p. 146-147°.
  Cryst. from water or ethanol.

Scopolamine, see Hyoscine.

Scopoletin, m.p. 206°.
  Cryst. from water or acetic acid.

Sebacic acid, m.p. 134.5°.

Purified via the disodium salt which, after crystn. from boiling water (charcoal), was again converted to the free acid. The free acid was cryst. repeatedly from hot distd. water and dried under vac.

Selenourea, m.p. 200° (decomp.).

Cryst. from water, under nitrogen.

Semicarbazide hydrochloride, m.p. 175°.

Cryst. from aq. 75% ethanol and dried under vac. over $CaSO_4$.

Sennoside A.
Sennoside B.

Cryst. from aq. acetone.

S-Serine, m.p. 228° (decomp.), $[\alpha]_D^{25}$ +14.5° (in 1 M HCl), $[\alpha]_{546}^{20}$ +16 (c = 5 in 5 N HCl).

Likely impurity: glycine. Cryst. from water by adding 4 volumes of ethanol. Dried. Stored in a desiccator.

Serotonin creatinine sulphate.

Cryst. (as monohydrate) from water.

Shikimic acid, m.p. 190.5°.
Sinomenine hydrochloride, m.p. 231°.

Cryst. from water.

Sitosterols.

Cryst. from ethanol.

β-Sitosterol, m.p. 136-137°, $[\alpha]_{546}^{20}$ -42° (c = 2 in $CHCl_3$).

Cryst. from methanol. Purified by zone melting.

Skatole, see 3-Methylindole.

Skellysolve A is essentially n-pentane, b.p. 28-30°.
           B is essentially n-hexane, b.p. 60-68°.
           C is essentially n-heptane, b.p. 90-100°.
           D is mixed heptanes, b.p. 75-115°.
           E is mixed octanes, b.p. 100-140°.
           F is pet.ether, b.p. 30-60°.
           G is pet.ether, b.p. 40-75°.
           H is hexanes and heptanes, b.p. 69-96°.
           L is essentially octanes, b.p. 95-127°.
   For methods of purification, see Petroleum Ether.

Smilagenin, m.p. 185°.
   Cryst. from acetone.

Solanidine, m.p. 218-219°.
   Cryst. from $CHCl_3$-methanol.

Solanine.
   Cryst. from aq. 85% ethanol.

Solasodine, m.p. 202°.
   Cryst. (as monohydrate) from aq. 80% ethanol.

Solasonine, m.p. 279°.
   Cryst. from aq. 80% dioxane.

Solochrome violet R.
   Converted to the monosodium salt by pptn. with a sodium acetate-
acetic acid buffer of pH 4, then purified as described for Chlorazole
sky blue FF. Dried at 110°. Is hygroscopic. [Coates and Rigg, TFS,
57, 1088 (1961).]

Sorbic acid, m.p. 134°.
   Cryst. from boiling water.

Sorbitol, m.p. 89-93° (hemihydrate), 110-111° (anhydr.).
   Cryst. (as hemihydrate) several times from ethanol-water (1:1),
then dried by fusing and storing over $MgSO_4$.

Spirilloxanthin, m.p. 216-218°, λ_{max} 463, 493, 528 nm, $E_{1cm}^{1\%}$ = 2680 (493 nm), in pet.ether (b.p. 40-70°).

Cryst. from $CHCl_3$-pet.ether, acetone-pet.ether, benzene-pet.ether or benzene. Purified by chromatography on a column of $CaCO_3$-Ca(OH)$_2$ mixture or deactivated alumina. [Polgar et al., Arch. Biochem. Biophys., 5, 243 (1944).] Stored in the dark, in inert atmosphere, at -20°.

Squalene, f.p. -5.4°, b.p. 213°/1 mm, d_4^{25} 0.8670, n_D^{20} 1.4905.

Cryst. repeatedly from acetone (1.4 ml. of acetone per ml) by cooling in a Dry-Ice bath, washing the crystals with cold acetone, then freezing the squalene from solvent under vac. The squalene was further purified by passage through a column of silica gel. It has also been chromatographed on activated alumina, using pet.ether as eluant. Dauben et al. [JACS, 74, 4321 (1952)] purified squalene via its hexachloride.

See also Capstack et al. [JBC, 240, 3258 (1965)], Krishna et al. [Arch. Biochem. Biophys., 114, 200 (1966)].

Starch.

Defatted by Soxhlet extraction with ethyl ether or 95% ethanol. For fractionation of starch into "amylose" and "amylopectin" fractions, see Lansky, Kooi and Schoch [JACS, 71, 4066 (1949)].

Stearic acid, m.p. 71.4°.

Cryst. from acetone, acetonitrile, ethanol, aq. methanol, methyl ethyl ketone or pet.ether (b.p. 60-90°), or by fractional pptn. by dissolving in hot 95% ethanol and pouring into distd. water, with stirring. The ppte., after washing with distd. water, was dried under vac. over P_2O_5. Also purified by zone melting.

Stearyl alcohol, m.p. 61°, b.p. 153-154°/0.3 mm.

Cryst. from thiophen-free benzene, and dried under vac. over P_2O_5.

Stigmasterol, m.p. 170°, $[\alpha]_D^{22}$ -51° (in $CHCl_3$).

Cryst. from hot ethanol. Dried in vac. over P_2O_5 for 3 hr at 90°. Purity checked by proton magnetic resonance spectrum.

cis-Stilbene, b.p. 145°/12 mm.

Purified by chromatography on alumina using hexane. (The final product contains about 0.1% of the trans isomer.)

trans-Stilbene, m.p. 125.9°.

Purified by vac. distn. (The final product contains about 1% of the cis isomer.) Cryst. from ethanol. Purified by zone melting.

Strychnine, m.p. 268°.

Cryst. as the hydrochloride from water, then neutralized with ammonia.

Styphnic acid, m.p. 179-180°.

Cryst. from ethyl acetate. [Explodes violently on rapid heating.)

Styrene, b.p. 41-42°/18 mm, 145.2°/760 mm, n_D^{20} 1.5469, n_D^{25} 1.5441.

Difficult to purify and keep pure. Usually contains added inhibitor (such as a trace of hydroquinone). Washed with aq. 5% NaOH to remove inhibitors (e.g. tert.-butyl catechol), then with water, dried for several hours with $MgSO_4$ and distd at 25° under reduced pressure in the presence of an inhibitor (such as 0.005% p-tert.-butylcatechol). Can be stored at -78°. It can also be stored and kept anhydrous with Linde type 5A molecular sieve, CaH_2, $CaSO_4$, BaO or sodium, being fractionally distd. just before use.

Styrene glycol, m.p. 67-68°.

Cryst. from pet.ether.

Styrene oxide, b. p. 84-86°/16.5 mm.

Fractional distn. at reduced pressure does not remove phenyl-acetaldehyde. If this material is present, the styrene oxide is treated with hydrogen under 3 atmospheres pressure in the presence of platinum oxide. The aldehyde, but not the oxide, is reduced (to β-phenylethanol) and separation is now readily achieved by fractional distn. [Schenck and Kaizerman, JACS, 75, 1636 (1953).]

Suberic acid, m.p. 141-142°.

Cryst. from acetone.

Succinamide, m.p. 254°.
    Cryst. from hot water.

Succinic acid, m.p. 185-185.5°.
    Washed with ethyl ether. Cryst. from acetone, distd. water, or
tert.-butyl alcohol.Dried under vac. over P_2O_5 or conc. H_2SO_4. Also
purified by conversion to the disodium salt which, after crystn. from
boiling water (charcoal), is treated with mineral acid to regenerate
the succinic acid. The acid is then recryst. and vac. dried.

Succinic anhydride, m.p. 119-120°.
    Cryst. from redistd. acetic anhydride or $CHCl_3$, then filtered,
washed with ethyl ether and dried under vac.

Succinimide, m.p. 124-125°.
    Cryst. from ethanol (1 ml/g) or water.

Succinonitrile, m.p. 57.9°, b.p. 108°/1 mm, 267°/760 mm.
    Purified by vac. sublimation, also cryst. from acetone.

Sucrose.
    Cryst. from water.

Sucrose diacetate hexaisobutyrate.
    Melted and, while molten, treated with $NaHSO_3$ and charcoal, then
filtered.

D-Sucrose octaacetate, m.p. 83-85°, $[\alpha]_{546}^{20}$ +70° (c = 1 in $CHCl_3$).
    Cryst. from ethanol.

Sudan III, CI 248.
    Cryst. from ethanol, ethanol-water or benzene-absolute ethanol
(1:1).

Sudan IV, m.p. 184°.
    Cryst. from ethanol-water or acetone-water.

Sudan yellow, m.p. 135°.
    Cryst. from ethanol.

Sulphaguanidine, m.p. 189–190°.
 Cryst. from hot water (7 ml/g).

Sulphamethazine, m.p. 178–179°.
 Cryst. from aq. dioxane.

Sulphanilamide, m.p. 166°.
 Cryst. from water or ethanol.

Sulphanilic acid.
 Cryst. (as dihydrate) from boiling water. Dried at 105° for 2–3
hr, then over 90% H_2SO_4 in a vac. desiccator.

Sulphapyridine, m.p. 193°.
 Cryst. from 90% acetone and dried at 90°.

2-Sulphobenzoic acid, m.p. 68–69°.
 Cryst. from water.

o-Sulphobenzoic anhydride, m.p. 128°, b.p. 184–186°/18 mm.
 Cryst. from dry benzene. Can be distd. under vac.

Sulpholane, m.p. 28.5°, b.p. 153–154°/18 mm, 285°/760 mm,
$[n]_D^{30}$ 1.4820.
 Dried by passage through a column of molecular sieve. Dist. at
reduced pressure through a column packed with stainless steel helices.
Again dried with molecular sieve and distd. [Cram et al., JACS, 83,
3678 (1961); Coetzee, PAC, 49, 211 (1977).]
 Also, stirred at 50° and small portions of solid $KMnO_4$ were added
until the colour persisted during 1 hr. Dropwise addition of methanol
then destroyed the excess $KMnO_4$, the soln. was filtered, freed from
potassium ions by passage through an ion-exchange column and dried
under vacuum.

5-Sulphosalicylic acid.
 Cryst. from water. Alternatively, converted to the monosodium
salt which was cryst. from water and washed with a little water,
ethanol and then ethyl ether.

Syringaldehyde, m.p. 113°.
    Cryst. from pet.ether.

D-Tagatose, m.p. 134-135°.
    Cryst. from aq. ethanol.

Tartaric acid, m.p. 169.5-170° (2S,3S-form, natural) $[\alpha]_{546}^{0}$ -15°
(c = 10 in H_2O); m.p. 208° (2RS,3RS-form).
    Cryst. from distd. water or benzene-ethyl ether containing 5% of
pet.ether (b.p. 60-80°)(1:1). Soxhlet extraction with ethyl ether
has been used to remove an impurity absorbing at 265 nm.

meso-Tartaric acid, m.p. 139-141°.
    Cryst. from water, washed with cold methanol and dried at 60°
under vac.

Taurocholic acid, m. p. 125° (decomp.).
    Cryst. from ethanol-ethyl ether.

Terephthalaldehyde, m.p. 116°, b.p. 245-248°/771 mm.
    Cryst. from water.

Terephthalic acid. Sublimes >300° without melting.
    Purified via the sodium salt which, after crystn. from water, was
reconverted to the acid by acidification with mineral acid.

Terephthaloyl chloride, m.p. 80-82°.
    Cryst. from dry hexane.

o-Terphenyl, m.p. 57-58°.
m-Terphenyl, m.p. 88-89°.
    Cryst. from ethanol. Purified by chromatography of CCl_4 solns. on
alumina, with pet.ether as eluant, followed by crystn. from pet.ether
(b.p. 40-60°) or pet.ether-benzene. Can also be distd. under vac.

p-Terphenyl, m.p. 212.7°.

   Cryst. from nitrobenzene or trichlorobenzene. Purified by chromatography on alumina in a darkened room, using pet.ether, and then crystallizing from pet.ether (b.p. 40-60°) or pet.ether-benzene.

Terpin hydrate, m.p. 105.5° (cis), 156-158° (trans).

   Cryst. from water or ethanol.

2,2',2"-Terpyridyl, m.p. 89-91°.

   Cryst. from ethyl ether, or from pet.ether, then aq. methanol, followed by vac. sublimation at 90°.

Terramycin. Sinters at 182°, melts at 184-185° (decomp.).

   Cryst. (as dihydrate) from water or aq. methanol.

Terreic acid, m.p. 127-127.5°.

   Cryst. from benzene or hexane. Sublimed in vac.

Testosterone, m.p. 155°.

   Cryst. from aq. acetone.

Testosterone propionate, m.p. 118-122°.

   Cryst. from aq. ethanol.

2,3,4,6-Tetraacetyl-α,d-glucopyranosyl bromide, m.p. 88-89°

$[\alpha]_D^9$ +199.3° (c = 3 in CHCl$_3$).

   Cryst. from isopropyl ether.

Tetra-n-amylammonium iodide, m.p. 127.6-128.8°.

   Cryst. from ethanol and dried at 35° under vac.

Tetrabenazine, m.p. 127-128°.

   Cryst. from methanol.

3',3",5',5"-Tetrabromo-m-cresolsulphonephthalein, see Bromocresol green.

1,1,2,2-Tetrabromoethane, f.p. 0.0°, b.p. 243.5°, n_D^{20} 1.63533.

    Washed successively with conc. H_2SO_4 (three times) and water (three times), dried with K_2CO_3 and $CaSO_4$ and distd.

Tetrabromophenolphthalein ethyl ester.

    Cryst. from benzene, dried at 120° and kept under vac.

3',3",5',5"-Tetrabromophenolsulphonephthalein, see Bromophenol blue.

Tetra-n-butylammonium bromide, m.p. 119.6°.

    Cryst. from benzene (5 ml/g) at 80° by adding hot n-hexane (three volumes) and allowing to cool. Dried over P_2O_5 or $Mg(ClO_4)_2$, under vac. The salt is hygroscopic. It can also be cryst. from ethyl acetate or dry acetone by adding ethyl ether.

Tetra-n-butylammonium chloride.

    Cryst. from acetone by addition of ethyl ether. Very hygroscopic. (All manipulations carried out in drybox.)

Tetra-n-butylammonium hydrogen sulphate, m.p. 171-172°.

    Cryst. from acetone.

Tetra-n-butylammonium iodide, m.p. 146°.

    Cryst. from benzene-pet.ether (see entry for corresp. bromide), acetone, ethyl acetate, ethanol-ethyl ether, nitromethane, aq. ethanol or water. Dried at room temperature under vac. Has also been dissolved in methanol-acetone (1:3, 10 ml/g), filtered and allowed to stand at room temperature to evaporate to about half its original volume. Distd. water (1 ml/g) was then added, and the ppte. was filtered off and dried.

Tetrabutylammonium nitrate, m.p. 119°.

    Cryst. from benzene (7 ml/g) or ethanol.

Tetra-n-butylammonium perchlorate.

    Cryst. from ethanol, or from ethyl ether-acetone mixture. Dried in vac. at room temperature.

Tetra-n-butylammonium picrate, m.p. 89°.
    Cryst. from ethanol and dried under vac.

Tetrabutylammoniumtetrabutylborate (Butyl$_4$N$^+$butyl$_4$B$^-$), m.p. 109.5°.
    Dissolved in methanol and cryst. by adding distilled water.

1,2,3,4-Tetrachloroaniline, m.p. 45-46°, b.p. 254°/760 mm.
1,2,3,5-Tetrachloroaniline, m.p. 51°, b.p. 246°/760 mm.
2,3,4,5-Tetrachloroaniline, m.p. 119-120°.
2,3,5,6-Tetrachloroaniline, m.p. 107-108°.
    Cryst. from ethanol.

1,2,4,5-Tetrachloroaniline, m.p. 139.5-140.5°, b. p. 240°/760 mm.
    Cryst. from ethanol, ether, benzene, benzene-ethanol or carbon
disulphide.

Tetrachloro-o-benzoquinone, m.p. 130°.
    Cryst. from glacial acetic acid.

1,1,2,2-Tetrachlorodifluoroethane, f.p. 26.0°, b.p. 92.8°/760 mm.
    Purified as for Trichlorotrifluoroethane.

sym-Tetrachloroethane, b.p. 146.2°, n$_D^{15}$ 1.49678.
    Stirred, on a steam-bath, with conc. H$_2$SO$_4$ until a fresh portion
of acid remained colourless. The organic phase was then separated,
distd. in steam, dried (CaCl$_2$ or K$_2$CO$_3$), and fractionally distd.

Tetrachloroethylene, b.p. 121.2°, d^{15} 1.63109, d^{30} 1.60640,
n$_D^{15}$ 1.50759, n$_D^{20}$ 1.50566.
    It decomposes under similar conditions to CHCl$_3$, to give phosgene
and trichloroacetic acid. Inhibitors of this reaction include ethanol,
ethyl ether and thymol (effective at 2-5 p.p.m.). Tetrachloroethylene
should be distd. under vac. (to avoid phosgene formation), and stored
in the dark out of contact with air. It can be purified by washing
with 2 M HCl until the aq. phase no longer becomes coloured, then
with water, drying with Na$_2$CO$_3$, Na$_2$SO$_4$, CaCl$_2$ or P$_2$O$_5$, and fractionally distilling just before use. 1,1,2-Trichloroethane and 1,1,1,2-
tetrachloroethane can be removed by counter-current extraction with
ethanol-water.

2,3,4,6-Tetrachloronitrobenzene, m.p. 42°.
2,3,5,6-Tetrachloronitrobenzene, m.p. 99-100°.
   Cryst. from ethanol.

2,3,4,5-Tetrachlorophenol, m.p. 116-117°.
2,3,4,6-Tetrachlorophenol, m.p. 70°, b. p. 150°/15 mm.
2,3,5,6-Tetrachlorophenol, m.p. 115°.
   Cryst. from ligroin.

Tetrachlorophthalic anhydride, m.p. 255°.
   Cryst. from $CHCl_3$ or benzene, then sublimed.

2,3,4,6-Tetrachloropyridine, m.p. 74-75°, b.p. 130-135°/16-20 mm.
   Cryst. from 50% ethanol.

Tetracosane, m.p. 54°, b.p. 243-244°/15 mm.
   Cryst. from ether.

Tetracosanoic acid, m.p. 87.5-88°.
   Cryst. from acetic acid.

1,2,4,5-Tetracyanobenzene.
   Cryst. from ethanol.

Tetracyanomethylene, m.p. 198-199° (sealed tube).
   Cryst. from chlorobenzene, dichloroethane, or methylene dichloride.
Stored at 0° in a desiccator over NaOH pellets. (It slowly evolves
HCN on exposure to moist air.) Can also be sublimed at 120° under vac.

Tetracycline, m.p. 172-174° (decomp.), $[\alpha]_{546}^{20}$ +270° (c = 1 in MeOH).
   Cryst. from toluene.

Tetradecane-1,14-dicarboxylic acid, m.p. 126°.
   Cryst. from 95% ethanol.

1-Tetradecanol, m.p. 39-39.5°, b.p. 170-173°/20 mm, 160°/10 mm.
   Cryst. from aq. ethanol. Purified by zone refining.

Tetradecyl ether.
   Vac. distd., then cryst. repeatedly from methanol-benzene.

Tetraethoxymethane, b.p. 159°.
   Dried with Na_2SO_4 and distd.

Tetraethylammonium bromide.
   Cryst. from ethanol, ethanol-ethyl ether (1:2), ethyl acetate,
water or boiling methanol-acetone (1:3) by adding an equal volume
of acetone and allowing to cool. Dried and stored over P_2O_5.

Tetraethylammonium chloride.
   Cryst. from ethanol by adding ethyl ether, from warm water by
adding ethanol and ethyl ether, or from dimethylacetamide.

Tetraethylammonium iodide.
   Cryst. from acetone-methanol, ethanol-water, dimethylacetamide
or ethyl acetate-ethanol (19:1). Dried under vac.

Tetraethylammonium perchlorate.
   Cryst. repeatedly from water and dried at 70° under vac.

Tetraethylene glycol dimethyl ether, b.p. 105°/1 mm.
   Stood with CaH_2, $LiAlH_4$ or sodium, and distd. as needed.

Tetraethylenepentamine, b.p. 169-171°/0.05 mm.
   Distd. under vac. Purified via its pentachloride, nitrate or
sulphate. Jonassen, Frey and Schaafsma [JPC, 61, 504 (1957)] cooled
a soln. of 150 g of the base in 300 ml of 95% ethanol, and added
dropwise 180 ml of conc. HCl, keeping the temperature below 20°.
The white ppte. was filtered, cryst. three times from ethanol-water,
then washed with ethyl ether and dried by suction. Reilley and
Holloway [JACS, 80, 2917 (1958)], starting with a similar soln.
cooled to 0°, added slowly (keeping the temperature below 10°) a
soln. of 4½ gram-moles of HNO_3 in 600 ml of aq. 50% ethanol (also
cooled to 0°). The ppte. was filtered by suction, recryst. five
times from aq. 5% HNO_3, then washed with acetone and absolute ethanol and dried at 50°. [For purification via the sulphate, see

Tetraethylenepentamine (continued)

Reilley and Vavoulis [AC, 31, 243 (1959)], and for an additional
purification step using Schiff base formation with benzaldehyde, see
Jonassen et al., [JACS, 79, 4279 (1957).]

Tetrafluoro-1,3-dithietane.

Purified by preparative gas chromatography.

2,2,3,3-Tetrafluoropropanol, b.p. 106-106.5°.

Tetrafluoropropanol (450 ml) was added to a soln. of 2.25 g.
NaHSO₃ in 90 ml of water, shaken vigorously and stood for 24 hr. The
fraction distilling at or above 99° was refluxed for 4 hr with 5-6 g
of KOH and then rapidly distd., followed by a final fractional distn.
[Kosower and Wu, JACS, 83, 3142 (1961).] Alternatively, shaken with
alumina for 24 hr, dried overnight with anhydrous K_2CO_3 and distd.,
taking the middle fraction (b.p. 107-108°).

Tetra-n-heptylammonium bromide.

Cryst. from n-hexane.

Tetra-n-heptylammonium iodide.
Tetra-n-hexylammonium chloride.

Cryst. from ethanol.

Tetrahydrofuran, b.p. 65.4°, n_D^{25} 1.4040.

Refluxing with, and distilling from, $LiAlH_4$ removes water, peroxides, inhibitor and other impurities. Peroxides can also be removed
by passage through a column of activated alumina, or by treatment
with aq. ferrous sulphate and sodium bisulphate, followed by solid
KOH. In both cases, the solvent is then dried and fractionally distd.
from sodium. Lithium wire or vigorously stirred molten potassium have
also been used for this purpose. CaH_2 has also been used as a drying
agent.

Several methods are available for obtaining the solvent almost
anhydrous. Ware [JACS, 83, 1296 (1961)] dried vigorously with sodium-
potassium alloy until a characteristic blue colour was evident in the
solvent at Dry-Ice-Cellosolve temperatures. The solvent was kept in
contact with the alloy until distd. for use. Worsfold and Bywater

Tetrahydrofuran (continued)

[JCS, 5234 (1960)], after refluxing and distilling from P_2O_5 and KOH, in turn, refluxed the solvent with sodium-potassium alloy and fluorenone until the green colour of the disodium salt of fluorenone was well established. [Alternatively, instead of fluorenone, benzophenone, which forms a blue ketyl, can be used.] The tetrahydrofuran was then fractionally distd., degassed and stored above CaH_2. p-Cresol or hydroquinone inhibit peroxide formation.

1,2,3,4-Tetrahydronaphthalene, see Tetralin.

RS-Tetrahydropalmatine, m.p. 148-149°.
    Cryst. from methanol by addition of water.

Tetrahydropyran, b.p. 88.0°, n_D^{20} 1.4202.
    Dried with CaH_2, then passed through a column of silica gel to remove olefinic impurities and fractionally distd. Freed from peroxides and moisture by refluxing with sodium, then distilling from $LiAlH_4$. Alternatively, peroxides can be removed by treatment with aq. ferrous sulphate and sodium bisulphate, followed by solid KOH, and fractional distn. from sodium.

Tetrahydrothiophen, b.p. 120.9°/760 mm, n_D^{20} 1.5289.
    Crude material was purified by crystn. of the Hg(II) chloride complex to a constant m.p. It was then regenerated, washed, dried, and fractionally distd. [Whitehead et al., JACS, 73, 3632 (1951).]

Tetrahydroxy-p-benzoquinone.
    Cryst. from water.

Tetralin, b.p. 65-66°/5 mm, 207.6°/760 mm, n_D^{20} 1.5413.
    Washed with successive portions of conc. H_2SO_4 until the acid layer no longer became coloured, then washed with aq. 10% Na_2CO_3, and distd. water. Dried ($CaSO_4$ or Na_2SO_4), filtered, refluxed and fractionally distd. at reduced pressure from sodium or BaO. Can also be purified by repeated fractional freezing.
    Bass [JCS, 3498 (1964)] freed tetralin, purified as above, from naphthalene and other impurities by conversion to ammonium tetralin-

Tetralin (continued)

6-sulphonate. Conc. H_2SO_4 (150 ml) was added slowly to stirred
tetralin (272 ml) which was then heated on a water bath for about
2 hr to give complete solution. The warm mixture, poured into aq.
NH_4Cl soln. (120 g in 400 ml), gave a white cryst. ppte. which, after
filtering off, was recryst. from boiling water, washed with 50% aq.
ethanol and dried at 100°. Evapn. of its boiling aq. soln. on a water
bath removed traces of naphthalene. The pure salt (229 g) was mixed
with conc. H_2SO_4 (266 ml) and steam distd. from an oil-bath at 165-
170°. An ether extract of the distillate was washed with aq. Na_2CO_3
(10%) and water, dried with Na_2SO_4,and the ether was evapd., prior
to distn. of tetralin from sodium. Tetralin has also been purified
via barium tetralin-6-sulphonate, conversion to the sodium salt and
decomposition in 60% H_2SO_4 using superheated steam.

Tetralin hydroperoxide, m.p. 56°.
    Cryst. from hexane.

Tetramethylammonium bromide. Sublimes with decomp. $>230^\circ$.
    Cryst. from ethanol, ethanol-ethyl ether, methanol-acetone, water
or from acetone-methanol (4:1) by adding an equal volume of acetone.
Dried at 110° under reduced pressure.

Tetramethylammonium chloride. Decomp. $>230^\circ$.
    Cryst. from ethanol, ethanol-$CHCl_3$, ethanol-ethyl ether, acetone-
ethanol (1:1), isopropyl alcohol or water. Traces of the free amine
can be removed by washing with $CHCl_3$.

Tetramethylammonium hydroxide. Pentahydrate m.p. 63° (decomp.).
    Freed from chloride ions by passage through an ion-exchange col-
umn (Amberlite IRA-400, prepared in its OH form by passing 2 M NaOH
until the effluent was free from chloride ions, then washed with
distd. water until neutral). A modification, to obtain carbonate-free
hydroxide, uses the method of Davies and Nancollas [Nature, 165, 237
(1950)].

Tetramethylammonium iodide. Decomp. >230°.

Cryst. from aq. 50% ethanol, ethanol-ethyl ether, ethyl acetate, water or from acetone-methanol (4:1) by adding an equal volume of acetone. Dried in a vac. desiccator.

1,2,4,5-Tetramethylbenzene, see Durene.

N,N,N',N'-Tetramethylbenzidine, m.p. 195.4-195.6°.

Cryst. from pet.ether, then from pet.ether-benzene. Dried under vac. in an Abderhalden pistol.

Tetramethylcyclobutan-1,3-dione, m.p. 114.5-114.9°.

Cryst. from benzene and dried under vac. over P_2O_5 in an Abderhalden pistol.

p,p'-Tetramethyldiaminophenylmethane, m.p. 89-90°.

Cryst. from ethanol (2 ml/g).

Tetramethylene glycol, m.p. 18.7°.

Fractionally cryst. from its melt, then distd. under vac.

Tetramethylene sulphone, m.p. 27°, b.p. 130°/6.5 mm.

Vac. distd. from NaOH pellets. Hygroscopic.

Tetramethylene sulphoxide.

Shaken with BaO for 4 days, then distd. from CaH_2 under reduced pressure.

N,N,N',N'-Tetramethylethylenediamine, b.p. 122°, n_D^{25} 1.4153.

Partially dried with molecular sieves (Linde type 4A), and distd. in vac. from butyllithium. This treatment removes all traces of primary and secondary amines and water. [Hay, McCabe and Robb, JCS Faraday Trans. I, 68, 1 (1972).] Or, dried with KOH pellets. Refluxed for 2 hr with one-sixth its weight of n-butyric anhydride (to remove primary and secondary amines) and fractionally distd. Refluxed with fresh KOH, and distd. under nitrogen. [Cram and Wilson, JACS, 85, 1245 (1963).] Also distd. from sodium.

<u>Tetramethylethylenediamine dihydrochloride</u>.
   Cryst. from 98% ethanol-conc. HCl (3:1).

<u>Tetramethylguanidine</u>.
   Refluxed over granulated BaO, then fractionally distd.

<u>3,4,7,8-Tetramethyl-1,10-phenanthroline</u>, m.p. 275-276°.
   Cryst. from benzene (charcoal).

<u>Tetramethyl-p-phenylenediamine</u>, m.p. 51°, b.p. 260°/760 mm.
   Cryst. from water.

<u>N,N,N',N'-Tetramethyl-p-phenylenediamine dihydrochloride</u>.
   Cryst. from isopropyl or <u>n</u>-butyl alcohols, satd. with HCl.
Treated with aq. NaOH to give the free base which was filtered,
dried and sublimed.

<u>Tetramethylthiuram disulphide</u>, m.p. 104°.
   Cryst. (three times) from boiling $CHCl_3$, then from boiling $CHCl_3$,
adding ethanol dropwise to initiate pptn. and allowing to cool.
Finally pptd. from cold $CHCl_3$ by adding ethanol (which retains the
monosulphide in soln.) [Ferington and Tobolsky, <u>JACS</u>, <u>77</u>, 4510 (1955).]

<u>1,1,3,3-Tetramethylurea</u>, f.p. -1.2°, b.p. 175.2°/760 mm.
   Dried over BaO and distd. under nitrogen.

<u>Tetramethyluric acid</u>, m.p. 228°.
   Cryst. from water.

<u>Tetranitromethane</u>, m.p. 13°.
   Shaken with dil. NaOH, washed, steam distd., dried with Na_2SO_4 and
fractionally cryst. by partial freezing. The melted crystals were
dried with $MgSO_4$ and fractionally distd. under reduced pressure.
Shaken with a large volume of dil. NaOH until no absorption attribut-
able to the nitroform anion was observable in the water. Then washed
with distd. water, and distd. at room temperature by passing a stream
of air or nitrogen through the liquid and condensing in a trap at -80°.
Can be dried with $MgSO_4$ or Na_2SO_4, fractionally cryst. from the melt,
and fractionally distd. under reduced pressure.

Tetra(p-nitrophenyl)ethylene.
   Cryst. from dioxane and dried at 150°/0.1 mm.

Tetrapentylammonium bromide.
   Cryst. from benzene or from acetone-ether and dried in vac. at 40-50°.

Tetraphenylethylene, m.p. 223-224°, b.p. 415-425°/760 mm.
   Cryst. from dioxane from ethanol-benzene. Sublimed under high vac.

Tetraphenylhydrazine.
   Cryst. from 1:1 CHCl$_3$-toluene. Stored in refrigerator, in the dark.

trans-1,1,4,4-Tetraphenyl-2-methylbutadiene.
   Cryst. from ethanol.

Tetraphenylphosphonium chloride.
   Cryst. from acetone. Dried at 70° under vac.

Tetra-n-propylammonium bromide.
   Cryst. from ethyl acetate-ethanol (9:1), acetone or methanol. Dried at 110° under reduced pressure.

Tetra-n-propylammonium iodide.
   Cryst. from ethanol, ethanol-ethyl ether or ethanol-water, and dried in a vac. desiccator.

Tetra-n-propylammonium perchlorate.
   Cryst. from acetonitrile-water (1:4 v/v), and dried under vac. over P$_2$O$_5$.

1,2,3,4-Tetrazole, m.p. 156°.
   Cryst. from ethanol.

TETREN, see Tetraethylenepentamine.

Thapsic acid, see 1,14-Hexadecanedioic acid.

Thebaine, m.p. 193° $[\alpha]_D^{25}$ -219° (in ethanol).
  Sublimed at 170-180°.

3-Thenoic acid, m.p. 137-138°.
  Cryst. from water.

Thenoyltrifluoroacetone.
  Cryst. from hexane or benzene. (Aq. solns. slowly decompose.)

2-Thenylamine, b.p. 78.5°/15 mm.
  Distd. under reduced pressure (nitrogen), from BaO, through a
column packed with glass helices.

Theobromine, m.p. 337°.
Theophylline, m.p. 270-274°.
  Cryst. from water.

Thevetin. Softens at 194°, m.p. 210°.
  Cryst. (as trihydrate) from isopropyl alcohol. Dried at 100°/10 mm
to give the hemihydrate (very hygroscopic).

Thianthrene, m.p. 158°.
  Cryst. from acetone (charcoal) or acetic acid.

ε-[2-(4-Thiazolidone)]hexanoic acid, m.p. 140°.
  Cryst. from water, acetone or methanol.

Thietane, b.p. 93.8-94.2°/752 mm, d_4^{23} 1.0283, n_D^{23} 1.5059.
  Purified by preparative gas chromatography on a dinonyl phthalate
column.

Thioacetamide, m.p. 112-113°.
  Cryst. from absolute ethyl ether or benzene. Stored over P_2O_5.
(Develops an obnoxious odour on keeping, and absorption at 269 nm
decreases, hence freshly cryst. before use.)

Thioacetanilide, m.p. 75-76°.
  Cryst. from water and dried in a vac. desiccator.

Thiobarbituric acid, m.p. 235° (decomp.)
  Cryst. from water.

Thiobenzanilide, m.p. 101.5-102°.
  Cryst. from methanol at Dry-Ice temperature.

Thiocarbanilide, see sym-Diphenylthiourea.

Thiochrome, m.p. 227-228°.
  Cryst. from CHCl$_3$.

Thiocresol, see Toluenethiol.

6,8-Thioctic acid, m.p. 45-47.5° (R-form); 60-61° (RS-form).
  Cryst. from cyclohexane.

Thiodiglycollic acid, m.p. 129°.
β,β'-Thiodipropionic acid, m.p. 134°.
  Cryst. from water.

Thioflavine T.
  Cryst. from benzene-methanol (1:1).

Thioformamide, m.p. 29°.
  Cryst. from ethyl acetate or ether-pet.ether.

Thioglycollic acid, b.p. 95-96°/8 mm.
  Mixed with an equal volume of benzene, the benzene then being
distd. to dehydrate the acid. After heating to 100° to remove most
of the benzene, the residue was distd. under vac. and stored in
closed bottles at 3°. [Eshelman et al., AC, 32, 844 (1960).]

Thioguanosine.
  Cryst. (as hemihydrate) from water.

Thioindigo, m.p. >280°.

Adsorbed on silica gel from CCl_4-benzene (3:1), eluted with benzene, cryst. from $CHCl_3$ and dried at 60-65°. [Wyman and Brode, JACS, 73, 1487 (1951). This paper also gives details of purification of other thioindigo dyes.]

Thiomalic acid, m.p. 153-154°.

Extracted from aq. soln. several times with ethyl ether.

β-Thionaphthol, m.p. 81°, b.p. 153.5°/15 mm, 286°/760 mm.

Cryst. from ethanol.

Thionin, ε_{590nm} = 6.2 x 10^4 l./mole.

Standard Biological Stain material is highly impure. It can be cryst. from water or 50% aq. ethanol, then chromatographed on alumina using $CHCl_3$ as eluant. [Shepp, Chaberek and MacNeil, JPC, 66, 2563 (1962).]

Thiophane, b.p. 40.3°/39.7 mm.

Distd. from sodium.

2-Thiophenaldehyde, b.p. 106°/30 mm.

Washed with 50% HCl and distd. under reduced pressure just before use.

Thiophen, f.p. -38.5°, b.p. 84.2°, n_D^{20} 1.52890, n_D^{30} 1.5223.

The simplest purification procedure is to dry with solid KOH, or reflux with sodium, and fractionally distil through a glass helices-packed column. More extensive treatments include an initial wash with aq. HCl, then water, drying with $CaSO_4$ or KOH, and passage through columns of activated silica gel or alumina. Fawcett and Rasmussen [JACS, 67, 1705 (1945)] washed thiophen successively with 7 M HCl, 4 M NaOH, and distd. water, dried with $CaCl_2$ and fractionally distd. Benzene was removed by fractional crystn. by partial freezing, and the thiophen was degassed and sealed in Pyrex flasks. [A method is described for recovering the thiophen from the benzene-enriched portion.]

Thiophene-2-acetic acid, m.p. 76°.
Thiophene-3-acetic acid, m.p. 79-80°.
   Cryst. from ligroin.

Thiophene-2-carboxylic acid, m.p.129-130°.
   Cryst. from water.

Thiophenol, see Benzenethiol.

Thiosalicylic acid, m.p. 164-165°.
   Cryst. from hot ethanol (4 ml/g), after adding hot distd. water
(8 ml/g) and boiling with decolourizing carbon.

Thiosemicarbazide, m.p. 181-183°.
   Cryst. from water.

1-Thiosorbitol, m.p. 92-93°.
   Cryst. from ethanol.

Thiouracil, m.p. 240° (decomp.).
   Cryst. from water or ethanol.

Thiourea, m.p. 179°.
   Cryst. from absolute ethanol, methanol or water. Dried under vac.
over H_2SO_4.

Thioxanthene-9-one, m.p. 209°.
   Cryst. from chloroform and sublimes in vacuo.

Thiram, see Bis-(dimethylthiocarbamyl)disulphide.

S-Threonine, m.p. 251-253°, $[\alpha]_D^{26}$ -28.4° (in water).
   Likely impurities: allo-threonine, glycine. Cryst. from water by
adding 4 volumes of ethanol. Dried. Stored in desiccator.

Thymidine, m.p. 185°.
   Cryst. from ethyl acetate.

Thymine, m.p. 326°.

Cryst. from ethyl acetate or water. Purified on preparative (2 mm thick) thin layer chromatographic plate of silica gel, eluting with ethyl acetate-isopropanol-water (75:16:9, v/v ; R_f = 0.75). Spot located by uv lamp, cut from plate, placed in methanol, shaken and filtered through millipore filter, then rotary evapd. [Infante et al., JCS Faraday Trans I, 69, 1586 (1973).]

S-Thyroxine, m.p. 235°, $[\alpha]_D^{22}$ +26° (in ethanol-1 M aq. HCl; 1:1).

Likely impurities: tyrosine, iodotyrosine, iodothyroxines, iodide. Dissolved in dilute ammonia at room temperature, then cryst. by adding dil. acetic acid to pH 6.

Tiglic acid, m.p. 63.5-64°, b.p. 198.5°.

Cryst. from water.

RS-α-Tocopherol, $E_{1cm}^{1\%}$ 74.2 at 292 nm, in methanol.

Dissolved in anhydrous methanol (15 ml/g) cooled to -6° for 1 hr, then chilled in a Dry Ice-acetone bath, crystn. being induced by scratching with a glass rod.

Tolan, see Diphenylacetylene.

o-Tolidine, m.p. 131-132°.

Dissolved in benzene, percolated through a column of activated alumina and cryst. from benzene-pet.ether.

p-Tolualdehyde, b.p. 199-200°.

Steam distd., dried with $CaSO_4$ and fractionally distd.

o-Toluamide, m.p. 141°.

Cryst. from hot water (10 ml/g) and dried in air.

Toluene, b.p. 110.6°, d^{10} 0.87615, d^{25} 0.86231, n_D^{20} 1.49693, n_D^{25} 1.49413.

Dried with $CaCl_2$, CaH_2 or $CaSO_4$, and dried further by standing with sodium, P_2O_5 or CaH_2. Can be fractionally distd. from sodium or P_2O_5. Unless specially purified, toluene is likely to be contaminated with methylthiophens and other sulphur-containing impurities. These can be

Toluene (continued)

removed by shaking with conc. H_2SO_4, but the temperature must be kept below 30° if sulphonation of the toluene is to be avoided. A typical procedure consists of shaking toluene twice with cold conc. H_2SO_4 (100 ml of acid per l.), once with water, once with aq. 5% $NaHCO_3$ or NaOH, again with water, then drying successively with $CaSO_4$ and P_2O_5, with final distn. from P_2O_5. Alternatively, the treatment with $NaHCO_3$ can be replaced by a boiling under reflux with 1% sodium amalgam. Sulphur compounds can also be removed by prolonged shaking of the toluene with mercury, or by two distns. from $AlCl_3$, the distillate then being washed with water, dried with K_2CO_3 and stored with sodium wire. Other purification procedures include refluxing and distn. of sodium-dried toluene from diphenylpicrylhydrazyl, and from $SnCl_2$ (to ensure freedom from peroxides). It has also been co-distd. with 10% by volume of methyl ethyl ketone, and again fractionally distd. [Brown and Pearsall, JACS, 74, 191 (1952).] For removal of carbonyl-containing impurities, see Benzene.

Toluene-2,4-diamine, m.p. 99°, b.p. $148-150^{\circ}/8$ mm, $292^{\circ}/760$ mm.

Recryst. from water containing a very small amount of sodium dithionate (to prevent air oxidation), and dried under vac.

o-Toluenesulphonamide, m.p. 155.5°.
p-Toluenesulphonamide, m.p. 138°.

Cryst. from hot water, then from ethanol.

p-Toluenesulphonic acid, m.p. 38° (anhydr.), m.p. $105-107^{\circ}$ (mono-hydrate).

Purified by pptn. from a satd. aq. soln. at 0° by introducing HCl gas. Also cryst. from conc. HCl, then cryst. from dil. HCl (charcoal) to remove benzenesulphonic acid. Also cryst. from ethanol-water. Dried in a vac. desiccator over solid KOH and $CaCl_2$. p-Toluenesulph-onic acid can be dehydrated by azeotropic distn. with benzene or by heating at 100° for 4 hr under water-pump vac. The anhydrous acid can be cryst. from benzene, $CHCl_3$, ethyl acetate, anhydrous methanol, or from acetone by adding a large excess of benzene. It can be dried under vac. at 50°.

p-Toluenesulphonyl chloride, m.p. 69°, b.p. 146°/15 mm.

Cryst. from benzene-pet.ether in the cold, from pet.ether (b.p. 40-60°) or benzene. Its soln. in ethyl ether has been washed with aq. 10% NaOH until colourless, then dried with Na_2SO_4 and cryst. by cooling in powdered Dry Ice. It has also been purified by dissolving in benzene, washing with aq. 5% NaOH, then dried with K_2CO_3 or $MgSO_4$, and distd. under reduced pressure.

α-Toluenethiol, see Benzylmercaptan.

p-Toluenethiol, m.p. 43.5-44°.
  Cryst. from pet.ether (b.p. 40-70°).

Toluhydroquinone, m.p. 128-129°.
  Cryst. from ethanol.

o-Toluic acid, m.p. 102-103°.
  Cryst. from benzene (2.5 ml/g) and dried in air.

 m-Toluic acid, m.p. 111-113°.
  Cryst. from water.

p-Toluic acid, m.p. 178.5-179.5°.
  Cryst. from water, water-ethanol (1:1), methanol-water or benzene.

o-Toluidine, f.p. -16.3°, b.p. 200.3°, n_D^{20} 1.57246, n_D^{25} 1.56987.

In general, methods similar to those for purifying aniline can be used, e.g. distn. from zinc dust, at reduced pressure, under nitrogen. Berliner and May [JACS, 49, 1007 (1927)] purified via the oxalate. Twice-distd. o-toluidine was dissolved in four times its volumes of ethyl ether and the equivalent amount of oxalic acid needed to form the dioxalate was added as its soln. in ethyl ether. (If p-toluidine is present, its dioxalate pptes. and can be removed by filtration.) Evapn. of the ether soln. gave crystals of o-toluidine dioxalate. They were filtered off, recryst. five times from water containing a small amount of oxalic acid (to prevent hydrolysis), then treated with dil. aq. Na_2CO_3 to liberate the amine which was separated, dried ($CaCl_2$) and distd. under reduced pressure.

m-Toluidine, f.p. -30.4°, b.p. 203.4°, n_D^{20} 1.56811, n_D^{25} 1.56570.

Can be purified as for Aniline. Twice-distd. m-toluidine was converted to the hydrochloride using a slight excess of HCl, and the
salt was fractionally cryst. from 95% ethanol (five times), and from
distd. water (twice), rejecting, in each case, the first material
that crystd. The amine was regenerated and distd. as for o-Toluidine.
[Berliner and May, JACS, 49, 1007 (1927).]

p-Toluidine, m.p. 44.8°, b.p. 200.6°, $n^{59.1}$ 1.55324.

In general, methods similar to those for purifying aniline can be
used. It can be separated from the o- and m-isomers by fractional
crystn. from its melt. p-Toluidine has been cryst. from hot water
(charcoal), ethanol, benzene, pet.ether or ethanol-water (1:4), and
dried in a vac. desiccator. It can also be sublimed at 30° under vac.
For further purification, use has been made of the oxalate, the
sulphate and acetylation. The oxalate, formed as described for
o-toluidine, was filtered, washed and recryst. three times from hot
distd. water. The base was regenerated with aq. Na_2CO_3 and recryst.
three times from distd. water. [Berliner and May, JACS, 49, 1007
(1927).] Alternatively, p-toluidine was converted to its acetyl
derivative which, after repeated crystn. from ethanol, was hydrolyzed
by refluxing (50 g) in a mixture of 500 ml of water and 115 ml of
conc. H_2SO_4 until a clear soln. was obtained. The amine sulphate was
isolated, suspended in water, and NaOH was added. The free base was
distd. twice from zinc under vac. The p-toluidine was then recryst.
from pet.ether and dried in a vac. desiccator. [Berliner and
Berliner, JACS, 76, 6179 (1954).]

Toluidine blue.

Cryst. from hot water (18 ml/g) by adding one and a half volumes
of alcohol and chilling on ice.

p-Toluidine hydrochloride, m.p. 245.9-246.1°.

Cryst. from methanol containing a few drops of conc. HCl. Dried
under vac. over paraffin chips.

2-p-Toluidinylnaphthalene-6-sulphonate.

Cryst. twice from 2% aq. KOH and dried under high vac. for 4 hr at room temperature. Tested for purity by thin-layer chromatography on silica gel with isopropanol as solvent.

o-Tolunitrile, b.p. 205.2°, n$_D^{20}$ 1.5279.

Fractionally distd., washed with conc. HCl or 50% H$_2$SO$_4$ at 60° until the smell of isonitrile had gone (this also removed any amines), then washed with satd. NaHCO$_3$ and dilute NaCl solns., then dried with K$_2$CO$_3$ and redistd.

m-Tolunitrile, b.p. 209.5-210°/773 mm, n$_D^{20}$ 1.5250.

Dried with MgSO$_4$, fractionally distd., then washed with aq. acid to remove possible traces of amines, dried and redistd.

p-Tolunitrile, m.p. 29.5°.

Melted, dried with MgSO$_4$, fractionally cryst. from its melt, then fractionally distd. under reduced pressure in a 6-in. spinning-band column. [Brown, JACS, 81, 3232 (1959).] Can also be cryst. from benzene-pet.ether (b.p. 40-60°).

p-Toluquinone, m.p. 68°.

Cryst. from ethanol, dried rapidly and stored under vac.

p-Toluyl-o-benzoic acid, m.p. 138-139°.

Cryst. from toluene.

p-Tolylacetic acid, m.p. 90.8-91.3°.

Cryst. from water.

4-o-Tolylazo-o-toluidine, see 2-Amino-5-azotoluene.

p-Tolyl carbinol, m.p. 61°, b.p. 116-118°/20 mm, 217°/760 mm.

Cryst. from pet.ether (b.p. 80-100°, 1 g/g). Can also be distd. under reduced pressure.

Tolyl diphenyl phosphate, n$_D^{25}$ 1.5758.

Vac. distd., then percolated through a column of alumina. Finally, passed through a packed column maintained at 150° to remove traces

Tolyl diphenyl phosphate (continued)
of volatile impurities in a countercurrent stream of nitrogen at
reduced pressure. [Dobry and Keller, JPC, 61, 1448 (1947).]

p-Tolyl disulphide, m.p. 45-46°.
   Purified by chromatography on alumina using hexane as eluant, then
cryst. from methanol. [Kice and Bowers, JACS, 84, 2384 (1962).]

p-Tolylsulphonylmethylnitrosamide, m.p. 62°.
   Cryst. from benzene-pet.ether.

p-Tolylurea, m.p. 181°.
   Cryst. from ethanol-water (1:1).

Traumatic acid, m.p. 165-166° (trans).
   Cryst. from ethanol or acetone.

α,α'-Trehalose, m.p. 96.5-97.5° (dihydrate), 203° (anhydr.).
   Cryst. (as the dihydrate) from aq. ethanol. Dried at 130°.

TREN, see Tris(2-aminoethyl)amine.

1,2,3-Triaminopropane trihydrochloride, m.p. 250°.
   Cryst. from aq. alcohol.

1,2,4-Triazole, m.p. 121°.
   Cryst. from aq. ethanol.

Tribenzylamine, m.p. 93-94°.
   Cryst. from absolute ethanol or pet.ether. Dried under vac. over
P_2O_5 at room temperature.

2,4,6-Tribromoacetanilide, m.p. 232°.
   Cryst. from ethanol.

2,4,6-Tribromoaniline, m.p. 120°.
   Cryst. from methanol.

sym-Tribromobenzene, m.p. 122°.

Cryst. from glacial acetic acid-water (4:1), then washed with chilled ethanol and dried in air.

Tribromochloromethane, m.p. 55°.

Melted, washed with aq. $Na_2S_2O_3$, dried with BaO and fractionally cryst. from its melt.

2,4,6-Tribromophenol, m.p. 94°.

Cryst. from ethanol or pet.ether. Dried under vac. over P_2O_5 at room temperature.

Tri-n-butylamine, b.p. 68°/3 mm, 120°/44 mm, d_4^{20} 0.7788, n_D^{20} 1.4294.

Purified by fractional distn. from sodium under reduced pressure. Pegolotti and Young [JACS, 83, 3251 (1961)] heated the amine over-night with an equal volume of acetic anhydride, on a steam-bath. The amine layer was separated and heated with water for 2 hr on the steam-bath (to hydrolyze any remaining acetic anhydride). The soln. was cooled, solid K_2CO_3 was added to neutralize any acetic acid that had been formed, and the amine was separated, dried (K_2CO_3) and distd. at 44 mm pressure. Davis and Nakshbendi [JACS, 84, 2085 (1962)] treated the amine with one-eighth of its weight of benzenesulphonyl chloride in aq. 15% NaOH at 0-5°. The mixture was shaken intermitt-ently and allowed to warm to room temperature. After a day, the amine layer was washed with aq. NaOH, then water and dried with KOH. It was further dried with CaH_2 and distd. under vac.

Tri-n-butylamine hydrobromide, m.p. 75.2-75.9°.

Cryst. from ethyl acetate.

sym-Tri-tert.-butylbenzene, m.p. 73.4-73.9°.

Cryst. from ethanol.

2,4,6-Tri-tert.-butylphenol, m.p. 131°, b.p. 103°/1 mm, 147°/10 mm, 278°/760 mm.

Cryst. from n-hexane.

Tributyl phosphate, b.p. 121-124°/3 mm, n_D^{25} 1.4222.

Washed successively with 0.2 M HNO_3 (three times), 0.2 M NaOH (three times) and water (three times), then fractionally distd. under vac. [Yoshida, J. Inorg. Nuclear Chem., 24, 1257 (1962).] Has also been purified via its uranyl nitrate addition compound, obtained by saturating the crude phosphate with uranyl nitrate. This compound was cryst. three times from n-hexane by cooling to -40°, and then decomposed by washing with Na_2CO_3 and water. Hexane was removed by steam distn. and the water was then evapd. under reduced pressure. [Siddall and Dukes, JACS, 81, 790 (1959).]

Tricarballylic acid, m. p. 166°.
    Cryst. from ethyl ether.

Trichloroacetamide, b.p. 85.7°.
    Its xylene soln. was dried with P_2O_5, then fractionally distd.

Trichloroacetanilide, m.p. 95°.
    Cryst. from benzene.

Trichloroacetic acid, m.p. 59.4-59.8°.
    Purified by fractional crystn. from its melt, then cryst. repeatedly from dry benzene and stored over conc. H_2SO_4 in a vac. desiccator. Can also cryst. from $CHCl_3$ or cyclohexane, and dry over P_2O_5 or $Mg(ClO_4)_2$ in a vac. desiccator. Trichloroacetic acid can be fractionally distd. under reduced pressure from $MgSO_4$. Layne, Jaffé and Zimmer [JACS, 85, 435 (1963)] dried trichloroacetic acid in benzene by distilling off the benzene-water azeotrope, then cryst. the acid from the remaining benzene soln. Manipulations were carried out under dry nitrogen.

2,3,4-Trichloroaniline, m.p. 67.5°, b.p. 292°/774 mm.
2,4,5-Trichloroaniline, m.p. 96.5°, b.p. 270°/760 mm.
2,4,6-Trichloroaniline, m.p. 78.5°, b.p. 262°/746 mm.
    Cryst. from ligroin.

1,2,3-Trichlorobenzene, m.p. 52.6°.
    Cryst. from ethanol.

1,2,4-Trichlorobenzene, m.p. 17°, b.p. 210°.

Separated from a mixture of isomers by washing with fuming H_2SO_4, then water, drying with $CaSO_4$ and slowly fractionally distilling. [Jensen, Marino and Brown, JACS, 81, 3303 (1959).]

1,1,1-Trichloro-2,2-bis(p-chlorophenyl)ethane, m.p. 108.5-109°.
Cryst. from 95% ethanol.

3,4,5-Trichloro-o-cresol, m.p. 77°.
2,3,5-Trichloro-p-cresol, m.p. 66-67°.
Cryst. from pet.ether.

1,1,1-Trichloroethane, f.p. -32.7°, b.p. 74.0°, n_D^{20} 1.4385.
1,1,2-Trichloroethane, f.p. -36.3°, b.p. 113.6°.

Washed successively with conc. HCl (or conc. H_2SO_4), aq. 10% K_2CO_3 (or Na_2CO_3), aq. 10% NaCl, dried with $CaCl_2$ or Na_2SO_4, and fractionally distd.

Trichloroethylene, f.p. -88°, b.p. 87.2°, $n_D^{21.7}$ 1.4767.

Undergoes decomposition in a similar way to $CHCl_3$, giving HCl, CO, $COCl_2$ and organic products. It reacts with KOH, NaOH and 90% H_2SO_4, and forms azeotropes with water, methanol, ethanol and acetic acid. Purified by washing successively with 2 M HCl, water and 2 M K_2CO_3, then dried with K_2CO_3 and $CaCl_2$, and fractionally distd. immediately before use. It has also been steam distd. from 10% $Ca(OH)_2$ slurry, most of the water being removed from the distillate by cooling to -30° to -50° and filtering off the ice through chamois skin: the trichloroethylene was then fractionally distd. at 250 mm pressure and collected in a blackened container. [Carlisle and Levine, IEC, 24, 1164 (1932).]

2,4,5-Trichloro-1-nitrobenzene, m.p. 57°.
Cryst. from ethanol.

3,4,6-Trichloro-2-nitrophenol, m.p. 92-93°.
Cryst. from pet.ether.

2,4,5-Trichlorophenol, m.p. 67°.
Cryst. from ethanol or pet.ether.

2,4,6-Trichlorophenol, m.p. 67-68°.
  Cryst. from benzene, ethanol or ethanol-water.

3,4,5-Trichlorophenol, m.p. 100°.
  Cryst. from pet.ether-benzene mixture.

(2,4,5-Trichlorophenoxy) acetic acid, m.p. 153°.
  Cryst. from benzene.

1,1,1-Trichloro-2-(2,2,2-trichloro-1-hydroxyethoxy)-2-methylpropane,
see Chloralacetone chloroform.

1,1,2-Trichlorotrifluoroethane, b.p. 47.6°/760 mm.
  Washed with water, then with weak alkali. Dried with $CaCl_2$ or
H_2SO_4 and distd. [Locke et al., JACS, 56, 1726 (1934).]

Tridecanoic acid, m.p. 44.5-45.5°, b.p. 199-200°/24 mm.
  Cryst. from acetone.

7-Tridecanone, m.p. 33°, b.p. 255°/766 mm.
  Cryst. from ethanol.

TRIEN, see Triethylenetetramine.

Triethanolamine hydrochloride, m.p. 177°.
  Cryst. from ethanol. Dried at 80°.

1,1,1-Triethoxyethane, b.p. 144°.
  Dried with Na_2SO_4 and distd.

Triethoxyphosphine oxide, b.p. 215°/760 mm, 146°/112 mm,
98-98.5°/8-10 mm, d_0^{19} 1.0725, $n_D^{17.1}$ 1.40674.
  Dried over Drierite, and distd. under reduced pressure.

Triethylamine, b.p. 89.4°, d_4^{20} 0.7280, n_D^{20} 1.4005.
  Dried with $CaSO_4$, $LiAlH_4$, Linde type 4A molecular sieve, CaH_2,
KOH or K_2CO_3, then distd., either alone or from BaO, sodium, P_2O_5
or CaH_2. It has also been distd. from zinc, under reduced pressure,

Triethylamine (continued)

under nitrogen. To remove traces of primary and secondary amines,
triethylamine has been refluxed with acetic anhydride, benzoic
anhydride or phthalic anhydride, then distd., refluxed with CaH_2
(ammonia-free) or KOH (or dried with activated alumina), and again
distd. Another purification involved refluxing for 2 hr with
p-toluenesulphonyl chloride, then distn. Grovenstein and Williams
[JACS, 83, 412 (1961)] treated triethylamine (500 ml) with benzoyl
chloride (30 ml), filtered off the ppte., and refluxed the liquid
for 1 hr with a further 30 ml of benzoyl chloride. After cooling, the
liquid was filtered, distd., and allowed to stand for several hours
with KOH pellets. It was then refluxed with, and distd. from,
stirred molten potassium. Triethylamine has been converted to its
hydrochloride, cryst. from ethanol (to m.p. 254°), then liberated
with aq. NaOH, dried with solid KOH and distd. from sodium under
nitrogen.

1,2,4-Triethylbenzene, b.p. 96.8-97.1°/12.8 mm, n_D^{20} 1.4951.

sym-Triethylbenzene, b.p. 102-102.5°/17 mm, n_D^{20} 1.5015.

 For separation from a commercial mixture, see Dillingham and Reid,
JACS, 60, 2606 (1938).

Triethylenediamine, m.p. 156-157° (sealed tube).

 Cryst. from 95% ethanol, pet.ether or methanol-ethyl ether (1:1).
Dried under vac. over $CaCl_2$ and BaO. Can be sublimed under vac.

Triethylene glycol, b.p. 115-117°/0.1 mm, 278.3°/760 mm, d_4^{15} 1.1274,
n_D^{15} 1.4578.

 Dried with $CaSO_4$ for 1 week, then repeatedly and very slowly fractionally distd. under vac. Stored in a vac. desiccator over P_2O_5.
It is very hygroscopic.

Triethylene glycol dimethyl ether, b.p. 225°.

 Refluxed with, and distd. from, sodium hydride.

Triethylenetetramine (TRIEN), b.p. 157°/20 mm.

 Dried with sodium, then distd. under vac. Further purification has
been via the nitrate or the chloride. For example, Jonassen and

Triethylenetetramine (TRIEN) (continued)

Strickland [JACS, 80, 312 (1958)] separated TRIEN from admixture
with TREN (38%) by soln. in ethanol, cooling to approximately 5° in
an ice-bath and adding conc. HCl dropwise from a burette, keeping
the temperature below 10°, until all of the white crystalline ppte.
of TREN.HCl had formed and been removed. Further addition of HCl
then pptd. thick creamy white TRIEN.HCl which was cryst. several
times from hot water by adding an excess of cold ethanol. The crystals were finally washed with acetone, then ether and dried in a vac.
desiccator.

Triethylenetetramine tetrahydrochloride, m.p. $266-270^{\circ}$.

   Cryst. repeatedly from hot water by pptn. with cold ethanol or
from ethanolic-HCl. Washed with acetone and absolute ethanol and
dried in a vac. oven at 90°.

Tri-(2-ethylhexyl)phosphate, b.p. $219^{\circ}/5$ mm, d_4^{25} 0.92042, n_D^{20}
1.44464.

   TEHP, in an equal volume of ethyl ether, was shaken with aq. 5%
HCl and the organic phase was filtered to remove traces of pyridine
(used as a solvent during manufacture) as its hydrochloride. This
layer was shaken with aq. Na_2CO_3, then water and the ether was distd.
off at room temperature. The ester was filtered, dried for 12 hr at
$100^{\circ}/15$ mm, and again filtered, then shaken intermittently for 2 days
with activated alumina (100 g/l.). It was decanted through a fine
sintered-glass disc (with exclusion of moisture), and distd. under
vac. [French and Muggleton, JCS, 5064 (1957).] Benzene can be used
as a solvent (to give 0.4 M soln.) instead of ether.

Triethyl phosphate, b.p. $98-98.5^{\circ}/8-10$ mm, $161^{\circ}/188$ mm.

   Dried by refluxing with solid BaO and fractionally distd. under
vac. Stood with sodium and distd., then stored in a receiver protected from light and moisture.

Triethyl phosphite, b.p. $52^{\circ}/12$ mm, d_4^{25} 0.9610, n_D^{25} 1.4108.

   Treated with sodium (to remove water and any dialkyl phosphonate),
then decanted and distd. under reduced pressure, with protection
against moisture.

Triethyl silane.
   Refluxed over molecular sieve, then redistd.

Trifluoroacetic acid, f.p. -15.3°, b.p. 72.4°, n_D^{20} 1.2850.
   Refluxed over KMnO$_4$ for 24 hr. Slowly distd. Made 0.05% in trifluoroacetic anhydride (to diminish water content) and distd.
[Conway and Novak, JPC, 81, 1459 (1977).] Can also be refluxed with,
and distd. from, P$_2$O$_5$. Further purified by fractional crystn. by
partial freezing and again distd.

Trifluoroacetic anhydride, b.p. 38-40°/760 mm.
   Distd. from KMnO$_4$, then slowly distd. at 39.5°.

1,1,1-Trifluoro-2-bromoethane.
   Washed with water, dried, and distd.

Trifluoroethanol, b.p. 72.4°/738 mm.
   Dried with CaSO$_4$ and a little NaHCO$_3$ (to remove traces of acid).

3-Trifluoromethyl-4-nitrophenol, m.p. 81°.
   Cryst. from benzene or from pet.ether-benzene mixture.

α,α,α-Trifluorotoluene, b.p. 102.5°, d_4^{30} 1.1739, n_D^{30} 1.4100.
   Purified by repeated treatment with boiling aq. Na$_2$CO$_3$ (until no
test for chloride ion was obtained), dried with K$_2$CO$_3$, then with
P$_2$O$_5$ and fractionally distd.

Triglycyl glycine.
   Cryst. from distd. water (optionally, by the addition of ethanol).

Triglyme. Decomp. >220°.
   Distd. from LiAlH$_4$.

Trigonelline, m.p. 218° (decomp.).
   Cryst. (as monohydrate) from aq. ethanol, then dried at 100°.

2,3,4-Trihydroxybenzoic acid, m.p. 207-208°.
2,4,6-Trihydroxybenzoic acid, m.p. 205-212° (decomp.).
   Cryst. from water.

4',5,7-Trihydroxyflavone, m.p. 345-350°.
  Cryst. from aq. pyridine.

3,4,5-Triiodobenzoic acid, m.p. 289-290°.
  Cryst. from aq. ethanol.

3,4,5-Triiodobenzyl chloride, m.p. 138°.
  Cryst. from CCl_4-pet.ether (charcoal).

3,3',5-Triiodo-S-thyronine, m.p. 236-237° (decomp.), $[\alpha]_D^{29.5}$ +21.5°
(in ethanol-1 M aq. HCl, 2:1).
  Likely impurities: as in thyroxine. Dissolved in dil. ammonia at
room temperature, then cryst. by adding dil. acetic acid to pH 6.

Triisoamyl phosphate, b.p. 143°/3 mm.
Triisobutyl phosphate, b.p.119-129°/8-12 mm, 192°/760 mm.
  Purified by repeated crystn. from hexane of its addition compound
with uranyl nitrate. (See Tributyl phosphate.)[Siddall, JACS, 81,
4176 (1959).]

Triisooctyl thiophosphate.
  Purified by passage of its soln. in CCl_4 through a column of
activated alumina.

Triisopropyl phosphite, b.p. 58-59°/7 mm, n_D^{25} 1.4082.
  Distd. from sodium, under vac., through a column packed with glass
helices. (This removes any dialkyl phosphonate.)

1,2,3-Triketohydrindene hydrate, see Ninhydrin.

Trimellitic acid, m.p. 218-220°.
  Cryst. from acetic acid or aq. ethanol.

Trimethallyl phosphate, b.p. 134.5-140°/5 mm, n_D^{25} 1.4454.
  Purified as for Triisoamyl phosphate.

Trimethylacetic acid, see Pivalic acid.

Trimethylamine, b.p. 3.5°.

Dried by passage of the gas through a tower filled with solid KOH. Water and impurities containing labile hydrogen were removed by treatment with freshly sublimed, ground P_2O_5. Has been refluxed with acetic anhydride, and then distd. through a tube packed with HgO and BaO. [Comyns, JCS, 1557 (1955).] For more extensive purification, trimethylamine has been converted to the hydrochloride, cryst., and regenerated by treating the hydrochloride with excess aq. 50% KOH, the gas passing through a $CaSO_4$ column into a steel cylinder containing sodium ribbon. After 1-2 days, the cylinder was cooled to -78° and hydrogen and air were removed by pumping. [Day and Felsing, JACS, 72,1698 (1950).]

Trimethylamine hydriodide, m.p. 263°.

Cryst. from methanol.

Trimethylamine hydrochloride.

Cryst. from $CHCl_3$, ethanol or n-propyl alcohol, and dried under vac. Has also been cryst. from benzene-methanol and dried under vac. over paraffin. It is hygroscopic.

2,4,6-Trimethylbenzoic acid, m.p. 155°.

Cryst. from water, or ligroin.

Trimethylbenzyl ammonium hydroxide (Triton B).

Prepared anhydrous by prolonged drying over P_2O_5 in a vac. desiccator.

2,2,5-Trimethylhexane, b.p. 124.1°, n_D^{20} 1.39971, n_D^{25} 1.39727.

Extracted with conc. H_2SO_4, washed with water, dried and fractionally distd.

Trimethylolethane, see 2-Hydroxymethyl-2-methylpropane-1,3-diol.

Trimethylolpropane, m.p. 57-59°.

Cryst. from acetone and ether.

2,2,3-Trimethylpentane, b.p. 109.8°, n_D^{20} 1.40295, n_D^{25} 1.40064.

Purified by azeotropic distn. with 2-methoxyethanol, which was subsequently washed out with water. The trimethylpentane was then dried and fractionally distd. [Forziati et al., J. Res. Nat. Bur. Stand., 36, 129 (1946).]

2,2,4-Trimethylpentane, b.p. 99.2°, n_D^{20} 1.39145, n_D^{25} 1.38898,

Distd. from sodium, passed through a column of silica gel or activated alumina (to remove traces of olefines), and again distd. from sodium. Extracted repeatedly with conc. H_2SO_4, then agitated with aq. $KMnO_4$, washed with water, dried ($CaSO_4$) and distd. Purified by azeotropic distn. with ethanol, which was subsequently washed out with water, and the trimethylpentane was dried and fractionally distd. [Forziati et al., J. Res. Nat. Bur. Stand., 36, 129 (1946).] Also purified by fractional crystn.

2,3,5-Trimethylphenol, m.p. 95-96°, b.p. 233°/760 mm.
  Cryst. from water or pet.ether.

2,4,5-Trimethylphenol, m.p. 70.5-71.5°.
  Cryst. from water.

2,4,6-Trimethylphenol, m.p. 69°, b.p. 220°/760 mm.
  Cryst. from water and sublimes in vacuo.

3,4,5-Trimethylphenol, m.p. 107°, b.p. 248-249°/760 mm.
  Cryst. from pet.ether.

Trimethylphenylammonium benzenesulphonate.
  Cryst. repeatedly from methanol (charcoal).

2,2,4-Trimethyl-6-phenyl-1,2-dihydroquinoline, m.p. 102°.
  Vac. distd., then cryst. from absolute ethanol.

Trimethyl phosphite, b.p. 22°/23 mm, 111-112°/760 mm, n_D^{20} 1.4095.
  Treated with sodium (to remove water and any dialkyl phosphonate), then decanted and distd. with protection against moisture.

<u>2,4,6-Trimethylpyridine</u>, b.p. 60.7°/13 mm, 170.4°/760 mm, d_4^{25} 0.9100, n_D^{20} 1.4981, n_D^{25} 1.4959.

Commercial samples may be grossly impure, likely contaminants including 3,5-dimethylpyridine, 2,3,6-trimethylpyridine and water. Brown, Johnson and Podall [<u>JACS</u>, <u>76</u>, 5556 (1954)] fractionally distd. 2,4,6-trimethylpyridine at reduced pressure through a 40-cm Vigreux column and added to 430 ml of the distillate slowly, with cooling to 0°, 45 g of BF_3-ethyl etherate. The mixture was again distd., and an equal volume of dry benzene was added to the distillate. Dry HCl was passed into the soln., which was kept cold in an ice-bath, and the hydrochloride was filtered off. It was recryst. from absolute ethanol (1.5 ml/g) to m.p. 286-287° (sealed tube). The free base was regenerated by treatment with aq. NaOH, then extracted with benzene, dried ($MgSO_4$) and distd. under reduced pressure. Sisler <u>et al</u>. [<u>JACS</u>, <u>75</u>, 446 (1953)] pptd. trimethylpyridine as its phosphate from a soln. of the base in methanol by adding 85% H_3PO_4, shaking and cooling. The free base was regenerated as above. Garrett and Smythe [<u>JCS</u>, 763 (1903)] purified <u>via</u> the $HgCl_2$ complex.

<u>Trimethylsulphonium iodide</u>, m.p. 215-220° (decomp.).
   Cryst. from ethanol.

<u>1,3,7-Trimethyluric acid</u>, m.p. 345° (decomp.).
<u>1,3,9-Trimethyluric acid</u>, m.p. 347°.
   Cryst. from water.

<u>1,7,9-Trimethyluric acid</u>, m.p. 345°.
   Cryst. from water or ethanol or sublimes <u>in vacuo</u>.

<u>Trimyristin</u>, m.p. 56.5°.
   Cryst. from ether.

<u>Trineopentyl phosphate</u>.
   Cryst. from hexane.

<u>2,4,6-Trinitroanisole</u>, m.p. 68°.
   Cryst. from ethanol or methanol. Dried under vac.

sym-Trinitrobenzene, m.p. 122-123°.

Cryst. from glacial acetic acid, $CHCl_3$, CCl_4, ethanol, aq. ethanol or ethanol-benzene, after (optionally) heating with dil. nitric acid. Air dried. Fused and cryst. under vac.

2,4,6-Trinitrobenzoic acid, m.p. 227-228°.

Cryst. from distd. water. Dried in a vac. desiccator.

2,4,6-Trinitro-m-cresol, m.p. 107.0-107.5°.

Cryst. successively from water, aq. ethanol and benzene-cyclo-hexane, then dried at 80° for 2 hr. [Davis and Paabo, J. Res. Nat. Bur. Stand., 64 A, 533 (1960).]

2,4,7-Trinitro-9-fluorenone, m.p. 176°.

Cryst. from nitric acid-water (3:1), washed with water and dried under vac. over P_2O_5.

2,4,6-Trinitroresorcinol, m.p. 177-178°.

Cryst. from water containing HCl.

2,4,6-Trinitrotoluene, m.p. 81.0-81.5°.

Cryst. from benzene and ethanol. Then fused and allowed to cryst-allize under vac. Gey, Dalbey and Van Dolah [JACS, 78, 1803 (1956)] dissolved TNT in acetone and added to cold water (1:2:15), filtered the ppte., washed solvent-free and stirred with five parts of aq. 8% Na_2SO_3 at 50-60° for 10 min. It was filtered, washed with cold water until the effluent was colourless and air dried. The product was dissolved in five parts of hot CCl_4, washed with warm water until the washings were colourless and TNT was recovered by cooling and filtering. It was cryst. from 95% ethanol and carefully dried over H_2SO_4.

2,4,6-Trinitro-m-xylene, m.p. 182.2°.

Cryst. from methyl ethyl ketone.

Tri-n-octylamine, b.p. 365-367°/760 mm.

Converted to the amine hydrochloride etherate which was cryst. four times from ethyl ether at -30°. Neutralization of this salt

Tri-n-octylamine (continued)
regenerated the free amine. [Wilson and Wogman, JPC, 66, 1552
(1962).] Distd. at 1-2 mm pressure.

Tri-n-octylammonium chloride.
   Cryst. from ethyl ether, then from n-hexane.

Tri-n-octylphosphine oxide, m.p. 53.5°.
   Mason, McCarty and Peppard [J. Inorg. Nuclear Chem., 24, 967
(1962)] stirred an 0.1 M soln. in benzene with an equal volume of 6 M
HCl at 40° in a sealed flask for 48 hr, then washed the benzene soln.
successively with water (twice), 5% aq. Na_2CO_3 (three times) and
water (six times). The benzene and water were then evapd. under
reduced pressure at room temperature. Zingaro and White [J. Inorg.
Nuclear Chem., 12, 315 (1960)] treated a pet.ether soln. with aq.
$KMnO_4$ (to oxidize any phosphinous acids to phosphinic acids), then
with sodium oxalate, H_2SO_4 and HCl (to remove any manganese compounds).
The pet.ether soln. was slurried with activated alumina (to remove
phosphinic acids) and recryst. from pet.ether at -20°. Can also be
cryst. from absolute ethanol.

Trioxane, m.p.62°.
   Cryst. from water and dried over $CaCl_2$. Cryst. from sodium-dried
ethyl ether. Purified by zone melting.

s-Trioxane, m.p. 64°, b.p. 114.5°/759 mm.
   Cryst. from ether.

Tripalmitin, m.p. 66.4°.
   Cryst. from acetone, ether or ethanol.

Triphenylamine, m.p. 127.3-127.9°.
   Cryst. from ethanol or from benzene-absolute ethanol, then ethyl
ether. Sublimed under vac.

Triphenylcarbinol, see Triphenylmethanol.

Triphenylene, m.p. 198°.
   Cryst. by zone melting.

1,2,3-Triphenylguanidine, m.p. 144°.
Cryst. from ethanol or ethanol-water, and dried under vac.

Triphenylmethane, m.p. 92-93°.
Cryst. from ethanol or benzene (with one molecule of benzene of crystn. which is lost on exposure to air or by heating on a water bath). Can also be sublimed under vac. It can be given a preliminary purification by refluxing with tin and glacial acetic acid, then filtered hot through a glass sinter disc, and pptd. by addition of cold water.

Triphenylmethanol, m.p. 163°.
Cryst. from ethanol, CCl_4 (4 ml/g), or pet.ether (b.p. 60-70°). Dried at 90°.

Triphenyl phosphate, m.p. 49.5-50°.
Cryst. from ethanol.

Triphenylphosphine, m.p. 80-81°.
Cryst. from hexane, methanol or 95% ethanol. Dried at 65°/<1 mm over $CaSO_4$ or P_2O_5.

Triphenylphosphine oxide, m.p. 152.0°.
Cryst. from absolute ethanol. Dried in vac.

Triphenyl phosphite, b.p. 181-189°/1 mm.
Its ethereal soln. was washed successively with aq. 5% NaOH, distd. water and satd. aq. NaCl, then dried with Na_2SO_4 and distd. under vac. after evaporating the ethyl ether.

2,3,5-Triphenyltetrazolium chloride, m.p. 243° (decomp.).
Cryst. from ethanol or $CHCl_3$, and dried at 105°.

Tri-n-propylamine, b.p. 156.5°.
Dried with KOH and fractionally distd.

2,2',2"-Tripyridine, m.p. 80-82°.
Cryst. from pet.ether (b.p. 40-60°).

Tris(2-aminoethyl)amine, b.p. $114°/15$ mm, $263°/744$ mm.

For a separation from a mixture containing 62% TRIEN, see entry under Triethylenetetramine.

Tris(2-aminoethyl)amine trihydrochloride, m.p. $300°$ (decomp.).

Cryst. several times by dissolving in a minimum of hot water and precipitating with excess cold ethanol. The ppte. was washed with acetone, then ethyl ether and dried in a vac. desiccator.

Tris(2-biphenylyl)phosphate, m.p. $115.5-117.5°$.

Cryst. from methanol containing a little acetone.

Tris(2-ethylhexyl)phosphate.

Purified by partial crystn. of its addition compound with uranyl nitrate. [Siddall, JACS, 81, 4176 (1959).]

TRIS Buffer, see Trishydroxymethylaminomethane.

Trishydroxymethylaminomethane, m.p. $172°$.

Tris can ordinarily be obtained in highly pure form suitable for use as an acidimetric standard. If only impure material is available, cryst. from 20% aq. ethanol. Dry in a vac. desiccator over P_2O_5 or $CaCl_2$.

Alternatively, dissolved in twice its weight of water at $55-60°$, filtered, concd. to half its original volume and poured slowly, with stirring, into about twice the volume of ethanol. The crystals which separated on cooling to $3-4°$ were filtered off, washed with a little methanol, air dried by suction, then finely ground and dried in a vac. desiccator over P_2O_5. Has also been cryst. from water, methanol or aq. methanol.

Trishydroxymethylaminomethane hydrochloride.

Cryst. from 50% ethanol, then from 70% ethanol. Tris-hydrochloride is also available commercially in a highly pure state. Otherwise, cryst. from 50% ethanol, then from 70% ethanol, and dry below $40°$ to avoid risk of decomposition.

1,1,1-Tris(hydroxymethyl)ethane, m.p. 200°.
    Dissolved in hot tetrahydrofuran, filtered and pptd. with hexane.
Dried in vac.

N-Tris(hydroxymethyl)methyl-2-aminoethanesulphonic acid (TES).
    Cryst. from hot ethanol containing a little water.

N-Tris(hydroxymethyl)methylglycine (Tricine).
    Cryst. from ethanol and water.

Tris(hydroxymethyl)nitromethane, see 2-(Hydroxymethyl)-2-nitro-
propane-1,3-diol.

s-Trithiane, m.p. 220°.
    Cryst. from acetic acid.

Tri-p-tolylamine, m.p. 116.5°.
    Cryst. from ethanol.

Tritolyl phosphate, b.p. 232-234°/4 mm, d_4^{25} 1.16484, n_D^{20} 1.56703.
    Dried with $CaCl_2$, then distd. under vac. and percolated through a
column of alumina. Passage through a packed column maintained at
150°, with a countercurrent stream of nitrogen at reduced pressure,
removed residual traces of volatile impurities.

Triuret.
    Cryst. from aq. ammonia.

3-Tropanol, m.p. 63°, b.p. 229°/760 mm.
    Distd. in steam and cryst. from ether.

RS-Tropic acid, m.p. 118°.
    Cryst. from water or benzene.

Tropine, m.p. 63°.
    Cryst. from ethyl ether. Hygroscopic. See also 3-Tropanol.

Tropolone, m.p. 49-50°.
    Cryst. from hexane or pet.ether and sublimed at 40°/4 mm.

Tryptamine, m.p. 116°.
  Cryst. from benzene.

S-Tryptophan, m.p. 278°, $[\alpha]_D^{20}$ -33.4°(in EtOH).
  Cryst. from water-ethanol, washed with anhydrous ethyl ether and
dried at room temperature under vac. over P_2O_5.

Tryptophol, m.p. 59°.
  Cryst. from ethyl ether-pet.ether.

d-Tubocurarine chloride, m.p. 274-275°, $[\alpha]_D^{22}$ +215° (in water).
  Cryst. from water.

Turanose, m.p. 157°.
  Cryst. from water by addition of ethanol.

Tyramine, m.p. 164-165°.
  Cryst. from benzene or ethanol.

Tyramine hydrochloride, m.p. 269°.
  Cryst. from ethanol by addition of ethyl ether.

S-Tyrosine, m.p. 290-295° (decomp.), $[\alpha]_D^{25}$ -10.0° (in 5 M HCl).
  Likely impurities: S-cystine, ammonium salt. Dissolved in dil.
ammonia, then cryst. by adding dil. acetic acid to pH 5. Also cryst.
from water or ethanol-water, and dried at room temperature under vac.
over P_2O_5.

Umbelliferone, m.p. 225-228°.
  Cryst. from water.

Undecan-1-ol, m.p. 16.5°.
Undec-10-enoic acid, m.p. 25-25.5°.
  Purified by repeated fractional crystn. from its melt.

Uracil, m.p. 335° (decomp.).
  Cryst. from water.

Uramil, m.p. >400° (decomp.).
  Cryst. from water.

Urea, m.p. 132.7-132.9°.

Recryst. twice from conductivity water using centrifugal drainage and keeping the temperature below 60°. The crystals were dried under vac. at 55° for 6 hr. Levy and Magoulas [JACS, 84, 1345 (1962)] prepared a 9 M soln. in conductivity water (keeping the temperature below 25°) and, after filtering through a medium-porosity glass sinter, added an equal volume of absolute ethanol. The mixture was stood at -27° for 2-3 days and filtered cold. The ppte. was washed with a small amount of ethanol and dried in air. Crystn. from 70% ethanol between 40° and -9° has also been used. Ionic impurities have been removed by treating the conc. aq. soln. at 50° with Amberlite MB-1 cation- and anion-exchange resin, and allowing to crystallize. [Benesch, Lardy and Benesch, JBC, 216, 663 (1955).] Also cryst. from methanol and dried under vac. at room temperature.

Urea nitrate, m.p. 152° (decomp.).
    Cryst. from dil. HNO_3.

Uric acid.
    Cryst. from hot distd. water.

Uridine, m.p. 165°, $[\alpha]_D^{20}$ +4.0 (in water).
    Cryst. from aq. 75% methanol.

Uridylic acid, m.p. 198.5°.
    Cryst. from methanol.

Urocanic acid, m.p. 225°.
    Cryst. from water and dried at 100°.

Ursodeoxycholic acid, m.p. 203°.
    Cryst. from ethanol.

Usnic acid, m.p. 204°, $[\alpha]_{546}^{20}$ +630° (c = 0.7 in $CHCl_3$).
    Cryst. from acetone, methanol or benzene.

Ustilagic acid, m.p. 146-147°.
    Cryst. from ethyl ether.

Vaccenic acid, m.p. 43-44°.
  Cryst. from acetone.

n-Valeraldehyde, b.p. 103°, n_D^{25} 1.40233.
  Purified via the bisulphite complex. [Birrell and Trotman-
Dickenson, JCS, 2059 (1960).]

n-Valeramide, m.p. 115-116°.
  Cryst. from ethanol.

Valeric acid, b.p. 186.4°, n_D^{20} 1.4080.
  Water was removed by distilling, in a Vigreux column, until the
boiling point reached 183°. A few crystals of $KMnO_4$ were added, and
after refluxing, the distn. was continued. [Andrews and Keefer,
JACS, 83, 3708 (1961).]

Valeronitrile, b.p. 142.3°, n_D^{15} 1.39913, n_D^{30} 1.39307.
  Washed with half its volume of conc. HCl (twice), then with satd.
aq. $NaHCO_3$, dried with $MgSO_4$ and fractionally distd. from P_2O_5.

S-Valine, m.p. 315°, $[\alpha]_D^{20}$ 26.7° (in 6 M HCl).
  Cryst. from water by addition of ethanol.

Vanillin, m.p. 83°.
  Cryst. from water.

Veratraldehyde, m.p. 42-43°.
  Cryst. from ethyl ether, pet.ether, CCl_4 or toluene.

Veratric acid, m.p. 180-181°.
  Cryst. from aq. ethanol.

Veratrole, see o-Dimethoxybenzene.

Vicine.
  Cryst. from water or aq. 85% ethanol, and dried at 135°.

Vinyl acetate, b.p. 72.3°.
  Inhibitors such as hydroquinone, and other impurities, are removed
by drying vinyl acetate with $CaCl_2$ and fractionally distilling under

Vinyl acetate (continued)

nitrogen, then refluxing briefly with a small amount of benzoyl peroxide and again distilling under nitrogen. Stored in the dark at 0°.

Vinyl butoxyethyl ether.

Washed with aq. 1% NaOH, dried with CaH_2, then refluxed with and
distd. from, sodium.

N-Vinylcarbazole, m.p. 66°.

Cryst. repeatedly from methanol in amber glassware. Vac. sublimed.

Vinylene carbonate, m.p. 22°.

Purified by zone melting.

1-Vinylnaphthalene, b.p. $124-125^{\circ}/15$ mm.

Fractionally distd. under reduced pressure on a spinning-band
column, dried with CaH_2 and again distd. under vac. Stored in sealed
ampoules in a freezer.

2-Vinylnaphthalene, see Naphthylethylene.

2-Vinylpyridine, b.p. $79-82^{\circ}/29$ mm.

Steam distd., then dried with $MgSO_4$ and distd. under vac.

Vinyl stearate, m.p. 35°, b.p. $166^{\circ}/1.5$ mm.

Vac. distd. under nitrogen, then cryst. from acetone (3 ml/g) or
ethyl acetate, at 0°.

Vioform, see 5-Chloro-7-iodo-8-hydroxyquinoline.

Violanthrene.

Purified by vac. sublimation in a muffle furnace.

Violuric acid, see 5-Isonitrosobarbituric acid.

Visnagin, m.p. $142-145^{\circ}$.

Cryst. from water.

Vitamin-A acetate, m.p. 57°.
    Cryst. from methanol.

Vitamin-A alcohol.
    Purified by chromatography on alumina. [See Ganguly et al., Arch.
Biochem. Biophys., 38, 275 (1952).]

Vitamin-A aldehyde, m.p. 61-64°.
    Cryst. from pet.ether.

Vitamin B_{12}, darkens at 210-220° but does not melt below 300°,
$[\alpha]_{656}^{23}$ -59°.
    Cryst. from de-ionized water and dried under vac. over $Mg(ClO_4)_2$.

Vitamins D_2 and D_3.
    Converted to their 3,5-dinitrobenzoyl esters and cryst. repeatedly
from acids. The esters were then saponified and the free vitamins
were isolated. [Laughland and Phillips, AC, 28, 817 (1956).]

Vitamin K_1.
    Cryst. from acetone or ethanol at -70°.

Warfarin, m.p. 161°.
    Cryst. from ethanol.

Xanthatin, m.p. 114.5-115°.
    Cryst. from methanol.

Xanthene, m.p. 100.5°, b.p. 310-312°/760 mm.
    Cryst. from benzene or ethanol.

9-Xanthenone, see Xanthone.

Xanthine.
    Pptd. by the addition of conc. ammonia to its soln. in hot 2 M HCl
(after treatment with charcoal), then cryst. from distd. water.

Xanthone, m.p. 175.6-176.4°.

  Cryst. from ethanol (25 ml/g) and dried at 100°.

Xanthophyll, m.p. 190°.

  Cryst. from ethyl ether by addition of methanol.

Xanthopterin, m.p. >410°.

  Cryst. by acidifying an ammoniacal soln., and collecting by centrifugation followed by washing with ethanol, ether and drying at 100° in vacuo.

Xanthorhamnin, m.p. 195°, $[\alpha]_D$ +3.75° (in EtOH).

  Cryst. from a mixture of ethyl and isopropyl alcohols, air dried, then dried for several hours at 110°.

Xanthosine.

  Cryst. from ethanol or water (as dihydrate).

Xanthydrol, m.p. 123-124°.

  Cryst. from ethanol and dried at 40-50°.

Xylene (mixed isomers).

  Usual impurities are ethylbenzene, paraffins, traces of sulphur compounds and water. It is not practicable to separate the m-, and p-isomers of xylene by fractional distn., although, with a sufficiently efficient still, o-xylene can be fractionally distd. from a mixture of isomers. Purified (and dried) by fractional distn. from LiAlH$_4$, P$_2$O$_5$, CaH$_2$ or sodium. This treatment can be preceded by shaking successively with conc. H$_2$SO$_4$, water, aq. 10% NaOH, water and mercury, and drying with CaCl$_2$ for several days. Xylene can be purified by azeotropic distn. with 2-ethoxyethanol or 2-methoxyethanol, the distillate being washed with water to remove the alcohol, then dried and fractionally distd.

o-Xylene, f.p. -25.2°, b.p. 84°/14 mm, 144.4°/760 mm, d^{20} 0.88020, d^{25} 0.87596, n_D^{20} 1.50543, n_D^{25} 1.50292.

  The general purification methods listed under Xylene are applicable. Clarke and Taylor [JACS, 45, 831 (1923)] sulphonated o-xylene

o-Xylene (continued)

(4.4 kg) by stirring for 4 hr with 2.5 l. of conc. H_2SO_4 at 95°. After cooling, and separating the unsaponified material, the product was diluted with 3 l. of water and neutralized with 40% NaOH. On cooling, sodium o-xylenesulphonate separated and was recryst. from half its weight of water. [A further crop of crystals was obtained by concentrating the mother liquor to one-third of its volume.) The salt was dissolved in the minimum amount of cold water, then mixed with the same volume of conc. H_2SO_4 and heated to 110°. o-Xylene was regenerated and steam distd. It was then dried and redistd.

m-Xylene, f.p. -47.9°, b.p. 139.1°, d^{20} 0.86417, d^{25} 0.85990, n_D^{20} 1.49721, n_D^{25} 1.49464.

The general purification methods listed under Xylene are applicable. The o- and p-isomers can be removed by their selective oxidation when a m-xylene sample containing them is boiled with dil. HNO_3 (one part conc. acid to three parts water). After washing with water and alkali, the product can be steam distd., then distd. and purified by sulphonation. [Clarke and Taylor, JACS, 45, 831 (1923).] m-Xylene is selectively sulphonated when a mixture of xylenes is refluxed with the theoretical amount of 50-70% H_2SO_4 at $85-95^\circ$ under reduced pressure. By using a still resembling a Dean and Stark apparatus, water in the condensate can be progressively withdrawn while the xylene is returned to the reaction vessel. Subsequently, after cooling, then adding water, unreacted xylenes are distd. off at reduced pressure. The m-xylene sulphonic acid is subsequently hydrolyzed by steam distn. up to 140°, the free m-xylene being washed, dried with silica gel and again distd.

p-Xylene, f.p. 13.3°, b.p. 138.3°, d^{20} 0.86105, d^{25} 0.85669, n_D^{20} 1.49581, n_D^{25} 1.49325.

The general purification methods listed under Xylene are applicable. p-Xylene can readily be separated from its isomers by crystn. from such solvents as methanol, ethanol, isopropanol, acetone, butanone, toluene, pentane or pentene. It can be further purified by fractional crystn. by partial freezing.

Xylenol, see Dimethylphenol.

Xylenol orange.

Purified by ion-exchange chromatography using DEAE-cellulose, eluting with 0.1 M NaCl soln. Cresol red, semixylenol orange and iminodiacetic acid bands elute first. [Sato, Yokoyama and Momoki, Anal. Chim. Acta, 94, 317 (1977).]

Xylidine, see Dimethylaniline.

α-D-Xylose, m.p. 146-147°, $[\alpha]_D^{20}$ -18.8° (c = 4, in water).

Purified by slow crystn. from aq. 80% ethanol or ethanol, then dried at 60° under vac. over P_2O_5. Stored in a vac. desiccator over $CaSO_4$.

α-Yohimbine, m.p. 278° (decomp.).

Cryst. from ethanol, and dried to remove ethanol of crystn.

δ-Yohimbine, see Ajmalicine.

CHAPTER 4

PURIFICATION OF INDIVIDUAL INORGANIC AND METAL-ORGANIC CHEMICALS

The commonest method of purification of inorganic species is by recrystallization, usually from water. However, especially with salts of weak acids or of cations other than the alkaline and alkaline earth metals, care must be taken to minimize the effect of hydrolysis. This can be achieved, for example, by crystallizing acetates in the presence of dilute acetic acid. Nevertheless, there are many inorganic chemicals that are too insoluble or are hydrolyzed by water so that no general purification method can be given. It is convenient that many inorganic substances have large temperature coefficients for their solubility in water, but in other cases recrystallization is still possible by partial solvent evaporation.

This chapter uses the same abbreviations, and follows the same arrangement, as the preceding one.

Acetarsol, see N-Acetyl-4-hydroxy-m-arsanilic acid.

N-Acetyl-4-hydroxy-m-arsanilic acid.
   Cryst. from water.

Alizarin red S.
   Commercial samples contain large amounts of sodium and potassium sulphate and chloride. Can be purified by passing through a Sephadex G-10 column, followed by elution with water, then 50% aq. ethanol. [King and Pruden, Analyst, 93, 601 (1968).]

Alumina (neutral).
   Stirred with hot 2 M HNO_3, either on a steam-bath for 12 hr (changing the acid every hour) or three times for 30 min, then washed with hot dist. water until the washings were at pH 4, followed by

Alumina (neutral)(continued)

three washings with hot methanol. The product was dried at 270°.
[Angyal and Young, JACS, 81, 5251 (1959).] For preparation of
alumina for chromatography, see Chapter 1.

Aluminium ammonium sulphate.

   Cryst. from warm water by cooling in ice.

Aluminium bromide, b.p. $114^{\circ}/10$ mm.

   Refluxed, and then distd. from, pure aluminium chips in a stream
of nitrogen into a flask containing more of the chips. It was then
distd. under vacuum into ampoules. [Tipper and Walker, JCS, 1352
(1959).] Anhydrous conditions are essential, and the white to very
light brown solid distillate can be broken into lumps in a dry-box
(under nitrogen). Fumes in moist air.

Aluminium tert.-butoxide.

   Cryst. from benzene. Can also sublime at 180°.

Aluminium chloride (anhydr.).

   Sublimed several times in an all-glass system under nitrogen at
30-50 mm pressure. Has also been sublimed in a stream of dry HCl and
has been subjected to a preliminary sublimation through a section of
granular aluminium metal. [For manipulative details, see Jensen,
JACS, 79, 1226 (1957).] Fumes in moist air.

Aluminium fluoride (anhydr.).

   Technical material may contain up to 15% alumina, with minor impurities such as aluminium sulphate, cryolite, silica and iron oxide.
Reagent grade AlF_3 (hydrated) contains only traces of impurities but
its water content is very variable (may be up to 40%). It can be
dried by calcining at $600-800^{\circ}$ in a stream of dry air (some hydrolysis occurs), followed by vac. distn. at low pressure in a graphite
system, heated to approximately 925° (condenser at 900°). [Henry and
Dreisbach, JACS, 81, 5274 (1959).]

Aluminium isopropoxide, m.p. 119°, b.p. $94^{\circ}/0.5$ mm, $135^{\circ}/10$ mm.

   Distd. under vac. Hygroscopic.

Aluminium nitrate, $Al(NO_3)_3.9H_2O$.

Cryst. from dil. HNO_3, and dried by passing dry nitrogen through the crystals for several hours at $40°$.

Aluminium potassium sulphate.

Cryst. from weak aq. H_2SO_4 (approx. 0.5 ml/g).

Aluminium rubidium sulphate.

Cryst. from weak aq. H_2SO_4 (approx. 2.5 ml/g).

Aluminium sulphate.

Cryst. from hot dil. H_2SO_4 (1 ml/g) by cooling in ice.

Ammonia (aqueous).

Obtained metal-free by saturating distd. water, in a cooling bath, with ammonia (tank) gas. Alternatively, can use isothermal distn. by placing a dish of conc. aq. ammonia and a dish of pure water in an empty desiccator and leaving for several days.

Ammonia (gas).

Dried by passage through porous BaO, or over alumina followed by glass wool impregnated with sodium (prepared by soaking the glass wool in a soln. of sodium in liquid ammonia, and evaporating off the ammonia). Can be rendered oxygen-free by passage through a soln. of potassium in liquid ammonia.

Ammonia (liquid).

Dried, and stored, with sodium in a steel cylinder, then distd. and condensed by means of liquid air, the non-condensable gases being pumped off.

Ammonium acetate.

Cryst. twice from anhydrous acetic acid, and dried under vac. for 24 hr at $100°$. [Proll and Sutcliffe, TFS, 57, 1078 (1961).]

Ammonium bisulphate.

Cryst. from water at room temp. (1 ml/g) by adding ethanol and cooling.

Ammonium bromide.
    Cryst. from 95% ethanol.

Ammonium chloride.
    Cryst. several times from conductivity water (1.5 ml/g) between 90°
and 0°.

Ammonium chromate.
    Cryst. from weak aq. ammonia (about 2.5 ml/g) by cooling from room
temp.

Ammonium dichromate.
    Cryst. from weak aq. HCl (about 1 ml/g).

Ammonium dihydrogen arsenate.
    Cryst. from water (about 1 ml/g).

Ammonium dihydrogen orthophosphate.
    Cryst. from water (0.7 ml/g) between 100° and 0°.

Ammonium fluoroborate.
    Cryst. several times from conductivity water (1 ml/g) between 100°
and 0°.

Ammonium fluorosilicate.
    Cryst. from water (2 ml/g).

Ammonium hexachloroiridate (IV).
    Pptd. several times from aq. soln. by satn. with ammonium chloride.
(This removes any palladium or rhodium.) Then washed with ice-cold
water and dried over conc. H_2SO_4 in a vac. desiccator. If osmium or
ruthenium is present, it can be removed as the tetroxide by heating
with conc. nitric acid, followed by conc. $HClO_4$, until most of the
acid has been driven off. (This treatment is repeated.) The near-dry
residue is dissolved in a small amount of water and added to excess
$NaHCO_3$ soln. and bromine water. On boiling, iridic (but not platinic)
hydroxide is pptd. It is redissolved in HCl and pptd. several times,
then dissolved in HBr and treated with HNO_3 and HCl to convert the

<u>Ammonium hexachloroiridate (IV)</u> (continued)
bromides to chlorides. Satn. with ammonium chloride and cooling
precipitates ammonium hexachloroiridate which is filtered off and
purified as above. [Woo and Yost, <u>JACS</u>, <u>53</u>, 884 (1931).]

<u>Ammonium hypophosphite</u>.
   Cryst. from hot ethanol.

<u>Ammonium iodate</u>.
   Cryst. from water (8 ml/g) between 100° and 0°.

<u>Ammonium iodide</u>.
   Cryst. from ethanol by addition of ethyl iodide. Very hygroscopic.
Stored in the dark.

<u>Ammonium metavanadate</u>.
   Cryst. from conductivity water (20 ml/g).

<u>Ammonium molybdate</u>.
   Cryst. from water (2.5 ml/g) by partial evapn. in a desiccator.

<u>Ammonium nitrate</u>.
   Cryst. twice from distd. water (1 ml/g) by adding ethanol, or from
warm water (0.5 ml/g) by cooling in an ice-salt mixture. Dried in
air, then under vac.

<u>Ammonium oxalate</u>.
   Cryst. from water (10 ml/g) between 50° and 0°.

<u>Ammonium perchlorate</u>.
   Cryst. twice from distd. water (2.5 ml/g) between 80° and 0°, and
dried in a vac. desiccator over P_2O_5. (Drying at 110° might lead to
slow decomposition to chloride.) <u>Potentially explosive</u>.

<u>Ammonium reineckate</u>.
   Cryst. from water, between 30° and 0°, working by artificial light.
(Solns. of reineckate decompose slowly at room temperature in the
dark, more rapidly at higher temperatures or in diffuse sunlight.)

Ammonium selenate.
   Cryst. from water at room temperature by adding ethanol and cooling.

Ammonium sulphamate.
   Cryst. from water at room temperature (1 ml/g) by adding ethanol and cooling.

Ammonium sulphate.
   Cryst. twice from hot water containing 0.2% EDTA to remove metal ions, then finally from distd. water. Dried in a desiccator for two weeks over $Mg(ClO_4)_2$.

Ammonium tetrafluoroborate.
   Cryst. from water.

Ammonium thiocyanate.
   Cryst. three times from dil. $HClO_4$, to give material optically transparent at wavelengths longer than 270 nm. Has also been cryst. from absolute methanol and from acetonitrile.

Ammonium tungstate.
   Cryst. from warm water by adding ethanol and cooling.

n-Amylmercuric chloride, m.p. $110°$.
   Cryst. from ethanol.

Antimony trichloride, m.p. $73°$.
   Dried over P_2O_5 or by mixing with toluene or xylene and distilling (water is carried off with the organic solvent), then distd. twice under dry nitrogen at 50 mm, degassed and sublimed twice, under vac. into ampoules. Can be cryst. from CS_2. Deliquescent. Fumes in moist air.

Antimony trifluoride, m.p. $292°$.
   Cryst. from methanol to remove oxide and oxyfluoride, then sublimed under vac. in an aluminium cup on to a water-cooled copper condenser. [Woolf, _JCS_, 279 (1955).]

Antimony triiodide, m.p. 167°.
   Sublimed under vac.

Antimony trioxide.
   Dissolved in just enough dil. HCl, filtered, and six volumes of
water were added to ppte. a basic antimonous chloride (free from
iron and Sb_2O_5). The ppte. was redissolved in dil. HCl, and added
slowly, with stirring, to a boiling soln. (containing slight excess)
of Na_2CO_3. The oxide was filtered off, washed with hot water, then
boiled and filtered, the process being repeated until the filtrate
gave no test for chloride. The product was dried in a vac. desiccator.
[Schuhmann, JACS, 46, 52 (1924).]

Argon.
   Rendered oxygen-free by passage over reduced copper at 450°, then
dried with $CaSO_4$, $Mg(ClO_4)_2$, or Linde molecular sieve. Other purification steps include passage through Ascarite (asbestos impregnated
with sodium hydroxide), through finely divided uranium at about 800°
and through a -78° cold trap.
   Alternatively, passed over CuO pellets at 300° to remove hydrogen
and hydrocarbons, over calcium chips at 600° to remove oxygen and,
finally, over titanium chips at 780° to remove nitrogen. Also
purified by freeze-pump-thaw cycles and by passage over sputtered
sodium.

o-Arsanilic acid, m.p. 153°.
p-Arsanilic acid, m.p. 232°.
   Cryst. from water or ethanol-ether.

Arsenazo.
   A satd. aq. soln. of the free acid is slowly added to an equal vol.
of conc. HCl. The orange ppte. is filtered, washed with acetonitrile
and dried for 1-2 hr at 110°. [Fritz and Bradford, AC, 30, 1021 (1958).]

Arsenic.
   Heated under vac. at 350° to sublime oxides, then sealed in a Pyrex
tube under vac. and sublimed at 600°, the arsenic condensing in the
cooler portions of the tube. Stored under vac. [Shih and Peretti,
JACS, 75, 608 (1953).]

Arsenic tribromide, b.p. 89°/11 mm, 221°/760 mm.
   Distd. under vac.

Arsenic trichloride, b.p. 130.0°.
   Refluxed with arsenic for 4 hr, then fractionally distd. The middle
fraction was stored with sodium wire for two days, then again distd.
[Lewis and Sowerby, JCS, 336 (1957).]

Arsenic triiodide, m.p. 146°.
   Cryst. from acetone.

Arsenious oxide.
   Cryst. from dil. HCl (1:2), washed, dried and sublimed. Analytical
reagent grade material is suitable for use as an analytical standard
after it has been dried by heating at 105° for 1-2 hr or has been
left in a desiccator for several hours over conc. H_2SO_4.

3-(2-Arsenophenylazo)-4,5-dihydroxy-2,7-naphthalene disulphonic acid,
trisodium salt, see Arsenazo.

Barium (metal).
   Cleaned by washing with ethyl ether to remove adhering paraffin,
then filed in an argon-filled glove box, washing first with ethanol
containing 2% conc. HCl, then with dry ethanol. Dried under vac. and
stored under pure argon. [Addison, Coldrey and Halstead, JCS, 3868
(1962).] Has also been purified by double distn. under 10 mm argon
pressure.

Barium acetate.
   Cryst. twice from anhydrous acetic acid and dried under vac. for
24 hr. at 100°.

Barium bromate.
   Cryst. from hot water (20 ml/g).

Barium bromide.
   Cryst. from water (1 ml/g) by partial evapn. in a desiccator.

Barium chlorate.
   Cryst. from water (1 ml/g) between 100° and 0°.

Barium chloride.
   Twice cryst. from water (2 ml/g) and oven dried to constant weight.

Barium dithionate.
   Cryst. from water.

Barium ferrocyanide.
   Cryst. from hot water (100 ml/g).

Barium formate.
   Cryst. from warm water (4 ml/g) by adding ethanol and cooling.

Barium hydroxide.
   Cryst. from water (1 ml/g).

Barium hypophosphite.
   Pptd. from aq. soln. (3 ml/g) by adding ethanol.

Barium iodate.
   Cryst. from a large volume of hot water by cooling.

Barium iodide.
   Cryst.from water (0.5 ml/g) by partial evapn. in a desiccator.

Barium nitrate.
   Cryst. twice from water (4 ml/g) and dried overnight at 110°.

Barium nitrite.
   Cryst. from water (1 ml/g) by cooling in ice-salt.

Barium perchlorate.
   Cryst. twice from water.

Barium propionate.
   Cryst. from warm water (50 ml/g) by adding ethanol and cooling.

Barium thiocyanate.
  Cryst. from water (2.5 ml/g) by partial evapn. in a desiccator.

Benzenearsonic acid, see Phenylarsonic acid.

Benzenestibonic acid.
  Cryst. from acetic acid, or from ethanol-CHCl$_3$ mixture by addition
of water.

Beryllium acetate (basic), m.p. 285-286°.
  Cryst. from CHCl$_3$.

Beryllium potassium fluoride.
  Cryst. from hot water (25 ml/g).

Beryllium sulphate.
  Cryst. from weak aq. H$_2$SO$_4$.

2,2'-Bipyridineruthenous dichloride hexahydrate.
  Cryst. from water.

Bis(2,9-dimethyl-1,10-phenanthroline) copper(I) perchlorate.
  Cryst. from acetone.

Bis-(ethyl)titanium(IV) chloride, $(C_2H_5)_2TiCl_2$.

Bis-(ethyl)zirconium(IV) chloride, $(C_2H_5)_2ZrCl_2$.
  Cryst. from boiling toluene.

Bismuth, m.p. 271-273°.
  Melted in an atmosphere of dry helium and filtered through dry
Pyrex wool to remove any bismuth oxide present. [Mayer, Yosim and
Topol, JPC, 64, 238 (1960).]

Bismuth trichloride, m.p. 233.6^o.
  Sublimed under high vac.; or dried under a current of HCl gas,
followed by fractional distn., once under HCl and once under argon.

Borax, see Sodium borate.

Boric acid.

Cryst. three times from water (3 ml/g) between 100° and 0°, after filtering through sintered glass. Dried to constant weight over metaboric acid in a desiccator.

Boron trichloride, b.p. $0^{\circ}/476$ mm.

Purified (from chlorine) by passage through two mercury-filled bubblers, then fractionally distd. under vac. A more extensive purification forms the nitrobenzene addition compound by passage of the gas over nitrobenzene in a vac. system at 10°. Volatile impurities are removed from the crystalline yellow solid by pumping at -20°, and the BCl_3 is recovered by warming the addition compound to 50°. Passage through a trap at -78° removes entrained nitrobenzene, the BCl_3 finally condensing in a trap at -112°. [Brown and Holmes, JACS, 78, 2173 (1956).]

Boron trifluoride, b.p. $111.8^{\circ}/300$ mm.

The usual impurities - bromine, BF_5, HF and non-volatile fluorides - are readily separated by distn. Brown and Johannesen [JACS, 72, 2934 (1950)] passed BF_3 into benzonitrile at 0° until the latter was satd. Evacuation to 10^{-5} mm then removed all traces of SiF_4 and other gaseous impurities. [A small amount of the BF_3-benzonitrile addition compound sublimed and was collected in a U-tube cooled to -80°.) Pressure was raised to 20 mm by admitting dry air, and the flask containing the BF_3 addition compound was warmed with hot water. The BF_3 evolved was passed through a -80° trap (to condense any benzonitrile) into a tube cooled in liquid air. The addition compound with anisole can also be used. For drying, BF_3 can be passed through H_2SO_4 saturated with boric oxide. Fumes in moist air.

Boron trifluoride ethyl etherate, b.p. $67^{\circ}/43$ mm.

Treated with a small quantity of ethyl ether (to ensure an excess of this component), and then distd. under reduced pressure, from CaH_2. Fumes in moist air.

Bromine, b.p. 59°.

Refluxed with solid KBr and distd., dried by shaking with an equal volume of conc. H_2SO_4, then separated and distd. The H_2SO_4 treatment can be replaced by direct distn. from BaO or P_2O_5. A more extensive

Bromine (continued)

purification [Hildenbrand et al., JACS, 80, 4129 (1958)] is to reflux approx. 1 l. of bromine for 1 hr with a mixture of 16 g CrO_3 in 200 ml of conc. H_2SO_4 (to remove organic material). The bromine is distd. into a clean dry, glass-stoppered bottle, and chlorine is removed by dissolving about 25 g of freshly fused CsBr in 500 ml of the bromine and standing overnight. To remove HBr and water, the bromine was then distd. back and forth through a train containing alternate tubes of MgO and P_2O_5. Highly toxic.

Bromine pentafluoride.

Purified via its KF complex, as described for chlorine trifluoride. Highly toxic.

n-Butylmercuric chloride, m.p. 130°.

Cryst. from ethanol.

Cacodylic acid.

Cryst. from warm ethanol (3 ml/g) by cooling and filtering. Dried in vac. desiccator over $CaCl_2$.

Cadmium.

Oxide has been removed by filtering the molten metal, under vac., through quartz wool.

Cadmium acetate.

Cryst. twice from anhydrous acetic acid and dried under vac. for 24 hr at 100°.

Cadmium bromide.

Cryst. from water (0.6 ml/g) between 100° and 0°, and dried at 110°.

Cadmium chloride.

Cryst. from water (1 ml/g) by addition of ethanol and cooling.

Cadmium fluoride.

Cryst. by dissolving in water at room temp. (25 ml/g) and heating to 60°.

Cadmium iodide.
    Cryst. from ethanol (2 ml/g) by partial evapn.

Cadmium lactate.
    Cryst. from water (10 ml/g) by partial evapn. in a desiccator.

Cadmium nitrate.
    Cryst. from water (0.5 ml/g) by cooling in ice-salt.

Cadmium potassium iodide.
    Cryst. from ethanol by partial evapn.

Cadmium salicylate.
Cadmium sulphate.
    Cryst. from distd. water by partial evapn. in a desiccator.

Caesium aluminium sulphate.
    Cryst. from hot water (3 ml/g).

Caesium bromide.
    Cryst. from water (0.8 ml/g) by partial evapn. in a desiccator.

Caesium carbonate.
    Cryst. from ethanol (10 ml/g) by partial evapn.

Caesium chloride.
    Cryst. from acetone-water, or from water (0.5 ml/g) by cooling in
$CaCl_2$-ice. Dried at 78° under vac.

Caesium chromate.
    Cryst. from water (1.4 ml/g) by partial evapn. in a desiccator.

Caesium fluoride.
    Cryst. from aq. soln. by adding ethanol.

Caesium iodide.
    Cryst. from warm water (1 ml/g) by cooling to -5°.

Caesium nitrate.
   Cryst. from water (0.6 ml/g) between $100°$ and $0°$.

Caesium perchlorate.
   Cryst. from water (4 ml/g) between $100°$ and $0°$.

Caesium sulphate.
   Cryst. from water (0.5 ml/g) by adding ethanol and cooling.

Calcium (metal).
   Cleaned by washing with ether to remove adhering paraffin, then filed in an argon-filled glove box and washed with ethanol containing 2% of conc. HCl. Then washed with dry ethanol, dried in a vacuum and stored under pure argon. [Addison, Coldrey and Halstead, JCS, 3868 (1962).]

Calcium acetate.
   Cryst. from water (3 ml/g) by partial evapn. in a desiccator.

Calcium benzoate.
   Cryst. from water (10 ml/g) between $90°$ and $0°$.

Calcium bromide.
   Cryst. from ethanol or acetone.

Calcium butyrate.
   Cryst. from water (5 ml/g) by partial evapn. in a desiccator.

Calcium carbamate.
   Cryst. from aq. ethanol.

Calcium chloride (hydrate).
   Cryst. from ethanol.

Calcium formate.
   Cryst. from water (5 ml/g) by partial evapn. in a desiccator.

Calcium hydroxide.

Heat analytical grade calcium carbonate to 1000° during one hour. Allow the resulting oxide to cool and add slowly to water. Heat the suspension to boiling, cool and filter through a sintered glass funnel of medium porosity (to remove soluble alkaline impurities). Dry the solid at 110° and crush to a uniformly finely granular condition.

Calcium iodate.

Cryst. from water (100 ml/g).

Calcium iodide.

Dissolved in acetone, which was then diluted and evapd. This drying process was repeated twice, then the CaI_2 was cryst. from acetone-ethyl ether and stored over P_2O_5. Very hygroscopic when anhydrous. [Cremlyn et al., JCS, 528 (1958).]

Calcium isobutyrate.

Cryst. from water (3 ml/g) by partial evapn. in a desiccator.

Calcium lactate.

Cryst. from warm water (10 ml/g) by cooling to 0°.

Calcium nitrate.

Cryst. four times from water (0.4 ml/g) by cooling in a $CaCl_2$-ice freezing mixture. The tetrahydrate was dried over conc. H_2SO_4 and stored over P_2O_5, to give the anhydrous salt.

Calcium nitrite.

Cryst. from hot water (1.4 ml/g) by adding ethanol and cooling.

Calcium pantothenate.

Cryst. from methanol.

Calcium permanganate.

Cryst. from water (3.3 ml/g) by partial evapn. in a desiccator.

Calcium propionate.

Cryst. from water (2 ml/g) by partial evapn. in a desiccator.

Calcium salicylate.
    Cryst. from water (3 ml/g) between 90° and 0°.

(p-Carbamylphenylarsylenedithio)diacetic acid.
    Cryst. from methanol or ethanol.

Carbon dioxide. Sublimes at -78.5°.
    Passed over CuO wire at 800° to oxidize CO and other reducing
impurities (such as H_2), then over copper dispersed on kieselguhr at
180° to remove oxygen. Drying at -78° removed water vapour. Final
purification was by vac. distn. at liquid-nitrogen temperature to
remove non-condensable gases. [Anderson, Best and Dominey, JCS, 3498
(1962).]
    Sulphur dioxide can be removed at 450° using hot silver wool
combined with a plug of platinized quartz wool. Halogens are removed
by using magnesium, zinc or copper, heated to 450°.

Carbon disulphide, see entry in Chapter 3.

Carbon monoxide, b.p. -191.5°.
    Iron carbonyl is a likely impurity in CO stored under pressure in
steel tanks: it can be decomposed by passage of the gas through a
hot porcelain tube at $350-400^\circ$. Passage through alkaline pyrogallol
soln. removes oxygen (and CO_2). Removal of CO_2 and water are effected
by passage through soda-lime or asbestos impregnated with NaOH,
followed by $Mg(ClO_4)_2$. Carbon monoxide can be condensed and distd.
at -195°.

Carbon tetrachloride, see entry in Chapter 3.

Carbonyl sulphide.
    Purified by scrubbing through three consecutive fritted washing
flasks containing conc. NaOH soln. at 0°. Then freeze-pumped repeatedly and distd. through a trap packed with glass wool and cooled to
-130° (using an n-pentane slurry).

Celite 545 (diatomaceous earth).

Stood overnight in conc. HCl after stirring well, then washed with distd. water until neutral and free of chloride ions. Washed with methanol and dried at 50°.

Ceric ammonium nitrate.

Ceric ammonium nitrate (125 g) is warmed with 100 ml of dil. HNO_3 (1:3 v/v) and 40 g of NH_4NO_3 until dissolved, and filtered hot through a sintered-glass funnel. The solid which separates on cooling in ice is filtered off on a sintered funnel (at the pump) and air is sucked through the solid for 1-2 hr to remove most of the nitric acid. Finally, the solid is dried at 80-85°.

Cerous acetate.

Cryst. twice from anhydrous acetic acid, then pumped dry under vac. at 100° for 8 hr.

Chloramine-T.

Cryst. from hot water (2 ml/g). Dried in a desiccator over $CaCl_2$. (Protect from sunlight.)

Chlorine.

Passed in succession through aq. $KMnO_4$, dil. H_2SO_4, conc. H_2SO_4, and a drying tower containing $Mg(ClO_4)_2$. Or, washed with water, dried over P_2O_5 and distd. from bulb to bulb. Highly toxic.

Chlorine trifluoride, b.p. 12.1°.

Impurities include chloryl fluoride, chlorine dioxide and hydrogen fluoride. Passed first through two U-tubes containing NaF to remove HF, then through a series of traps in which the liquid is fractionally distd. Can be purified via the KF complex, $KClF_4$, formed by adding excess ClF_3 to solid KF in a stainless steel cylinder in a dry-box and shaking overnight. After pumping out the volatile materials, pure ClF_3 is obtained by heating the bomb to 100-150° and condensing the evolved gas in a -196° trap. [Schack, Dubb and Quaglino, Chem. Ind., 545 (1967).] Highly toxic.

Chloroform, see entry in Chapter 3.

p-Chloromercuribenzoate.

Its suspension in water is stirred with enough 1 M NaOH to dissolve most of it: a small amount of insoluble matter is removed by centrifugation. The chloromercuribenzoate is then pptd. by adding 1 M HCl and centrifuged off. The pptn. is repeated twice. Finally, the ppte. is washed three times with distd. water (by centrifuging), then dried in a thin layer under vac. over P_2O_5. [Boyer, JACS, 76, 4331 (1954).]

Chlorosulphonic acid.
Distd. in an all-glass apparatus, taking the fraction boiling at 156-158°.

Chromazurol S.
Purified by paper chromatography using n-butanol, glacial acetic acid and water (7:1:3). First and second spots extracted.

Chromic chloride (anhydr.).
Sublimed in a stream of dry HCl. Alternatively, the impure chromic chloride (100 g) was added to 1 l. of 10% aq. $K_2Cr_2O_7$ and several millilitres of conc. HCl, and the mixture was brought to a gentle boil with constant stirring for 10 min. (This removed a reducing impurity.) The solid was separated, and washed by boiling with successive 1 l. lots of distd. water until the wash water no longer gave a test for chloride ion, then dried at 110°. [Poulsen and Garner, JACS, 81, 2615 (1959).]

Chromium ammonium sulphate.
Cryst. from a satd. aq. soln. at 55° by cooling slowly with rapid mechanical stirring. The resulting fine crystals were filtered on a Büchner funnel, partly dried on a porous plate, then equilibrated for several months in a vac. desiccator over crude chromium ammonium sulphate (partially dehydrated by heating at 100° for several hours before use). [Johnston, Hu and Horton, JACS, 75, 3922 (1953).]

Chromium potassium sulphate.
Cryst. from hot water (2 ml/g) by cooling.

Chromium trioxide.

    Cryst. from water (0.5 ml/g) between 100° and -5°, or from water-conc. HNO_3 (1:5). Dried in a vac. desiccator over NaOH pellets.

Chromyl chloride, b.p. 115.7°.

    Purified by distn. under reduced pressure.

Cobaltous acetate.

    Cryst. several times as the tetrahydrate from 50% aq. acetic acid. Converted to the anhydrous salt by drying at 80°/1 mm for 60 hr.

Cobaltous ammonium sulphate.

    Cryst. from boiling water (2 ml/g) by cooling. Washed with ethanol.

Cobaltous bromide.

    Cryst. from water (1 ml/g) by partial evapn. in a desiccator.

Cobaltous chloride.

    A satd. aq. soln. at room temperature was fractionally cryst. by standing overnight. The first half of the material that cryst. in this way was used in the next crystn. The process was repeated several times, water being removed in a dry-box using air filtered through glass wool and dried over $CaCl_2$. [Hutchinson, JACS, 76, 1022 (1954).] Has also been cryst. from dil. aq. HCl.

Cobaltous nitrate.

    Cryst. from water (1 ml/g) or ethanol (1 ml/g), by partial evapn.

Cobaltous perchlorate.

    Cryst. from warm water (0.7 ml/g) by cooling.

Cobaltous sulphate.

    Cryst. three times from conductivity water (1.3 ml/g) between 100° and 0°.

Cupric acetate.

    Cryst. twice from warm dil. acetic acid solns. (5 ml/g) by cooling.

Cupric ammonium chloride.
   Cryst. from weak aq. HCl (1 ml/g).

Cupric benzoate.
   Cryst. from hot water.

Cupric bromide.
   Cryst. twice by dissolving in water (140 ml/g), filtering to remove any Cu_2Br_2, and concentrating under vac. at $30°$ until crystals appeared. The cupric bromide was then allowed to crystallize by leaving the soln. in a vac. desiccator containing P_2O_5. [Hope, Otter and Prue, JCS, 5226 (1960).]

Cupric chloride.
   Cryst. from hot dil. aq. HCl (0.6 ml/g) by cooling in $CaCl_2$-ice. Dehydrated by heating on a steam-bath under vac.

Cupric lactate.
   Cryst. as the monohydrate from boiling water (3 ml/g) by cooling.

Cupric nitrate.
   Cryst. from weak aq. HNO_3 (0.5 ml/g) by cooling from room temperature.

Cupric oleate.
   Cryst. from ethyl ether.

Cupric perchlorate.
   Cryst. from distd. water.

Cupric sulphate.
   After adding 0.02 g of KOH to a litre of nearly satd. aq. soln., it was left for two weeks, then the ppte. was filtered on to a fibreglass filter with pore diameter of 5-15 microns. The filtrate was heated to $90°$ and allowed to evaporate until some $CuSO_4.5H_2O$ had cryst. The soln. was then filtered hot and cooled rapidly to give crystals which were freed of mother liquor by filtering under suction. [Geballe and Giauque, JACS, 74, 3513 (1952).] Alternatively, cryst. from water (0.6 ml/g) between $100°$ and $0°$.

Cuprous chloride.

Dissolved in strong HCl, pptd. by dilution with water and filtered off. Washed with ethanol and ethyl ether, then dried and stored in a vac. desiccator. [Österlöf, Acta Chem. Scand., 4, 375 (1950).]

Decaborane, m.p. 99.7-100°.

Purified by vac. sublimation, followed by crystn. from methyl-cyclohexane, methylene dichloride, or dry olefin-free n-pentane, the solvent being subsequently removed by storing the crystals in a vac. desiccator containing $CaCl_2$.

Deuterium.

Passed over activated charcoal at -195°. [MacIver and Tobin, JPC, 64, 451 (1960).] Purified by diffusion through nickel. [Pratt and Rogers, JCS Faraday Trans I, 92, 1589 (1976).]

Deuterium oxide, f.p. 3.8°/760 mm, b.p. 101.4/760 mm.

Distd. from alkaline $KMnO_4$. [de Giovanni and Zamenhof, BJ 87, 79 (1963).]

Dextran sulphuric acid ester, sodium salt.

Cryst. from ethanol-ethyl ether.

Diammonium hydrogen orthophosphate.

Cryst. from water (1 ml/g) between 70° and 0°.

Diethyl aluminium chloride, m.p. -75.5°, b.p. 106.5-108°/24.5 mm.

Distd. from excess dry NaCl (to remove ethyl aluminium dichloride) in a 50-cm column containing a heated nichrome spiral.

Dinitrogen tetroxide, N_2O_4, m.p. -11.2°, b.p. 21.1°.

Purified by oxidation at 0° in a stream of oxygen until the blue colour changed to red-brown. Distd. from P_2O_5, then solidified on cooling in a deep-freeze (giving nearly colourless crystals). Oxygen can be removed by alternate freezing and melting.

Diphenylmercury, m.p. 125.5-126°.

Sublimed, then cryst. from nitromethane or ethanol. If phenyl-mercuric halides are present they can be converted to phenylmercuric

Diphenylmercury (continued)

hydroxide which, being much more soluble, remains in the alcohol or benzene used for crystn. Thus, crude material (10 g) is dissolved in warm ethanol (about 150 ml) and shaken with moist Ag_2O (about 10 g) for 30 min, then heated under reflux for 30 min and filtered hot. Concn. of the filtrate by evapn. gives diphenylmercury, which is recryst. from benzene. [Blair, Bryce-Smith and Pengilly, JCS, 3174 (1959).]

Diphenylsilanediol, m.p. 148° (decomp.).

Cryst. from $CHCl_3$-methyl ethyl ketone.

Disodium benzene-1,3-disulphonate.

Cryst. twice from methanol or water, and dried under vac.

Disodium calcium ethylenediaminetetraacetate.

Dissolved in a small amount of water, filtered and pptd. with excess ethanol. Dried at 80°.

Disodium dihydrogen ethylenediaminetetraacetate.

Analytical reagent grade material can be used as primary standard after drying at 80°. Commercial grade material can be purified by crystn. from water or by preparing a 10% aq. soln. at room temperature, then adding ethanol slowly until a slight permanent ppte. is formed, filtering, and adding an equal volume of ethanol. The ppte., filtered off on a sintered-glass funnel, is washed with acetone, followed by ethyl ether, and dried in air overnight to give the dihydrate. Drying at 80° for at least 24 hr converts it to the anhydrous form.

Disodium 1,8-dihydroxynaphthalene-3,6-disulphonate.

Cryst. from water.

Disodium ethylenebis[dithiocarbamate].

Cryst. (as hexahydrate) from aq. ethanol.

Disodium-β-glycerophosphate.

Cryst. from water.

Disodium hydrogen orthophosphate (anhydrous).

Cryst. twice from warm water, by cooling. Air dried, then oven dried overnight at 130°. Hygroscopic: should be dried before use.

Disodium magnesium ethylenediaminetetraacetate.

Dissolved in a small amount of water, filtered and pptd. with an excess of methanol. Dried at 80°.

Disodium phenylphosphate.

Dissolved in a minimum amount of methanol, filtering off an insoluble residue of inorganic phosphate, then pptd. by adding an equal volume of ethyl ether. Washed with ethyl ether and dried. [Tsuboi, Biochim. Biophys. Acta, 8, 173 (1952).]

Disodium succinate.

Cryst. twice from water. Freed from other metal ions by passage of an 0.1 M soln. through a column of Dowex ion-exchange resin A-1, sodium form.

Di-p-tolylmercury, m.p. 244-246°.

Cryst. from xylene.

Eosin (as disodium salt).

Dissolved in water and pptd. by addition of dil. HCl. The ppte. was washed with water, cryst. from ethanol, then dissolved in the minimum amount of dil. NaOH soln., and evapd. to dryness on a water-bath. The purified disodium salt was then cryst. twice from ethanol. [Parker and Hatchard, TFS, 57, 1894 (1961).]

Ethylarsonic acid, m.p. 99.5°.

Cryst. from ethanol.

Ethylmercuric chloride, m.p. 193-194°.

Mercuric chloride can be removed by suspending ethylmercuric chloride in hot distd. water, filtering with suction in a sintered-glass crucible and drying. Then cryst. from ethanol and sublimed under reduced pressure. Can also be cryst. from water.

Ethylmercuric iodide, m.p. 186°.

    Cryst. once from water (50 ml/g).

Ferric acetylacetonate, m.p. 179°.

    Cryst. from 95% ethanol and dried for 1 hr at 120°.

Ferric ammonium oxalate.

    Cryst. from water (0.5 ml/g) between 80° and 0°.

Ferric ammonium sulphate.

    Cryst. from aq. ethanol.

Ferric chloride (anhydr.).

    Sublimed at 200° in an atmosphere of chlorine. Stored in a weighing bottle inside a desiccator.

Ferric chloride (decahydrate).

    An aq. soln., satd. at room temperature, was cooled to -20° for several hours. Pptn. was slow, even with scratching and seeding, and it was generally necessary to stir overnight. The presence of free HCl retards the pptn. [Linke, JPC, 60, 91 (1956).]

Ferric perchlorate.

    Cryst. twice from conc. $HClO_4$, the first time in the presence of a small amount of H_2O_2 to ensure that the iron was fully oxidized. [Sullivan, JACS, 84, 4256 (1962).]

Ferrocene, m.p. 173-174°.

    Cryst. from methanol or ethanol.

Ferrous ammonium sulphate.

    A soln. in warm water (1.5 ml/g) was cooled rapidly to 0°, and the resulting fine crystals were filtered at the pump, washed with cold distd. water and pressed between sheets of filter paper to dry.

Ferrous chloride.

    A 550 ml round-bottom Pyrex flask was connected, via a glass tube fitted with a medium porosity sintered-glass disc, to a similar flask. To 240 g of $FeCl_2.4H_2O$ in the first flask was added conduct-

Ferrous chloride (continued)

ivity water (200 ml), 38% HCl (10 ml), and pure electrolytic iron
(8-10 g). A stream of purified nitrogen was passed through the assembly, escaping through a mercury trap. The salt was dissolved by heating which was continued until complete reduction has occurred. By
inverting the apparatus and filtering (under nitrogen pressure)
through the sintered-glass disc, unreacted iron was removed. After
cooling and crystn., the unit was again inverted and the crystals of
ferrous chloride were filtered free of the mother liquor by applied
nitrogen pressure. Partial drying by overnight evacuation at room
temperature gave a mixed hydrate which, on further evacuation on a
water bath at 80°, lost water of hydration and its absorbed HCl
(with vigorous effervescence) to give a white powder, $FeCl_2.2H_2O$.
[Gayer and Wootner, JACS, 78, 3944 (1956).]

Ferrous perchlorate.

Cryst. from $HClO_4$.

Ferrous sulphate.

Cryst. from 0.4 M H_2SO_4.

Fluorine, b.p. -129.2°.

Passed through a bed of NaF at 100° to remove HF. [For descriptions
of stills used in fractional distn., see Greenberg et al., JPC, 65,
1168 (1961); Stein, Rudzitis and Settle, Purification of Fluorine by
Distillation, Argonne National Laboratory, ANL-6364 (1961)(from Office
of Technical Services, U.S. Dept. of Commerce, Washington 25).]
Highly toxic.

Fluoroboric acid.

Cryst. several times from conductivity water.

Gallium.

Dissolved in dil. HCl and extracted into ethyl ether. Pptn. with
H_2S removed many metals, and a second extraction with ethyl ether
freed Ga more completely, except from Mo, Th(III) and Fe which were
largely removed by pptn. with NaOH. The soln. was then electrolyzed
in 10% NaOH with a Pt anode and cathode (2-5 A at 4-5 V) to deposit
Ga, In, Zn and Pb, from which Ga was obtained by fractional crystn.

Gallium (continued)

of the melt. [Hoffman, J. Res. Nat. Bur. Stand., 13, 665 (1934).]

Also purified by heating to b.p. in 0.5-1 N HCl, then heating to 40° in water and pouring the molten Ga with water under vac. on to a glass filter (30-50 μ pore size), to remove any unmelted metals or oxide film. The Ga was then fractionally crystd. from the melt under water.

Germanium.

Copper contamination on the surface and in the bulk of single crystals of germanium can be removed by immersion in molten alkali cyanide under nitrogen. The germanium is placed in dry cyanide powder in a graphite holder in a quartz or porcelain boat. The boat was then inserted into a heated furnace which, after a suitable time, was left to cool to room temperature. At 750°C, a 1 mm thickness requires about 1 min, whereas 0.5 cm needs about ½ hr. The boat was removed and the samples were taken out with plastic-coated tweezers, carefully rinsed in hot water and dried in air. [Wang, JPC, 60, 45 (1956).]

Graphite.

Treated with hot 1:1 HCl. Filtered, washed, dried, powdered and heated in an evacuated quartz tube at 1000° until a high vacuum was obtained. Cooled and stored in an atmosphere of helium. [Craig, Van Voorhis and Bartell, JPC, 60, 1225 (1956).]

Helium.

Dried by passage through a column of Linde molecular sieve and calcium sulphate, then passed through an activated-charcoal trap cooled in liquid nitrogen, to adsorb nitrogen, argon, xenon and krypton. Passed over CuO pellets at 300° to remove hydrogen and hydrocarbons, over calcium chips at 600° to remove oxygen, and then over titanium chips at 780° to remove nitrogen.

Hexamminecobalt(III) chloride.

Cryst. from warm water (8 ml/g) by cooling.

Hexammineruthenium(III) chloride.

Cryst. twice from 1 M HCl.

Hydrazine (anhydr.), f.p. 1.5-2.0°, b.p. 113-113.5°.

Hydrazine hydrate is dried by refluxing with an equal weight of NaOH pellets for 3 hr, then distd. from fresh NaOH or BaO in a current of dry nitrogen.

Hydrazine dihydrochloride.

Cryst. from aq. ethanol and dried under vac. over $CaSO_4$.

Hydrazine sulphate.

Cryst. from water. Dried at 140°.

Hydriodic acid, b.p. 127°.

Iodine can be removed from aq. HI, probably as the amine hydrogen triiodide, by three successive extractions using a 4% soln. of Amberlite LA-2 (a long-chain aliphatic amine) in CCl_4, toluene or pet.ether (10 ml per 100 ml of acid). [Davidson and Jameson, Chem. & Ind., 1686 (1963).] Extraction with tributyl phosphate in $CHCl_3$ or other organic solvents is also suitable. Alternatively, a De-acidite FF anion-exchange resin column, converted to its OH$^-$ form using 2 M NaOH, then into its I$^-$ form by passing dil. KI soln., can be used. Passage of an HI soln. under CO_2 through such a column removes poly-iodide. The column can be regenerated with NaOH. [Irving and Wilson, Chem. & Ind., 653 (1964).] The earlier method was to reflux with red phosphorus and distil in a stream of nitrogen. The colourless product was stored in ampoules in the dark. [Bradbury, JACS, 74, 2709 (1952).] Fumes in moist air.

Hydrobromic acid.

A soln. about 48% (w/w) HBr was distd. twice from a little red phosphorus, and the middle half of the distillate was taken. (The azeotrope at 760 mm contains 47.8% (w/w) HBr.) [Hetzer, Robinson and Bates, JPC, 66, 1423 (1962).] Free bromine can be removed by Irving and Wilson's method for hydriodic acid (see above), except that the column is regenerated by washing with an ethanolic soln. of aniline or styrene. Hydrobromic acid can also be purified by aerating with H_2S, distilling and collecting the fraction boiling at 126 ± 1°.

Hydrochloric acid.

Readily purified by fractional distn. as constant b.p. acid, following dilution with water. The constant-boiling fraction contains 1 mole of HCl in the following weights of distillate at the stated pressures - 179.555 g (730 mm); 179.766 (740); 179.979 (750); 180.193 (760); 180.407 (770). [Foulk and Hollingsworth, JACS, 45, 1220 (1923).]

Hydrofluoric acid.

Freed from lead (Pb $\leq$0.002 p.p.m) by co-pptn. with SrF_2, by addition of 10 ml of 10% $SrCl_2$ soln. per kilogram of the conc. acid. After the ppte. has settled, the supernatant is decanted through a filter in a hard-rubber or paraffined-glass vessel. [Rosenqvist, Amer. J. Sci., 240, 358 (1942).]

Pure aq. HF solns. (up to 25 M) can be prepared by isothermal distn. in polyethylene, polypropylene or platinum apparatus. [Kwestroo and Visser, Analyst, 90, 297 (1965).] Highly toxic.

Hydrogen.

Usually purified by passage through a suitable absorption train. Carbon dioxide is removed with KOH pellets, soda-lime or NaOH pellets. Oxygen is removed with a "De-oxo" unit or by passage over copper heated to 450-500°, copper on kieselguhr at 250° or platinized asbestos at 200-350°. Passage over a mixture of MnO_2 and CuO ("Hopcalite") oxidizes any CO to CO_2 (which is removed as above). Hydrogen can be dried by passage through dried silica-alumina at -195°, through a Dry-Ice trap followed by a liquid-nitrogen trap packed with glass wool, through $CaCl_2$ tubes, or through $Mg(ClO_4)_2$ or P_2O_5. Other purification steps include passage through a hot palladium thimble [Masson, JACS, 74, 4731 (1952)], through an activated-charcoal trap at -195°, and through a non-absorbent cotton-wool filter or small glass spheres coated with a thin layer of silicone grease.

Hydrogen bromide (anhydr.).

Dried by passage through $Mg(ClO_4)_2$ towers. This procedure is hazardous, see M.J. Stoss and G.B. Zimmermann, Ind. Eng. Chem., 17, 70 (1939). Shaken with mercury, distd. through a -78° trap and condensed at -195°/10^{-5} mm. Fumes in moist air.

Hydrogen chloride (anhydr.).

Passed through conc. H_2SO_4, then over activated charcoal and silica gel. Fumes in moist air.

Hydrogen cyanide (anhydr.), b.p. 25.7°.

Cylinder HCN may contain stabilizers against explosive polymerization, together with small amounts of H_3PO_4, H_2SO_4, SO_2, and water. It can be purified by distn. from P_2O_5, then frozen in Pyrex bottles at Dry-Ice temperature for storage. Highly poisonous.

Hydrogen fluoride (anhydr.), b.p. 19.4°.

Can be purified by trap-to-trap distn., followed by drying over CoF_2 at room temperature and further distn. Alternatively, it can be absorbed on NaF to form $NaHF_2$ which is then heated under vac. at 150° to remove volatile impurities. The HF is regenerated by heating at 300° and stored with CoF_3 in a nickel vessel, being distd. as required. (Water content ≤0.01%.) To avoid contact with base metal, use can be made of nickel, polychlorotrifluoroethylene and gold-lined fittings. [Hyman, Kilpatrick and Katz, JACS, 79, 3668 (1957).] Highly toxic.

Hydrogen iodide (anhydr.), b.p. -35.5°.

After removal of free iodine from aq. hydriodic acid (q.v.), the soln. is frozen, then covered with P_2O_5 and allowed to melt under vac. The gas evolved is dried by passage through P_2O_5 on glass wool. It can be freed from iodine contamination by repeated fractional distn. at low temperatures. Fumes in moist air.

Hydrogen peroxide.

The 30% material has been steam distd. using distd. water. Gross and Taylor [JACS, 72, 2075 (1950)] made 90% H_2O_2 approx. 0.001 M in NaOH and then distd. under its own vapour pressure, keeping the temperature below 40°, the receiver being cooled with a Dry-Ice-isopropyl alcohol mush. The 98% material has been rendered anhydrous by repeated fractional crystn. in all-quartz vessels. Explosive in contact with organic material.

Hydrogen sulphide, b.p. -59.6°.

Washed, then passed through a train of flasks containing satd. Ba(OH)$_2$ (two), water (two), and dil. HCl. [Goates et al., JACS, 73, 707 (1951).] Highly poisonous.

Hydroxylamine, m.p. 33.1°, b.p. 56.5°/22 mm.

Cryst. from n-butanol at -10°, collected by vac. filtration and washed with cold ethyl ether.

Hydroxylamine hydrochloride, m.p. 151°.

Cryst. from aq. 75% ethanol or boiling methanol, and dried under vac. over CaSO$_4$ or P$_2$O$_5$. Has also been dissolved in a minimum of water and satd. with HCl; after three such crystns. it was dried under vac. over CaCl$_2$ and NaOH.

Hydroxylamine sulphate, m.p. 170° (decomp.).

Cryst. from boiling water (1.6 ml/g) by cooling to 0°.

4-Hydroxy-3-nitrobenzenearsonic acid.

Cryst. from water.

Hypophosphorous acid, m.p. 26.5°.

Phosphorous acid is a common contaminant of commercial 50% hypophosphorous acid. Jenkins and Jones [JACS, 74, 1353 (1952)] purified this material by evaporating about 600 ml in a 1 l. flask, at 40°, under reduced pressure (in nitrogen), to a volume of about 300 ml. After the soln. cooled, it was transferred to a wide-mouthed Erlenmeyer flask which was stoppered and left in a Dry-Ice-acetone bath for several hours to freeze (if necessary, with scratching of the wall). When the flask was then left at about 5° for 12 hr, about 30-40% of the contents became liquid and were filtered off through Whatman No.44 filter paper. The solid, after pressing dry on the paper, was allowed to stand in the cold in a crystallizing dish until about 20-30% of it liquefied, and again filtered. This process was repeated, then the solid was stored over Mg(ClO$_4$)$_2$ in a vac. desiccator in the cold. Subsequent crystn. from n-butanol by dissolving it at room temperature and then cooling in an ice-salt bath at -20° did not appear to purify it further.

Indium.

Before use, the metal surface can be cleaned with dil. nitric acid, followed by a thorough washing with water and an alcohol rinse.

Indium sulphate.

Cryst. from dil. aq. H_2SO_4.

Iodine, m.p. 113.6°.

Usually purified by vac. sublimation. Preliminary purifications include grinding with 25% by weight of KI, blending with 10% BaO and subliming; subliming with CaO; grinding to a powder and treating with successive portions of water to remove dissolved salts, then drying; and crystn. from dry benzene. Barrer and Wasilewski [TFS, 57, 1140 (1961)] dissolved iodine in conc. KI and distd. it, then steam distd. three times, washing with distd. water. Organic material was removed by sublimation in a current of oxygen over platinum at about 700°, the iodine being finally sublimed under vac.

Iodine monobromide, m.p. 42°.
Iodine monochloride, m.p. 27.2°.

Purified by repeated fractional crystn. from its melt.

Iodine pentafluoride, m.p. -8.0°, b.p. 97°.

Rogers et al. [JACS, 76, 4843 (1954)] removed dissolved iodine from IF_5 by agitating with a mixture of dry air and ClF_3 in a fluorothene beaker using a magnetic stirrer. The mixture was transferred to a still and the more volatile impurities were pumped off as the pressure was reduced below 40 mm. The still was gradually heated (kept at 40 mm) to remove the ClF_3 before the IF_5 distd. Stevens [JOC, 26, 3451 (1961)] pumped IF_5 under vac. from its cylinder, trapping it at -78°, then allowing it to melt in a stream of dry nitrogen.

Iodine trichloride, m.p. 33°, b.p. 77° (decomp.).

Purified by sublimation at room temperature.

Iron (wire).

Cleaned in conc. HCl, rinsed in de-ionized water, then reagent grade acetone and dried under vac.

Iron pentacarbonyl, b.p. 102.5°/760 mm.

Distd. under vac., the middle cut being redistd. twice and stored in a bulb protected from light. (Photosensitive.)

Lead(II) acetate.

Cryst. twice from anhydrous acetic acid and dried under vac. for 24 hr at 100°.

Lead(II) bromide.

Cryst. from water containing a few drops of HBr (25 ml of water per gram $PbBr_2$) between 100° and 0°.

Lead(II) chloride.

Cryst. from distd. water at 100° (33 ml/g) after filtering through sintered-glass and adding a few drops of HCl, by cooling. After three crystns. the solid was dried under vac. or under anhydrous HCl vapour by heating slowly to 400°.

Lead(II) formate.

Cryst. from aq. formic acid.

Lead(II) iodide.

Cryst. from a large volume of water.

Lead monoxide.

Higher oxides were removed by heating under vac. at 550° with subsequent cooling under vac. [Ray and Ogg, JACS, 78, 5994 (1956).]

Lead nitrate.

Pptd. twice from a hot (60°) conc. aq. soln. by adding nitric acid. The ppte. was sucked dry in a sintered-glass funnel, then transferred to a crystallizing basin which was covered by a clock glass and left in an electric oven at 110° for several hours. [Beck, Singh and Wynne-Jones, TFS, 55, 331 (1959).]

Lead tetraacetate.

Dissolved in hot glacial acetic acid, any lead oxide being removed by filtration. White crystals of lead tetraacetate separated on cooling. Store in a vac. desiccator over P_2O_5 and KOH for 24 hr

Lead tetraacetate (continued).
before use.

Lissapol C (mainly sodium salt of cetyl oleyl alcohol sulphate).
Lissapol LS (mainly sodium salt of anisidine sulphate).
   Refluxed with 95% ethanol, then filtered to remove insoluble inorganic electrolytes. The alcohol soln. was then concentrated and the
residue was poured into dry acetone. The ppte. was filtered off,
washed in acetone and dried under vac. [Biswas and Mukherji, JPC, 64,
1 (1960).]

Lithium (metal).
   After washing with pet.ether to remove storage oil, lithium was
fused at 400° and then forced through a 10-micron stainless-steel
filter with argon pressure. It was again melted in a dry-box, skimmed,
and poured into an iron distn. pot. After heating under vac. to 500°,
cooling and returning to the dry-box for a further cleaning of its
surface, the lithium was distd. at 600° using an all-iron distn.
apparatus. [Gunn and Green, JACS, 80, 4782 (1958).]

Lithium acetate.
   Cryst. from ethanol (5 ml/g) by partial evapn.

Lithium aluminium hydride.
   Extracted with ethyl ether, and, after filtering, the solvent was
removed under vac. The residue was dried at 60° for 3 hr, under high
vac. [Ruff, JACS, 83, 1788 (1961).] Ignites in the presence of a
small amount of water.

Lithium benzoate.
   Cryst. from ethanol (13 ml/g) by partial evapn.

Lithium borohydride.
   Cryst. from ethyl ether, and pumped free of ether at 90-100° during
2 hr. [Schaeffer, Roscoe and Stewart, JACS, 78, 729 (1956).]

Lithium bromide.
   Cryst. several times from water or absolute ethanol, then dried under
high vac. for 2 days at room temperature, followed by drying at 100°.

Lithium carbonate.
Cryst. from water. Its solubility decreases as the temperature is raised.

Lithium chloride.
Cryst. from water (1 ml/g) or methanol and dried for several hours at 130°. Other metal ions can be removed by a preliminary crystallization from hot aq. 0.01 M disodium EDTA. Has also been cryst. from conc. HCl, fused in an atmosphere of dry HCl gas, cooled under dry nitrogen and pulverized in a dry-box. Kolthoff and Bruckenstein [JACS, 78, 1 (1956)] prepared a satd. soln. in almost boiling acetic acid, then cooled to 50° and added a twentyfold excess of benzene. After washing thoroughly with benzene, the ppte. was dried at 50° under vac. Runner and Wagner [JACS, 74, 2529 (1952)] pptd. with ammonium carbonate, washed the Li_2CO_3 five times by decantation and finally with suction, then dissolved in HCl. The LiCl soln. was evapd. slowly with continuous stirring in a large evaporating dish, the dry powder being stored (while still hot) in a desiccator over $CaCl_2$.

Lithium formate.
Cryst. from hot water (0.5 ml/g) by chilling.

Lithium hydroxide.
Cryst. from hot water (3 ml/g) as the monohydrate. Dehydrated at 150° in a stream of CO_2-free air.

Lithium iodate.
Cryst. from water and dried in a vac. oven at 60°.

Lithium iodide.
Cryst. from hot water (0.5 ml/g) by cooling in $CaCl_2$-ice, or from acetone. Dried under vac. over P_2O_5 for 1 hr at 60° and then at 120°.

Lithium nitrate.
Cryst. from water or ethanol. If it is cryst. from water keeping the temperature above 70°, formation of trihydrate is avoided. The anhydrous salt is dried at 120° and stored in a vac. desiccator over $CaSO_4$.

Lithium nitrite.
   Cryst. from water by cooling from room temperature.

Lithium perchlorate.
   Cryst. from water or 50% aq. methanol. Rendered anhydrous by heating the trihydrate at 170-180° in an air oven. It can then be recryst. from acetonitrile and again dried at 170°.

Lithium salicylate.
   Cryst. from ethanol (2 ml/g) by partial evapn.

Lithium sulphate.
   Cryst. from water (4 ml/g) by partial evapn.

Magnesium acetate.
   Cryst. from anhydrous acetic acid, then dried under vac. for 24 hr at 100°.

Magnesium ammonium chloride.
   Cryst. from water (6 ml/g) by partial evapn. in a desiccator over KOH.

Magnesium ammonium sulphate.
   Cryst. from water (1 ml/g) between 100° and 0°.

Magnesium benzoate.
   Cryst. from water (6 ml/g) between 100° and 0°.

Magnesium bromide.
   Cryst. from ethanol.

Magnesium chloride.
   Cryst. from hot water (0.3 ml/g) by cooling.

Magnesium iodate.
   Cryst. from water (5 ml/g) between 100° and 0°.

Magnesium iodide.
   Cryst. from water (1.2 ml/g) by partial evapn. in a desiccator.

Magnesium lactate.
   Cryst. from water (6 ml/g) between 100° and 0°.

Magnesium nitrate.
   Cryst. from water (2.5 ml/g) by partial evapn. in a desiccator.

Magnesium perchlorate.
   Cryst. from water. Coll, Nauman and West [JACS, 81, 1284 (1959)] removed traces of an unspecified contaminant by washing with small portions of ethyl ether. Explosive in contact with organic materials.

Magnesium succinate.
   Cryst. from water (0.5 ml/g) between 100° and 0°.

Magnesium sulphate.
   Cryst. from warm water (1 ml/g) by cooling.

Manganous acetate.
   Cryst. from water acidified with acetic acid.

Manganous ammonium sulphate.
   Cryst. from water (2 ml/g) by partial evapn. in a desiccator.

Manganous chloride.
   Cryst. from warm water (0.3 ml/g) by cooling.

Manganous ethylenebis[dithiocarbamate].
   Cryst. from ethanol.

Manganous lactate.
   Cryst. from water.

Manganous sulphate.
   Cryst. from water (0.9 ml/g) at 54-55° by evaporating about two-thirds of the water.

Mercuric acetate.
   Cryst. from glacial acetic acid.

Mercuric bromide, m.p. 238.1°.

Cryst. from hot satd. alcoholic soln., dried and kept at 100° for several hours under vac., then sublimed.

Mercuric chloride.

Cryst. twice from distd. water, dried at 70° and sublimed under high vac.

Mercuric cyanide.

Cryst. from water.

Mercuric iodide.

Cryst. from ethanol or methanol, and washed repeatedly with distd. water. Has also been mixed thoroughly with excess 0.001 N iodine soln., filtered, washed with cold distd. water, rinsed with alcohol and ethyl ether, and air dried.

Mercuric oxide.

Dissolved in $HClO_4$ and pptd. with NaOH soln.

Mercury, b.p. 356.7°.

After air had been bubbled through mercury for several hours to oxidize metallic impurities, it was filtered to remove the coarser particles of oxide and dirt, then sprayed through a 4-ft column containing 10% HNO_3. It was washed with distd. water, dried with filter paper and distd. under vac.

Metanil yellow.

Salted out from water three times with sodium acetate, then repeatedly extracted with ethanol. [McGrew and Schneider, JACS, 72, 2547 (1950).]

Methylarsonic acid, m.p. 161°.

Cryst. from absolute ethanol.

Methylmercuric chloride, m.p. 167°.

Cryst. from absolute ethanol (20 ml/g).

Methyl orange, see Sodium p-(p-dimethylaminobenzeneazo)-benzenesulphonate.

Molybdenum trichloride.
   Boiled with 12 M HCl, washed with absolute ethanol and dried in a
vac. desiccator.

Molybdenum trioxide.
   Cryst. from water (50 ml/g) between 70° and 0°.

Monocalcium phosphate monohydrate.
   Cryst. from a near-satd. soln. in 50% aq. reagent grade phosphoric
acid at 100° by filtering through fritted glass and cooling to room
temperature. The crystals were filtered off and this process was
repeated three times using fresh acid. For the final crystn. the
soln. was cooled slowly with constant stirring to give thin plate
crystals that were filtered off on fritted glass, washed free of acid
with anhydrous acetone and dried in a vac. desiccator. [Egan,
Wakefield and Elmore, JACS, 78, 1811 (1956).]

Neodymium oxide.
   Dissolved in $HClO_4$, pptd. as the oxalate with doubly recryst.
oxalic acid, washed free of soluble impurities, dried at room temperature and ignited in platinum at higher than 850° in a stream of
oxygen. [Tobias and Garrett, JACS, 80, 3532 (1958).]

Nickel ammonium sulphate.
   Cryst. from water (3 ml/g) between 90° and 0°.

Nickel bromide.
   Cryst. from dil. HBr (0.5 ml water per g) by partial evapn. in a
desiccator.

Nickel chloride.
   Cryst. from dil. HCl.

Nickel nitrate.
   Cryst. from water (0.3 ml/g) by partial evapn. in a desiccator.

Nickel potassium sulphate, see Potassium nickel sulphate.

Nickel sulphate.
    Cryst. from warm water (0.25 ml/g) by cooling.

Nitric acid.
    Obtained colourless (approx. 92%) by direct distn. of fuming nitric
acid under reduced pressure at 40-50° with an air leak at the head of
the fractionating column. Stored in a desiccator kept in a refrigerator. Nitrite-free nitric acid can be obtained by vac. distn. from urea.

Nitric oxide.
    Bubbling through 10 M NaOH, or passage through asbestos impregnated
with NaOH, removes NO_2. The gas is dried with solid NaOH pellets or
by passing through silica gel cooled to -78°, followed by fractional
distn. from a liquid-nitrogen trap. This purification does not eliminate nitrous oxide. Other gas scrubbers sometimes used include one
containing conc. H_2SO_4 and another containing mercury.

p-Nitrobenzenediazonium fluoroborate.
    Cryst. from water.

Nitrogen, b.p. -195.8°.
    Cylinder nitrogen can be freed of oxygen by passage through
Fieser's soln. (which comprises 2 g sodium anthraquinone-β-sulphonate
and 15 g sodium hydrosulphite dissolved in 100 ml of 20% KOH [Fieser,
JACS, 46, 2639 (1924)]) followed by scrubbing with satd. lead
acetate soln. (to remove any H_2S generated by the Fieser's soln.),
conc. H_2SO_4 (to remove moisture), then soda-lime (to remove any
H_2SO_4 and CO_2). Alternatively, after passage through Fieser's soln.,
nitrogen can be dried by washing with a soln. of the metal ketyl from
benzophenone and sodium in absolute ethyl ether. (If ether vapour in
the nitrogen is undesirable, the ketyl from liquid sodium-potassium
alloy under xylene can be used.)
    Another method for removing oxygen is to pass the nitrogen through
a long column tightly packed with copper turnings, the surface of
which is constantly renewed by scrubbing it with ammonia (S.G. 0.880)
soln. The gas is then passed through a column packed with glass beads
moistened with conc. H_2SO_4 (to remove ammonia), through a column
packed with KOH pellets (to remove any H_2SO_4 and to dry the nitrogen),

Nitrogen (continued)

and finally through a glass trap packed with chemically clean glass
wool immersed in liquid nitrogen.

A typical dry purification method consists of a mercury bubbler (as
trap), followed by a small column of silver and gold turnings to
remove any mercury vapour, towers containing anhydrous $CaSO_4$, dry
molecular sieve or $Mg(ClO_4)_2$, a tube filled with fine copper turnings
and heated to $400°$ by an electric furnace, a tower containing soda-
lime or NaOH on asbestos, and finally a plug of glass wool as a
filter. Variations include tubes of silica gel, palladized asbestos
heated to $300°$, traps containing activated charcoal cooled in a
Dry-Ice bath, copper on kieselguhr heated to $250°$, and copper and
iron filings at $400°$.

Nitrophenolarsonic acid.

Nitroso-R-salt.

   Cryst. from water.

Nitrosyl chloride, b.p. $-5.5°$.

   Fractionally distd. at atmospheric pressure in an all-glass, low
temperature still, taking the fraction boiling at $-4°$ and storing in
sealed tubes.

Nitrous oxide, b.p. $-88.5°$.

   Washed with conc. alkaline pyrogallol soln., to remove O_2, CO_2 and
NO_2, then dried by passage through columns of P_2O_5 or Drierite, and
collected in a trap cooled in liquid nitrogen. Further purified by
freeze-pump-thaw and distn. cycles under vac. [Ryan and Freeman,
JPC, 81, 1455 (1977).]

Orange II.

   Extracted with a small volume of water, then cryst. by dissolving
in boiling water, cooling to about $80°$, adding two volumes of ethanol
and cooling. When cold, the ppte. is filtered off, washed with a
little alcohol and dried in air. It can be salted out from aq. soln.
with sodium acetate, then repeatedly extracted with ethanol. Meggy
and Sims [JCS, 2940 (1956)], after crystallizing the sodium salt
twice from warm water, dissolved it in cold water (11 ml/g) and
added conc. HCl (0.5 ml/g) to ppte. the dye acid which was separated

Orange II (continued)

at the centrifuge, redissolved and again pptd. with acid. After
washing the ppte. three times with 0.5 M acid it was dried over
NaOH, recryst. twice from absolute ethanol, washed with a little
ethyl ether, dried over NaOH and stored over conc. H_2SO_4 in the dark.

Orange RO.

Salted out three times with sodium acetate, then repeatedly
extracted with ethanol.

Perchloric acid (explosive).

The 72% acid has been purified by double distn. from silver oxide
under vacuum: this frees the acid from metal contamination. Anhydrous
acid can be obtained by adding gradually 400-500 ml of oleum (20%
fuming H_2SO_4) to 100-120 ml of 72% $HClO_4$ in a reaction flask cooled
in an ice-bath. The pressure is reduced to 1 mm (or less), with the
reaction mixture at 20-25°. The temperature is gradually raised
during 2 hr to 85°, the distillate being collected in a receiver
cooled in Dry Ice. (A protective screen should be used.)[For details
of the distn. apparatus, see Smith, JACS, 75, 184 (1953).]

Phenylarsonic acid, m.p. 155-158° (decomp.).

Cryst. from water (3 ml/g) between 90° and 0°.

Phenylmercuric hydroxide, m.p. 195-203°.

Cryst. from dil. aq. NaOH.

Phenylmercuric nitrate.

Cryst. from water.

Phosgene, b.p. 8.2°/756 mm.

Dried with Linde type 4A molecular sieve, degassed and distd. under
vac. Highly toxic.

Phosphonitrilic chloride (tetramer).

Purified by zone melting, then cryst. from pet.ether (b.p. 40-60°).

Phosphonitrilic chloride (trimer), m.p. 112.8°.

Purified by zone melting, by crystn. from pet.ether or benzene, and by sublimation.

Phosphoric acid, m.p. 42.3°.

Pyrophosphate can be removed from phosphoric acid by diluting with distd. water and refluxing overnight. By cooling to 11° and seeding with crystals obtained by cooling a few millilitres in a Dry-Ice-acetone bath, 85% orthophosphoric acid can be cryst. as $H_3PO_4 \cdot \frac{1}{2}H_2O$. The crystals are separated using a sintered-glass filter.

Phosphorus (red).

Boiled for 15 min with distd. water, allowed to settle and washed several times with boiling water. Transferred to a Büchner funnel, washed with hot water until the washings are neutral, then dried at 100° and stored in a desiccator.

Phosphorus (white).

Purified by melting under a dil. H_2SO_4-dichromate mixture and allowing to stand for several days in the dark at room temperature. (It remains liquid, and the initial milky appearance due to insoluble, oxidizable material gradually disappears.) The phosphorus can then be distd. under vac. in the dark. [Holmes, TFS, 58, 1916 (1962).] Other methods include extraction with dry CS_2 followed by evapn. of the solvent, or washing with 6 M HNO_3, then water, and drying under vac. Poisonous.

Phosphorus oxychloride, b.p. 105.5°.

Distd. under reduced pressure to separate from the bulk of the HCl and the phosphoric acid, the middle fraction being distd. into ampoules containing purified mercury. These ampoules were sealed and stored in the dark for 4-6 weeks with occasional shaking to facilitate reaction of any free chlorine with the mercury. The $POCl_3$ was then again fractionally distd. and stored in sealed ampoules in the dark until used. [Herber, JACS, 82, 792 (1960).] Lewis and Sowerby [JCS, 336 (1957)] refluxed their distd. $POCl_3$ with sodium wire for 4 hr, then removed the sodium and again distd.

Phosphorus pentabromide, m.p. <100°, b.p. 106° (decomp.).

Dissolved in pure nitrobenzene at 60°, filtering off any insoluble residue on to sintered glass, then cryst. by cooling. Washed with dry ethyl ether and removed the ether in a current of dry nitrogen. (All manipulations were carried out in a dry-box.) [Harris and Payne, JCS, 3732 (1958).] Fumes in moist air.

Phosphorus pentachloride.

Sublimed at 160-170° in an atmosphere of chlorine. The excess chlorine was then displaced by dry nitrogen gas. All subsequent manipulations were performed in a dry-box. [Downs and Johnson, JACS, 77, 2098 (1955).] Fumes in moist air.

Phosphorus pentasulphide, m.p. 277-283°.

Purified by extraction and crystn. with CS_2, using a Soxhlet extractor. Liberates H_2S in moist air.

Phosphorus pentoxide.

Sublimed at 250° under vac. into glass ampoules. Fumes in moist air.

Phosphorus sesquisulphide, P_4S_3, m.p. 172°.

Extracted with CS_2, filtered and evapd. to dryness. Placed in water, and steam was passed through for an hour. The water was then removed, the solid was dried, followed by crystn. from CS_2. [Rogers and Gross, JACS, 74, 5294 (1952).]

Phosphorus trichloride, b.p. 76°.

Heated under reflux to expel dissolved hydrogen chloride, then distd. It has been further purified by vac. fractionation several times through a -45° trap into a receiver at -78°.

12-Phosphotungstic acid.

A few drops of conc. HNO_3 were added to 100 g of phosphotungstic acid dissolved in 75 ml of water, in a separating funnel, and the soln. was extracted with ethyl ether. The lowest of the three layers, which contained a phosphotungstic acid-ether complex, was separated, washed several times with 2 M HCl, then with water and again extracted with ether. Evapn. of the ether, under vac., with mild heating on a water-

12-Phosphotungstic acid (continued)
bath gave crystals which were dried under vac. and ground.
[Matijević and Kerker, JACS, 81, 1307 (1959).]

Potassium (metal), m.p. 62.3°.

Oil was removed from the surface by immersion for long periods in n-hexane and pure ethyl ether. The surface oxide was next removed by scraping under ether and the potassium was melted under vac. It was then allowed to flow through metal constrictions into tubes that could be sealed, followed by distn. under vac. in the absence of mercury vapour. (See Sodium.) Explosive in water.

Potassium acetate.

Cryst. three times from water-ethanol (1:1) drying to constant weight in a vac. oven, or cryst. from anhydrous acetic acid and pumped dry under vac. for 30 hr at 100°.

Potassium p-aminobenzoate.

Cryst. from ethanol.

Potassium antimony tartrate.

Cryst. from water (3 ml/g) between 100° and 0°. Dried at 100°.

Potassium benzoate.

Cryst. from water (1 ml/g) between 100° and 0°.

Potassium bicarbonate.

Cryst. from water at 65-70° (1.25 ml/g) by filtering, then cooling to 15°. During all operations, carbon dioxide is passed through the stirred mixture. The crystals, sucked dry at the pump, are washed with distilled water, dried in air and then over sulphuric acid in an atmosphere of carbon dioxide.

Potassium biiodate.

Cryst. three times from hot water (3 ml/g), stirring continuously during each cooling. After drying at 100° for several hours, the crystals are suitable for use in volumetric analysis.

Potassium bisulphate.
   Cryst. from water (1 ml/g) between 100° and 0°.

Potassium borohydride.
   Cryst. from liquid ammonia.

Potassium bromate.
   Cryst. from distd. water (2 ml/g) between 100° and 0°. To remove
bromide contamination, a 5% soln. in distd. water, cooled to 10°, has
been bubbled with gaseous chlorine for 2 hr, then filtered and extracted with reagent grade CCl_4 until colourless and odourless. After
evaporating the aq. phase to about half its volume, it was cooled
again slowly to about 10°. The cryst. $KBrO_3$ was separated, washed
with 95% ethanol and vac. dried. [Boyd, Cobble and Wexler, JACS, 74,
237 (1952).]

Potassium bromide.
   Cryst. from distd. water (1 ml/g) between 100° and 0°. Washed with
95% ethanol, followed by ethyl ether. Dried in air, then heated at
115° for 1 hr, pulverized and heated in a vac. oven at 130° for 4 hr.
Has also been cryst. from aq. 30% ethanol, or ethanol, and dried over
P_2O_5 under vac. before heating in an oven.

Potassium carbonate.
   Cryst. from water (0.3 ml/g) between 100° and 0°.

Potassium chlorate.
   Cryst. from water (1.8 ml/g) between 100° and 0°, filtering the
crystals on to sintered glass.

Potassium chloride.
   Dissolved in conductivity water, filtered, and satd. with chlorine
(generated from A.R. HCl and $KMnO_4$). Excess chlorine was boiled off,
and the KCl was pptd. by HCl (generated by dropping conc. A.R. HCl
into conc. H_2SO_4). The ppte. was washed with water, dissolved in
conductivity water at $90-95^\circ$, and cryst. by cooling to about -5°. The
crystals were drained at the centrifuge, dried in a vac. desiccator
at room temperature, then fused in a platinum dish under nitrogen,
cooled and stored in a desiccator. Potassium chloride has also been

Potassium chloride (continued)
sublimed in a stream of prepurified nitrogen gas and collected by
electrostatic discharge. [Craig and McIntosh, Canad. J. Chem., 30,
448 (1952).]

Potassium chloroplatinate.
   Cryst. from water (20 ml/g) between 100° and 0°.

Potassium chloroplatinite.
   Cryst. from aq. 0.75 M HCl (20 ml/g) between 100° and 0°. Washed
with ice-cold water and dried.

Potassium chromate.
   Cryst. from conductivity water (soly. 0.6 g/ml at 20°), and dried
between 135° and 170°.

Potassium cobalticyanide.
   Cryst. from water to remove trace of HCN.

Potassium cobaltous sulphate.
   Cryst. from water (1 ml/g) between 50° and 0°.

Potassium cyanate.
   Common impurities include ammonia and bicarbonate ion (from hydrolysis). Purified by preparing a satd. aq. soln. at 50°, neutralizing
with acetic acid, filtering, adding two volumes of ethanol and keeping
for 3-4 hr in an ice-bath. (More alcohol can lead to co-pptn. of
$KHCO_3$.) Filtered, washed with ethanol and dried quickly in a vac.
desiccator. The process was repeated. [Vanderzee and Myers, JPC, 65,
153 (1961).]

Potassium cyanide.
   A satd. soln. in water-ethanol (1:3) at 60° was filtered and cooled
to room temperature. Absolute ethanol was added, with stirring, until
crystn. ceased. The soln. was again allowed to cool to room temperature
(during 2-3 hr) then the crystals were filtered off, washed with absolute ethanol, and dried, first at $70-80^\circ$ for 2-3 hr, then at 105°
               own, Adisesh and Taylor, JPC, 66, 2426 (1962).]
               .ed by vac. melting and zone refining. Highly poisonous.

Potassium dichromate.

   Cryst. from water (1 ml/g) between 100° and 0° and dried under vac.
at 156°.

Potassium dihydrogen citrate.

   Cryst. from water. Dried at 80°.

Potassium dihydrogen orthophosphate.

   Dissolved in boiling distd. water (2 ml/g), kept on a boiling water-
bath for several hours, then filtered through paper pulp to remove any
turbidity. Cooled rapidly with constant stirring, and the crystals
were separated on to hardened filter paper, using suction, washed
twice with ice-cold water, once with 50% ethanol and dried at 105°.
Alternative crystns. are from water, then 50% ethanol, and again
water, or from conc. aq. soln. by addition of ethanol. Freed from
traces of copper by extracting its aq. soln. with diphenylthiocarbazone in CCl_4, followed by repeated extraction with CCl_4 to remove
traces of diphenylthiocarbazone.

Potassium dithionate.

   Cryst. from water (1.5 ml/g) between 100° and 0°.

Potassium ethyl xanthate.

   Cryst. from absolute ethanol, ligroin-ethanol or acetone by addition
of ether. Washed with ethyl ether, then dried in a desiccator.

Potassium ferricyanide.

   Cryst. repeatedly from hot water (1.3 ml/g). Dried under vac. in a
desiccator.

Potassium ferrocyanide.

   Cryst. repeatedly from distd. water, never heating above 60°.
Prepared anhydrous by drying at 110° or over P_2O_5 in a vac. desiccator.
To obtain the trihydrate, it is necessary to equilibrate in a desiccator over satd. aq. soln. of sucrose and NaCl. Can also be pptd. from
a satd. soln. at 0° by adding an equal volume of cold 95% ethanol,
standing for several hours, then centrifuging and washing with cold
95% ethanol. Finally sucked air dry with a water-pump. The anhydrous
salt can be obtained by drying in a platinum boat at 90° in a slow

Potassium ferrocyanide (continued)
stream of nitrogen. [Loftfield and Swift, JACS, 60, 3083 (1938).]

Potassium fluoroborate.
   Cryst. several times from conductivity water (15 ml/g).

Potassium fluorosilicate.
   Cryst. several times from conductivity water (100 ml/g) between
100° and 0°.

Potassium hexachloroiridate(IV).
   Cryst. from a hot aq. soln. containing a few drops of HNO_3.

Potassium hexachloroosmate(IV).
   Cryst. from hot dil. aq. HCl.

Potassium hexacyanochromate(III).
   Cryst. from water.

Potassium hexafluorophosphate.
   Cryst. from alkaline aq. soln., using polyethylene vessels, or from
95% ethanol, and dried in a vac. desiccator over NaOH.

Potassium hydrogen fluoride.
Potassium hydrogen R-glucarate.
   Cryst. from water.

Potassium hydrogen malate.
   A satd. aq. soln. at 60° was decolorized with activated charcoal,
and filtered. The filtrate was cooled in water-ice bath and the salt
was pptd. by addition of ethanol. After being cryst. five times from
ethanol-water mixtures, it was dried overnight at 130° in air.
[Eden and Bates, J. Res. Nat. Bur. Stand., 62, 161 (1959).]

Potassium hydrogen oxalate.
   Cryst. from water by dissolving 20 g in 100 ml water at 60° containing 4 g of potassium oxalate, filtering and allowing to cool to 25°.
The crystals, after washing three or four times with water, are
allowed to dry in air.

Potassium hydrogen phthalate.
   Cryst. first from a dil. aq. soln. of K_2CO_3, then from water
(3 ml/g) between 100° and 0°. Before being used as a standard in
volumetric analysis, analytical grade potassium hydrogen phthalate
should be dried at 120° for 2 hr, then allowed to cool in a desiccator.

Potassium hydrogen saccharate.
   Cryst. from hot water.

Potassium hydrogen d-tartrate.
   Cryst. from water (17 ml/g) between 100° and 0°. Dried at 110°.

Potassium hydroxide (solution).
   Its carbonate content can be reduced by rinsing KOH sticks rapidly
with water prior to dissolving them in boiled out distd. water.
Alternatively, a slight excess of satd. $BaCl_2$ or $Ba(OH)_2$ can be added
to the soln. which, after shaking well, is left so that the $BaCO_3$
ppte. will separate out. Davies and Nancollas [Nature, 165, 237
(1950)] rendered KOH solns. carbonate free by ion exchange using a
column of Amberlite IR-400 in the hydroxyl form.

Potassium iodate.
   Cryst. twice from distd. water (3 ml/g) between 100° and 0°, dried
for 2 hr at 140° and cooled in a desiccator. Analytical reagent grade
material dried in this way is suitable for use as an analytical
standard.

Potassium iodide.
   Cryst. from distd. water (0.5 ml/g) by filtering the near-boiling
soln. and cooling. To minimize oxidation to iodine, the crystn. can
be carried out under nitrogen and the salt is dried under vac. over
P_2O_5 at $70-100^{\circ}$. Before drying, the crystals can be washed with
ethanol or with acetone followed by pet.ether. Has also been cryst.
from water-ethanol.

Potassium isoamyl xanthate.
   Cryst. twice from acetone-ethyl ether. Dried in a desiccator for
two days and stored under refrigeration.

Potassium metaperiodate, see Potassium periodate.

Potassium nickel sulphate.
  Cryst. from water (1.7 ml/g) between 75° and 0°.

Potassium nitrate.
  Cryst. from hot water (0.5 ml/g) by cooling. Dried for 12 hr under vac. at 70°.

Potassium nitrite.
  A satd. soln. at 0° can be warmed and partially evapd. under vac., the crystals so obtained being filtered from the warm soln. (This procedure is designed to reduce the level of nitrate impurity and is based on the effects of temperature on solubility. The soly. of KNO_3 in water is 13 g/100 ml at 0°, 247 g/100 ml at 100°; for KNO_2 the corresponding figures are 280 g/100 ml and 413 g/100 ml.)

Potassium oleate.
  Cryst. from ethanol (1 ml/g).

Potassium oxalate.
  Cryst. from hot water.

Potassium perchlorate.
  Cryst. from boiling water (5 ml/g) by cooling. Dried under vac. at 105°.

Potassium periodate.
  Cryst. from distd. water.

Potassium permanganate.
  Cryst. from hot water (4 ml/g at 65°), then dried and vac. desiccated over $CaSO_4$. Phillips and Taylor [JCS, 4242 (1962)] cooled an aq. soln. of $KMnO_4$, satd. at 60°, to room temperature in the dark, and filtered through a no.4 porosity sintered-glass filter funnel. The soln. was allowed to evaporate in air in the dark for 12 hr, and the supernatant liquid was decanted from the crystals, which were dried as quickly as possible with filter paper.

Potassium peroxydisulphate.

Cryst. twice from distd. water and dried at 40° in a vac. desiccator.

Potassium perphosphate.

Dissolved in ethylene glycol-water (1:1) at 60°, added water slowly at near 55°, and then filtered. Product washed with methanol-water (1:1) and then pure methanol.

Potassium perrhenate.

Cryst. from water (7 ml/g), then fused in a platinum crucible in air at 750°.

Potassium persulphate.

Cryst. from warm water (10 ml/g) by cooling. Dried under vac. at 50°.

Potassium p-phenolsulphonate.

Cryst. several times from distd. water at 90°, after treatment with charcoal, by cooling to approx. 10°. Dried at $90-110^\circ$.

Potassium picrate.

Cryst. from water (4 ml/g) between 100° and 15°.

Potassium propionate.

Cryst. from water (30 ml/g) or 95% ethanol.

Potassium reineckate.

Cryst. from KNO_3 soln., then from warm water. [Adamson, JACS, 80, 3183 (1958).]

Potassium selenocyanate.

Dissolved in acetone, filtered and pptd. by adding ethyl ether.

Potassium sodium tartrate.

Cryst. from warm water (1.5 ml/g) by cooling to 0°.

Potassium sulphate.

Cryst. from distd. water (4 ml/g) between 100° and 0°.

Potassium d-tartrate.
   Cryst. from distd. water (0.4 ml/g) between 100° and 0°.

Potassium tetraphenylborate.
   Pptd. from a soln. of KCl acidified with dil. HCl, then cryst. twice from acetone, washed thoroughly with water and dried at 110°. [Findeis and de Vries, AC, 28, 1899 (1956).]

Potassium tetroxalate.
   Cryst. from water below 50°. Dried below 60° at atmospheric pressure.

Potassium thiocyanate, m.p. 172°.
   Cryst. from water if much chloride ion is present in the salt, otherwise from ethanol or methanol (optionally by addition of ethyl ether). Filtered on to a Büchner funnel without paper, and dried in a desiccator at room temperature before being heated for 1 hr at 150°, with a final 10-20 min at 200° to remove the last traces of solvent. [Kolthoff and Lingane, JACS, 57, 2126 (1935).] Stored in the dark.

Potassium thiosulphate.
   Cryst. from warm water (0.5 ml/g) by cooling in ice-salt mixture.

Potassium tungstate.
   Cryst. from hot water (0.7 ml/g).

Praseodymium trichloride.
   Its 1 M soln. in 6 M HCl was passed twice through a Dowex-1 anion-exchange column. The eluate was evapd. in a vac. desiccator to about half its volume and allowed to crystallize. [Katzin and Gulyas, JPC, 66, 494 (1962).]

Praseodymium oxide.
   Dissolved in acid, pptd. as the oxalate and ignited at 650°.

Reinecke salt, see Ammonium reineckate.

Rubidium bromide.
   Cryst. from near-boiling water (0.5 ml/g) by cooling to 0°.

Rubidium chlorate.

Cryst. from water (1.6 ml/g) by cooling from 100°.

Rubidium chloride.

Cryst. from water (0.7 ml/g) by cooling to 0° from 100°.

Rubidium nitrate.

Cryst. from hot water (0.25 ml/g) by cooling to room temperature.

Rubidium perchlorate.

Cryst. from hot water (1.6 ml/g) by cooling to 0°.

Rubidium sulphate.

Cryst. from water (1.2 ml/g) between 100° and 0°.

Ruthenium dioxide.

Freed from nitrates by boiling in distd. water and filtering. A more complete purification is based on fusion in a $KOH-KNO_3$ mix to form the soluble ruthenate and perruthenate salts. The melt is dissolved in water, and filtered, then acetone is added to reduce the ruthenates to the insoluble hydrated oxide which, after making a slurry with paper pulp, is filtered and ignited in air to form the anhydrous oxide. [Campbell, Ortner and Anderson, AC, 33, 58 (1961).]

Selenious acid.

Cryst. from water.

Selenium, m.p. 217.4°.

Dissolved in small portions in hot conc. nitric acid (2 ml/g) filtered and evapd. to dryness to give selenious acid which was then dissolved in conc. HCl. Passage of SO_2 into the soln. pptd. selenium (but not tellurium) which was filtered off and washed with conc. HCl. This purification process was repeated. The selenium was then converted twice to the selenocyanate by treating with a 10% excess of 3M aq. KCN, heating for half an hour on a sand-bath and filtering. Addition of an equal weight of chopped ice to the cold soln., followed by an excess of cold, conc. HCl, with stirring (in a well-ventilated fume hood because HCN is evolved) pptd. selenium as a powder, which, after washing with water until odourless, and then with methanol, was

Selenium (continued)

heated in an oven at 105°, then by fusion for 2 hr under vac. It was cooled, crushed and stored in a desiccator. [Tideswell and McCullough, JACS, 78, 3036 (1956).]

Selenium dioxide, m.p. 340°.

Purified by sublimation, or by soln. in nitric acid, pptn. of selenium which, after standing for several hours or boiling, is filtered off, then re-oxidized by nitric acid and cautiously evapd. to dryness below 200°. The dioxide is dissolved in water and again evapd. to dryness.

Silica gel.

Before use as a drying agent, silica gel is heated in an oven, then cooled in a desiccator. Conditions in the literature range from heating at 110° for 15 hr to 250° for 2-3 hr. Silica gel has been purified by washing with hot acid (in one case successively with aqua regia, conc. HNO_3, then conc. HCl; in another case digested overnight with hot conc. H_2SO_4), followed by exhaustive washing with distd. water (one week in a Soxhlet apparatus has also been used), and prolonged oven drying. Alternatively, silica gel has been extracted with acetone until all soluble material was removed, then dried in a current of air, washed with distd. water and oven dried. Silica gel has also been washed successively with water, M HCl, water, and acetone, then activated at 110° for 15 hr.

Silicon tetrachloride.

Distd. under vac.

12-Silicotungstic acid.

Extracted with ethyl ether from a soln. acidified with HCl. The ethyl ether was evapd. under vac., and the free acid was cryst. twice. [Matijević and Kerker, JPC, 62, 1271 (1958).]

Silver (metal).

For purification by electrolysis, see Craig et al. [J. Res. Nat. Bur. Stand., 64 A, 381 (1960).]

Silver acetate.

Shaken with acetic acid for three days, the process being repeated with fresh acid, the solid then being dried in a vac. oven at 40° for 48 hr. Has also been cryst. from water containing a trace of acetic acid, and air dried.

Silver bromate.

Cryst. from hot water (80 ml/g).

Silver bromide, m.p. 432°.

Purified from Fe, Mn, Ni and Zn by zone melting in a quartz vessel under vac.

Silver chlorate.

Cryst. from hot water (2 ml/g).

Silver iodate.

Washed with warm dil. HNO_3, then water and dried at 100°, or cryst. from an ammoniacal soln. by adding nitric acid, filtering, washing with water and drying at 100°.

Silver nitrate.

Cryst. from hot conductivity water by slow addition of freshly distd. ethanol. Before being used as a standard in volumetric analysis, analytical reagent grade silver nitrate should be finely powdered, dried at 120° for 2 hr, then cooled in a desiccator.

Silver nitrite.

Cryst. from hot conductivity water (70 ml/g) in the dark. Dried in the dark under vac.

Silver perchlorate.

Refluxed with benzene (6 ml/g) in a flask fitted with a Dean and Stark distilling receiver until all the water was removed azeotropically (about 4 hr). The soln. was cooled and diluted with dry pentane (4 ml/g of $AgClO_4$). The pptd. $AgClO_4$ was filtered off and dried in a desiccator over P_2O_5 at 1 mm for 24 hr. [Radell, Connolly and Raymond, JACS, 83, 3958 (1961).] Has also been cryst. from perchloric acid.

Silver sulphate.

Cryst. by dissolving in hot conc. H_2SO_4 containing a trace of HNO_3, cooling and diluting with water. The ppte. was filtered off, washed and dried at 120°.

Sodium (metal), m.p. 97.5°.

The metal was placed on a coarse grade of sintered-glass filter, melted under vac. and forced through the filter using argon. The Pyrex apparatus was then re-evacuated and sealed off below the filter, so that the sodium could be distd. at 460° through a side arm and condenser into a receiver bulb which was then sealed off. [Gunn and Green, JACS, 80, 4782 (1958).] Explodes in water.

Sodium acetate.

Cryst. from acetic acid and pumped under vac. for 10 hr at 120°. Alternatively, cryst. from aq. ethanol, as the trihydrate. This material can be converted to the anhydrous salt by heating slowly in a porcelain, nickel or iron dish, so that the salt liquefies. Steam is evolved and the mass again solidifies. Heating is now increased so that the salt again melts. (If it is heated too strongly, the salt chars.) After several minutes, the salt is allowed to solidify and cool to a convenient temperature before being powdered and bottled.

Sodium alginate.

Freed from heavy metal impurities by treatment with ion-exchange resins (Na-form) or with a dil. soln. of the sodium salt of EDTA. Also dissolved in 0.1 M NaCl, centrifuged and fractionally pptd. by gradual addition of ethanol or 4 M NaCl. The resulting gels were centrifuged off, washed with aq. ethanol or acetone, and dried under vac. [Büchner, Cooper and Wassermann, JCS, 3974 (1961).]

Sodium p-aminobenzoate.

Cryst. from water.

Sodium p-aminosalicylate.

Cryst. from water at room temperature (2 ml/g) by adding acetone and cooling.

Sodium ammonium hydrogen phosphate.
   Cryst. from hot water (1 ml/g).

Sodium amylpenicillin.
   Cryst. from moist acetone or moist ethyl acetate.

Sodium anthraquinone-1,5-disulphonate.
   Separated from insoluble impurities by continuous extraction with water. Cryst. twice from hot water and dried under vac.

Sodium anthraquinone-β-sulphonate.
   Cryst. from hot water (4 ml/g) or from water by addition of ethanol. Dried under vac. over $CaCl_2$ or in an oven at 70°. Stored in the dark.

Sodium antimonyl tartrate.
   Cryst. from water.

Sodium arsenate.
   Cryst. from water (2 ml/g).

Sodium azide.
   Cryst. from hot water or from water by the addition of absolute ethanol or acetone. Dried under vac. for several hours in an Abderhalden pistol. Highly poisonous.

Sodium barbitone.
   Cryst. from warm water (3 ml/g) by adding an equal volume of ethanol and cooling to 5°. Dried under vac. over P_2O_5.

Sodium benzenesulphonate.
   Cryst. from ethanol or aq. 70-100% methanol, and dried under vac. at 80-100°.

Sodium benzoate.
   Cryst. from ethanol (12 ml/g).

Sodium benzylpenicillin.
   Cryst. from methanol-ethyl acetate.

Sodium bicarbonate.

Cryst. from hot water (6 ml/g). The solid should not be heated above 40°.

Sodium bisulphite.

Cryst. from hot water (1 ml/g). Dried at 100° under vac. for 4 hr.

Sodium borate.

Most of the water of hydration was removed from the decahydrate by evacuation at 25° for three days, followed by heating to 100° and evacuation with a high-speed diffusion pump. The dried sample was then heated gradually to fusion (above 966°), and allowed to cool gradually to 200° before being transferred to a desiccator containing P_2O_5. [Grenier and Westrum, JACS, 78, 6226 (1956).]

Sodium borate (decahydrate).

Cryst. from water (3.3 ml/g) keeping below 55° to avoid formation of the pentahydrate. Filtered at the pump, washed with water and equilibrated for several days in a desiccator containing an aq. soln. satd. wtih respect to sucrose and sodium chloride. Borax can be prepared more quickly (but its water content is somewhat variable) by washing the recryst. material at the pump with water, followed by 95% ethanol, then ethyl ether, and air drying at room temperature for 12-18 hr on a clock glass.

Sodium borohydride.

After adding $NaBH_4$ (10 g) to freshly distd. diglyme (120 ml) in a dry three-necked flask fitted with a stirrer, nitrogen inlet and outlet, the mixture was stirred for 30 min at 50° until almost all of the solid had dissolved. Stirring was stopped, and, after the solid had settled, the supernatant liquid was forced under nitrogen pressure through a sintered glass filter into a dry flask. [The residue was centrifuged to obtain more of the soln. which was added to the bulk.] The soln. was cooled slowly to 0° and then decanted from the white needles that separated. The crystals were dried by pumping for 4 hr to give anhydrous $NaBH_4$. Alternatively, after the filtration at 50° the soln. was heated at 80° for 2 hr to give a white ppte. of substantially anhydrous $NaBH_4$ which was collected on

Sodium borohydride (continued)

a sintered-glass filter under nitrogen, then pumped at 60° for 2 hr.
[Brown, Mead and Subba Rao, JACS, 77, 6209 (1955).]

NaBH$_4$ has also been cryst. from isopropylamine by dissolving it in
the solvent at reflux, cooling, filtering and allowing the soln. to
stand in a filter flask connected to a Dry-Ice-acetone trap. After
most of the solvent was passed over into the cold trap, crystals
were removed with forceps, washed with dry ethyl ether and dried
under vac. Somewhat less pure crystals were obtained more rapidly by
using Soxhlet extraction with only a small amount of solvent and
extracting for about 8 hr. The crystals that formed in the flask were
filtered off, then washed and dried as before. [Stockmayer, Rice and
Stephenson, JACS, 77, 1980 (1955).] Other solvents used for crystn.
include water and liquid ammonia.

Sodium bromate.

   Cryst. from hot water (1.1 ml/g).

Sodium bromide.

   Cryst. from water (0.86 ml/g) between 50° and 0°, and dried at 140°
under vac. (This purification may not eliminate chloride ion.)

Sodium 4-bromobenzenesulphonate.

   Cryst. from methanol, ethanol or distd. water.

Sodium cacodylate.

   Cryst. from aq. ethanol.

Sodium carbonate.

   Cryst. from water as the decahydrate which was redissolved in water
to give a near-satd. soln. By bubbling in CO_2, NaHCO$_3$ was pptd. It
was filtered, washed and ignited for 2 hr at 280°. [MacLaren and
Swinehart, JACS, 73, 1822 (1951).] Before being used as a volumetric
standard, analytical grade material should be dried by heating at
$260-270^\circ$ for ½ hr and allowed to cool in a desiccator.

   For preparation of primary standard material, see PAC, 25, 459
(1969).

Sodium cetyl sulphate.
    Cryst. from methanol.

Sodium chlorate.
    Cryst. from hot water (0.5 ml/g).

Sodium chloride.
    Cryst. from satd. aq. soln.   (2.7 ml/g) by passing in HCl gas, or
by adding ethanol or acetone. Can be freed from bromide and iodide
impurities by adding chlorine water to an aq. soln. and boiling for
some time to expel free bromine and iodine. Traces of iron can be
removed by prolonged boiling of solid NaCl with 6 M HCl, the crystals
then being washed with ethanol and dried at about 100°. Sodium
chloride has been purified by sublimation in a stream of pre-purified
nitrogen and collected by electrostatic discharge. [Ross and Winkler,
JACS, 76, 2637 (1954).] For use as a primary analytical standard,
analytical reagent grade sodium chloride should be finely powdered,
dried in an electric furnace at $500-600^\circ$ in a platinum crucible, and
allowed to cool in a desiccator. For most purposes, however, drying
at $110-120^\circ$ is satisfactory.

Sodium chlorite.
    Cryst. from hot water and stored in a cool place. Has also been
cryst. from methanol by countercurrent extraction with liquid
ammonia. [Curti and Locchi, AC, 29, 534 (1957).]

Sodium 4-chlorobenzenesulphonate.
Sodium 4-chloro-m-toluenesulphonate.
    Cryst. twice from methanol and dried under vac.

Sodium chromate.
    Cryst. from hot water (0.8 ml/g).

Sodium creatinephosphate.
    Cryst. from water-ethanol.

Sodium cyanate.
    Cryst. from water.

Sodium p-cymenesulphonate.

Dissolved in water, filtered and evapd. to dryness. Cryst. twice from absolute ethanol and dried at 110°.

Sodium deoxycholate.

Cryst. from ethanol. Dried in oven.

Sodium deoxycholate solution.

Freed from particulate matter by centrifuging at 100,000 g for 2 hr and from most of the soluble components by repeated extraction with acid-washed charcoal.

Sodium 2,5-dichlorobenzenesulphonate.

Cryst. from methanol, and dried under vac.

Sodium 5,5-diethylbarbiturate, see Sodium barbitone.

Sodium di(ethylhexyl)sulphosuccinate (Aerosol-OT).

Dissolved in methanol and inorganic salts which pptd. were filtered off. Water was added and the soln. was extracted several times with hexane. The residue was evapd. to 1/5 original volume, benzene was added and azeotropic distn. was continued until no water remained. Solvent was then evapd. The white solid was crushed and dried in vac. over P_2O_5 for 48 hr. [El Seoud and Fendler, JCS Faraday Trans I, I, 71, 452 (1975).]

Sodium diethyloxalacetate.

Extracted several times with boiling ethyl ether (until the solvent remained colourless) and then residue was dried in air.

Sodium dihydrogen orthophosphate.

Cryst. from warm water (0.5 ml/g) by chilling.

Sodium 2,4-dihydroxyphenylazobenzene-4'-sulphonate.

Cryst. from absolute ethanol.

Sodium 2,2'-dihydroxy-1-naphthaleneazobenzene-5'-sulphonate.

Purified by pptn. of the free acid from aq. soln. using conc. HCl, washing and extracting with ethanol in a Soxhlet extractor. The acid

Sodium 2,2'-dihydroxy-1-naphthaleneazobenzene-5'-sulphonate (continued)
pptd. on evapn. of the ethanol and was reconverted to the sodium salt.

Sodium p-(p-dimethylaminobenzeneazo)-benzenesulphonate.
   Cryst. from water.

Sodium p-dimethylaminoazobenzene-o'-carboxylate.
Sodium p-dimethylaminoazobenzene-p'-carboxylate.
   Pptd. from aq. soln. as the free acid which was cryst. from 95%
ethanol, then reconverted to the sodium salt.

Sodium 2,4-dimethylbenzenesulphonate.
Sodium 2,5-dimethylbenzenesulphonate.
   Cryst. from methanol and dried under vac.

Sodium N,N-dimethylsulphanilate.
   Cryst. from water.

Sodium dithionate.
   Cryst. from hot water (1.1 ml/g) by cooling.

Sodium dodecanesulphonate.
   Cryst. from water or ethanol, and dried in vac.

Sodium dodecyl sulphate.
   Cryst. from ethanol or methanol. Purified by foaming [See Cockbain
and McMullen, TFS, 47, 322 (1951)] or by liquid-liquid extraction.
[See Harrold, J. Colloid Sci., 15, 280 (1960).]

Sodium ethylmercurithiosalicylate.
   Cryst. from ethanol-ethyl ether.

Sodium ferricyanide.
   Cryst. from hot water (1.5 ml/g) or by pptn. from 95% ethanol.

Sodium ferrocyanide.
   Cryst. from hot water (0.7 ml/g), until free of ferricyanide as
shown by absence of Prussian Blue formation with ferrous sulphate.

Sodium fluoride.

Cryst. from water by partial evapn. in a vac. desiccator. Or,
dissolved in water, and approximately half of it pptd. by addition of
ethanol. Ppte. dried in air oven at 130° for one day, and then
stored in desiccator over NaOH.

Sodium fluoroborate.

Cryst. from hot water (50 ml/g) by cooling to 0°. Alternatively,
purified from insoluble material by dissolving in a minimum amount
of water, then fluoride ion was removed by adding conc. lanthanum
nitrate in excess. After removing the lanthanum fluoride by centrifuging, the supernatant was passed through a cation-exchange column
(Dowex 50, sodium form) to remove any remaining lanthanum. [Anbar
and Guttman, JPC, 64, 1896 (1960).]

Sodium fluorosilicate.

Cryst. from hot water (40 ml/g) by cooling.

Sodium formate (anhydr.).

A satd. aq. soln. at 90° (0.8 ml water per gram) was filtered and
allowed to cool slowly. (The final temperature was above 30° to
prevent formation of the hydrate.) After two such crystns. the
crystals were dried in an oven at 130°, then under high vac.
[Westrum, Chang and Levitin, JPC, 64, 1553 (1960).]

Sodium glycollate.

Pptd. from aq. soln. by ethanol, and air-dried.

Sodium hydrogen diglycollate.

Cryst. from hot water (7.5 ml/g) by cooling to 0° with constant
stirring, the crystals being filtered off on to a sintered-glass
funnel and dried at 110° overnight.

Sodium hydrogen oxalate.

Cryst. from hot water (5 ml/g) by cooling.

Sodium hydrogen succinate.

Cryst. from water. Dried at 110°.

Sodium hydrogen d-tartrate.

   Cryst. from warm water (10 ml/g) by cooling to 0°.

Sodium hydroxide (anhydr.).

   Common impurities are water and sodium carbonate. Sodium hydroxide
can be purified by dissolving 100 g in 1 l. of pure dry ethanol,
filtering the soln. under vac. through a fine sintered-glass disc to
remove insoluble carbonates and halides. (This and subsequent operations should be performed in a dry, CO_2-free box.) The soln. is
concentrated under vac., using mild heating, to give a thick slurry
of the mono-alcoholate which is transferred to a coarse sintered-
glass disc and pumped free of mother liquor. After washing the crystals several times with purified alcohol to remove traces of water,
they are vac. dried, with mild heating, for about 30 hr to decompose
the alcoholate, leaving a fine white cryst. powder. [Kelly and
Snyder, JACS, 73, 4114 (1951).]

Sodium hydroxide solutions (caustic).

   Carbonate ion can be removed by passage through an anion-exchange
column (such as Amberlite IRA-400) in the hydroxide form. The column
should be freshly prepared from the chloride form by slow prior
passage of sodium hydroxide soln. until the effluent gives no test
for chloride ion. After use, the column can be regenerated by washing with dil. HCl, then water. Similarly, other metal ions are removed when a 1 M (or more dilute) NaOH soln. is passed through a
column of Dowex ion-exchange A-1 resin in its sodium form.

   Alternatively, carbonate contamination can be reduced by rinsing
analytical reagent quality sticks of NaOH rapidly with water, then
dissolving them in distd. water, or by preparing a conc. aq. soln.
of NaOH and drawing off the clear supernatant liquid. (Insoluble
Na_2CO_3 is left behind.) Carbonate can also be removed by adding a
slight excess of conc. $BaCl_2$ or $Ba(OH)_2$ to a sodium hydroxide soln.,
shaking well and allowing the $BaCO_3$ ppte. to settle. If the presence
of barium in the soln. is unacceptable, an electrolytic purification
can be used. For example, sodium amalgam is prepared by the electrolysis of 3 l. of 30% NaOH with 500 ml of pure mercury for cathode,
and a platinum anode, passing 15 faradays at 4 A, in a thick-walled
polyethylene bottle. The bottle is then fitted with inlet and outlet

Sodium hydroxide solutions (caustic)(continued)
tubes, the spent soln. being flushed out by CO_2-free nitrogen. The
amalgam is then washed thoroughly with a large volume of deionized
water (with the electrolysis current switched on to minimize loss of
sodium). Finally, a clean steel rod is placed in contact in the soln.
with the amalgam (to facilitate hydrogen evolution), reaction being
allowed to proceed until a suitable conc. is reached, before being
transferred to a storage vessel and diluted as required. [Marsh and
Stokes, Aust. J. Chem., 17, 740 (1964).]

Sodium 2-hydroxy-4-methoxybenzophenone-5-sulphonate.
   Cryst. from methanol and dried under vac.

Sodium p-hydroxyphenylazobenzene-p'-sulphonate.
   Cryst. from 95% ethanol.

Sodium iodate.
   Cryst. from hot water (3 ml/g) by cooling.

Sodium iodide.
   Cryst. from water-ethanol soln. and dried for 12 hr under vac. at
70°. Alternatively, dissolved in acetone, filtered and cooled to -20°,
the resulting yellow crystals being filtered off and heated in a vac.
oven at 70° for 6 hr to remove acetone. The sodium iodide was then
cryst. from very dil. NaOH, dried under vac. and stored in a vac.
desiccator. [Verdin, TFS, 57, 484 (1961).]

Sodium isopropylxanthate.
   Cryst. from ligroin-ethanol.

Sodium laurate.
   Cryst. from methanol.

Sodium lauryl sulphate.
   Purified by dissolving in hot 95% ethanol (14 ml/g), filtering and
cooling, then drying in a vac. desiccator.

Sodium mandelate.
   Cryst. from 95% ethanol.

Sodium metanilate.

Sodium metaperiodate, $NaIO_4$.

   Cryst. from hot water.

Sodium metasilicate.

   Cryst. from aq. 5% NaOH soln.

Sodium 3-methyl-1-butanesulphonate.

   Cryst. from 90% methanol.

Sodium molybdate.

   Cryst. from hot water (1 ml/g) by cooling to 0°.

Sodium β-naphthalenesulphonate.

   Cryst. from hot 10% NaCl soln. and dried in a steam oven.

Sodium 2-naphthylamine-5,7-disulphonate.

   Cryst. from water (charcoal) and dried in a steam oven.

Sodium nitrate.

   Cryst. from hot water (0.6 ml/g) by cooling to 0°, or from conc.
aq. soln. by addition of methanol. Dried under vac. at 140°.

Sodium nitrite.

   Cryst. from hot water (0.7 ml/g) by cooling to 0°, or from its own
melt. Dried over P_2O_5.

Sodium oleate.

   Cryst. from ethanol. Dried in oven.

Sodium oxalate.

   Cryst. from hot water (16 ml/g) by cooling to 0°. Before use as a
volumetric standard, analytical reagent quality sodium oxalate should
be dried for 2 hr at 120° and allowed to cool in a desiccator.

Sodium palmitate.

   Cryst. from ethanol. Dried in oven.

Sodium perchlorate.

Because its solubility in water is high (2.1 g/ml at 15°) and it has a rather low temperature coefficient of solubility, sodium perchlorate is usually cryst. from acetone, methanol, water-ethanol or dioxane-water (33 g dissolved in 36 ml of water and 200 ml of dioxane). After filtering and crystallizing, the solid is dried under vac. at 140-150° to remove solvent of crystn. Basic impurities can be removed by crystn. from hot acetic acid, followed by heating at 150°. If $NaClO_4$ is pptd. from distd. water by adding $HClO_4$ to the chilled soln., the ppte. contains some free acid.

Sodium p-phenolsulphonate.

Cryst. from hot water (1 ml/g) by cooling to 0°, or from methanol. Dried under vac.

Sodium phenoxide.

Washed with ethyl ether then heated under vac. to 200° to remove any free phenol.

Sodium phenylacetate.

Its aq. soln. was evapd. to crystn. on a steam-bath, the crystals being washed with absolute ethanol and dried under vac. at 80°.

Sodium o-phenylphenolate.

Cryst. from acetone and dried under vac. at room temperature.

Sodium phosphoamidate.

Dissolved in water below 10°, and acetic acid added dropwise to pH 4.0 so that the monosodium salt was pptd. The ppte. was washed with water and ethyl ether, then air dried. Addition of one equivalent of NaOH to the soln. gave the disodium salt, the soln. being adjusted to pH 6.0 before use. [Rose and Heald, BJ, 81, 339 (1961).]

Sodium phytate.

Cryst. from water.

Sodium piperazine-N,N'-bis(2-ethanesulphonate)(monohydrate).

Cryst. from water and ethanol.

Sodium pyrophosphate (decahydrate).
  Cryst. from warm water and air dried at room temperature.

Sodium selenate.
Sodium selenite.
  Cryst. from water.

Sodium silicate solution.
  Purified by contact filtration with activated charcoal.

Sodium succinate.
  Cryst. from hot water (1.2 ml/g) by cooling to 0°. Dried at 125°.

Sodium sulphanilate.
  Cryst. from water.

Sodium sulphate (decahydrate).
  Cryst. from water at 30° (1.1 ml/g) by cooling to 0°. (Sodium
sulphate becomes anhydrous at 32°.)

Sodium sulphide.
  Some purification of the hydrated salt can be achieved by selecting
large crystals and removing the surface layer (contaminated with
oxidation products) by washing with distd. water. Other metal ions
can be removed from sodium sulphide solns. by passage through a
column of Dowex ion-exchange A-1 resin, sodium form. The hydrated
salt can be rendered anhydrous by heating in a stream of hydrogen or
of nitrogen until water is no longer evolved. (The resulting cake
should not be heated to fusion because it is readily oxidized.)

Sodium sulphite.
  Cryst. from warm water (0.5 ml/g) by cooling to 0°.

Sodium R-tartrate.
  Cryst. from warm dil. aq. NaOH by cooling.

Sodium tetrametaphosphate.

Cryst. twice from water at room temperature by adding ethanol (300 g of $Na_4P_4O_{12}H_2O$, 2 1. of water, and 1 1. of ethanol), washing first with 20% ethanol then with 50% ethanol. Air dried. [Quimby, JPC, 58, 603 (1954).]

Sodium tetraphenylborate.

Cryst. from acetone-hexane or $CHCl_3$, or from ethyl ether-cyclohexane (3:2) by warming the soln. to ppte. the compound. Dried in vac. at 80°.

Sodium thioantimonate.

Cryst. from warm water (2 ml/g) by cooling to 0°.

Sodium thiocyanate.

Cryst. from water; or from methanol using ethyl ether for washing, then drying at 130°. (The latter purification removes material reacting with iodine.) Sodium thiocyanate solns. can be freed from traces of iron by repeated batch extractions with ethyl ether.

Sodium thioglycollate.

Cryst. from 60% ethanol (charcoal).

Sodium thiosulphate.

Cryst. from ethanol-water solns. or from water (0.3 ml/g) below 60° by cooling to 0°, and dried at 35° under vac.

Sodium p-toluenesulphinate.

Cryst. from water (to constant u.v. spectrum), and dried under vac., or extracted with hot benzene, then dissolved in ethanol-water and heated with decolorizing charcoal. Filtered and cooled to give crystals of the dihydrate.

Sodium p-toluenesulphonate.

Dissolved in distd. water, filtered to remove insoluble impurities and evapd. to dryness. Then cryst. from methanol or ethanol, and dried at 110°. Its solubility in ethanol is not high (maximum 2.5%) so that Soxhlet extraction with ethanol may be preferable. Sodium p-toluenesulphonate has also been cryst. from acetic acid, washed thoroughly with ethyl ether and dried under vac. at 50°.

Sodium trifluoroacetate.

Pptd. from ethanol by adding dioxane, then cryst. several times from hot absolute ethanol. Dried at 120-130°/1 mm.

Sodium 2,2',4-trihydroxyazobenzene-5'-sulphonate.

Purified by precipitating the free acid from aq. soln. using conc. HCl, then washing and extracting with ethanol in a Soxhlet extractor. Evaporation of the ethanol left the purified acid.

Sodium trimetaphosphate.

Pptd. from an aq. soln. at 40° by adding ethanol. Air dried.

Sodium 2,4,6-trimethylbenzenesulphonate.

Cryst. twice from methanol and dried under vac.

Sodium triphosphate.

Purified by repeated pptn. from aq. soln. by slow addition of methanol. Air dried.

Sodium tripolyphosphate.

A soln. of anhydrous sodium tripolyphosphate (840 g) in water (3.8 1.) was filtered, and methanol (1.4 1.) was added with vigorous stirring to ppte. $Na_5P_3O_{10}.6H_2O$. The ppte. was collected on a filter, air dried by suction, then left to dry in air overnight. It was crystd. twice more in this way, using a 13% aq. soln. (w/w), and leaching the crystals with 200-ml portions of water. [Watters, Loughran and Lambert, JACS, 78, 4855 (1956).] Similarly, ethanol can be added to ppte. the salt from a filtered 12-15% aq. soln., the final soln. containing about 25% ethanol (v/v). Air drying should be at a relative humidity of 40-60%. Heat and vac. drying should be avoided. [Quimby, JPC, 58, 603 (1954).]

Sodium tungstate.

Cryst. from hot water (0.8 ml/g) by cooling to 0°.

Sodium m-xylenesulphonate.
Sodium p-xylenesulphonate.

Dissolved in distd. water, filtered, then evapd. to dryness. Cryst. twice from absolute ethanol and dried at 110°.

Stannic chloride.

Refluxed with clean mercury or P_2O_5 for several hours, then distd. under (reduced) nitrogen pressure into a receiver containing P_2O_5. Finally redistd. Alternatively, distd. from tin metal under vac. in an all-glass system and sealed off in large ampoules. Fumes in moist air.

Stannic iodide.

Cryst. from anhydrous $CHCl_3$, dried under vac. and stored in a vac. desiccator.

Stannic oxide, SnO_2.

Refluxed repeatedly with fresh hydrochloric acid until the acid showed no tinge of yellow. The oxide was then dried at 110°.

Stannous chloride (anhydr.).

Analytical reagent grade stannous chloride dihydrate is dehydrated by adding slowly to vigorously stirred, redistd. acetic anhydride (120 g salt per 100 g of anhydride). (In a fume cupboard.) After about an hour, the anhydrous $SnCl_2$ is filtered on to a sintered glass or Büchner funnel, washed free from acetic acid with dry ethyl ether (2 x 30 ml), and dried under vac. It is stored in a sealed container. [Stephen, JCS, 2786 (1930).]

Strontium acetate.

Cryst. from acetic acid, then dried under vac. for 24 hr at 100°.

Strontium bromide.

Cryst. from water (0.5 ml/g).

Strontium chloride.

Cryst. from warm water (0.5 ml/g) by cooling to 0°.

Strontium chromate.

Cryst. from hot water (40 ml/g) by cooling.

Strontium hydroxide.

Cryst. from hot water (2.2 ml/g) by cooling.

Strontium lactate.

  Cryst. from aq. ethanol.

Strontium nitrate.

  Cryst. from hot water (0.5 ml/g) by cooling to 0°.

Strontium oxalate.

  Cryst. from hot water (20 ml/g) by cooling.

Strontium salicylate.

  Cryst. from hot water (4 ml/g) or ethanol.

Strontium tartrate.

  Cryst. from hot water.

Strontium thiosulphate.

  Cryst. from hot water (2 ml/g) by cooling to 0°.

Sulphamic acid, m.p. 205° (decomp.).

  Cryst. from water at 70° (300 ml per 125 g), after filtering, by
cooling a little and discarding the first batch of crystals (about
25 g) before standing in an ice-salt mixture for 20 min. The crystals
were filtered by suction, washed with a small quantity of ice water,
then twice with cold ethanol and finally with ethyl ether. Air dried
for 1 hr, then stored in a desiccator over $Mg(ClO_4)_2$. [Butler, Smith
and Audrieth, IECAE, 10, 690 (1938).]

  For preparation of primary standard material, see PAC, 25, 459
(1969).

Sulphamide, m.p. 91.5°.

  Cryst. from absolute ethanol.

Sulphur, m.p. between 112.8° and 120°, depending on form.

  Murphy, Clabaugh and Gilchrist [J. Res. Nat. Bur. Stand., 64A, 355
(1960)] have obtained sulphur of about 99.999 moles per cent purity
by the following procedure: Roll sulphur was melted and filtered
through a coarse-porosity glass filter funnel into a 2 l. round-
bottomed Pyrex flask with two necks. Conc. H_2SO_4 (300 ml) was added to
the sulphur (2.5 kg), and the mixture was heated to 150°, stirring

<u>Sulphur</u> (continued)

continuously for 2 hr. Over the next 6 hr, conc. HNO_3 was added in about 2 ml portions at 10-15 min intervals to the heated mixture. It was then allowed to cool to room temperature and the acid was poured off. The sulphur was rinsed several times with distd. water, then remelted, cooled, and rinsed several times with distd. water, this process being repeated four or five times to remove most of the acid entrapped in the sulphur. An air-cooled reflux tube (about 40 cm long) was attached to one of the necks of the flask, and a gas delivery tube (the lower end about 1 in. above the bottom of the flask) was inserted into the other. While the sulphur was boiled under reflux, a stream of helium or nitrogen was passed through to remove any water, HNO_3 or H_2SO_4, as vapour. After 4 hr, the sulphur was cooled so that the reflux tube could be replaced by a bent air-cooled condenser. The sulphur was then distd., rejecting the first and the final 100 ml portions, and transferred in 200 ml portions to 400 ml glass cylind-rical ampoules (which were placed on their sides during solidification). After adding about 80 ml of water, displacing the air with nitrogen, and sealing the ampoule, it was heated to $125°$ and shaken to extract residual H_2SO_4. The ampoule was cooled, and the water was titrated with 0.02 M NaOH, the process being repeated until the acid content was negligible. Finally, entrapped water was removed by alternate evacuation to 10 mm Hg and refilling with nitrogen while the sulphur was kept molten.

Other purifications include crystn. from CS_2 (which is less satis-factory because the sulphur retains appreciable amounts of organic material), benzene or benzene-acetone, followed by melting and de-gassing. Has also been boiled with 1% MgO, then decanted, and dried under vac. at $40°$ for 2 days over P_2O_5. [For purification of S_6, "recryst. S_8" and "Bacon-Fanelli sulphur", see Bartlett, Cox and Davis, <u>JACS</u>, <u>83</u>, 103 109 (1961).]

<u>Sulphur dichloride</u>, b.p. $59°$/760 mm.

Twice distd. in the presence of a small amount of PCl_3 through a 12-in. Vigreux column, the fraction boiling between 55-61° being redistd. (in the presence of PCl_3), and the fraction distilling between 58-61° retained. (The PCl_3 is added to inhibit the decompos-ition of SCl_2 into S_2Cl_2 and Cl_2. The SCl_2 must be used as quickly as

Sulphur dichloride (continued)

possible after distn.; within 1 hr at room temperature the sample contains 4% S_2Cl_2. On long standing this reaches 16-18%.)

Sulphur dioxide, b.p. -10°.

Dried by bubbling through conc. H_2SO_4 and by passage over P_2O_5, mist being removed by passage through a glass-wool plug. Frozen with liquid air and pumped to a high vac. to remove dissolved gases.

Sulphuric acid.

Sulphuric acid, and also 30% fuming H_2SO_4, can be distd. in an all-Pyrex system, optionally from potassium persulphate. Also purified by fractional crystn. of the monohydrate from the liquid.

Sulphur monochloride, b.p. 138°/760 mm.

Purified by distn. below 60° from a mixture containing sulphur (2 %) and activated charcoal (1%), under reduced pressure (e.g. 50 mm). Stored in the dark in a refrigerator.

Telluric acid.

Cryst. once from nitric acid, then repeatedly from hot water (0.4 ml/g).

Tellurium, m.p. 450°.

Purified by zone refining and repeated sublimation to an impurity of less than 1 part in 10^8 (except for surface contamination by TeO_2). [Machol and Westrum, JACS, 80, 2950 (1958).] (Tellurium is volatile at 500°/0.2 mm.) Also purified by electrodeposition. [Mathers and Turner, Trans. Amer. Electrochem. Soc., 54, 293 (1928).]

Tellurium dioxide.

Dissolved in 5 M NaOH, filtered and pptd. by adding 10 M HNO_3 to the filtrate until the soln. was acid to phenolphthalein. After decanting the supernatant, the ppte. was washed five times with distd. water, then dried for 24 hr at 110°. [Horner and Leonhard, JACS, 74, 3694 (1952).]

Terbium oxide.

Dissolved in acid, pptd. as its oxalate and ignited at 650°.

Tetraethyllead

Its more volatile contaminants can be removed by exposure to a low pressure (by continuous pumping) for 1 hr at 0°. Purified by stirring with an equal volume of H_2SO_4 (S.G. 1.40), keeping the temperature below 30°, repeating this process until the acid layer is colourless. It is then washed with dil. Na_2CO_3 and distd. water, dried with $CaCl_2$ and fractionally distd. at low pressure under H_2 or N_2. [Calingaert, Chem. Rev., 2, 43 (1926).]

Tetramethylammonium triphenylborofluoride, m.p. 186°.

Cryst. from acetone or acetone-ethanol.

Tetramethylsilane.

Distd. from conc. H_2SO_4 (after shaking with it) or $LiAlH_4$, through a 5-ft vacuum-jacketed column packed with glass helices into an ice-cooled condenser, then percolated through silica gel to remove traces of halide.

Tetraphenylarsonium chloride, m.p. $261-263^\circ$.

A neutralized aq. soln. was evapd. to dryness. The residue was extracted into absolute ethanol, evapd. to small volume and pptd. by addition of absolute ethyl ether. It was again dissolved in a small volume of absolute ethanol or ethyl acetate and repptd. with ethyl ether.

Tetraphenylsilane, m.p. $231-233^\circ$.

Cryst. from benzene.

Tetraphenyltin, m.p. 226°.

Cryst. from $CHCl_3$, xylene or benzene-cyclohexane, and dried at $75^\circ/20$ mm.

Thallous bromide, m.p. 460°.

Thallous bromide (20 g) was refluxed for 2-3 hr with water (200 ml) containing 3 ml of 47% HBr. It was then washed until acid free, heated to 300° for 2-3 hr and stored in brown bottles.

Thallous carbonate.

Cryst. from hot water (4 ml/g) by cooling.

Thallous chlorate.

Cryst. from hot water (2 ml/g) by cooling.

Thallous chloride, m.p. 430°.

Cryst. from 1% HCl and washed until acid-free, or cryst. from hot water (50 ml/g), then dried at 140° and stored in brown bottles.

Thallous hydroxide.

Cryst. from hot water (0.6 ml/g) by cooling.

Thallous iodide.

Refluxed for 2-3 hr with water containing HI, then washed until acid-free and dried at 120°. Stored in brown bottles.

Thallous nitrate.

Cryst. from warm water (1 ml/g) by cooling to 0°.

Thallous perchlorate.

Cryst. from hot water (0.6 ml/g) by cooling.

Thallous sulphate.

Cryst. from hot water (7 ml/g) by cooling, then dried under vac. over P_2O_5.

Thionyl chloride, b.p. 77°.

Crude $SOCl_2$ can be freed from sulphuryl chloride, sulphur mono-chloride and sulphur dichloride by refluxing with sulphur and then fractionally distilling twice. (The SO_2Cl_2 is converted to SO_2 and sulphur chlorides. The S_2Cl_2 (b.p. 135.6°) is left in the residue, whereas SCl_2 (b.p. 59°) passes over in the forerun.) The usual purification is to distil from quinoline (50 g $SOCl_2$ to 10 g of quinoline) to remove acid impurities, followed by distn. from boiled linseed oil (50 g $SOCl_2$ to 20 g of oil.) Precautions must be taken to exclude moisture.

Thionyl chloride for use in organic syntheses can be prepared by distn. of technical $SOCl_2$ in the presence of diterpene (12 g/250 ml $SOCl_2$), avoiding overheating. Further purification is achieved by redistn. from linseed oil (1-2%).[Rigby, Chem. Ind., 1508 (1969).]

Thionyl chloride (continued)
Gas-chromatographically pure material is obtained by distn. from 10%
(w/w) triphenyl phosphite. [Friedman and Wetter, JCS (A), 36 (1967).]

Thorium chloride.
    Freed from anionic impurities by passing a 2 M soln. of $ThCl_4$ in
3 M HCl through a Dowex-1 anion-resin column. The eluate was partially
evapd. to give crystals which were filtered off, washed with ethyl
ether and stored in a desiccator over H_2SO_4 to dry. Alternatively, a
satd. soln. of $ThCl_4$ in 6 M HCl was filtered through asbestos and
extracted twice with ethyl, or isopropyl, ether (to remove iron), then
evapd. to a small volume on a hot plate. (Excess silica pptd. and was
filtered off through asbestos. The filtrate was cooled to 0° and satd.
with dry HCl gas.) It was shaken with an equal volume of ethyl ether,
agitating with HCl gas, until the mixture became homogeneous. On
standing, $ThCl_4 \cdot 8H_2O$ pptd. and was filtered off, washed with ethyl
ether, and dried. [Kremer, JACS, 64, 1009 (1942).]

Thorium sulphate.
    Cryst. from water.

RS-Thyroxine sodium salt.
    Cryst. from absolute ethanol and dried for 8 hr at $30^{\circ}/1$ mm.

Tin (powder).
    The powder was added to about twice its weight of 10% aq. NaOH and
shaken vigorously for 10 min. (This removed oxide film and stearic
acid or similar material sometimes added for pulverization.) It was
then filtered, washed with water until the washings were no longer
alkaline to litmus, rinsed with methanol and air dried. [Sisido,
Takeda and Kinugama, JACS, 83, 538 (1961).]

Titanium tetrachloride, b.p. 136.4°.
    Refluxed with mercury or a small amount of pure copper turnings to
remove the last traces of light colour (due to $FeCl_3$ and V(IV)
chloride), then distd. under nitrogen in an all-glass system, taking
precautions to exclude moisture. Clabaugh, Leslie and Gilchrist
[J. Res. Nat. Bur. Stand., 55, 261 (1955)] removed organic material

Titanium tetrachloride (continued)

by adding aluminium chloride hexahydrate as a slurry with an equal amount of water (the slurry being about one-fiftieth the weight of $TiCl_4$), refluxing for 2-6 hr while bubbling in chlorine, which was subsequently removed by passing a stream of clean dry air. The $TiCl_4$ was then distd., refluxed with copper and again distd., taking precautions to exclude moisture. Volatile impurities were then removed using a technique of freezing, pumping and melting.

Titanyl sulphate, $TiOSO_4 \cdot 2H_2O$.

Dissolved in water, filtered and cryst. three times from boiling 45% H_2SO_4, washing with ethanol to remove excess acid, then with ethyl ether. Air dried for several hours, then oven dried at 105-110°. [Hixson and Fredrickson, IEC, 37, 678 (1945).]

Tributyl borate, b.p. 232.4°, n_D^{20} 1.4092, n_D^{30} 1.4051.

The chief impurities are n-butyl alcohol and boric acid (from hydrolysis). It must be handled in a dry-box, and can readily be purified by fractional dist. under reduced pressure.

Trichloroborane.

Purified by condensing into a trap cooled in acetone-Dry Ice, where it was pumped for 15 min to remove volatile impurities. It was then warmed, recondensed and again pumped.

B-Trichloroborazine.

Purified by distn. from mineral oil.

Triethylaluminium.

Distd. under vac. in a 50-cm column containing a heated nichrome spiral, taking the fraction 112-114°/27 mm.

Triethylborane.

Distd. at 56-57°/220 mm.

Triethyl borate, b.p. 118.6°.

Dried with sodium, then distd.

Triethyltin hydroxide.

   Treated with HCl, followed by KOH, and filtered to remove diethyltin oxide. [Prince, JCS, 1783 (1959).]

Tri-n-hexylborane.

   Treated with hex-1-ene and 10% anhydrous ethyl ether for 6 hr at gentle reflux under nitrogen, then vac. distd. through an 18-in. glass helices-packed column under nitrogen taking the fraction b.p. $130^{o}/2.1$ mm to $137^{o}/1.5$ mm. The distillate still contained some di-n-hexylborane. [Mirviss, JACS, 83, 3051 (1961).]

Trimethyl borate, b.p. 65°.
   Dried with sodium, then distd.

Triphenylborane, m.p. 142-142.5^{o}, b.p. $203^{o}/15$ mm.
   Cryst. three times from benzene under nitrogen.

Triphenylchlorosilane, m.p. 97-99°.

   Likely impurities are tetraphenylsilane, small amounts of hexaphenyldisiloxane and traces of triphenylsilanol. Purified by dist. at 2 mm, then cryst. from ethanol-free $CHCl_3$, and from pet.ether (b.p. 30-60°) by cooling in a Dry Ice-acetone bath.

Triphenylchlorostannane, m.p. 104°.

   Cryst. repeatedly from pet.ether (b.p. 30-60°) or ethanol, then sublimed under vac.

Triphenylsilanol.

   Cryst. from benzene or ethyl ether-pet.ether (1:1).

Triphenyltin hydroxide.

   West, Baney and Powell [JACS, 82, 6269 (1960)] purified a sample which was grossly contaminated with tetraphenyltin and diphenyltin oxide by dissolving it in ethanol, most of the impurities remaining behind as an insoluble residue. Evapn.of the ethanol gave the crude hydroxide which was converted to triphenyltin chloride by grinding in a mortar under 12 M HCl, then evaporating the acid soln. The chloride, after crystn. from ethanol, had m.p. 104-105°. It was dissolved in ethyl ether and converted to the hydroxide by stirring with excess

Triphenyltin hydroxide (continued)

aq. ammonia. The ether layer was separated, dried, and evaporated to give triphenyltin hydroxide which, after crystn. from ethanol and drying under vac., was in the form of white crystals (m.p. 119-120°), which retained some cloudiness in the melt above 120°. The hydroxide retains water (0.1-0.5 moles of water per mole) tenaciously.

Tri-n-propyl borate.

   Dried with sodium and then distd.

Tris(2,2'-bipyridine)ruthenium(II) dichloride.

   Cryst. from water.

Trisodium citrate.

   Cryst. from warm water by cooling to 0°.

Trisodium 1,3,6-naphthalenetrisulphonate.

   The free acid was obtained by passage through an ion-exchange column and converted to the lanthanum salt by treatment with La$_2$O$_3$. This salt was cryst. twice from hot water. (The much lower solubility of La$_2$(SO$_4$)$_3$ and its retrograde temperature dependence allows a good separation from sulphate impurity.) The lanthanum salt was then passed through an appropriate ion-exchange column to obtain the free acid, the sodium or the potassium salt. [The sodium salt is hygro-scopic.) [Atkinson, Yokoi and Hallada, JACS, 83, 1570 (1961).]

Trisodium orthophosphate.

   Cryst. from warm dil. aq. NaOH (1 ml/g) by cooling to 0°.

Tritium.

   Purified from hydrocarbons and ^{3}He by diffusion through the wall of a hot nickel tube. [Landecker and Gray, Rev. Sci. Inst., 25, 1151 (1954).] Radioactive.

Tungsten (rod.).

   Cleaned with conc. NaOH soln., rubbed with very fine emery-paper until its surface was bright, washed with previously boiled and cooled conductivity water and dried with filter paper.

<u>Uranium hexafluoride</u>, b.p. $0°/17.4$ mm, m.p. $64.8°$.
  Purified by fractional distn. to remove HF.

<u>Uranium trioxide</u>.

  The oxide was dissolved in perchloric acid (to give a uranium content of 5%), and the soln. was adjusted to pH 2 by addition of dilute
ammonia. Dropwise addition of 30% H_2O_2, with rapid stirring, pptd.
U(VI) peroxide, the pH being held constant during the pptn. by addition of small amounts of the ammonia soln. (The H_2O_2 was added until
further quantities caused no change in pH.) After stirring for 1 hr,
the slurry was filtered through coarse filter paper in a Büchner
funnel and washed with 1% H_2O_2 acidified to pH 2 with $HClO_4$, then with
1% H_2O_2 and dried at $110°$. Finally, the cake was ground and heated at
$350°$ for three days in a large platinum dish. [Baes, JPC, 60, 878
(1956).]

<u>Uranyl nitrate</u>, m.p. $60.2°$ (hexahydrate), b.p. $118°$.

  Cryst. from warm water by cooling to $-5°$, taking only the middle
fraction of the solid which separated. Dried as the hexahydrate over
35-40% H_2SO_4 in a vac. desiccator.

<u>Vanadium</u> (metal).

  Cleaned by rapid exposure consecutively to HNO_3, HCl, HF, de-ionized
water and reagent grade acetone, then vac. desiccated.

<u>Vanadyl acetylacetonate</u>.

  Cryst. from acetone.

<u>Water</u>, b.p. $100.0°$.

  Conductivity water (specific conductance about 10^{-7}mho) can be
obtained by distilling water in a steam-heated tin-lined still, then,
after adding 0.25% of solid NaOH and 0.05% of $KMnO_4$, distilling once
more from an electrically heated Barnstead-type still, taking the
middle fraction into a Jena glass bottle. During these operations
suitable traps must be used to protect against entry of CO_2 and NH_3.
Water only a little less satisfactory for conductivity measurements
(but containing traces of organic materials) can be obtained by passing ordinary distd. water through a mixed bed ion-exchange column containing, for example, Amberlite resins IR 120 (cation exchange) and

Water (continued)

IRA 400 (anion), or Amberlite MB-1. This treatment is also a convenient one for removing traces of heavy metals. (The heavy metals
copper, zinc, lead, cadmium and mercury can be tested for by adding
pure conc. ammonia to 10 ml of sample and shaking vigorously with
1-2 ml 0.001% dithizone in CCl_4. Less than 0.1 µg of metal ion will
impart a faint colour to the CCl_4 layer.) For almost all laboratory
purposes, simple distn. yields water of adequate purity, and most of
the volatile contaminants such as ammonia and CO_2 are removed if the
first fraction of distillate is discarded.

Zinc (dust).

   Commercial zinc dust (1.2 kg) was stirred with 2% HCl (3 l.) for
1 min, (then the acid was removed by filtration), and washed in a
4 l. beaker with a 3 l. portion of 2% HCl, three 1 l. portions of
distd. water, two 2 l. portions of 95% ethanol, and finally with
2 l. of absolute ethyl ether. (The wash solns. were removed each
time by filtration.) The material was then dried thoroughly and if
necessary, any lumps were broken up in a mortar.

Zinc (metal).

   Fused under vac., cooled, then washed with acid to remove the oxide.

Zinc acetate.

   Cryst. (in poor yield) from hot water or, better, from ethanol.

Zinc acetylacetonate.

   Cryst. from hot 95% ethanol.

Zinc bromide.

   Heated to $300°$ under vac. (2×10^{-2} mm) for 1 hr, then sublimed.

Zinc caprylate.

   Cryst. from ethanol.

Zinc chloride, m.p. $283°$.

   The anhydrous material can be sublimed under a stream of dry HCl,
followed by heating to $400°$ in a stream of dry nitrogen.

                                                    [continued]

Zinc chloride (continued)

Purified by refluxing (50 g) in dioxane (400 ml) with 5 g zinc dust, filtering hot and cooling to ppte. $ZnCl_2$. Cryst. from dioxane and stored in a desiccator over P_2O_5. Hygroscopic: minimal exposure to the atmosphere is desirable.

Zinc diethyldithiocarbamate.
Zinc dimethyldithiocarbamate, m.p. 148°.
Zinc ethylenebis[dithiocarbamate].

Cryst. several times from hot toluene or from hot $CHCl_3$ by addition of ethanol.

Zinc formate.

Cryst. from water (3 ml/g).

Zinc iodide.

Heated to 300° under vac. (2×10^{-2} mm) for 1 hr, then sublimed.

Zinc RS-lactate.

Cryst. from water (6 ml/g).

Zinc perchlorate.

Cryst. from water.

Zinc phenol-p-sulphonate.

Cryst. from warm water by cooling to 0°.

Zinc sulphate.

Cryst. from aq. ethanol.

Zirconium tetrachloride.

Cryst. repeatedly from conc. HCl.

Zirconyl chloride.

Cryst. repeatedly from 8 M HCl as $ZrOCl_2 \cdot 8H_2O$. (The product was not free of hafnium.)

CHAPTER 5

GENERAL METHODS FOR THE PURIFICATION OF CLASSES OF COMPOUNDS

Chapters 3 and 4 list a large number of individual compounds, with a brief indication of how each one may be purified. For substances that are not included in these chapters the following general procedures may prove helpful.

If the laboratory worker does not know of a reference to the preparation of a commercially available substance, he may be able to make a reasonable guess at the method used from published laboratory syntheses. This information, in turn, can simplify the necessary purification steps by suggesting probable contaminants.

Impurities that amount to a few per cent can be observed by measuring some of the spectroscopic and physical properties. Thus a combination of u.v., i.r. and p.m.r. spectra, colour, gas or t.l.c. chromatographic properties and spot tests can give one a very good idea of what impurities are present. Purification methods can then be devised to remove these impurities, and a monitoring method will have already been established.

Physical methods of purification depend largely on the m.pt. or b.pt. of the material. For gases and low-boiling liquids use is commonly made of the freeze-pump-thaw method, by which the chemical to be purified is condensed by liquid air, solid CO_2, etc., in a trap. The solid distillate is pumped at low pressure to remove low-boiling impurities, and then it is allowed to thaw so that the chemical can be distilled off, leaving less volatile impurities in the residual fraction. Gas chromatography is also useful, especially for low-boiling point liquids. Liquids are usually purified by refluxing with drying agents, acids or bases, reducing agents, charcoal, etc., followed by fractional distillation under reduced pressure. For solids, general methods include fractional freezing of the melted material, taking the middle fraction. A related procedure is zone melting. Another procedure is sublimation of the solid under reduced pressure. The other commonly used method for purifying solids is by crystallization from a solution in a suitable solvent, by cooling with or without the prior addition of a solvent in which the solute is not very soluble.

Purification becomes meaningful only insofar as adequate tests of
purity are applied: the higher the degree of purity that is sought,
the more stringent must these tests be. If the material is an organic
solid, its melting point should first be taken and compared with the
recorded value. Also, as part of this preliminary examination, the
sample might be examined by paper (or thin layer) chromatography
(for the latter, see E. Demole, "Recent Progress in Thin-Layer
Chromatography", Chromatographic Reviews, 4, 26 (1962)) in several
different solvent systems and in high enough concentrations to
facilitate the detection of minor components. On the other hand, if
the substance is a liquid, its boiling point should be measured. If,
further, it is a high-boiling liquid, its chromatographic behaviour
should be examined. Liquids, especially the more volatile ones, can
be studied very satisfactorily by gas chromatography, preferably
using at least two different stationary phases. Application of these
tests at successive steps will give a good indication of whether or
not the purification is satisfactory and will also show when adequate
purification has been achieved.

The nature of the procedure adopted will depend to a large extent on
the quantity of purified material that is required. For example, for
small quantities (50-250 mg) of a pure liquid, preparative gas
chromatography is probably the best method. Two passes through a
suitable column may well be sufficient. Similarly, for small amounts
(100-500 mg) of an organic solid, column chromatography is likely to
be very satisfactory, the eluate being collected as a number of
separate fractions (about 5-10 ml each) which are examined by ultraviolet spectrophotometry, by paper or thin layer chromatography, or
by some other appropriate technique. (In finding by experiment a
suitable adsorbent and eluant, the book by Lederer and Lederer should
prove useful.) Preparative thin layer chromatography can be used
successfully for purifying up to 500 mg of solid.

Where larger quantities (upwards of 1 g) are required, most of the
impurities should be removed by preliminary treatments such as
solvent extraction, liquid-liquid partition, or conversion to a
derivative (preferably solid, see Chapter 2) which can be purified
by crystallization or fractional distillation before being reconverted
to the starting material. The substance is then crystallized or
distilled. If the final amounts must be in excess of 25 g, preparation
of a derivative is sometimes omitted because of the cost involved.
In all of the above cases, purification is likely to be more laborious
if the impurity is an isomer or a derivative with closely similar
physical properties.

In the general methods of purification described below, it is assumed
that the impurities belong essentially to a class of compound different from the one being purified. They are suggested for use in cases
where substances are not listed in Chapters 3 or 4. In such cases,
the experimenter is advised to employ them in conjunction with information given in these chapters for the purification of suitable anal-
ogues. Also, for a wider range of drying agents, solvents for extraction and solvents for recrystallization, the reader is referred to
Chapter 1.

CLASSES OF COMPOUNDS

Acetals. These are generally diethyl or dimethyl acetal derivatives of aldehydes. They are more stable to alkali than to acids. Their common impurities are the corresponding alcohol, aldehyde and water. Drying with sodium wire removes alcohols and water, and polymerizes aldehydes so that, after decantation, the acetal can be fractionally distilled. In cases where the use of sodium is too drastic, aldehydes can be removed by shaking with alkaline hydrogen peroxide solution and the acetal is dried with sodium carbonate or potassium carbonate. Residual water and alcohols (up to n-propyl) can be removed with Linde type 4A molecular sieve. The acetal is then filtered and fractionally distilled. Solid acetals (i.e. acetals of high molecular weight aldehydes) are generally low-melting and can be recrystallized from low-boiling petroleum ether, benzene or a mixture of both.

Acids. (a) Carboxylic: Liquid carboxylic acids are first freed from neutral and basic impurities by dissolving them in aqueous alkali and extracting with ethyl ether. (The pH of the solution should be at least three units above the pK_a of the acid.) The aqueous phase is then acidified to a pH at least three units below the pK_a of the acid and again extracted with ether. This extract is dried with magnesium sulphate or sodium sulphate and the ether is distilled off. The acid is fractionally distilled through an efficient column. It can be further purified by conversion to its methyl or ethyl ester (see p. 70) which is then fractionally distilled. Hydrolysis yields the original acid which is again purified as above.

Acids that are solids can also be purified in this way, except that distillation is replaced by repeated recrystallization (preferably from at least two different solvents such as water, alcohol or aqueous alcohol, benzene, benzene-peteroleum ether or acetic acid.) Water-insoluble acids can be partially purified by dissolution in N sodium hydroxide solution and precipitation with dilute mineral acid.

The separation and purification of naturally occurring fatty acids, based on distillation, salt solubility and low temperature crystallization, are described by K.S. Markley (Ed.), Fatty Acids, 2nd edn., part 3, chap. 20, Interscience, New York, 1964.

Aromatic carboxylic acids can be purified by conversion to their sodium salts, recrystallization from hot water, and reconversion to the free acids.

(b) Sulphonic: The low solubility of sulphonic acids in organic solvents and their high solubility in water makes necessary a treatment different from that for carboxylic acids. Sulphonic acids are strong, they have a tendency to hydrate, and many of them contain water of crystallization. The lower-melting and liquid acids can generally be purified with only slight decomposition by fractional distillation, preferably under reduced pressure. A common impurity is sulphuric acid, but this can be removed by recrystallization from concentrated aqueous solutions. The wet acid can then be dried by azeotropic removal of water with benzene, followed by distillation. The higher-melting acids, or acids that melt with decomposition, can be crystallized from water or, occasionally, from alcohol.

(c) <u>Sulphinic</u>: These acids are less stable, less soluble and less acidic than the corresponding sulphonic acids. The common impurities are the corresponding sulphonyl chlorides from which they have been prepared, and the thiosulphonates (neutral) and sulphonic acids into which they decompose. The first two of these can be removed by solvent extraction from an alkaline solution of the acid. On acidification of the alkaline solution the sulphinic acid crystallizes out, leaving the sulphonic acid behind. The lower molecular weight members are isolated as their metal (ferric) salts, but the higher members can be crystallized from water (made slightly acidic) or alcohol.

<u>Acid chlorides</u>. The corresponding acid and hydrogen chloride are the most likely impurities. Usually these can be removed by efficient fractional distillation. Where the acid chlorides are not readily hydrolyzed (e.g. phenyl sulphonyl chlorides) the compound can be freed from contaminants by dissolving it in a suitable solvent such as alcohol-free chloroform, dry benzene or light petroleum and shaking with dilute sodium bicarbonate solution. The organic phase is then washed with water, dried with sodium sulphate or magnesium sulphate, and distilled. This procedure is hazardous with readily hydrolyzable acid chlorides such as acetyl chloride and benzoyl chloride. Solid acid chlorides are satisfactorily crystallized from benzene, benzene-light petroleum, light petroleum, alcohol-free chloroform-toluene, and, occasionally, from dry ethyl ether. Hydroxylic or basic solvents are to be strictly avoided. All operations should be carried out in a fume cupboard because of the irritant nature of these compounds.

<u>Alcohols</u>. (a) <u>Monohydric</u>: The common impurities in alcohols are aldehydes or ketones, and water. [Ethanol in Chapter 3 is typical.) Aldehydes and ketones can be removed by adding a small amount of sodium metal and refluxing for 2 hr, followed by distillation. Water can be removed in a similar way but it is preferable to use magnesium metal instead of sodium because it forms a more insoluble hydroxide, thereby shifting the equilibrium more completely from metal alkoxide to metal hydroxide. The magnesium should be activated with iodine and the water content should be low, otherwise the magnesium will be deactivated.

Acidic materials can be removed by treatment with anhydrous Na_2CO_3, followed by a suitable drying agent, such as calcium hydride, and fractional distillation, using gas chromatography to establish the purity of the product. [Ballinger and Long, <u>JACS</u>, <u>82</u>, 795 (1960).] Alternatively, the alcohol can be refluxed with freshly ignited CaO for 4 hours and then fractionally distilled. [McCurdy and Laidler, <u>Can. J. Chem</u>., <u>41</u>, 1867 (1963).]

With higher-boiling alcohols it is advantageous to add some freshly prepared magnesium ethoxide solution (only slightly more than required to remove the water), followed by fractional distillation. Alternatively, in such cases, water can be removed by azeotropic distillation with benzene or toluene. Higher-melting alcohols can be purified by crystallization from methanol or ethanol, benzene-light petroleum or light petroleum. Sublimation in vacuum, molecular distillation and gas chromatography are also useful means of purification. For purification <u>via</u> a derivative, see p.66.

(b) Polyhydric: These alcohols are more soluble in water than are the monohydric ones. Liquids can be freed from water by shaking with type 4A Linde molecular sieves and can safely be distilled only under high vacuum. Carbohydrate alcohols can be crystallized from strong aqueous solution or, preferably, from mixed solvents such as ethanol-light petroleum or dimethyl formamide-benzene. Crystallization usually requires seeding and is extremely slow. Further purification can be effected by conversion to the acetyl derivatives which are much less soluble in water and which can be readily recrystallized. Hydrolysis of the acetyl derivatives, followed by removal of acetate and metal ions by ion-exchange chromatography, gives the purified material. On no account should solutions of carbohydrates be concentrated above $40°$. Ion exchange, charcoal or cellulose column chromatography has been used for the purification and separation of carbohydrates. (See Heftmann, Chromatography.)

Aldehydes. Common impurities found in aldehydes are the corresponding alcohols, aldols and water from self-condensation, and the corresponding acids formed by auto-oxidation. Acids can be removed by shaking with aqueous 10% sodium bicarbonate solution. The organic liquid is then washed with water, followed by dilute mineral acid, and again with water. It is dried with sodium sulphate or magnesium sulphate and then fractionally distilled. Water-soluble aldehydes must be dissolved in a suitable solvent such as ethyl ether before being washed in this way. Further purification can be effected via the bisulphite (see p.67) or the Schiff base formed with aniline or benzidine. Solid aldehydes can be dissolved in ethyl ether and purified as above. Alternatively, they can be steam distilled, then sublimed and crystallized from benzene or light petroleum.

Amides. Amides are stable compounds. The lower-melting members (such as acetamide) can be readily purified by fractional distillation. Most amides are solids which have low solubilities in water. They can be recrystallized from large quantities of water, ethanol, ethanol-ether, aqueous ethanol, chloroform-benzene, chloroform or acetic acid. The likely impurities are the parent acids of the alkyl esters from which they have been made. The former can be removed by thorough washing with aqueous ammonia followed by crystallization, whereas elimination of the latter is by trituration or crystallization from an organic solvent. Amides can be freed from solvent or water by drying below their melting points. These purifications can also be used for sulphonamides and acid hydrazides.

Amines. The common impurities found in amines are nitro compounds (if prepared by reduction), the corresponding halides (if prepared from them), coloured products, water, and (in aliphatic or strongly basic amines) the corresponding carbamates. Amines are dissolved in aqueous acid, the pH of the solution being at least three units below the pK_a value of the base to ensure almost complete cation formation. They are extracted with ethyl ether to remove neutral impurities and to decompose the carbamates. The solution is then made strongly alkaline and the amines are separated, extracted into a solvent (ethyl ether or benzene) or steam distilled. The liquid amines or extracts are dried with anhydrous potassium carbonate or sodium carbonate and then fractionally distilled. The latter process removes coloured impurities. Note that chloroform cannot be used as a solvent for primary amines because, in the presence of alkali,

Amines (continued)

poisonous carbylamines are formed. However, chloroform is a useful
solvent for the extraction of heterocyclic bases. In this case it
has the added advantage that while the extract is being freed from
the chloroform most of the moisture is removed with the solvent.

Alternatively, the amine may be dissolved in a suitable solvent
(such as toluene) and dry HCl gas is passed into the solution to
precipitate the amine hydrochloride. This is purified by recrystallization from a suitable solvent mixture (such as toluene-ethanol).
The free amine can be regenerated by adding sodium hydroxide.

Liquid amines can be further purified via their acetyl or benzoyl
derivatives (see p.69). Solid amines can be crystallized from water,
alcohol, benzene or benzene-light petroleum. Care should be taken in
handling large quantities of amines because their vapours are harmful
and they are readily absorbed through the skin.

Amino acids. Because of their zwitterionic nature, amino acids are
soluble in water. Their solubility in organic solvents rises as the
fat-soluble portion of the molecule increases. The likeliest impurities are traces of salts, heavy metal ions, proteins and other
amino acids. Purification from these is usually easy, by recrystallization from water or water-ethanol mixtures. The amino acid is
dissolved in the boiling solvent, decolorized if necessary by boiling
with 1 g acid-washed charcoal/100 g amino acid, then filtered hot,
chilled and stood several hours to crystallize. The crystals are
filtered off, washed with ethanol, then ether, and dried.

Amino acids have high melting or decomposition points and are best
examined for purity by paper chromatography. The spots are developed
with ninhydrin solution (see, for example, Lederer and Lederer,
Chromatography). Customary methods for the purification of small
quantities of amino acids obtained from natural sources (i.e. 1-5 g)
are ion-exchange chromatography (see p. 29) or countercurrent
distribution (see p.39).

A useful source of details such as likely impurities, stability and
tests for homogeneity of amino acids is Specifications and Criteria
for Biochemical Compounds, 3rd edn., 1972, National Academy of
Sciences, U.S.A.

Anhydrides. The corresponding acids, resulting from hydrolysis, are
the most likely impurities. Distillation from phosphorus pentoxide,
followed by fractional distillation, is usually satisfactory. With
high-boiling or solid anhydrides, another method involves refluxing
for ½-1 hr with acetic anhydride, followed by fractional distillation.
Acetic acid distils first, then acetic anhydride, and finally the
required anhydride. Where the anhydride is a solid, removal of acetic
acid and acetic anhydride at atmospheric pressure is followed by
heating under vacuum. The solid anhydride is then either crystallized
as for acid chlorides or (in some cases) sublimed in a vacuum. A
preliminary purification when large quantities of acid are present in
a solid anhydride (such as phthalic anhydride) can sometimes be
effected by preferential solvent extraction of the (usually) more
soluble anhydride from the acid (e.g. with chloroform in the case of
phthalic anhydride). All operations should be carried out in a fume

Anhydrides (continued)
cupboard because of the lachrimatory properties of liquid anhydrides.

Carotenoids. Carotenoids are decomposed by light, air and solvents,
so that degradation products are likely as impurities. Chromatography
and absorption spectra permit the ready detection of coloured impurities, and separations are possible using solvent distribution,
chromatography or crystallization. Thus, in partition between immiscible solvents, xanthophyll remains in 90% methanol while carotenes
pass into the petroleum ether phase. For small amounts of material,
thin-layer or paper chromatography may be used, while column chromatography is suitable for larger amounts. Colourless impurities may
be detected by infrared, n.m.r. or mass spectrometry. The more
common separation procedures are described by P. Karrer and E. Jucker
in Carotenoids, E.A. Braude (translator), Elsevier, New York, 1950.

Purity can be assayed by chromatography (on thin-layer plates,
kieselguhr paper or columns), by visible spectrum, or n.m.r.

Esters. The most common impurities are the corresponding acid and
hydroxy compound (i.e. alcohol or phenol) and water. A liquid ester
from a carboxylic acid is washed with 2 N sodium carbonate or sodium
hydroxide to remove acidic material, then shaken with calcium
chloride to remove methyl or ethyl alcohols (if it is a methyl or
ethyl ester). It is dried with potassium carbonate or magnesium
sulphate, and distilled. Fractional distillation then removes residual traces of hydroxy compounds. This method does not apply to esters
of inorganic acids (e.g. dimethyl sulphate) which are more readily
hydrolyzed in aqueous solution when heat is generated in the neutralization of the excess acid. In such cases, several fractional distillations, preferably under vacuum, are usually sufficient.

Solid esters are easily crystallizable materials. It is important
to note that esters of alcohols must be recrystallized either from
non-alcoholic solvents (e.g. benzene) or from the alcohol from which
the ester is derived. Thus methyl esters should be crystallized from
methanol or methanol-benzene, but not from ethanol or n-butanol, in
order to avoid alcohol exchange and contamination of the ester with
a second ester. Useful solvents for crystallization are the corresponding alcohols or aqueous alcohols, benzene, benzene-light petroleum, chloroform and chloroform-toluene. Carboxylic acid esters
derived from phenols are more difficult to hydrolyze and exchange,
hence any alcoholic solvent can be used freely. Sulphonic acid esters
of phenols are even more resistant to hydrolysis: they can safely be
crystallized not only from the above solvents but also from acetic
acid or aqueous acetic acid.

Fully esterified phosphoric and phosphonic acids differ only in
detail from the above-mentioned esters. Their major contaminants are
alcohols or phenols, phosphoric or phosphonic acids (from hydrolysis),
and (occasionally) basic material, such as pyridine, which is used in
their manufacture. Water-insoluble esters are washed thoroughly and
successively with dilute acid (e.g. 0.2 N sulphuric acid), water,
0.2 N sodium hydroxide and water. After drying with calcium chloride
they are fractionally distilled. Water-soluble esters should first
be dissolved in a suitable organic solvent and, in the washing
process, water should be replaced by saturated aqueous sodium chloride.

Esters (continued)

These esters can be further purified through their uranyl adducts (see p.72). Traces of water or hydroxy compounds can be removed by percolation through, or shaking with, activated alumina (about 100 g/l. of liquid or solution), followed by filtration and fractional distillation in a vacuum. For high molecular weight esters (which cannot be distilled without some decomposition) it is advisable to carry out distillation at as low a pressure as possible. Solid esters can be recrystallized from benzene or light petroleum. Alcohols can be used for recrystallizing phosphoric or phosphonic esters of phenols.

Ethers. The purification of ethyl ether (given in Chapter 3) is typical for liquid ethers. The most common contaminants are the alcohols or hydroxy compounds from which the ethers are prepared, their oxidation products (e.g. aldehydes), peroxides and water. Peroxides, aldehydes and alcohols can be removed by shaking with alkaline potassium permanganate solution for several hours, followed by washing with water, concentrated sulphuric acid, then water. After drying with calcium chloride, the ether is distilled. It is then dried with sodium or with lithium aluminium hydride, redistilled and given a final fractional distillation. The drying process should be repeated if necessary.

Alternative methods for removing peroxides include leaving the ether to stand in contact with iron filings or copper powder, shaking the ether with a solution of iron(II) sulphate acidified with sulphuric acid, shaking with a copper-zinc couple [Fierz-David, Chimia, 1, 246 (1947)], passage through a column of activated alumina, and refluxing with phenothiazine. Cerium(III) hydroxide has also been used. [Ramsay and Aldridge, JACS, 77, 2561 (1955).]

A simple test for ether peroxides is to add 10 ml of the ether to a stoppered cylinder containing 1 ml of freshly prepared 10% solution of potassium iodide containing a drop of starch indicator. No colour should develop during one minute. Alternatively, a 1% solution of iron(II) ammonium sulphate, 0.1 M in sulphuric acid and 0.01 M in potassium thiocyanate should not increase appreciable in red colour when shaken with two volumes of the ether.

As a safety precaution against explosion (in case the purification has been insufficiently thorough) at least a quarter of the total volume of ether should remain in the distilling flask when the distillation is discontinued. To minimize peroxide formation, ethers should be stored in dark bottles and, if they are liquids, they should be left in contact with type 4A Linde molecular sieve, in a cold place, over sodium amalgam. The rate of ether formation depends on storage conditions and is accelerated by heat, light, air and moisture. Peroxide formation is decreased in the presence of diphenylamine as stabilizer.

Ethers that are solids (e.g. phenyl ethers) can be steam distilled from an alkaline solution which will hold back any phenolic impurity. After the distillate is made alkaline with sodium carbonate, the insoluble ether is collected either by extraction (e.g. with chloroform, ethyl ether or benzene) or by filtration. It is then crystallized from alcohols, alcohol-light petroleum, light petroleum, benzene

Ethers (continued)
or mixtures of these solvents, sublimed in a vacuum and recrystallized.

Halides. Aliphatic halides are likely to be contaminated with
halogen acids and the alcohols from which they have been prepared,
whereas in aromatic halides the impurities are usually aromatic
hydrocarbons, amines or phenols. In both groups the halogen atom is
less reactive than it is in acid chlorides. Purification is by
shaking with concentrated hydrochloric acid, followed by washing
successively with water, 5% sodium carbonate or bicarbonate, and
water. After drying with calcium chloride, the halide is distilled
and then fractionally distilled using an efficient column. For a
solid halide the above purification is carried out by dissolving it
in a suitable solvent such as benzene. Solid halides can also be
purified by chromatography using an alumina column and eluting with
benzene or light petroleum. They can be recrystallized from benzene,
light petroleum, benzene-light petroleum or benzene-chloroform-light
petroleum. Care should be taken in handling organic halogen compounds
because of their toxicity.

Liquid aliphatic halides are obtained alcohol-free by distillation
from phosphorus pentoxide. They are stored in dark bottles to
prevent oxidation and, in some cases, the formation of phosgene.

Hydrocarbons. Gaseous hydrocarbons are best freed from water and
gaseous impurities by passage through suitable adsorbents and (if
olefinic material is to be removed) oxidants such as alkaline
potassium permanganate solution, followed by fractional cooling (see
p.47 for cooling baths) and fractional distillation at low temperature. To effect these purifications and also to store the gaseous
sample, a vacuum line is necessary.

Impurities in hydrocarbons can be characterized and evaluated by
gas chromatography and mass spectrometry. The total amount of
impurities present can be estimated from the shape of the thermo-
metric freezing curve.

Liquid aliphatic hydrocarbons are freed from aromatic impurities
by shaking with concentrated sulphuric acid whereby the aromatic
compounds are sulphonated. Shaking is carried out until the sulphuric
acid layer remains colourless for several hours. The hydrocarbon is
then freed from the sulphuric acid and the sulphonic acids by
separating the two phases and washing the organic one successively
with water, 2 N sodium hydroxide, and water. It is dried with
calcium chloride or sodium sulphate, and then distilled. The distillate is dried with sodium wire, P_2O_5, or metallic hydrides, or
passage through a dry silica gel column, or preferably, and more
safely, with molecular sieve (see p. 40) before being finally
fractionally distilled through an efficient column. If the hydrocarbon is contaminated with olefinic impurities, shaking with
aqueous alkaline potassium permanganate is necessary prior to the
above purification. Alicyclic and paraffinic hydrocarbons can be
freed from water, nonhydrocarbon and aromatic impurities by passage
through a silica-gel column before the final fractional distillation.
This may also remove isomers. (For the use of a chromatographic
method to separate mixtures of aromatic, paraffinic and alicyclic

Hydrocarbons (continued)
hydrocarbons, see Meir, J. Res. Nat. Bur. Stand., 34, 453 (1945)).
Another method of removing branched-chain and unsaturated hydrocarbons
from straight-chain hydrocarbons depends on the much faster reaction
of the former with chlorosulphonic acid.

Isomeric materials which have closely similar physical properties
can be serious contaminants in hydrocarbons. With aromatic hydrocarbons, e.g. xylenes and alkyl benzenes, advantage is taken of
differences in ease of sulphonation. If the required compound is
sulphonated more readily, the sulphonic acid is isolated, recrystallized (e.g. from water), and decomposed by passing superheated steam
through a flask containing the acid. The sulphonic acid undergoes
hydrolysis and the liberated hydrocarbon distils with the steam. It
is separated from the distillate, dried, distilled and then fractionally distilled. For small quantities (10-100 mg) vapour phase chromatography is the most satisfactory way of obtaining a pure sample.
(For column packings, see pp. 35,55).

Azeotropic distillation with methanol or 2-ethoxyethanol has been
used to obtain highly purified saturated hydrocarbons and aromatic
hydrocarbons such as xylenes and isopropylbenzene.

Carbonyl-containing impurities can be removed from hydrocarbons
(and other oxygen-lacking solvents such as $CHCl_3$ and CCl_4) by passage
through a column of Celite 545 (100 g) mixed with concentrated
sulphuric acid (60 ml). After first adding some solvent and about
10 g of granular Na_2SO_4, the column is packed with the mixture and a
final 7-8 cm Na_2SO_4 is added to the top. [Hornstein and Crowe, Anal.
Chem., 34, 1037 (1962).] Alternatively, Celite impregnated with
2,4-dinitrophenylhydrazine can be used.

With solid hydrocarbons such as naphthalene, preliminary purification is by sublimation in vacuum (or high vacuum if the substance
is high-melting), followed by zone melting and finally by chromatography (e.g. on alumina) using a low-boiling liquid hydrocarbon
eluant. These solids can be recrystallized from alcohols, alcohol-
light petroleum or from liquid hydrocarbons (including benzene) and
dried below their melting points. Aromatic hydrocarbons that have
been purified by zone melting include terphenyl, biphenyl, naphthalene, fluoranthene, anthracene, pyrene, perylene and phenanthrene.

Olefinic compounds have a very strong tendency to polymerize and
commercially available materials are generally stabilized, e.g. with
hydroquinone. When distilling compounds such as vinylpyridine or
styrene the stabilizer remains behind and the purified olefinic
material is more prone to polymerization. The most common impurities
are higher-boiling dimeric or polymeric compounds. Vacuum distillation in a nitrogen atmosphere not only separates monomeric from polymeric materials but in some cases also depolymerizes the impurities.
The distillation flask should be charged with a polymerization
inhibitor and the purified material should be used immediately or
stored in the dark, mixed with a small amount of stabilizer (e.g.
0.1% of hydroquinone).

A general method for purifying chlorohydrocarbons uses repeated
shaking with concentrated sulphuric acid until no further colour

Hydrocarbons (again continued)
develops in the acid, then washing with a solution of sodium bicarbonate, followed by water. After drying with calcium chloride, the
chlorohydrocarbon is fractionally distilled. Finally, it is fractionally crystallized to constant melting point.[Barton and Howlett,
JCS, 155 (1949).]

Imides. Imides (e.g. phthalimide) can be purified by conversion to
their potassium salts by reaction in ethanol with ethanolic potassium
hydroxide. The imides are regenerated when the salts are hydrolyzed
with dilute acid. Like amides, imides readily crystallize from
alcohols and, in some cases (e.g. quinolinic imide), from glacial
acetic caid.

Imino compounds. These substances contain the —C = NH group and,
because they are strong, unstable bases, they are kept as their more
stable salts, such as the hydrochlorides. (The free base usually
hydrolyzes to the corresponding oxo compound and ammonia.) Like amine
hydrochlorides, the salts are purified by solution in alcohol containing a few drops of hydrochloric acid. After treatment with charcoal, and filtering, ethyl ether (or light petroleum if ethanol is
used as the alcohol) is added until crystallization sets in. The
salts are dried and kept in a vacuum desiccator.

Ketones. Ketones are more stable to oxidation than aldehydes and
can be purified from oxidizable impurities by refluxing with potassium permanganate until the colour persists, followed by shaking with
sodium carbonate (to remove acidic impurities) and distilling.
Traces of water can be removed with type 4A Linde molecular sieve.
Ketones which are solids can be purified by crystallization from
alcohol, benzene, or light petroleum, and are usually sufficiently
volatile for sublimation in vacuum. Ketones can be further purified
as their bisulphite, semicarbazone or oxime derivatives (see p. 71).
The bisulphite addition compounds are formed only by aldehydes and
methyl ketones but they are readily split with dilute acid or
alkali.

Nitriles. All purifications should be carried out in an efficient
fume cupboard because of the poisonous nature of the compounds.

Nitriles are usually prepared either by reacting the corresponding halide or diazonium salt with a cyanide salt or by dehydrating
an amide. Hence, possible contaminants are the respective halide or
alcohol (from hydrolysis), phenolic compounds, amines or amides.
Small quantities of phenols can be removed by chromatography on
alumina. More commonly, purification of liquid nitriles or solutions
of solid nitriles in a solvent such as ethyl ether is by shaking with
dilute aqueous sodium hydroxide, followed by washing successively
with water, dilute acid and water. After drying with sodium sulphate,
the solvent is distilled off. Liquid nitriles are best distilled from
a small amount of phsophorus pentoxide which, besides removing water,
dehydrates any amine to the nitrile. About one quarter of the nitrile
should remain in the distilling flask at the end of the distillation.
This purification also removes alcohols and phenols. Solid nitriles
can be crystallized from ethanol, benzene or light petroleum, or from
a mixture of these solvents. They can also be sublimed under vacuum.
Preliminary purification by steam distillation is usually possible.

Nitriles (continued)

Strong alkali or heating with dilute acids may lead to hydrolysis of the nitrile.

Nitro compounds. Aliphatic nitro compounds are acidic. They are freed from alcohols or alkyl halides by standing for a day with concentrated sulphuric acid, then washed with water, dried with magnesium sulphate followed by calcium sulphate and distilled. The principal impurities are isomeric or homologous nitro compounds. In cases where the nitro compound was originally prepared by vapour-phase nitration of the aliphatic hydrocarbon, fractional distillation should separate the nitro compound from the corresponding hydrocarbon. Fractional crystallization was more effective than fractional distillation. [Coetzee and Cunningham, JACS, 87, 2529 (1965).]

The impurities present in aromatic nitro compounds depend on the aromatic portion of the molecule. Thus, benzene, phenol or anilines are probable impurities in nitrobenzene, nitrophenol and nitro-anilines, respectively. Purification should be carried out accordingly. Isomeric compounds are likely to remain as impurities after the preliminary purifications to remove basic or acidic contaminants. For example, o-nitrophenol may be found in samples of p-nitrophenol. Usually, the o-nitro compounds are more steam-volatile than the p-isomers, and can be separated in this way. Polynitro impurities in mononitro compounds can be readily removed because of their relatively lower solubilities. With acidic or basic nitro compounds which cannot be separated in the above manner, advantage may be taken of their differences in pK_a values. The compounds can thus be purified by preliminary extractions with several sets of aqueous buffers of known pH (see, for example p.56) from a solution of the substance in a suitable solvent such as ethyl ether. This method is more satisfactory and less laborious the larger the difference between the pK_a value of the impurity and the required compound. Heterocyclic nitro compounds require similar treatment to the nitroanilines. Neutral nitro compounds can be steam distilled.

Phenols. Because phenols are weak acids, they can be freed from neutral impurities by dissolution in aqueous N sodium hydroxide and extraction with a solvent such as ethyl ether or by steam distillation to remove the nonacidic material. The phenol is recovered by acidification of the aqueous phase with 20% sulphuric acid, and either extracted with ether or steam distilled. In the second case the phenol is extracted from the steam distillate after saturating it with sodium chloride. A solvent is unnecessary when large quantities of liquid phenols are purified. The phenol is fractionated by distillation under reduced pressure, preferably in an atmosphere of nitrogen to minimize oxidation. Solid phenols can be crystallized from benzene, light petroleum or a mixture of these solvents, and can be sublimed under vacuum. Purification can also be effected by fractional crystallization or zone melting. For further purification of phenols as their acetyl or benzoyl derivatives, see p.71.

Quinones. Quinones are neutral compounds which are usually coloured. They can be separated from acidic or basic impurities by extraction of their solutions in organic solvents with aqueous basic or acidic solutions, respectively. Their colour is a useful property in their purification by chromatography through an alumina column

Quinones (continued)

with, e.g. benzene as eluant. They are volatile enough for vacuum sublimation although, with high-melting quinones, a high vacuum is necessary. p-Quinones are stable compounds and can be recrystallized from water, ethanol, aqueous ethanol, benzene, light petroleum or glacial acetic acid. o-Quinones, on the other hand, are readily oxidized. They should be handled in an inert atmosphere, preferably in the absence of light.

Salts (organic). (a) With metal ions: Water-soluble salts are best purified by preparing a concentrated aqueous solution to which, after decolorizing with charcoal and filtering, ethanol or acetone is added so that the salts crystallize. They are collected, washed with aqueous alcohol or aqueous acetone, and dried. In some cases, water-soluble salts can be recrystallized satisfactorily from alcohols. Water-insoluble salts are purified by Soxhlet extraction, first with organic solvents and then with water, to remove soluble contaminants. The purified salt is recovered from the thimble.

(b) With organic ions: Organic salts (e.g. trimethylammonium benzoate) are usually purified by recrystallization from polar solvents (e.g. water, alcohol or dimethyl formamide). If the salt is too soluble in a polar solvent, its concentrated solution should be treated dropwise with a miscible nonpolar solvent (see p.46) until crystallization begins.

Sulphur compounds. (a) Disulphides can be purified by extracting acidic or basic impurities with aqueous base or acid, respectively. However, they are somewhat sensitive to strong alkali which slowly cleaves the disulphide bond. The lower-melting members can be fractionally distilled under vacuum. The higher members can be crystallized from alcohol, benzene or glacial acetic acid.

(b) Sulphones are neutral and extremely stable compounds that can be distilled without decomposition. They are freed from acidic and basic impurities in the same way as disulphides. The low molecular weight members are quite soluble in water but the higher members can be recrystallized from water, alcohol, aqueous alcohol or glacial acetic acid.

(c) Sulphoxides are odourless, rather unstable compounds, and should be distilled under vacuum in an inert atmosphere. They are water-soluble but can be extracted from aqueous solution with a solvent such as ethyl ether.

(d) Thioethers are neutral stable compounds that can be freed from acidic and basic impurities as described for disulphides. They can be crystallized from organic solvents and distil without decomposition.

(e) Thiols are stronger acids than the corresponding hydroxy compounds but can be purified in a similar manner. However, care must be exercised in handling thiols to avoid the oxidation to disulphides. For this reason, purification is best carried out in an inert atmosphere in the absence of oxidizing agents. Similarly, thiols should be stored out of contact with air. They can be distilled without change, and the higher-melting thiols (which are

Sulphur compounds (continued)

(e) Thiols (continued)

usually more stable) can be crystallized, e.g. from water or dilute alcohol. They oxidize readily in alkaline solution but can be separated from the disulphide which is insoluble in this medium. All operations with thiols should be carried out in an efficient fume cupboard because of their unpleasant odour and their toxicity.

(f) Thiosulphonates (disulphoxides) are neutral and are somewhat light-sensitive compounds. Their most common impurities are sulphonyl chlorides (neutral) or the sulphinic acids from which they are usually prepared. The former can be removed by partial freezing or crystallization, and the latter by shaking with dilute alkali. Thiosulphonates decompose slowly in dilute, or rapidly in strong, alkali to form disulphides and sulphonic acids. Thiosulphonates also decompose on distillation but they can be steam distilled. The solid members can be recrystallized from water, alcohols or glacial acetic acid.

Index

For individual organic chemicals, listed alphabetically,
see Chapter 3, beginning on page 74,
and for inorganic and metal-organic compounds
see Chapter 4, beginning on page 466.